# Ocean Circulation Theory

Springer
*Berlin*
*Heidelberg*
*New York*
*Barcelona*
*Budapest*
*Hong Kong*
*London*
*Milan*
*Paris*
*Santa Clara*
*Singapore*
*Tokyo*

Joseph Pedlosky

# Ocean Circulation Theory

With 167 Figures

Springer

Dr. JOSEPH PEDLOSKY
Department of Physical Oceanography
Woods Hole Oceanographic Institution
Woods Hole, MA 02543, USA



ISBN 3-540-60489-8 Springer-Verlag Berlin Heidelberg New York

Library of Congress Cataloging-in-Publication Data. Pedlosky, Joseph. Ocean circulation theory/ Joseph Pedlosky. p. cm. Includes index. ISBN 3-540-60489-8 (hardcover). – ISBN 0-387-60489-8 (hardcover) 1. Ocean circulation. I. Title. GC228.5.P427 1996 551.47 – dc20 96-13298



Cover design: Springer-Verlag, E. Kirchner

Typesetting: Scientific Publishing Services (P) Ltd, Madras

SPIN: 10509713 32/3136/SPS – 5 4 3 2 1 0 – Printed on acid-free paper

# Preface

The waters of the earth are gathered in shallow, irregular, interconnecting basins. Heated by the sun and driven by the wind, the oceans circulate endlessly.

The general circulation of the ocean is the persistent pattern of this flow on the scale of the basins. It is the heart and soul of physical oceanography, and the explanation and prediction of the flow from the principles of fluid dynamics is the chief goal of the physical oceanographer. Along with the pattern of the motion, the associated fields of pressure, temperature, salinity, and density are also necessary components of a complete theory for the ocean circulation since they are dynamically linked to the motion of the oceans.

The physical problem posed by the general circulation of the oceans is a difficult one both experimentally and theoretically. From the point of view of theoretical fluid mechanics the difficulty springs fundamentally from the recirculating character of the circulation. The fluid is gathered into a single, though highly contorted basin. Aside from a relatively small amount of water which enters as precipitation and leaves the ocean to enter the atmosphere due to evaporation, the water remains in its basin and never leaves. As is well known, the hardest problems in fluid mechanics are just those of recirculating fluid flows since there is no location in the fluid that can be taken as a starting point at which the properties of the flow can be prescribed a priori. Dissipation, no matter how small, has all the time it needs to affect the circulation in fundamental ways. When we realize that the ocean circulation is vigorous enough that its dynamics are also nonlinear, the difficulty becomes even more obvious.

Nevertheless, considerable progress has been achieved in the past 15 years in understanding the oceanic general circulation. Problems which stubbornly defied solution for decades, such as the theory of the thermocline, have been clarified to a considerable degree. At the same time, problems which had been considered "solved," such as the homogeneous model of wind-driven circulation, have displayed new and unforeseen complexities and richness in the underlying dynamics.

In the middle of the past decade Bill Young and I introduced an advanced graduate course in the MIT/WHOI Joint Program in Physical Oceanography whose purpose was to introduce our students to the recent advances in the theory of the ocean circulation. When Bill left MIT I continued the course, and the present book has as its basis the material presented in that course.

Neither that course nor this book are meant to be an exhaustive review of all aspects of ocean circulation theory. Time, space, and personal interest lead to a selection of material which is meant to be fundamental without being encyclopedic. Researchers will note important subjects not treated. It is, after all, a big field. The subjects which are treated have been chosen primarily because, in my view, they offer the best insight into the solid advances recently made in the field and are likely to remain as building blocks for future developments.

Chapters 1 and 2 review the simplest ideas of ocean circulation theory. Chapter 1 describes Sverdrup's theory of horizontal transport. This theory is still one of the foundations of all theories of the ocean circulation, and the chapter describes its development and recent attempts to gauge its validity. In Chapter 2 the theory of the wind-driven circulation of a fluid of uniform density is taken up. This is perhaps the most classical of all the problems in physical oceanography, and it is humbling and exciting to recognize both the inadequacy of our present understanding and the recently discovered dynamical complexity of this simplest of all nontrivial theories for the circulation. The discussion, of interest in itself, is also designed as a warning concerning the incomplete character of the theories of the stratified ocean circulation which form the remainder of the text.

Chapter 3 describes the quasi-geostrophic theory of the wind-driven circulation of a stratified ocean. The brilliant theoretical work of Rhines and Young in the early years of the past decade revolutionized our understanding of the vertical structure of the wind-driven circulation. The concepts of geostrophic contours, potential vorticity homogenization, and their role in shaping the pattern of the circulation form the core of this chapter.

Chapter 4 takes up the question of understanding the full density and velocity structure of the wind-driven circulation by going beyond the quasi-geostrophic theory to consider the important effect of the ventilation of the thermocline which occurs as oceanic density surfaces rise to intersect the oceanic mixed layer. The theory of the ventilated thermocline, on which I have worked with my colleagues Jim Luyten and Henry Stommel, provides a mechanism to explain both the vertical and horizontal structures of the density field in the thermocline and is the starting point of a synthesis of the theories of ventilation and recirculation.

The adiabatic theories of the thermocline are extended slightly in Chapter 5 to include the effects of heating and cooling in the gyres as represented by specified transformations of density within the body of the ocean. Although this a priori specification of the heating and cooling is a weakness of the theory, the resulting buoyancy-driven circulations form an important complement to the purely adiabatic motions of the theories of the previous two chapters.

Chapter 6 describes the extension of the ventilated thermocline theory to the equator where the fundamental dynamical approximation of geostrophic balance fails. The proper continuation of the *concept* of thermocline theory to the equator both automatically provides a theory for the equatorial undercurrent, one of the most fascinating of oceanic phenomena, and emphasizes the inescapable dynamic link between midlatitudes and the equatorial zone.

Chapter 7 treats the abyssal circulation of the ocean. This is a fundamental and relatively underdeveloped area of research. The classical theory of Stommel and Arons is reviewed, and two particularly important extensions to their theory are provided. Both the role of large-scale topography, manifested in the great oceanic ridge and fault systems, and the role of stratification, which is required to explain the observed layering of the abyssal circulation, are taken up in simple models.

My interest in the general circulation of the oceans was rekindled after my move to Woods Hole by the presence there of the most creative researcher in oceanography I have ever known, Henry Stommel. He exerted his influence on the field in many ways. Foremost, of course, was the continuing fountain of ideas and theories which flowed from an endlessly active and fruitful intelligence. At the same time he constantly encouraged his colleagues to follow his example as best they could and concentrate on coming to grips with important problems in physical oceanography. He was inspiring to work with, and it was a joy and privilege to be his colleague. Everywhere in physical oceanography and thus everywhere in this book the positive and benevolent influence of his ideas ought to be clear. His central vision was that of physical oceanography as a "hard" science. Thus he was moved to remind us (Future Prospects for Physical Oceanography, *Science* 1970, vol. 168, pp. 1531–1537) that, "Most physical oceanographers will agree that the chief concern of their field is to understand the actual hydrodynamical mechanisms at work," and "The physical scientist does not see the ocean as a source of wealth or as a jumble of geographic curiosities. He sees it as a hydrodynamic phenomenon: larger than his laboratory, smaller than a star." This book attempts to capture the spirit of that subtle challenge.

This book was written during a sabbatical year spent in Venice, Italy, during the period 1994–1995. I acknowledge with pleasure a J. William Fulbright fellowship from the Council for International Exchange of Scholars for the early period of the writing of the book. The continued support of the Doherty Foundation, which has generously supported my research and teaching at the Woods Hole Oceanographic Institution, is also gratefully acknowledged as well as the support of the National Science Foundation. The hospitality of the Director and staff of the Istituto per lo Studio della Dinamica delle Grandi Masse, in Venice, was invaluable in the preparation of the manuscript. Once again, the city of Venice and this institute of the Italian National Research Council provided me a situation of unparalleled tranquillity and inspiration.

I owe a debt beyond my ability to express it to all my colleagues in the Physical Oceanography Department at Woods Hole who, over the course of my years there, taught me much and always generously shared with me their insights into the physics of the ocean circulation.

Woods Hole, USA
May 1996

JOSEPH PEDLOSKY

# Contents

# 1 Sverdrup Theory

## 1.1 Introduction

Pressure maps of the flow in the upper ocean, such as that shown in Fig. 1.1.1 (a, Stommel et al. 1978; b, Reid 1994) indicate a large-scale clockwise circulation in the subtropical North Atlantic, i.e., in the region south of 45°N but north of a few degrees of the equator. Similarly deduced patterns of circulation have been found in each of the oceans in both northern and southern hemispheres, and the reader is referred to any number of introductory texts (for example, Pickard and Emery 1982) dealing with descriptive oceanography, for a review of the overall circulation patterns deduced from a painstaking accumulation of oceanographic data over the past 100 years. Such large-scale patterns of flow comprise the general circulation of the ocean, and this book presents some recent theoretical advances in our understanding of the dynamics of the circulation. We refer at times to observations pertinent to particular theoretical issues, but we do not attempt to discuss comprehensively the observational description of the general circulation.

In each basin of the great world oceans the overall pattern of flow is notably similar within the first 1–2 km in depth. A broad flow, sweeping clockwise across the entire breadth of the basin is returned within a remarkably narrow western boundary current; in the case of the North Atlantic this is the Gulf Stream, and its presence is apparent in the closely packed isolines of pressure in the western part of the gyre. Although it is difficult even today to give a precise numerical value for the magnitudes of the various parts of this upper ocean circulation, its broad outlines have become increasingly clear. Schmitz and McCartney (1993), for example, have attempted to synthesize available data and to assign transport values to the component parts of the circulation, as shown in Fig. 1.1.2. Each isoline in the figure is labeled by a circled number indicating the estimated transport on each such branch of the circulation. The units are in sverdrups (1 sverdrup = 1 million $m^3/s$). To obtain transports of this magnitude over scales of the order of a few thousand kilometers and a depth of about 1 km requires an average velocity in the region of the broad southward flow of the order of 1 cm/s

This glimpse of the general circulation of the ocean, limited as it is to the upper ocean and to the subtropical gyres, is still sufficient to alert us to one of

**Fig. 1.1.1. a** Isobars in the North Atlantic subtropical gyre as constructed by Stommel et al. (1978). The isolines are of dynamic height at 100 m with respect to 1500 m. To the degree the flow is in geostrophic balance the velocity (relative to 1500 m) would be along these isolines of pressure. Note the large clockwise gyre between 10°N and 40°N in the Atlantic basin. The western boundary current is manifest in the close-packed isobars along the east coast of North America. **b** A similar map for the entire Atlantic ocean by Reid (1994) showing the field of isobars at about 500 m depth

the most important physical features of the flow that has from the start influenced attempts to construct deductive theories for the circulation. As is clear from Fig. 1.1.2, the circulation simultaneously possesses several distinct spatial scales. At the very least there is the region of broad, slow flow in the oceanic *interior* (the region to the east of the boundary currents) and that of the narrow, swift flow in the western boundary current regimes. There may be other scales embedded in each region as well, but for the moment we can focus

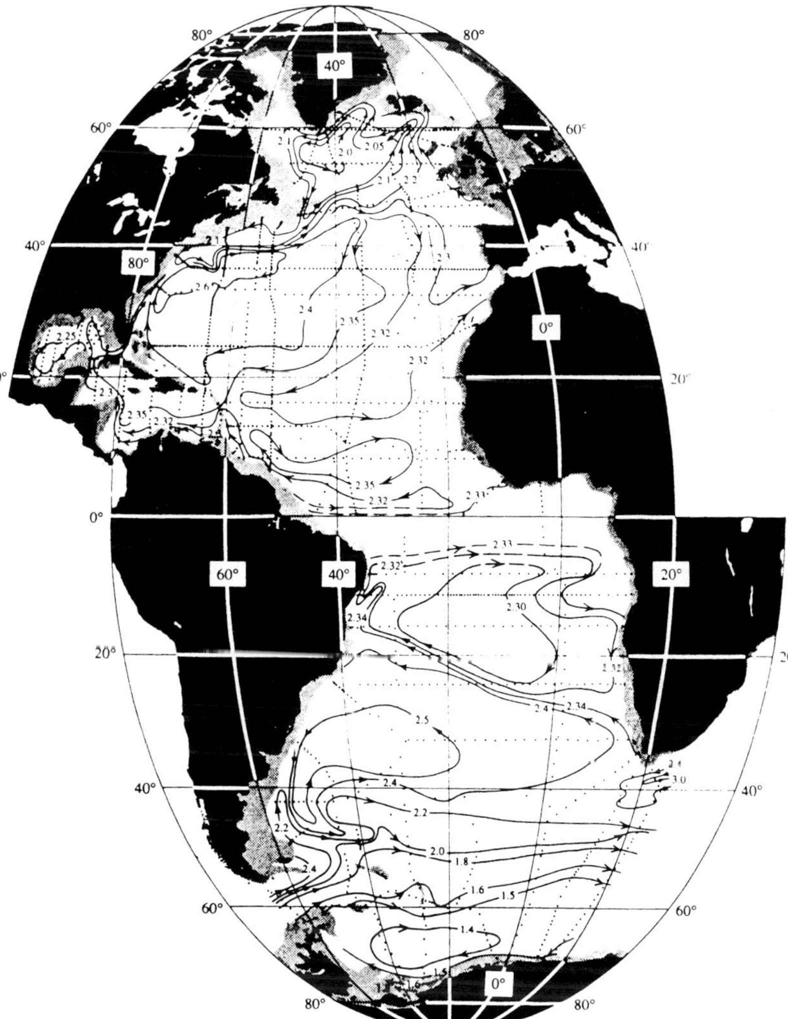


Fig. 1.1.1b

on just this fundamental disparity. From general fluid dynamical principles, we expect the dynamics of each of these regions to differ due to the presence of differing magnitudes of the fluid accelerations and of the spatial gradients of the dynamical fields in the two regions.

On the one hand, this promises increased complexity for the analysis of the dynamics while on the other hand the distinctive dynamics in the various regions provides the theoretician with an opportunity to exploit the special dynamic character of each region, profiting from simplifications possible in each of them. It then remains to fit the pieces of the solution together to reproduce the unitary natural picture, which is of course independent of the methodological expedients employed in analysis.

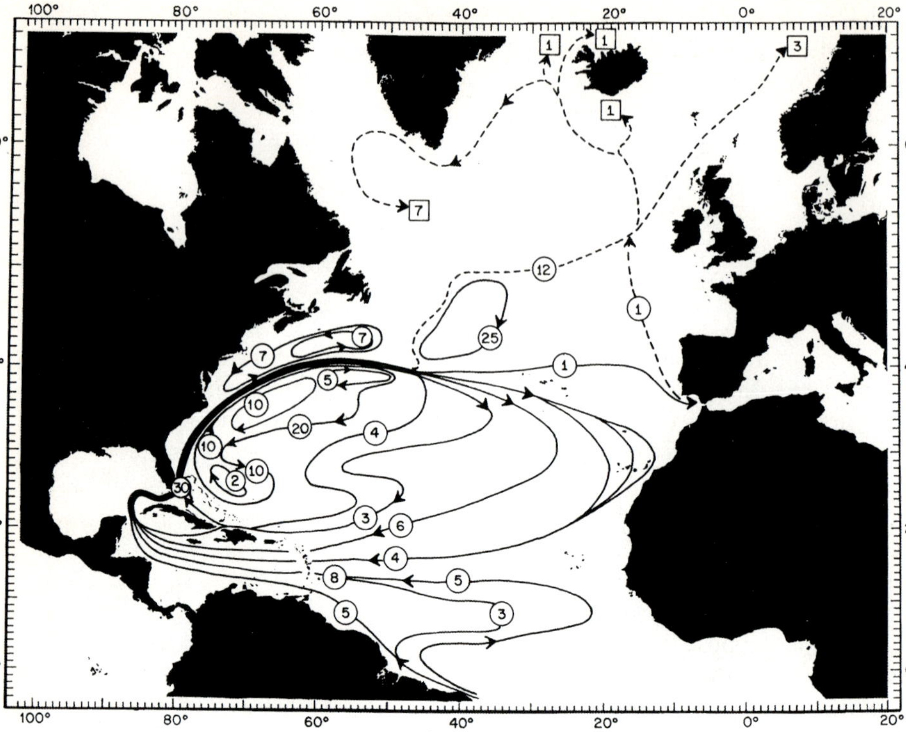


**Fig. 1.1.2.** Transport estimates of elements of the upper ocean circulation (waters warmer than 7°C) in the North Atlantic. *Circled numbers,* estimates of the number of Sverdrups carried in each branch of the circulation. (From Schmitz and McCartney 1993)

It is fair to say that the modern era in the development of the theory of the oceanic general circulation began with the pioneering paper of Sverdrup (1947). Although important work was carried out earlier by Ekman (1905), the paper by Sverdrup, although many of the arguments are nowadays modified or qualified, still forms the cornerstone of our attempts to construct a theory of the oceanic circulation.

This preliminary chapter presents a brief review of Sverdrup theory, and we also undertake a critical examination of its foundations and a discussion of its limitations. Our review is heuristic rather than formal, and the reader may wish to consult, for example, Pedlosky (1987) for a more formal though no more correct, development of the theory.

## 1.2 Scaling for Sverdrup Theory

Sverdrup theory is meant to apply to the broad reaches of the ocean outside the intense western boundary currents. It is therefore meant to serve as a model for the dynamics of the oceanic interior. The theory consists of two parts. The first is the derivation of a simplified vorticity balance for the interior region, and the second is a vertical average of this balance which, under appropriate conditions, leads to a remarkable relationship between the vertically averaged meridional flow and the local value of the wind stress on the ocean surface.

Our starting point is the momentum equation written in a frame of reference rotating with the earth at an angular velocity $\Omega$ ($7.3 \times 10^{-5}\text{s}^{-1}$), i.e.:

$$\frac{D\vec{u}}{Dt} + 2\vec{\Omega} \times \vec{u} = -\frac{\nabla p}{\rho} + \vec{g} + \frac{\Im}{\rho} \tag{1.2.1}$$

where $\vec{u}$ is the velocity seen in the rotating frame, $p$ is the pressure, $\rho$ is the density, $\vec{g}$ is the acceleration due to gravity, and $\Im$ is an implicit representation of turbulent momentum mixing effects of smaller scale motions which act, generally, to diffuse the momentum of the large scale flow. The momentum equation in the form (2.1.1) is far too general to give much insight into the dynamics of the ocean circulation. To proceed further we require approximations specific to the oceanic interior, and it is in the act of making these approximations in a systematic way that the dynamical character of Sverdrup theory really becomes apparent.

Let us suppose that $U$ is a characteristic value for the midocean horizontal velocity (i.e., the velocity tangent to the earth's surface) outside western boundary currents, and that $L$ is a characteristic horizontal scale of that circulation. Assuming that the large-scale flow pattern is essentially stationary over the circulation time, $L/U$, of the interior flow, the ratio of the relative acceleration to the Coriolis acceleration in (1.2.1), i.e., the ratio of the first to the second term in the momentum equation is:

$$\frac{UU/L}{2\Omega \sin\theta U} = \frac{U}{fL} \equiv R_0 \tag{1.2.2}$$

where $\theta$ is the latitude and is the complement of the angle between the rotation vector and the local normal to the earth's surface. The parameter $f$ is the Coriolis parameter:

$$f = 2\Omega \sin\theta. \tag{1.2.3}$$

The parameter $R_0$ is the Rossby number. For large-scale flows $R_0$ is very small. For example, if $U$, $L$, and $f$ are 1 cm/s, 1000 km and $10^{-4}$ $\text{s}^{-1}$, respectively (typical values at midlatitudes), then $R_0$ is $10^{-4}$.

If we consider the horizontal equation of motion, i.e. the equations for the motion tangent to the earth, the frictional force $\Im$ can be split into two parts. The first describes the vertical divergence of the horizontal stress, $\vec{\tau}$ acting on horizontal surfaces in the flow, while the second part is the horizontal divergence of the horizontal stress acting on vertical faces of the fluid element so that:

$$\Im = \frac{\partial \vec{\tau}}{\partial z} + \Im_H \tag{1.2.4}$$

where $z$ is a coordinate parallel to the local vertical direction. The second term on the right side of (1.2.4) is often written as:

$$\Im_H = \rho A_H \nabla^2 \vec{u} \tag{1.2.5}$$

as if the turbulent, horizontal momentum transfer by unresolved scales were equivalent to a molecular friction acting on the horizontal momentum with the augmented turbulent viscosity coefficient $A_H$. At this stage we need not justify the form of (1.2.5) since we introduce it only to assist in determining its likely insignificance for the interior motion. For if we use (1.2.5) to make an estimate of the importance of $\Im_H$ in the interior, we obtain as a ratio of this force to the Coriolis acceleration:

$$\frac{A_H U/L^2}{fU} = \frac{A_H}{fL^2} \equiv E_H \tag{1.2.6}$$

where $E_H$ is called the "horizontal" Ekman number. It is of course as difficult to estimate $E_H$ as it is to estimate the mixing coefficient $A_H$. A range of estimates has been given in the literature depending on which scales of motion are considered unresolved. However, if the values of $f$ and $L$ used above for the interior are retained to estimate (1.2.6), $A_H$ would have to exceed $10^{12}$ cm$^2$/s for $E_H$ to be 0(1). This is far greater than any current reasonable estimate. For example, Bryan (1987) has reported an equivalent mixing coefficient in current eddy resolving numerical *models* of the circulation of the order of $10^8$ cm$^2$/s while Brown and Owens (1981), estimating the effect of scales smaller than geostrophic eddies, i.e. due to the internal gravity wave field, have suggested a mixing coefficient no larger than $10^6$ cm$^2$/s. Thus we may take $E_H$ as a small parameter.

This leaves for the horizontal momentum equation, with both $R_0$ and $E_H$ less than unity, the approximate balance:

$$\rho f \hat{k} \times \vec{u} = -\nabla p + \frac{\partial \vec{\tau}}{\partial z}. \tag{1.2.7}$$

Here $\hat{k}$ is a unit vector in the local vertical direction.

Equation (1.2.7) would be the geostrophic balance between the horizontal pressure gradient and the horizontal Coriolis acceleration were it not for the last term representing the vertical flux of horizontal momentum by turbulent small-scale mixing. This turbulent mixing is generally thought to be small beneath the upper mixed layer, i.e., below a depth of 0(100 m), although it may become important in regions near the sea floor. Thus over the bulk of the water column and beneath the mixed layer the approximation (1.2.7) itself simplifies to the *geostrophic balance:*

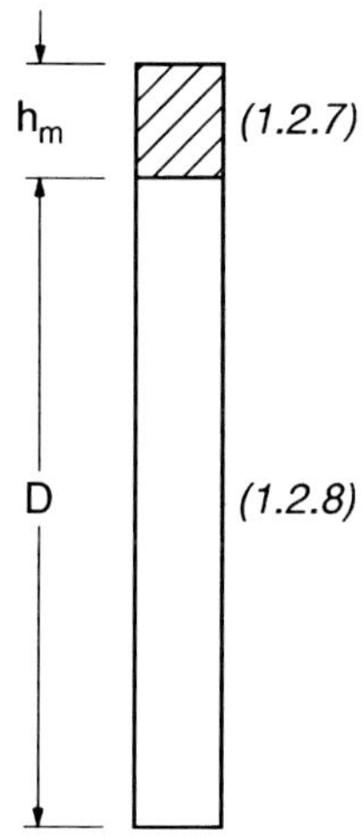


**Fig. 1.2.1.** Schematic presentation of a fluid column in the interior. In the clear region of thickness, $D$, below the wind-stirred layer of depth $h_m$, the geostrophic balance (1.2.8) holds. Above that region (*cross-hatched*) the turbulent stresses are important, and the geostrophic balance must be altered as given by (1.2.7)

$$\rho f\hat{k} \times \vec{u} = -\nabla p \tag{1.2.8}$$

which defines the geostrophic region of the interior flow. Figure 1.2.1 shows how the interior water column is divided between the geostrophic region and the mixed layer.

If the pressure gradient is eliminated by taking the vertical component of the curl of (1.2.7) we obtain:

$$\nabla \cdot \rho f\vec{u} = \frac{\partial}{\partial z}\hat{k} \cdot \nabla \times \vec{\tau} \tag{1.2.9}$$

where the divergence in (1.2.9) refers to the horizontal two-dimensional divergence operator.

The density variation in the ocean is very small, normally only a few parts per thousand. Thus in comparison with the velocity, which varies by its own order over the scale of the oceanic gyre, the density in (1.2.9) can be considered constant and taken out of the divergence operator so that (1.2.9) can be rewritten,

$$\rho\nabla\ f\vec{u} = \frac{\partial}{\partial z}\hat{k} \cdot \nabla \times \vec{\tau}, \quad \text{or} \quad \rho f\,\nabla \cdot \vec{u} + \rho\vec{u} \cdot \nabla f = \frac{\partial}{\partial z}\hat{k} \cdot \nabla \times \vec{\tau}. \tag{1.2.10}$$

Were we to consider the geostrophic region below the mixed layer (1.2.9) would still apply but now with a zero right-hand side.

To proceed further we need to consider the equation for conservation of mass. We approximate the ocean as an incompressible fluid, and thus the equation for the conservation of mass reduces to the condition of volume conservation, i.e.:

$$\nabla \cdot \vec{u} + \frac{\partial w}{\partial z} = 0. \tag{1.2.11}$$

Here we have neglected a term of order $w/R$, where $R$ is the earth's radius, in comparison with $\partial w/\partial z$, which yields an error of $0(D/R)$ which is very small. It is important to recall that $\vec{u}$ in (1.2.11) is the horizontal velocity, and that $\nabla$ is the two-dimensional version of the gradient operator. Using (1.2.11) to eliminate the horizontal divergence we obtain:

$$\rho\beta v = \rho f \frac{\partial w}{\partial z} + \frac{\partial}{\partial z}\hat{k} \cdot \nabla \times \vec{\tau} \tag{1.2.12}$$

where $\beta$ is the northward spatial derivative of $f$. That is:

$$\beta = \frac{1}{R}\frac{\partial f}{\partial \theta} = \frac{2\Omega \cos\theta}{R} \tag{1.2.13}$$

while below the mixed layer in the geostrophic region (1.2.12) becomes simply:

$$\beta v = f\frac{\partial w}{\partial z}. \tag{1.2.14}$$

In this approximation the vorticity equation is reduced to the particularly simple form (1.2.14). Its physical interpretation is simply that when a fluid column is stretched in the vertical direction, i.e., when $\partial w/\partial z > 0$, vorticity is produced at a rate equal to this stretching times the planetary vorticity, $f$, itself. In the midoceanic interior approximation the only vorticity which the fluid has is the planetary vorticity, $f$. The vorticity must increase if the stretching is positive. However, since the relative vorticity is negligible, the fluid elements, to increase their vorticity, must move northward where the planetary vorticity is greater in order that the planetary vorticity which they possess becomes greater. Thus $v$ must be positive. This balance of vorticity production by the stretching of planetary vorticity filaments $f$ and the accession of greater vorticity by meridional motion in the planetary vorticity gradient $\beta$, is called the *Sverdrup relation*. It is important to note that it is a local differential balance in the sense that it holds on each fluid element beneath the mixed layer.

In the mixed layer the effect of stretching is augmented by the direct differential torque due to the fluid turbulent stresses which also act to alter the planetary vorticity of the fluid, and this enters as an additional source on the right side of (1.2.12).

We discuss more completely in Section 1.4 further conditions for the validity of (1.2.14) beyond those that we have so far described. The reader may well ask why any further conditions beyond the smallness of $R_0$ and $E_H$, already described, are necessary. The need for further discussion is, in fact, a measure of the real subtlety of the balance, which at first glance appears so simple and even obvious. As we see below, it is actually neither.

A further balance can be obtained by integrating (1.2.12) over the entire water column. If, again, the slight variations of density are ignored in (1.2.12), the density may be replaced by its average value $\rho_0$ and thus:

$$\rho_0\beta\int v\,dz = \rho_0 f[w(top) - w(bottom) + \hat{k}\cdot\nabla\times\vec{\tau}(top) - \vec{\tau}(bottom)] \tag{1.2.15}$$

where the integral is over the *entire* water column and *top* and *bottom* refer to the top and bottom of the water column.

If we assume, as did Sverdrup (1947) in his pioneering paper, that the time average large-scale vertical velocity at the upper surface is zero (due to the strong gravitational stability of the free surface), and that the interaction of the fluid with the oceanic bottom is negligible because the bottom velocity is very weak and the accompanying stress is weak, then the balance in (1.2.15) involves, as a forcing term on the right side of the equation, only the stress at the oceanic surface. By continuity of stress at the upper surface this must be equal to the applied wind stress on the oceanic surface. To conform to standard notation we call this upper stress simply $\vec{\tau}$. The reader must be careful to distinguish this from the stress that appears in (1.2.4), for example, where $\vec{\tau}$ refers to an internal fluid stress conceptually unconnected to the wind. We introduce the further notational simplification:

$$\text{curl}\,\vec{\tau} \equiv \hat{k}\cdot\nabla\times\vec{\tau}$$

in terms of which we obtain the *Sverdrup balance*:

$$\beta V_S \equiv \beta\int_{-H}^{0} v\,dz = \text{curl}\left(\frac{\vec{\tau}}{\rho_0}\right) \tag{1.2.16}$$

where the interval $(-H, 0)$ refers to the entire water column, $D+h_m$.

This is an extraordinary result. The meridional transport of the entire water column is directly related and locally determined by the curl of the wind stress. Point by geographical point, the meridional transport per unit width depends only on the local wind stress curl. It is otherwise independent of the large-scale distribution of wind stress and is completely independent of the distribution of density stratification in the ocean. No consideration of the density structure of the ocean is required to achieve the result in (1.2.16). Except insofar as it depends of the smallness of the bottom velocity to neglect the interaction with the bottom, it is independent of the vertical distribution of the fluid velocity. It also gives no information about the distribution in the vertical of the fluid velocity. In this sense it is absolutely indifferent to the vertical structure of the motion field. This is both the strength and weakness of the Sverdrup balance.

If the vertical velocity at the top and bottom of the ocean is negligible on the large scale, the vertical integral of (1.2.11) yields:

$$\nabla \cdot \vec{U}_S = 0 \tag{1.2.17}$$

where:

$$\vec{U}_S = \int_{-H}^{0} \vec{u}\, dz. \tag{1.2.18}$$

Thus the total horizontal transport is horizontally nondivergent, and (1.2.17) implies that it may be derived from a stream function, thus:

$$\vec{U}_S = \hat{k} \times \nabla\Psi \tag{1.2.19}$$

where $\Psi$ is the stream function for the horizontal Sverdrup transport, i.e., the transport for the entire water column. In spherical polar coordinates (1.2.19) can be written:

$$U_S = -\frac{1}{R}\frac{\partial\Psi}{\partial\theta}, \quad V_S = \frac{1}{R\cos\theta}\frac{\partial\Psi}{\partial\phi} \tag{1.2.20}$$

where $U_S$ and $V_S$ are the zonal and meridional components of the transport, and $\phi$ is the longitude.

The Sverdrup balance (1.2.16) can then be rewritten as:

$$\frac{\beta}{R\cos\theta}\frac{\partial\Psi}{\partial\phi} = \text{curl}\left(\frac{\vec{\tau}}{\rho_0}\right). \tag{1.2.21}$$

This form of the Sverdrup balance illustrates the first difficulty associated with the theory. The equation for $\Psi$ is first order in $\phi$, and yet $\Psi$ must at least satisfy the condition of no normal flow at the two longitudes that span the basin at each latitude $\theta$ (see Fig. 1.2.2). Integration of (1.2.21) along a line of constant latitude as shown in the figure leads, in general, to a contradiction. The fundamental reason is that (1.2.21) is first order in $\phi$ while two boundary conditions must be applied, one at the east and one at the western boundary of the basin. Physically the reason for this difficulty is that in making approximations suitable for the midocean we have jettisoned many physical processes, some of which become important in local areas of the basin, in particular in the western boundary current, so that (1.2.21) is simply not valid

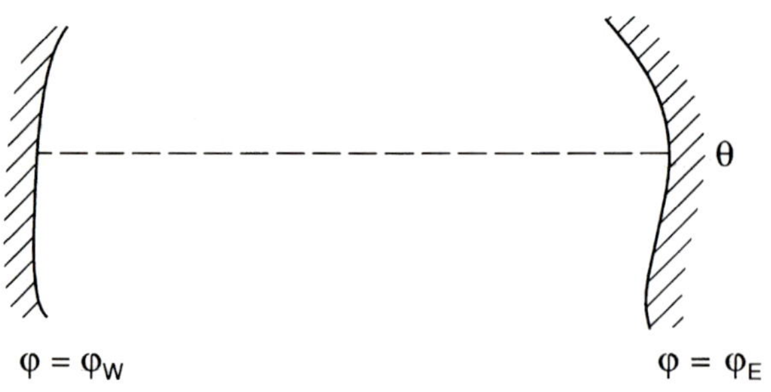


**Fig. 1.2.2.** An ocean basin showing both eastern and western boundaries on which the transport stream function must be a constant. Integrating the Sverdrup relation along the *dotted line* $\theta$ = constant generally leads to a contradiction

over the entire range of longitude of the basin. In the narrow region of the western boundary current the physics that we have retained for the Sverdrup relation must be supplemented by physics that we have neglected, either frictional dissipation or nonlinearity or both.

Sverdrup (1947), without comment, used (1.2.21) to satisfy the boundary condition on the eastern edge of the basin. No doubt he took the very reasonable view that the observed fact of the existence of strong western boundary currents precluded its application in the vicinity of the western boundary. It is important to realize that there is nothing in the Sverdrup balance itself that determines which boundary should be singled out for such treatment.

From a theoretical viewpoint it is necessary first to examine on which boundary it is possible to construct a boundary layer which connects smoothly with the Sverdrup interior solution and satisfies the conditions of no normal flow. Assuming that this can occur on only one boundary forces us to choose the other boundary as the one where the solution of (1.2.21) must satisfy the no normal flow condition. Since the theory of Stommel (1948) for the western intensification of the circulation, and the subsequent prescient discussion of Stewart (1964; both of which we take up in more detail in Chap. 2), it has become clear that if there is to be a Sverdrup interior, the Sverdrup theory must satisfy the boundary condition on the eastern boundary.

This then allows (1.2.21) to be integrated to yield:

$$\Psi = \frac{-1}{\rho_0 \beta} \int_{\phi}^{\phi_E} \operatorname{curl}(\vec{\tau}) R \cos\theta \, d\phi' \tag{1.2.22}$$

from which both $U_S$ and $V_S$ can be obtained using (1.2.20). If the zonal and meridional components of $\vec{\tau}$ are $\tau_\phi$ and $\tau_\theta$, respectively, the curl which appears in (1.2.22) can be written:

$$\operatorname{curl} \vec{\tau} = \frac{1}{R\cos\theta}\left(-\frac{\partial(\cos\theta\tau_\phi)}{\partial\theta} + \frac{\partial\tau_\theta}{\partial\phi}\right). \tag{1.2.23}$$

We note that in (1.2.22) the integral is over the linear *distance* interval from the point at $\phi$ to the eastern boundary.

## 1.3 The Geostrophic Sverdrup Relation

The Sverdrup balance is supposed to give the transport for the entire water column and it is clear from the derivation that the driving force for the transport is the wind, and that this force is transmitted to the water column through the action of the turbulent stresses acting in the mixed layer. Yet we know from observation that the circulation is not limited to the mixed layer even though the stress is. The geostrophic region, in which the turbulent stresses vanish, is also set in motion. How this happens is an important part of the story that we spend much time in developing below, but at this stage it is illuminating and also useful for our later discussion to understand at least how the geostrophic region considered as a whole is set in motion, even though the turbulent stresses vanish there (at least to lowest order).

We start by reconsidering the approximate momentum equation (1.2.7). Since the equation is linear in the velocity field, it is convenient to split the velocity field into two parts. The first is the *wind-driven* part, $\vec{u}_E$ defined by the relation:

$$\rho_0 f\hat{k} \times \vec{u}_E = \frac{\partial \vec{\tau}}{\partial z} \tag{1.3.1}$$

where we remind ourselves that here $\tau$ refers to the turbulent stress in the mixed layer. We assume that $\tau$ and thus also $\vec{u}_E$ vanish below the depth of the mixed layer $h_m$.

The geostrophic part of the velocity is similarly defined as $\vec{u}_G$ and satisfies:

$$\rho_0 f\hat{k} \times \vec{u}_G = -\nabla p \tag{1.3.2}$$

so that:

$$\vec{u} = \vec{u}_G + \vec{u}_E. \tag{1.3.3}$$

If (1.3.1) is integrated over the interval $(-h_m, 0)$ where the turbulent stresses are different from zero, we obtain an expression for the total horizontal transport associated with the wind stress, called the Ekman transport, namely:

$$\vec{U}_E = \int_{-h_M}^{0} \vec{u}_E \, dz = -\frac{\hat{k} \times \vec{\tau}}{\rho_0 f} \tag{1.3.4}$$

where the stress in (1.3.4) is the wind stress. Since the continuity equation is also linear in the velocity a similar decomposition can be made for the vertical velocity. Thus:

$$\frac{\partial w_E}{\partial z} = -\nabla \cdot \vec{u}_E \tag{1.3.5}$$

which when integrated over the mixed layer depth yields:

$$w_E(0) = -\nabla \cdot \vec{U}_E = -\text{curl}\left(\frac{\vec{\tau}}{\rho_0 f}\right) \tag{1.3.6}$$

where the upper surface is placed at $z = 0$. Note that the frictionally driven vertical velocity vanishes at the base of the mixed layer, by definition.

The total vertical velocity must vanish at the upper surface, thus:

$$w_G(0) = -w_E(0) = \text{curl}\left(\frac{\vec{\tau}}{\rho_0 f}\right) . \tag{1.3.7}$$

The convergence of the stress-driven mass flux in the upper mixed layer is compensated for by a vertical flux from the mixed layer into the geostrophic region below the mixed layer as given by (1.3.7). It is traditional to think of (1.3.7) as being the vertical mass flux at the base of the mixed layer produced by the wind-driven flow. It is, actually, the geostrophic vertical velocity at the surface. If $h_m$ is much less than $D$, the difference between these two interpretations is negligible, and this is the case when formal boundary layer theory is used to deal with the stress-driven flow of the mixed layer, for then, asymptotically, $h_m/D \ll 1$. If $h_m/D$ is small but not asymptotically small (as is the case in the natural ocean) the current treatment is necessary. Note that the actual vertical velocity at the base of the mixed layer is:

$$w(-h_m) = w_G(-h_M) = w_G(0) + \int_{-h_M}^{0} \frac{\beta v_G}{f} dz' . \tag{1.3.8}$$

Using (1.2.8) and the continuity equation, we obtain for the geostrophic velocity:

$$\beta v_G = f \frac{\partial w_G}{\partial z} \tag{1.3.9}$$

which when integrated over the entire water column yields the meridional transport of the geostrophic flow, i.e., again assuming no interaction with the bottom:

$$V_G = \int_{-D}^{0} V_G \, dz = \frac{f}{\beta} \text{curl}\left(\frac{\vec{\tau}}{\rho_0 f}\right). \tag{1.3.10}$$

We can think of the geostrophic flow being driven by the weak vertical velocity produced in the upper mixed layer by the convergence of the Ekman transport.

Since:

$$f \text{curl}\left(\frac{\tau}{\rho_0 f}\right) = \text{curl}\left(\frac{\vec{\tau}}{\rho_0}\right) + \hat{k} \cdot \left(\vec{\tau} \times \frac{\nabla f}{\rho_0 f}\right) = \beta V_S + \frac{\tau_\phi \beta}{\rho_0 f} \tag{1.3.11}$$

it follows that

$$\beta(V_G + V_E) = \beta V_S \tag{1.3.12}$$

so that the Sverdrup transport is split between the direct wind-driven transport of the mixed layer and the geostrophic transport which is forced by the stretching (or compression) of planetary vortex filaments by the vertical velocity pumped out of the mixed layer. The latter mechanism is a purely inertial one, caused indirectly by the stress in the mixed layer. The wind directly drives the flow in the mixed layer through the action of turbulent stresses. The ocean beneath is set into motion by the purely inertial effects of vortex tube stretching. The weak vertical velocity acting on the (relatively) large planetary vorticity is the principle source of motion in the geostrophic region below the mixed layer.

Generally the transport associated with the geostrophic portion of the Sverdrup flow exceeds that due to the Ekman transport since their ratio is:

$$\frac{V_G}{V_E} = O\left(\frac{\text{curl}(\vec{\tau}/\beta)}{\vec{\tau}/f}\right) = O\left(\frac{f}{\beta L}\right) = \frac{R}{L}\tan\theta. \tag{1.3.13}$$

This ratio is normally greater than 1 since the length scale, $L$, associated with the wind stress and the gyre scale is usually considerably smaller than the earth's radius $R$. At 45°N, where $\tan\theta$ is 1, the ratio is of the order of 6 if $L$ is 1000 km, and it is of order 3.5 at 30°N. In fact, in many cases theories for the oceanic circulation focus entirely on the geostrophic portion of the transport. This can lead to some confusion about the total mass balance if care is not taken. In addition, the ratio in (1.3.13) reverses in magnitude for small $\theta$ as the equator is approached. Indeed, a careful examination of (1.3.4) and (1.3.10) shows that as the equator is approached $V_G \to -V_E$.

## 1.4 On the Validity of Sverdrup Theory

The elements of the Sverdrup theory, while limited to a description of only the vertical integral of the motion field in the interior of the ocean, still form a central building block of all the theories of the circulation that we discuss in later chapters. It is therefore useful to examine more carefully the requirements for its validity and the evidence for its correctness.

Sverdrup theory, as we have presented it, consists of two parts. There is first the approximations leading to the Sverdrup vorticity balance, for example (1.2.14) for the region below the mixed layer, and the Sverdrup balance (1.2.16), for the vertically averaged meridional transport. The latter depends on the former and the requirements for the *Sverdrup balance* are consequently more severe. We examine each of these in turn.

First, let us consider the Sverdrup vorticity relation. As discussed in Section 1.2, at least both the Rossby number and horizontal Ekman number must be small, i.e.:

$$\begin{aligned} R_0 &= \frac{U}{fL} \ll 1 \\ E_H &= \frac{A_H}{fL^2} \ll 1. \end{aligned} \tag{1.4.1a, b}$$

The first of these is certainly met for the large-scale flow, and the second, we have argued, is most likely to be true. These conditions are required so that, below the mixed layer, the ocean is in geostrophic balance. That is, that the advection of momentum is then negligible compared to the Coriolis acceleration or equivalently, that the relative vorticity:

$$\zeta = \hat{k} \cdot \nabla \times \vec{u} = O(U/L) \tag{1.4.2}$$

be small compared to the planetary vorticity, $f$.

The Sverdrup relation is an approximation to the *vorticity equation* and comes from taking the curl of the momentum equation. This involves a comparison of the advection of the relative vorticity and the advection of the planetary vorticity rather than a comparison of the relative and planetary vorticities themselves. The relative vorticity, $\zeta$, varies on the length scale $L$. However the planetary vorticity, by (1.2.13), varies on the planetary scale $R$. As long as $R$ exceeds $L$, this implies that $f$ varies more slowly than $\zeta$. Thus, in order to neglect the advection of relative vorticity compared with the advection of planetary vorticity we must have:

$$\vec{u} \cdot \nabla\zeta = O\left(\frac{UU}{L^2}\right) \ll \vec{u} \cdot \nabla f = O(U\beta) \quad \text{or} \tag{1.4.3}$$

$$\varepsilon = \frac{U}{\beta L^2} \ll 1. \tag{1.4.4}$$

Since usually $\beta L < f$ this last condition is a more stringent condition than the smallness of the Rossby number. For $L = 1000$ km, $\beta = 2.1 \times 10^{-13}$ cm$^{-1}$ s$^{-1}$, and $U = 1$ cm/s, $\varepsilon$ is $5 \times 10^{-4}$, which is larger than $R_0$ but is still small enough to imply that we may neglect relative vorticity advection.

A more subtle requirement for the validity of the Sverdrup relation has been somewhat hidden by our scaling argument. When we make a scaling argument and attribute to the motion field a scale $L$, say, for the length scale of the motion, we are tacitly assuming that there is only one length scale that characterizes the flow. The same is true for the velocity scale $U$. If, as in the case of the ocean, there are several scales of motion coexisting in space, the argument becomes more subtle. Since the 1960s it has been evident that the ocean is populated by small- and meso-scale eddies. The eddies are variable in intensity, and their geographical distribution is nonuniform. The character of the eddy field has been the subject of intense observational and theoretical study, and its description is beyond the scope of this book. However, we need to recognize that the eddies are distributed over the whole oceanic basin, and that they possess scales for length and velocity $L_{eddy}$ and $U_{eddy}$, respectively, such that:

$$U_{eddy}/\beta L^2_{eddy} = O(1) \tag{1.4.5}$$

so that on the *eddy* scale the advection of relative vorticity cannot be ignored.

For the large scale flow of interest to us here, it is the collective, average effect of the eddies that is of interest. In the vorticity balance on the large scale this may disturb the Sverdrup relation if the *large*-scale divergence of the eddy flux of relative vorticity is as large as the advection of planetary vorticity. The former is:

$$\nabla \cdot \overline{\vec{u}\zeta} = O\left(\frac{U_{eddy}^2/L_{eddy}}{L}\right) \tag{1.4.6}$$

where the overbar refers to an average over the large-scale flow with length scale $L$. In (1.4.6) it is assumed that the relative vorticity flux has the local value $U_{eddy}(U_{eddy}/L_{eddy})$, and that this flux varies smoothly over the large-scale $L$. When compared with the large-scale advection of planetary vorticity which is $O(U\beta)$, this gives a ratio of the divergence of eddy flux of relative vorticity to the planetary advection of vorticity which is apparently of the order:

$$\frac{U_{eddy}^2/L_{eddy}L}{U\beta} = \left(\frac{U_{eddy}}{\beta L_{eddy}^2}\right)\frac{U_{eddy}L_{eddy}}{UL}. \tag{1.4.7}$$

If $L_{eddy}$ is of the order of 50 km, with $\beta$ as before an eddy velocity of only 5 cm/s would make the first bracket on the right-hand side of (1.4.7) of order 1. The second factor on the right-hand side of (1.4.7) is the ratio, assuming the large-scale and eddy flows are both in geostrophic balance, of the contribution to the dynamic pressure field of both the mean and the eddies. It is likely that this number is of order 1 as well since the geostrophic disturbance to the large scale pressure field by the eddy is of the order of the background field itself if the eddies spring from a local instability of the large-scale flow. This would seem to imply that the eddies could easily upset the Sverdrup relation of vorticity. However, our estimate of $E_H$ which is supposed to measure the effect of smaller scale motions on the vorticity dynamics of the large scale implied that the eddies would be negligible.

If nothing else, this illustrates the care which must be taken in making scaling estimates. Let us return to our estimate of the relative vorticity flux $\vec{u}\zeta$. We estimated its average over the large scale by multiplying its local value by the large scale area, $L^2$, and then dividing by the same area to obtain the estimate in the numerator of (1.4.6). However, each component of the vector vorticity flux for the eddies can itself be written as a divergence. This occurs because on the eddy scale the motion is quasi-geostrophic, and the eddy velocity field is horizontally nearly nondivergent (Pedlosky 1987; Chap. 7). On the smaller scales of the eddies a locally Cartesian coordinate frame is adequate so that we can write:

$$\vec{u}\zeta = \vec{u}\left(\frac{\partial v}{\partial x} - \frac{\partial u}{\partial y}\right) \tag{1.4.8}$$

where $x$ and $y$ are coordinates to the east and north, respectively. The $x$ and $y$ components of (1.4.8) can be written:

$$u\zeta = \frac{\partial}{\partial x}(uv) + \frac{1}{2}\frac{\partial}{\partial y}(v^2 - u^2), \quad v\zeta = -\frac{\partial}{\partial y}(uv) + \frac{1}{2}\frac{\partial}{\partial x}(v^2 - u^2) \tag{1.4.9}$$

so that when integrated over the large scale the vorticity flux, which is written in terms of perfect differentials, yields terms for the average vorticity flux that depend only on their values at the periphery of the integration region and hence are of order:

$$\begin{aligned} \overline{\vec{u}\zeta} - O\left(\frac{U^2_{eddy}}{L}\right) = \left(\frac{U^2_{eddy}}{L_{eddy}}\right)\frac{L_{eddy}}{L}, \quad \text{or} \\ \nabla \cdot \overline{\vec{u}\zeta} = O\left(\frac{U^2_{eddy}/L_{eddy}}{L}\right)\frac{L_{eddy}}{L}. \end{aligned} \tag{1.4.10}$$

This leads to a more accurate estimate for the eddy flux of vorticity on the large scale which is an order $L_{eddy}/L$ smaller than our previous estimate (1.4.6). This ratio of length scales is very small and we can then, as implied by our discussion of $A_H$, neglect the eddy flux on the large scale unless we are in regions of the general circulation possessing length scales considerably smaller than the gyre scale. Such regions do exist near the regions of strong currents on the western boundaries of the ocean, and we cannot expect the Sverdrup relation to be valid in such regions.

It seems reasonable that the Sverdrup relation, which is the local approximation to the vorticity equation would be valid outside regions of strongly varying eddy fields, i.e., in the eastern oceanic interior. The situation is far less clear for the *Sverdrup balance* (1.2.12) which is the approximation to the vertical integral of the vorticity equation in which all interactions with the bottom are neglected.

A priori there are two principal potential contributors to the bottom interaction that could upset the integrated Sverdrup balance. The first is due to the existence of a nonzero bottom stress while the second is due to a nonzero vertical velocity forced at the bottom by sloping topography. We can attempt to estimate the former as follows. From general boundary layer theory the effect of the bottom stress can be represented in terms of a vertical velocity pumped out of the bottom boundary layer as a consequence of that stress. Ekman layer theory for a viscous boundary layer on the bottom (Pedlosky 1987; Chap. 4) would give, as at least an estimate of that vertical velocity:

$$w_b = \frac{\delta_E}{2}\zeta_b \tag{1.4.11}$$

where $\delta_E$ is the thickness of the bottom boundary layer, and $\zeta_b$ is the vorticity of the flow just above the bottom boundary layer. This leads to an estimate for $w_b$ of the order of:

$$w_b = O\left(\frac{\delta_E U}{L}\right) \tag{1.4.12}$$

while using the Sverdrup relation for the geostrophic flow allows us to estimate $U$ in terms of the Ekman pumping, so that:

$$w_b = O\left(\frac{\delta_E}{D} \cdot \frac{f}{\beta L}\right) w_E \tag{1.4.13}$$

and thus, as long as $\delta_E/D$ is much smaller than $\beta L/f$, the effect of the bottom stress can be ignored. This condition is likely to be met on the large scale since $\beta L/f$ is of order 1/6 while $\delta_E/D$ is considerably smaller.

The more likely difficulty may be due to the role of topography. In the presence of a sloping bottom, i.e., in regions where $D$ is not constant, the vertical velocity produced by the bottom slope is of the order of:

$$w_b = -\vec{u}_b \cdot \nabla D \tag{1.4.14}$$

where $\vec{u}_b$ is the horizontal velocity just above the bottom boundary layer. If the depth changes by only 1 km in 1000 km of horizontal distance, a bottom velocity of only 0.1 cm/s would give rise to a vertical velocity of $O(10^{-4}$ cm/s), which is as large as the estimates of Ekman pumping velocity given for the subtropical gyres. Whether such persistent large-scale bottom velocity exists is unclear. However, the Sverdrup balance is clearly vulnerable to such weak flows.

Observational attempts to verify the Sverdrup theory have been scanty due to the difficulty of making long-term measurements directly of velocity in the interior of the ocean over the entire water column. Certainly, it is currently impossible to measure the vertical velocity directly to decide whether the Sverdrup vorticity equation (1.2.14) is locally valid. Most attempts have instead focused on the Sverdrup balance (1.2.16) in an effort to compare the transport deduced from the observed wind stress curl (a difficult quantity itself to produce) with independently produced estimates, usually from hydrographic data, of the gyre transport.

At the present time the discussion of the adequacy of the Sverdrup balance remains controversial. The first attempt to use the Sverdrup relation to calculate quantitatively the circulation was due that of Welander (1959). His map of the Sverdrup circulation bore a strong qualitative resemblance to the observed oceanic gyres, but it requires more careful analysis to go beyond a mere qualitative resemblance and discuss the quantitative adequacy of the theory. More recently Leetmaa et al. (1977) and Schmitz et al. (1992) have attempted to compare quantitatively the transport across 24°N in the North Atlantic calculated from the Sverdrup balance with the transport inferred from hydrographic data. In fact, the interior transport is not at all easy to calculate from hydrographic data. The observed density field in the ocean can be used to obtain the vertical shear of the horizontal velocity from the thermal wind relation (e.g., Pedlosky 1987), but the calculation of the velocity and hence the transport requires assumptions about the vertical structure of the velocity, in particular the position in depth where the velocity vanishes (or, equivalently, a value for the velocity at a particular depth is required). The location of this "level of no motion" is one of the most perplexing issues in observational oceanography, and its determination is often a matter requiring very subjective judgments. Instead, the comparison of the Sverdrup transport across 24°N calculated from the wind stress curl has been made with the more reliable direct measurements of the return flow in the Florida Current. This is the region where the Gulf Stream flows through the Straits of Florida, and the transport of the current can be measured relatively easily there. There is general agreement that the transport is about 30 sverdrups and represents the return flow which balances the *total net* mass transport southward in the interior. The estimated Sverdrup transport across the Atlantic at this latitude is also about 30 sverdrups (see Fig. 1.4.1), and thus the agreement in this sense is quite good.

Difficulties arise, however, when various authors have attempted to attribute to a particular depth range the wind-driven portion of the total circulation. In Fig. 1.4.1, for example, an attempt is made to compare the

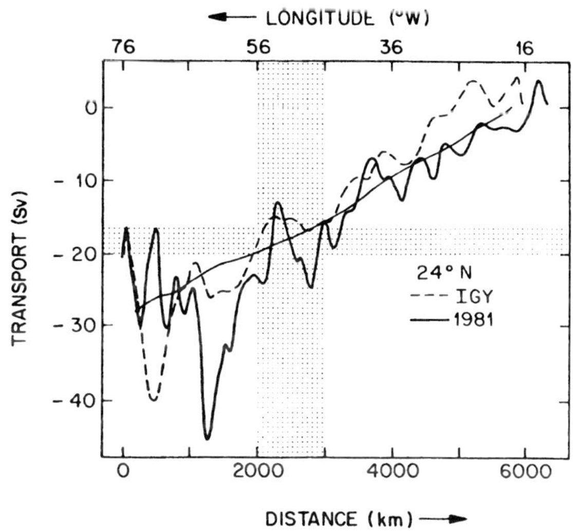


a

**Fig. 1.4.1a,b.** Estimated Sverdrup transport across 24°N (from Schmitz et al. 1992). **a** *Thin line*, Sverdrup transport as a function of longitude starting at zero on the eastern boundary and increasing to –30 sverdrups at the Atlantic's western edge. **b** As in **a** but with a slightly different wind field. The transport is proportional to the ordinate. The solid line is calculated from the wind stress, the circles from hydrographic observation

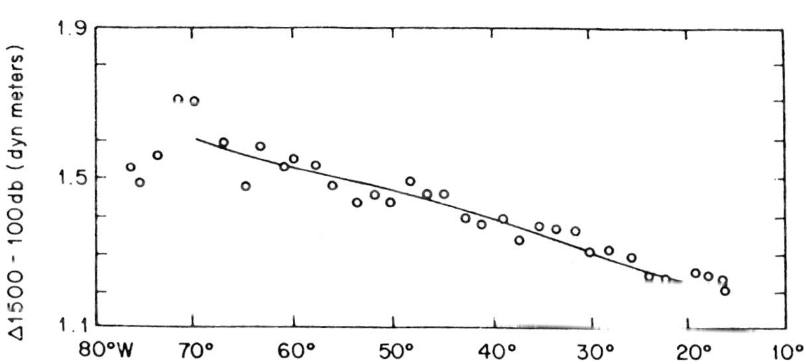


b

interior geostrophic transport, calculated from hydrographic data, for the region above the $\sigma_\theta = 27.4$ density surface (at about 900 m in depth) with the Sverdrup transport. The agreement is quite good in the eastern basin, i.e., east of 55°W, but the calculated transport is spatially quite variable to the west of that longitude. Only an extrapolation of the trend of the geostrophic transport yields the 30 sverdrups that would balance the Florida Current transport and match the calculation of Sverdrup transport. The issue of this extrapolation and the question of whether the wind-driven Sverdrup transport is sufficient to explain the observed transports has led to considerable controversy. If only the region east of 55°W is used as a reliable region for the calculation of the wind-driven circulation by the Sverdrup theory, the total transport that results is only about 17 sverdrups, much less than the 30 sverdrups needed. Wunsch and Roemmich (1985) have suggested, partly as a consequence of this imbalance, that the evidence for the agreement between the geostrophic wind-driven transport and the Sverdrup transport is fundamentally lacking. Schmitz et al. (1992) and Schmitz and McCartney (1993) have made an alternative suggestion in which only 17 sverdrups of the Florida Current are wind driven while 13 sverdrups flow northward in the current in compensation for a deep southward flow in the interior driven by deep water formation in the northern North Atlantic. Thus they suggest that actually only 17 sverdrups need be acounted for by the Sverdrup transport and consider this evidence of the validity of the Sverdrup balance. However, unless the southward flowing transport of 13 sverdrups occurs in a deep western boundary layer, east of Florida current but exempt from Sverdrup dynamics because it forms part of the general western boundary current system, the *total* interior transport in the interior would still have to account for 30 sverdrups, in agreement with the direct calculation using (1.2.22).

The total circulation can be driven by both wind stress and heating and cooling at the surface (or within the fluid). The virtue of the Sverdrup balance is that, if it is true, it tells us that the net, vertically averaged transport must be due only to the wind stress curl, and that any other part of the motion must have a zero vertical average. Unless the total flow vanishes below a certain depth, it is problematic to attempt to apply the Sverdrup relation to only a portion of the water column in the belief that it will capture only the wind-driven portion of the flow. Figure 1.4.2 shows a schematic of the problem. On the left is a schematic presentation of the velocity distribution in the interior. The upper fifth of the water column possesses a strong southward flow, and beneath it is a fairly weak flow which, however, distributed over the water column, has about the same transport. We can consider the flow as made up, as shown on the right of the figure, of a part that has a net transport (and is here arbitrarily limited to the upper part of the water column) and another part which is a purely internal mode, i.e., a part that has no net transport. If we attempt to identify the Sverdrup relation with the transport in the upper part of the water column, as shown on the left, we would seriously underestimate the transport. The Sverdrup balance should apply to the whole column and would

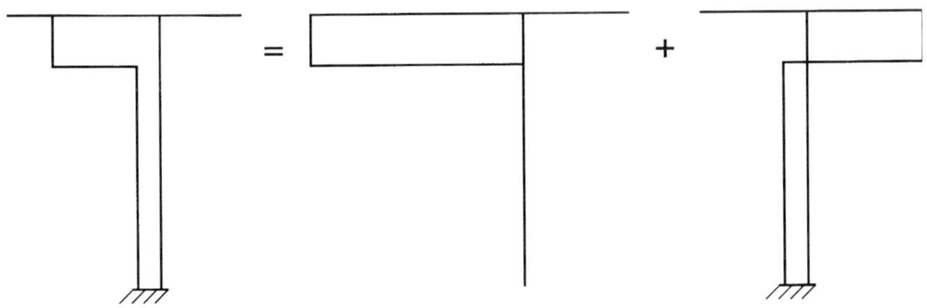

**Fig. 1.4.2.** Schematic presentation of the way in which the observed distribution of transport with depth can be considered a sum of a net component plus an internal mode. The Sverdrup balance should apply only to the net component, i.e., the vertical average over the total water column

give us the transport of the wind-driven portion as shown on the left of the second panel. If the Sverdrup balance is applied to only a portion of the water column, the vertical velocity in (1.2.15) at the "bottom" of the range of integration is not zero, and the connection of the transport to the wind driving will be lost. Unless a deep level can be determined where $w$ is zero, the Sverdrup balance must be applied to the whole water column.

Much of the disagreement in the literature over the validity of the Sverdrup balance is in fact connected to such attempts to apply the relation to only a portion of the water column. Wunsch and Roemmich (1985) have criticized earlier attempts to make such applications and present compelling arguments for the fragility of currently available supporting evidence beyond the overall transport balance previously noted. In particular they argue, as described above, that the integrated Sverdrup balance, as opposed to the Sverdrup relation, i.e., the approximate vorticity equation, is vulnerable to the effects of even small velocities interacting with the bottom slope.

Calculations of the pattern of Sverdrup transport more modern than Welander's nevertheless give rather good agreement with the overall *pattern* of flow deduced from hydrography (see Fig. 1.4.3, for example, which should be compared with Fig. 1.1.1a). However, in spite of this apparent agreement the reader must remain aware that even to this day precise, incontrovertible observational evidence for the validity of either the Sverdrup vorticity equation or the Sverdrup balance is lacking although the evidence in its favor is compelling. Nevertheless, the Sverdrup relation and balance are employed throughout this book. It currently forms the basis for all our present theories for the circulation.

There is even a more subtle issue associated with the validity of the Sverdrup balance. Let us even suppose that observational evidence clearly supports its validity. Could we maintain that we can theoretically justify or explain the Sverdrup balance even assuming, for example, that there were no interaction with the bottom? One might suppose that for a given wind stress we could calculate the velocity from Sverdrup theory and then verify a posteriori

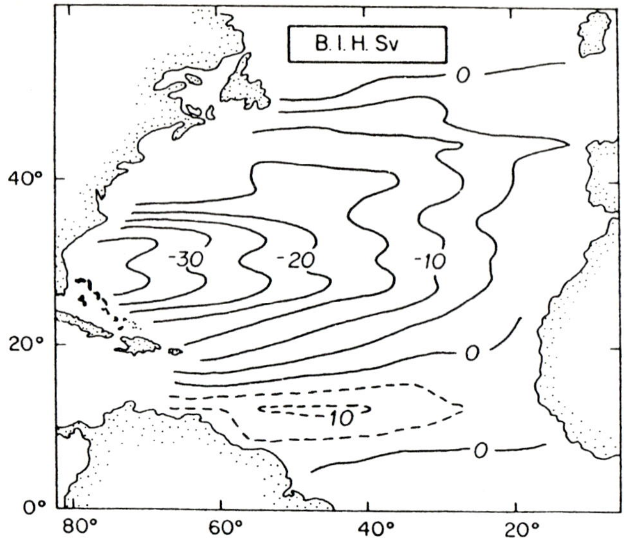


**Fig. 1.4.3.** Calculated Sverdrup transport from observed winds. This figure should be compared with Figure 1.1.1. (From Schmitz et al. 1992)

that the terms we have neglected are indeed small. This would not however constitute a verification of the theory. The Sverdrup solution for the interior is still only a partial solution to the circulation problem, and we must demonstrate that it can be consistently connected to a western boundary current, or a system of such currents to close the circulation and render the solution complete. The Sverdrup solution may appear to satisfy all the parameter inequalities required for its validity, but since the Sverdrup theory is incapable by itself in satisfying all the physical boundary conditions on the flow, *it fails even as a partial solution if it cannot be connected consistently to a western boundary layer that satisfies the physical boundary conditions*. This is neither an obvious nor a simple calculation to carry out. In the next chapter we examine a very simple model of the ocean circulation, the homogeneous model, in order to examine this point in detail. However, the reader should bear in mind that in addition to the absence of complete observational evidence for the validity of Sverdrup theory, the Sverdrup theory of the interior by itself cannot be considered theoretically justified without a discussion of the full problem. Before Sverdrup theory can be accepted as a justified theoretical deduction, the full problem which includes western boundary currents with their more complicated physics, hitherto neglected, must be understood.

# References

Brown, E.D. and Owens B., 1981: Observations of the horizontal interactions between the internal wave field and the mesoscale flow. *J. Phys. Ocean.* **11,** 1474–1480.

Bryan, K., 1987: Potential vorticity in models of the ocean circulation. *Q. J. R. Meteorol. Soc.* **113,** 713–734.

Ekman, V.W., 1905: On the influence of the earth's rotation on ocean currents. *Arkiv f. Matem., Astr. o. Fysik* (Stockholm) Bd. 2, No. 11, 53 pp.

Leetmaa, A., Niiler P. and Stommel H., 1977: Does the Sverdrup relation accoount for the mid-Atlantic circulation? *J. Mar. Res.* **35,** 1–10.

Pedlosky, J., 1987: Geophysical fluid dynamics. *Springer,* New York, 710 pp.

Pickard, G.L. and Emery W.J., 1982: Descriptive physical oceanography. *Pergamon Press,* Oxford, pp 249.

Reid, J.L., 1994: On the total geostrophic circulation of the North Atlantic Ocean: Flow patterns, tracers and transports. *Prog. Ocean.,* **33,** 1–92.

Schmitz, W.J., and McCartney M.S., 1993: On the North Atlantic circulation. *Rev. of Geophys.,* **31,** 29–43.

Schmitz, W.J., Thompson J.D. and Luyten J.R., 1992: The Sverdrup circulation for the Atlantic along 24°N. *J. Geophys. Res.* **97,** 7251–7256.

Stewart, R.W., 1964: The influence of friction on inertial models of the oceanic circulation. *Studies on oceanog.* Univ. of Wash. Press. 3–9

Stommel, H., 1948: The westward intensification of wind-driven ocean curents. *Trans. Amer. Geophys Union,* **29,** 202–206.

Stommel H, Niiler P. and Anati D., 1978: Dynamic topography and recirculation of the North Atlantic. *J. Mar. Res.,* **36,** 449–468.

Sverdrup, H.U., 1947: Wind-driven currents in a baroclinic ocean; with application to the equatorial currents of the eastern Pacific. *Proc. Natl. Acad. Sci. Sci.* **33,** 318–326.

Welander, P., 1959: On the vertically integrated mass transport in the oceans. In *The atmosphere and the sea in motion.* Ed. Bolin B. *Rockefeller Institute Press,* New York, pp 75–100.

Wunsch, C. and Roemmich D. 1985: Is the North Atlantic in Sverdrup balance? *J. Phys. Ocean.* **15,** 1876–1880.

# 2 Homogeneous Models of the Ocean Circulation

## 2.1 Introduction

The Sverdrup balance, which was derived in Chapter 1 for a baroclinic flow that does not interact with the bottom, is also valid for a homogenous (constant density) ocean if the bottom is flat, and if the bottom friction is small in the sense described in Section 1.4. Thus both a stratified and a homogeneous model would have the same Sverdrup transport for the same wind stress. This has encouraged the belief that the homogenous model, more tractable than a general stratified model because its horizontal velocity is independent of depth, might adequately represent at least the vertical average of the actual circulation. Although this might be true in the interior of the ocean under the conditions that the Sverdrup theory holds, it is less likely to be true in any region, such as the western boundary layer, where nonlinear effects are probably important. The vertical average of quadratic products representing nonlinear processes in the equations of motion does not equal the product of the vertical average of each of the terms. Hence even for the vertical average of the circulation the homogeneous model is expected to be seriously inaccurate.

We are interested in the homogeneous model actually less because we expect it to be able to simulate accurately any particular aspect of the actual circulation than because it presents in a more tractable form some of the same important dynamical issues as the general baroclinic problem. In particular, we are interested in understanding the relationship between the interior circulation and the dynamics of the western boundary current. Can the Sverdrup interior really always be joined to a western boundary current? There are, as we shall see in investigating this central question, some surprising links between the detailed dynamics of the western boundary current and the overall circulation pattern, even in the interior.

One of the most vexing problems in theoretical oceanography is the specification of a useful model for the dissipation of energy and vorticity, i.e., the parameterization of small-scale turbulent mixing unresolved by our explicit, large-scale dynamical equations. The hope has often been expressed that when the circulation is vigorous enough that nonlinearity is dominant, the effect of small-scale dissipation of unresolved motions is negligible for the dynamics, or at least unimportant, qualitatively. Confidence in the inconsequential nature of this dissipation reached its culmination, perhaps, in the 1960s. The very interesting pioneering paper of Carrier and Robinson (1962), in which the explicit role of dissipation is essentially ignored in favor of an attempt to construct a purely inertial model of the circulation, is a good example of such confidence. The hope in the inconsequential role for friction is partly related to our inability to specify with confidence either an appropriate mathematical representation of the role of small-scale friction or its magnitude. It is an understandable human inclination to hope that what we cannot do is unimportant.

Since most parameterizations of small-scale turbulent dissipation lead to mathematical representations, such as the diffusion of large-scale momentum, that raise the mathematical order of the governing equations, additional boundary conditions must also be specified. Various boundary conditions could be applied in addition to the kinematic condition of no normal flow, such as a condition of no slip on the tangential velocity, a condition on the boundary stress, and others which we discuss below. Several are plausible; none is rigorously required from first principles for the approximate equations governing only the large-scale flow. It is unclear how to chose between them.

Along with the hope that friction is physically inconsequential if only small enough, there is also frequently expressed the hope that the overall circulation is largely independent of the boundary condition chosen to accompany the raised order of the equations since there is rarely any determining argument that forces us to choose one or the other of these proposed boundary conditions. Often the conditions are chosen for convenience of calculation or from a subjective preference for the results eventuating from one or another of the choices.

Some reflection indicates that these hopes are all likely to be disappointed. The fact that the circulation of the ocean actually consists of an endless recirculation means that no matter how small the dissipation is, the fluid has substantial time to experience the action of dissipative forces. Stated differently, if the forcing by the wind continuously puts vorticity into the fluid, dissipation must continuously take it out. Since each fluid element is directly acted on by the wind during at least part of the time that it traverses the gyre, each fluid element receives an input of vorticity during its repetitive journey which it must somehow contrive to lose in the remaining part of its circuit around the gyre if a steady state is to be established. Each fluid element must therefore pass through regions where dissipation is important so that the input of vorticity and energy by the wind can be balanced by a dissipative drain of the same quantity. Since the input is order 1 (in the sense that the flow is driven by this input) the dissipation, *somewhere* must be order 1 for *each* fluid element no matter how small the dissipation is estimated to be for the large-scale motion.

The question that then arises is not whether this dissipation occurs. It must. The dynamical issue of importance is whether in contriving a way to rid itself of the vorticity and energy continuously added by the wind, the resulting circulation retains the character of the Sverdrup theory for the interior, and to what degree the nature of the total circulation depends on the details of the model of dissipation that we adopt.

This chapter considers the dynamics of the homogeneous model with this aspect of the dynamics in mind as a preliminary to our study of stratified models of the circulation. After a brief introduction and review of linear theory we confront the issues associated with the conjunction of nonlinearity and dissipation. The reader is referred to earlier reviews, such as Hendershott (1987) and Pedlosky (1987), for a more complete discussion of the formulation of the model and its relevant linear theory.

## 2.2 Formulation of the Homogeneous Model

There are actually at least two different ways to think about the homogeneous model. We might think of a single layer of fluid in which the motion is exactly two-dimensional and incompressible in which the rate of change of the total vorticity, planetary plus relative vorticity, is produced by the wind-stress curl acting as a body force in the layer, and in which the vorticity is dissipated on each fluid element by dissipation mechanisms of various forms, for example, proportional to the vorticity itself and/or due to the lateral diffusion of the vorticity. The exact two-dimensionality of such a model allows the flow to be written immediately in terms of a stream function $\psi$.

We take another approach that leads to a governing equation with the same structure but which is more consistent with the physical ideas discussed in Chapter 1 and the models described in later chapters.

We consider the model shown in Fig. 2.2.1. A homogeneous layer of fluid of constant thickness, $D$, lies beneath a mixed layer in which the turbulent stresses excited by the wind stress drive an Ekman transport as described in Chapter 1. Under the layer, and in contact with the bottom, lies another viscous boundary layer in which viscous stresses produced by the motion couple the fluid to the solid bottom.

We assume that the motion in all parts of the fluid is slow enough so that the flow is in geostrophic balance everywhere. Further, we assume that the north-south scale of the motion, though large enough to make the Rossby number small, is at the same time small enough that the $\beta$ plane approximation is valid. This allows us to use Cartesian coordinates, but more importantly it allows us to write the geostrophic balance for the momentum approximation in terms of a characteristic value, $f_0$, of the Coriolis parameter in the gyre. Thus if:

$$\beta L/f_0 \ll 1 \tag{2.2.1}$$

the geostrophic relations for the eastward and northward velocities in the layer of depth $D$ can be written, to $O(\beta L/f_0)$:

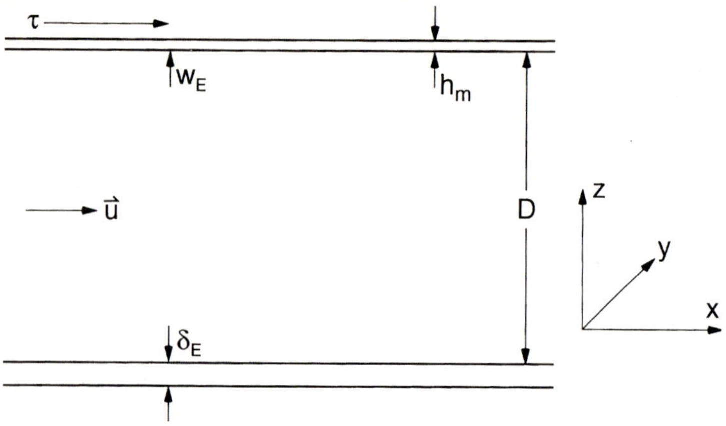


**Fig. 2.2.1.** Schematic cross section of the homogeneous model. The model decribes the motion in layer of thickness $D$ surmounted by a mixed layer in which the wind-driven turbulent stresses drive an Ekman transport whose divergence yields the Ekman vertical velocity $w_E$. The layer is bounded below by a viscous Ekman layer that couples the flow to the bottom and helps dissipate the motion

$$u = -\frac{\partial\psi}{\partial y}, \quad v = \frac{\partial\psi}{\partial x} \tag{2.2.2}$$

where:

$$\psi = \frac{P}{\rho f_0}. \tag{2.2.3}$$

The fluid pressure is $p$, its density is $\rho$, and $f_0$ is the Coriolis parameter at the central latitude of the gyre.

On the $\beta$ plane the vorticity equation takes a particularly simple form. Since the horizontal velocity is independent of depth, the only way in which the total vorticity can be altered in the geostrophic region between the upper and lower boundary layers is by vortex tube stretching of the total vorticity or by lateral diffusion of the relative vorticity. Thus if:

$$\zeta = \frac{\partial v}{\partial x} - \frac{\partial u}{\partial y} \tag{2.2.4}$$

is the relative vorticity, it satisfies the equation:

$$\frac{d(\zeta + \beta y)}{dt} = (f_0 + \beta y + \zeta)\frac{\partial w}{\partial z} + A_H\nabla^2\zeta. \tag{2.2.5}$$

The first term on the right side is the vorticity production by the stretching of the total vortex filaments by the vertical velocity while the second term is the lateral diffusion of vorticity. Since $\beta y \ll f_0$ by the $\beta$ plane approximation, and $\zeta \ll f_0$ by the geostrophic approximation:

$$f_0 + \beta y + \zeta \approx f_0 \tag{2.2.6}$$

in the stretching term in (2.2.5).

The density of the layer is constant and so the spatially variable part of the pressure field, which is in hydrostatic balance with the gravitational force, is independent of $z$. The horizontal, geostrophic velocity is therefore independent of $z$, and thus an integral of (2.2.5) over the layer thickness, $D$, yields:

$$\frac{d\zeta}{dt} + \beta v = \frac{f_0}{D}(w_E - w_B) + A_H \nabla^2 \zeta \tag{2.2.7}$$

where $w_B$ is the velocity pumped out of the lower boundary layer and $w_E$ is the Ekman velocity pumped out of (or into) the upper mixed layer.

We use standard Ekman layer theory to represent $w_B$ as (see Pedlosky 1987):

$$w_B = \frac{\delta_E}{2}\zeta \tag{2.2.8}$$

where $\delta_E$ is much less than $D$. $w_E$ can be written in terms of the wind stress as in Chapter 1 although it is convenient to leave it in its present form. It is important to remember, however, that it is given entirely in terms of the wind stress, and that its curl is independent of the motion and is thus a specified function of horizontal position and provides the forcing for the circulation.

If (2.2.2) is used to write the velocity and vorticity in terms of the stream function, we obtain the governing equation for the homogeneous model:

$$\frac{\partial}{\partial t}\nabla^2\psi + J(\psi, \nabla^2\psi) + \beta\frac{\partial\psi}{\partial x} = \frac{f_0}{D}w_E - r\nabla^2\psi + A_H\nabla^4\psi. \tag{2.2.9}$$

In (2.2.9) the second term on the left side is the Jacobian of the stream function with the relative vorticity $\zeta = \nabla^2\psi$ and represents the advection of relative vorticity by the motion field, i.e.:

$$u\frac{\partial\zeta}{\partial x} + v\frac{\partial\zeta}{\partial y} = -\frac{\partial\psi}{\partial y}\frac{\partial\nabla^2\psi}{\partial x} + \frac{\partial\psi}{\partial x}\frac{\partial\nabla^2\psi}{\partial y} \equiv J(\psi, \nabla^2\psi) \tag{2.2.10}$$

The second term on the right side of (2.2.9) represents the effect of the vorticity change due to vortex stretching produced by the vertical velocity pumped out of the bottom boundary layer, and we refer to this as the bottom friction term. The coefficient:

$$r = f_0\frac{\delta_E}{2D} \tag{2.2.11}$$

is an inverse time scale for vorticity decay due to bottom friction.

We would have obtained the same equation if we had chosen to think about the model as a purely two-dimensional model with the stress as a body force. Conceptually the development given above has the advantage of emphasizing the important physical fact that the vertical velocity, though weak, is not zero and is actually responsible for setting the geostrophic region of the flow into motion through the mechanism of stretching of the planetary vorticity filaments.

As noted in Chapter 1, we can expect the Ekman pumping to drive an anticyclonic circulation in the subtropical regions of the oceans and a cyclonic circulation poleward of that. For example, in both the northern Atlantic and Pacific oceans the Ekman pumping inferred from observations is downward in the latitudes south of about 45°N and upward north of that latitude (see Figs. 2.2.2, 2.2.3). Most of the theoretical work described in this chapter concentrates on the structure of the *subtropical* gyre, and therefore a restricted basin bounded on the north by a latitude meant to represent the position of the

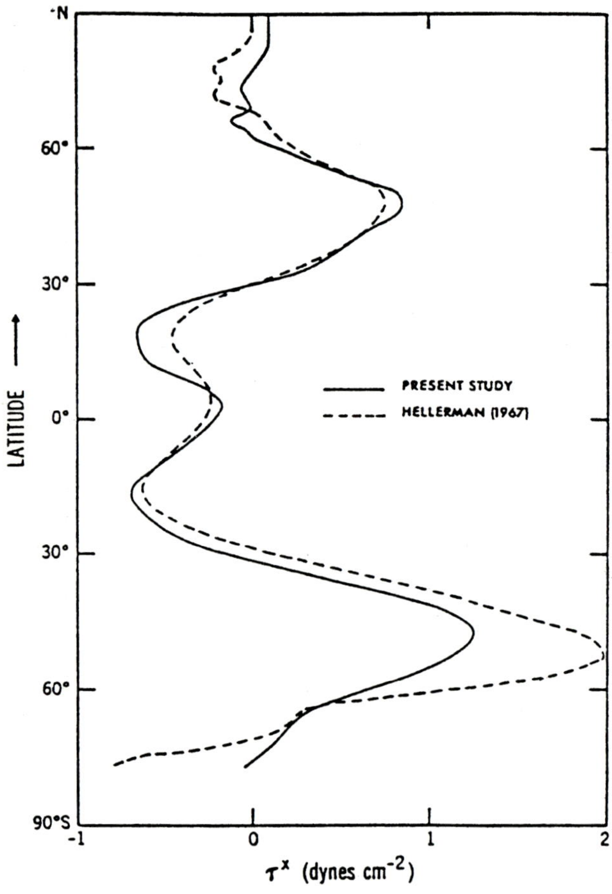


**Fig. 2.2.2.** Zonal mean of the eastward wind stress over the world's oceans. This stress drives the meridional transport in the upper Ekman layer. The convergence of this transport produces Ekman downwelling. The *Solid and dashed curves,* two estimates of the stress. (From Hellerman and Rosenstein 1983)

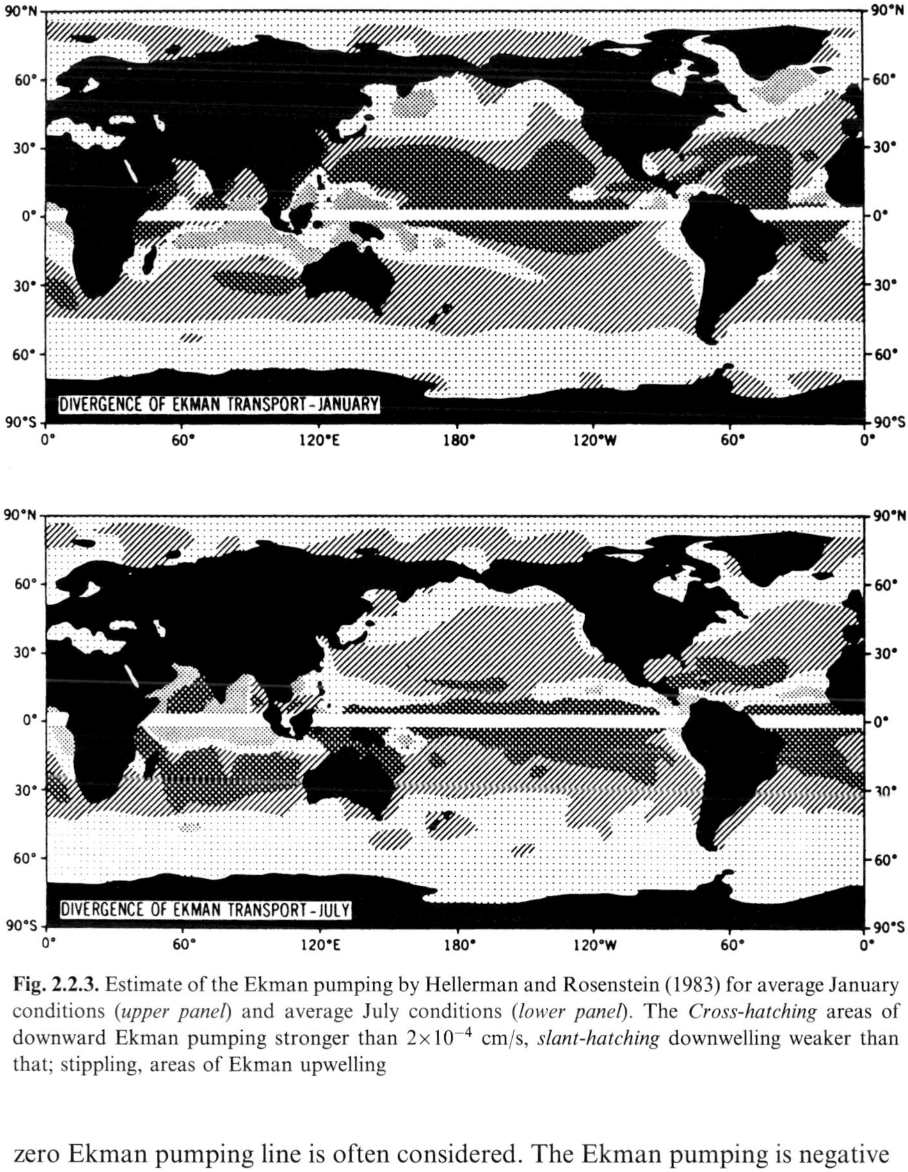


**Fig. 2.2.3.** Estimate of the Ekman pumping by Hellerman and Rosenstein (1983) for average January conditions (*upper panel*) and average July conditions (*lower panel*). The *Cross-hatching* areas of downward Ekman pumping stronger than $2\times10^{-4}$ cm/s, *slant-hatching* downwelling weaker than that; stippling, areas of Ekman upwelling

zero Ekman pumping line is often considered. The Ekman pumping is negative in this idealization of the subtropical gyre. If the latitudinal extent of the basin is $0 \leq y \leq L$, the Ekman pumping is frequently idealized in the calculations that we consider in the form:

$$w_E = -W_0 \sin\left(\frac{\pi y}{L}\right). \tag{2.2.12}$$

## 2.3 The Vorticity Equation in Nondimensional Form

It is illuminating to scale the variables of the problem to bring the vorticity equation into nondimensional form, for this brings out very clearly the key parameters of the problem. If the flow is contained in a basin of characteristic scale $L$, which we suppose also characterizes the spatial scale of variation of the Ekman pumping, we may provisionally associate $L$ with the scale of the motion, at least in the interior. We are, however, prepared for the fact that motion on smaller scales occurs naturally in the region of the western boundary current. We also provisionally suppose that the velocity, or transport in the interior, is such that the Sverdrup balance is achieved for the interior. That is, we expect a balance between the last term on the left side of (2.2.9) and the first term on the right side. If $U$ is a characteristic scale for the horizontal velocity, and $W_E$ is a characteristic scale for $w_E$, to balance those two terms we must have

$$U = W_E \frac{f_0}{\beta D}. \tag{2.3.1}$$

If $\psi$ is scaled with $UL$ and $x$ and $y$ are scaled with $L$ while $t$ is scaled with $1/\beta L$, the governing equation (2.2.9) becomes:

$$\frac{\partial}{\partial t}\nabla^2\psi + \varepsilon J(\psi, \nabla^2\psi) + \frac{\partial \psi}{\partial x} = w_E - \mu\nabla^2\psi + E\nabla^4\psi \tag{2.3.2}$$

where:

$$\begin{aligned} \varepsilon &= \frac{U}{\beta L^2} = \left(\frac{\delta_I}{L}\right)^2 \\ \mu &= \frac{r}{\beta L} = \left(\frac{\delta_s}{L}\right) \\ E &= \frac{A_H}{\beta L^3} = \left(\frac{\delta_M}{L}\right)^3 \end{aligned} \tag{2.3.3a, b, c}$$

and where each term in (2.3.2) is nondimensional. The three parameters of the steady circulation problem, $\varepsilon, \mu$, and $E$ measure the relative importance of nonlinearity, bottom friction, and lateral diffusion. Each of them has also been written as a ratio of length scales to some power. In each case the relevant parameter is a ratio of a boundary-layer scale to the scale of the interior motion, i.e., $L$. In the case of bottom friction $\delta_S$ is the scale of Stommel's boundary-layer thickness which appears in his theory (Stommel 1948) for the oceanic western intensification. The physics of lateral friction gives rise to a boundary layer first described by Munk (1950) which has a thickness $\delta_M$ while the inertial theory for the Gulf Stream described by Charney (1955) and Carrier and Robinson (1962) yields a length scale $\delta_I$.

Which of these possible boundary layers is appropriate for the problem depends on which of the boundary-layer thicknesses is the largest. As the driving increases in magnitude, eventually $\delta_I$ becomes larger than either of the viscous boundary-layer scales, and the physics of the boundary layer becomes nonlincar.

The ratio of the inertial advection of relative vorticity to the diffusion of vorticity is the Reynolds number, Re, where:

$$\mathrm{Re} = \frac{\varepsilon}{E} = \frac{\delta_I^2 L}{\delta_M^3} \tag{2.3.4}$$

where $L$ is the scale of the *motion*, and which might well be different than the basin scale. Thus if $L$ were actually equal to $\delta_I$, i.e., if we were discussing boundary-layer physics and our boundary layer were inertial, the Reynolds number on that length scale would be:

$$\mathrm{Re} = \left(\frac{\delta_I}{\delta_M}\right)^3 \equiv R_\delta \tag{2.3.5}$$

which for consistency should be greater than 1 if the boundary is fundamentally inertial. If the boundary layer is basically viscous and has the scale of the Munk layer, $\delta_M$, the boundary-layer Reynolds number becomes (for $L = \delta_M$):

$$\mathrm{Re} = \left(\frac{\delta_I}{\delta_M}\right)^2 \equiv \lambda \tag{2.3.6}$$

which should be less than 1 if the linear viscous physics is to dominate in the boundary layer. Nonlinearity affects the physics of the boundary layer and, as is seen below, the interior when Re is $O(1)$.

## 2.4 Boundary Conditions

Equation (2.3.2) requires the specification of boundary conditions on the stream function. On the boundary of the basin the normal velocity must be specified. If the basin has rigid lateral boundaries and is closed, the normal velocity must vanish on the boundary. Thus, on the boundary:

$$\vec{u} \cdot \hat{n} = 0 \tag{2.4.1}$$

where $\hat{n}$ is a unit vector normal to the boundary. Since:

$$\vec{u} = \hat{k} \times \nabla\psi \tag{2.4.2}$$

this is equivalent to the condition that $\psi$ be constant on the boundary. If there is a single boundary girdling the basin, we can choose that constant to be zero, so that (2.4.2) can be simplified to:

$$\psi = 0 \tag{2.4.3}$$

on the boundary.

The presence of viscous diffusion of vorticity implies the need for an additional dynamic boundary condition on the walls of the basin. If we interpret (2.3.2) as applying to all scales of motion with $A_H$ as the actual coefficient of viscosity, the standard condition of no slip on the boundary would be required. This would be:

$$\vec{u} \cdot \hat{t} = 0 \tag{2.4.4}$$

where $\hat{t}$ is a unit vector tangent to the boundary. In terms of $\psi$ this becomes:

$$\hat{n} \cdot \nabla\psi = 0. \tag{2.4.5}$$

It is not clear, however, that (2.4.5) is the correct boundary condition when our equations explicitly resolve only the large-scale motion. The no-slip condition is a reflection of the interaction of the fluid with a solid boundary on the microscopic level, and there is no reason why motion on scales of the order of hundreds or thousands of kilometers should mimic that interaction. We might even argue instead that the small-scale processes that occur right against the wall buffer the large-scale flow and allow it to slip smoothly along the wall. A condition of no tangential *stress* for the large-scale flow might be equally logical then. For a north-south wall, for example, this would require:

$$\frac{\partial v}{\partial x} + \frac{\partial u}{\partial y} = 0$$

at the wall. Since $u$ is already zero there by (2.4.1), it follows that $\partial v/\partial x$ itself must vanish there. We can write this boundary condition, which we call the *slip* condition, in a form that is independent of the coordinate system by recasting the condition in the form:

$$\zeta = \nabla^2\psi = 0 \tag{2.4.6}$$

on each boundary.

On the other hand, if the lateral diffusion term in (2.3.2) is considered to be a parameterization of small-scale mixing of vorticity by an unresolved eddy field, we could argue that we are actually specifying a relation between the eddy flux of total vorticity and the gradient of the large-scale vorticity field, i.e.:

$$-\overline{\vec{u}'\zeta'} = A_H\nabla(\zeta + \beta y) \tag{2.4.7}$$

whose divergence yields the diffusion term in (2.3.2). At a solid boundary the normal component of the eddy flux of vorticity must vanish since the eddy normal velocity must vanish there. This would imply that:

$$\hat{n} \cdot \nabla(\zeta + \beta y) = 0. \tag{2.4.8}$$

On western boundaries this would in particular require that:

$$\frac{\partial}{\partial x}\nabla^2\psi = 0 \tag{2.4.9}$$

or on a general boundary that:

$$\hat{n}\cdot\nabla\left(\nabla^2\psi\right) + \beta\hat{n}\cdot\hat{j} = 0 \tag{2.4.10}$$

where $\hat{j}$ is a unit vector in the $y$ direction. This boundary condition is called either *superslip* (when it applies only to the relative vorticity) or *hyperslip* when it is applied to the total vorticity (or its generalization to the potential vorticity as defined in subsequent chapters).

Which of (2.4.5), (2.4.6), and (2.4.10) should we apply? It is very difficult to say a priori. If the dissipation is small, i.e., if $E \ll 1$, does the choice make any difference for the overall character of the circulation? The answer is, as we see below, that the choice makes a great difference.

## 2.5 The Sverdrup Interior

If $E$, $\varepsilon$, and $\mu$ are each small, the dominant balance in (2.3.2) is, for the steady circulation problem is:

$$\frac{\partial\psi}{\partial x} = w_E \tag{2.5.1}$$

whose solution, which satisfies the condition of no normal flow on the eastern boundary, is:

$$\psi = -\int_x^{x_E} w_E(x', y)\,dx' \tag{2.5.2}$$

which is the nondimensional form of the Sverdrup balance discussed in Chapter 1, written here for the homogeneous model. In choosing to use the Sverdrup theory to satisfy the boundary condition on the eastern boundary we are assuming that no boundary current correction to the interior is possible on the eastern boundary to relieve the Sverdrup solution of the responsibility to satisfy the no normal flow condition. We are also *assuming*, in writing (2.5.2) as the interior solution, that a boundary layer solution can be found on the western boundary that enables us to satisfy the boundary conditions. It remains for us to verify this. For completeness we note that in *dimensional* units (2.5.2) is:

$$\psi = -\frac{f}{\beta D}\int_x^{x_e} w_E\,dx'. \tag{2.5.3}$$

## 2.6 The Western Boundary-Layer Equation

If the Sverdrup theory is to be valid in the interior, it is necessary that a boundary-layer solution for the western boundary region can be found which smoothly joins to the Sverdrup interior and at the same time satisfies all the boundary conditions on the western boundary. In the vicinity of the western boundary, which for simplicity we place at a constant value of $x$, i.e., at $x = x_w$, terms in the vorticity equation neglected in the interior become important as physical processes associated with the smaller length scales of the boundary-layer increase in magnitude. On the small scales in $x$ associated with the boundary layer the estimates of the viscous and nonlinear terms alter as $L$, the length scale of the motion, becomes less than the basin scale.

In reevaluating the importance of various processes and their mathematical representations it is sensible to suppose first of all that derivatives in the $x$ direction in the boundary layer become large compared with derivatives in the $y$ direction when operating on the same function. This may not always be true, but we start with this plausible hypothesis. Then in the vicinity of the western boundary if:

$$\frac{\partial}{\partial x} \gg \frac{\partial}{\partial y} \tag{2.6.1}$$

the vorticity equation in *dimensional form* (2.2.9) becomes in the steady state:

$$J\left(\psi, \frac{\partial^2 \psi}{\partial x^2}\right) + \beta \frac{\partial \psi}{\partial x} = \frac{f_0}{D} w_E - r \frac{\partial^2 \psi}{\partial x^2} + A_H \frac{\partial^4 \psi}{\partial x^4}. \tag{2.6.2}$$

In the boundary layer $\psi$ is a rapid function of $x$ so that one or the other of the terms which are small in the interior become order 1 in the boundary layer. This means that $\psi$ is actually a function of:

$$\xi = \frac{x}{\delta}$$

where $\delta$ is the dimensional boundary layer width so that the derivatives of $\psi$ can be estimated as being of the order of $\psi/\delta$.

In the boundary layer the order of magnitude of the streamfunction must be at least as large as its value in the interior, $\psi_I$, since it joins smoothly to it as the interior is approached. It could be larger if there is a strong recirculation in the region of the boundary layer. Thus in the western boundary-current region:

$$\beta \frac{\partial \psi}{\partial x} \geq O\left(\beta \frac{\psi_I}{\delta}\right) = O\left(\frac{L}{\delta} \frac{f_0}{D} w_E\right) \gg \frac{f_0}{D} w_E \tag{2.6.3}$$

so that the direct action of the forcing, i.e., the vorticity source due to the Ekman pumping, is negligible in (2.6.2) within the boundary layer. Physically and mathematically the boundary-layer current is produced only by the discrepancy between the boundary conditions at $x = x_W$ and the value of the streamfunction given by the Sverdrup solution there.

The boundary-layer vorticity equation can also be written as:

$$u\frac{\partial^2 v}{\partial x^2} + v\frac{\partial v}{\partial x\,\partial y} + \beta v = -r\frac{\partial v}{\partial x} + A_H\frac{\partial^3 v}{\partial x^3}. \tag{2.6.4}$$

Since the velocity is horizontally nondivergent (2.6.4) can be rewritten as:

$$\frac{\partial}{\partial x}\left[u\frac{\partial v}{\partial x} + v\frac{\partial v}{\partial y} + \beta\psi + rv - A_H\frac{\partial^2 v}{\partial x^2}\right] = 0$$

which when integrated in $x$, and after using the fact that for large $x$ (compared to $\delta$) the $\beta$ term becomes the dominant term in the equation, yields:

$$u\frac{\partial v}{\partial x} + v\frac{\partial v}{\partial y} + \beta\psi = \beta\psi_I(x_w, y) - rv + A_H\frac{\partial^2 v}{\partial x^2} \tag{2.6.5}$$

or, in terms of the streamfunction

$$J\left(\psi, \frac{\partial\psi}{\partial x}\right) + \beta\psi = \beta\psi_I - r\frac{\partial\psi}{\partial x} + A_H\frac{\partial^3\psi}{\partial x^3}. \tag{2.6.6}$$

The steady momentum equation in the $y$ direction is:

$$u\frac{\partial v}{\partial x} + v\frac{\partial v}{\partial y} + fu = -\frac{1}{\rho}\frac{\partial p}{\partial y} - rv + A_H\frac{\partial^2 v}{\partial x^2} \tag{2.6.7}$$

if we include the bottom friction term as a drag force in the momentum equation. Comparison of (2.6.5) and (2.6.7) allows us to associate the meridional pressure gradient in the boundary layer with the discrepancy between the interior stream function at $x = x_w$ and that in the boundary layer. More precisely:

$$\frac{1}{\rho}\frac{\partial p}{\partial y} = \beta[\psi - \psi_I(x_w, y)] - fu \tag{2.6.8}$$

so that on $x = x_w$, where $u$ and $\psi$ both vanish:

$$\frac{1}{\rho}\left(\frac{\partial p}{\partial y}\right)_{wall} = -\beta\psi_I(x_w, y). \tag{2.6.9}$$

Thus the pressure decreases northward along the western boundary as long as the interior stream function is positive just outside the western boundary current. If $\psi$ and $u$ are zero on $x = x_e$, and if the Sverdrup flow is southward, i.e., if $w_E < 0$, then by (2.5.1) $\psi_I$ is positive and the pressure therefore decreases northward along the wall. This has important consequences, as we see below for ability of the boundary layer to stay attached to the western wall.

## 2.7 The Linear Problem

### *Munk's Solution*

It is useful to consider the closure of the Sverdrup problem in the linear limit as a starting point for our general investigation of the relationship between the dynamics of the boundary current and the interior solution. The linear limit provides us with an easily accessible solution, and its relation to the interior is straightforward. The solution was first found by Munk (1950).

In the Munk solution nonlinearity in the boundary layer is completely ignored. This requires that the boundary-layer Reynolds number (2.3.6) is small. This requires in turn that the forcing be weak,or the horizontal turbulent mixing be large. It is not possible to obtain a realistic model of the western boundary current in such a limit (see, for example, the discussion in Pedlosky 1987), but the solution is of considerable theoretical and historical interest. Munk also intended his model to describe the vertical average of a baroclinic ocean with negligible bottom velocities, and the Munk model thus also ignores the bottom friction term in (2.6.2).

In this case the equation in the boundary layer becomes:

$$A_H \frac{\partial^4 \psi}{\partial x^4} - \beta \frac{\partial \psi}{\partial x} = 0. \tag{2.7.1}$$

For simplicity and without loss of generality we place the western boundary $x = 0$. The general solution that satisfies the condition that $\psi$ is zero on $x = 0$ and also tends to $\psi_I$ for $x \gg \delta_M$ is:

$$\psi = \psi_I(x, y)\left[1 - e^{-x/2\delta_M} \cos\left(\frac{\sqrt{3}x}{2\delta_M}\right)\right] + C(y)e^{-x/2\delta_M} \sin\left(\frac{\sqrt{3}x}{2\delta_M}\right). \tag{2.7.2}$$

The function $C(y)$ must be determined by the remaining the boundary condition, i.e., either (2.4.5), (2.4.6) or (2.4.10).

If the no slip condition is used, so that $v = 0$ at $x = 0$, then the solutions for $\psi, v$, and $\zeta$ become:

$$\psi = \psi_I\left[1 - e^{-x/2\delta_M}\left(\cos\frac{\sqrt{3}x}{2\delta_M} + \frac{1}{\sqrt{3}} \sin\frac{\sqrt{3}x}{2\delta_M}\right)\right]$$

$$v = \frac{2}{\sqrt{3}} \frac{\psi_I}{\delta_M} e^{-x/2\delta_M} \sin\left(\frac{\sqrt{3}x}{2\delta_M}\right) \tag{2.7.3a, b, c}$$

$$\zeta = \frac{\psi_I}{\delta_M^2} e^{-x/2\delta_M}\left[\cos\left(\frac{\sqrt{3}x}{2\delta_M}\right) - \frac{1}{\sqrt{3}} \sin\left(\frac{\sqrt{3}x}{2\delta_M}\right)\right].$$

Figure 2.7.1 shows the complete circulation pattern in the case when $w_E$ has the form $W_0 \sin(\pi y/L)$. The Munk layer thickness has been chosen to be

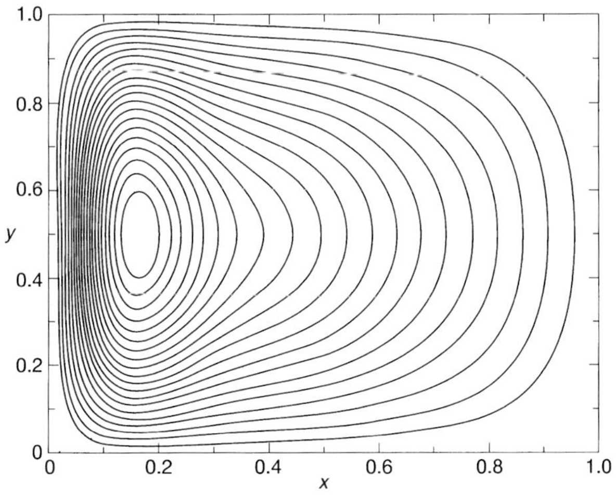


**Fig. 2.7.1.** Contours of the stream function for Munk's model with a no-slip boundary condition. The Munk boundary-layer thickness, $\delta_M$, is chosen to be 1/20 of the basin width. The circulation is clockwise if the Ekman pumping is negative, but since the problem is linear, a reversal of the sign of the Ekman pumping would leave the pattern of flow the same and merely reverse its direction

$L/20$ where $L$ is the basin width. The interior circulation is the Sverdrup flow as can be seen from (2.7.3a). The term $\psi$ approaches the interior streamfunction exponentially fast as a function of $x/\delta_M$. The Sverdrup solution is undisturbed for distances greater than $O(\delta_M)$, and the southward flow in the Sverdrup interior is returned in a western boundary current which smoothly attaches to the interior. It is left as an exercise for the reader to show that no such boundary layer can be found on the eastern boundary which simultaneously satisfies the conditions of no normal flow and no slip. This justifies our use (in fact it requires our use) of the Sverdrup solution to satisfy the no normal flow condition on the eastern boundary.

Figure 2.7.2 shows the northward velocity (scaled by $\psi_I/\delta_M$) in the boundary layer. The velocity is zero at the wall, reaches its maximum at $x = 1.20\,\delta_M$ and then slowly diminishes. For large $x/\delta_M$, $v$ becomes slightly negative before merging with the interior flow so that a boundary layer recirculation is set up in the edge of the boundary layer as it merges with the interior. That is, some of the flow in the boundary layer is not joined to the interior but recirculates endlessly within the region of the western boundary current. However, this recirculation is very weak and is not even noticeable in the streamline contours of Fig 2.7.1.

Also shown in the figure is the relative vorticity, $\zeta$, in the boundary layer. This is given by $\partial v/\partial x$ within the boundary layer approximation. Since $v$ is zero at the western boundary and rises to its maximum at $x = 1.20\,\delta_M$, the vorticity is positive at the western boundary and decreases to zero at the position of maximum $v$. The diffusive flux of vorticity in the $x$ direction is $-A_H(\partial\zeta/\partial x)$,

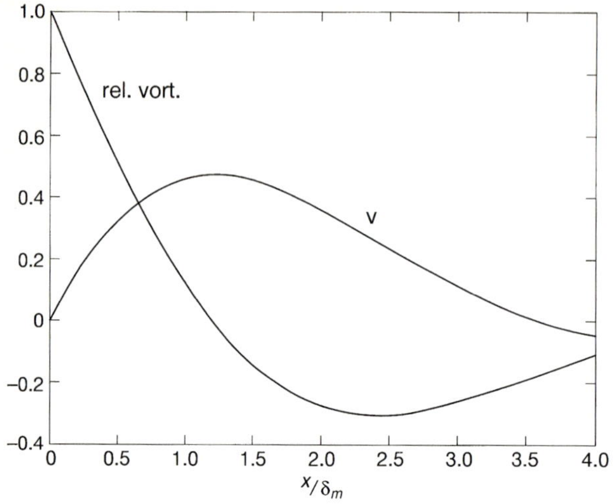


**Fig. 2.7.2.** Profiles of $v$ (scaled by $\psi_I/\delta_M$) and vorticity (scaled by $\psi_I/\delta_M^2$) for the Munk boundary layer with no slip boundary conditions

that is, the flux of relative vorticity flows down the vorticity gradient. Thus, at the wall positive (cyclonic) vorticity is diffusing *into* the basin, and this is happening to fluid elements all along their trajectory in the western boundary current. This is necessary to achieve the vorticity balance that we shall discuss more completely in following sections, but we can already appreciate the process in the Munk model. Fluid which enters the western boundary layer from the interior in the south, say at $y = y_s$, has a smaller total vorticity, $f_0 + \beta y_s$, than fluid leaving the boundary layer in the northern part of the gyre at $y = y_n$ and entering the interior where the planetary vorticity is $f_0 + \beta y_n$. The action of the wind in driving the fluid southward continuously contributes negative vorticity to the fluid, and the resulting southward motion is the Sverdrup theory's response to that negative vorticity input. As we have seen, the wind vorticity source is negligible in the boundary layer (the flow is too swift for the direct vorticity input to matter). However, when the fluid rejoins the interior in the north, it must have the same planetary vorticity (which is its total vorticity in the interior) as its environment, i.e., $f_0 + \beta y_n$. Thus in the boundary layer the fluid must gain positive vorticity in the amount $\beta(y_n - y_s)$. This vorticity enters diffusively from the western boundary.

In fact if (2.6.2) is integrated across the basin and nonlinearity and bottom friction are ignored, as is appropriate in the Munk theory, we obtain, at *each* latitude $y$.

$$0 = \int_0^x \frac{f_0}{D} w_E(x', y)dx' - A_H \frac{\partial \zeta}{\partial x}(0) \quad \text{or}$$

$$A_H \frac{\partial \zeta}{\partial x}(0) = -\beta \psi_I(0, y). \tag{2.7.4a,b}$$

In deriving (2.7.4) we have used the fact that the integral across the basin of $v$ must vanish to conserve mass and that the gradient of vorticity at the eastern boundary is utterly negligible in comparison with its value on the western boundary. The result (2.7.4a) shows us that *in the linear model* the flux of vorticity into the western boundary layer at each latitude exactly balances the vorticity input by the wind across the basin at the same latitude. The vorticity balance is attained locally at each latitude. It is important to emphasize that this is a consequence of ignoring advective effects. Note that the viscous flux required is proportional to the discrepancy between the value of the interior streamfunction at the western boundary and the value (zero) the actual streamfunction must attain there.

If instead of using (2.4.5) as the second boundary condition to determine $C(y)$ in (2.7.2) we were to use the no-stress, or slip, condition (2.4.6), the solutions for $\psi, v$, and $\zeta$ would be:

$$\begin{aligned} \psi &= \psi_I\left[1 - e^{-x/2\delta_M}\left(\cos\frac{\sqrt{3}x}{2\delta_M} + \frac{1}{\sqrt{3}}\sin\frac{\sqrt{3}x}{2\delta_M}\right)\right] \\ v &= \frac{2}{\sqrt{3}}\frac{\psi_I}{\delta_M}e^{-x/2\delta_M}\left(\cos\frac{\sqrt{3}x}{2\delta_M} + \frac{1}{\sqrt{3}}\sin\frac{\sqrt{3}x}{2\delta_M}\right) \\ \zeta &= -\frac{2}{\sqrt{3}}\frac{\psi_I}{\delta_M^2}e^{-x/2\delta_M}\sin\frac{\sqrt{3}x}{2\delta_M}. \end{aligned} \tag{2.7.5}$$

The northward velocity and the vorticity are shown in Fig. 2.7.3. Now the northward velocity has its maximum at $x = 0$. The vorticity is zero there and negative for $x > 0$. The vorticity gradient is again negative at $x = 0$, and there is, as in the case of the no-slip condition, a flux of positive relative vorticity into

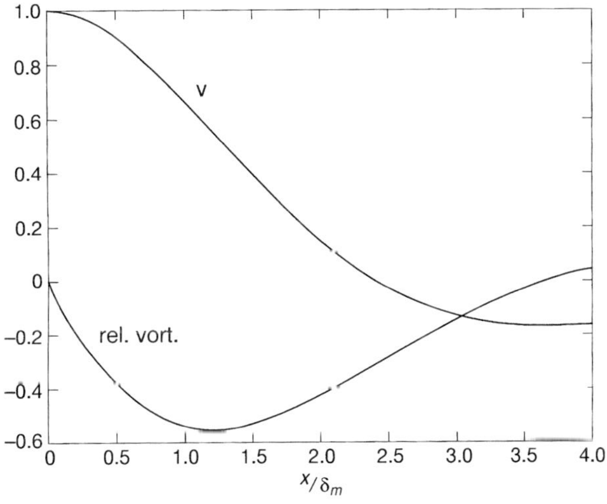


**Fig. 2.7.3.** As in Fig. 2.7.2 but now with slip (no-stress) conditions

the basin. Indeed, since (2.7.4) is equally valid for this boundary condition, the flux of vorticity entering is *exactly the same as in the case of the no-slip boundary condition.* Although the fluid slips along the boundary, the vorticity gradient at $x = 0$ is the same in each case, namely$-\psi_I(0, y)/\delta_M^3$. It is important to note that since $\delta_M^3 = A_H/\beta$, the *flux of vorticity into the basin by diffusion is independent of $A_H$,* which is the physical content of (2.7.4b).

On the other hand, if we tried to apply the boundary condition (2.4.10), there would be *no boundary layer solution possible,* as (2.7.4) shows us directly, for if (2.4.10) is used in (2.7.4), a contradiction is reached immediately. This is the first indication that the circulation is sensitive to the boundary conditions applied.

It might be thought that this sensitivity is an artifact of the linear solution that imposes the local balance (2.7.4). This is not the case, as we see below.

### *Stommel's Solution*

We include here for the sake of completeness the solution to the linear problem when lateral friction is ignored, and *only* bottom friction is kept. This is Stommel's (1948) original model for the western intensification. Note that in this case the order of the problem is lowered, and the boundary layer equation becomes:

$$r\frac{\partial^2\psi}{\partial x^2} + \beta\frac{\partial\psi}{\partial x} = 0 \tag{2.7.6}$$

which can satisfy only the no normal flow condition on $x = 0$ and the condition that the solution merges smoothly with $\psi_I$ when $x \gg \delta_S = r/\beta$.

The solution in this case is:

$$\begin{aligned} \psi &= \psi_I(x, y)\left(1 - e^{-x/\delta_S}\right) \\ v &= \frac{\psi_I}{\delta_S} e^{-x/\delta_S}. \end{aligned} \tag{2.7.7a,b}$$

In the absence of horizontal diffusion of vorticity the input of vorticity by the wind via the Ekman pumping must be balanced by a draining away of vorticity in the boundary layer through bottom friction. Analysis similar to that which leads to (2.7.4) requires, in the case of bottom friction, that:

$$v(0, y) = \frac{f_0}{rD}\int_0^{x_e} w_E(x', y)dx'. \tag{2.7.8}$$

If (2.5.3) and (2.7.7b) are used, we can see that (2.7.8) is satisfied automatically. Thus, in linear theory there is no difficulty for sufficient dissipation to occur locally in the boundary layer. A vorticity balance is always achieved locally between the interior source and the boundary layer dissipation which allows the flow to traverse a north-south path in the boundary layer, dissipate just the right amount of anomalous vorticity, and smoothly rejoin the Sverdrup interior.

## 2.8 The Inertial Boundary Layer

The dependence of the structure of Munk's model and Stommel's model on largely indeterminate values of friction coefficients and the difficulty of tuning the frictional models to obtain a sufficiently narrow current to be realistic (50–100 km) without yielding a Reynolds number greater than 1 encouraged Charney (1955) and Morgan (1956) to attempt to find models of the western boundary currents that are purely inertial. Our physical discussion above indicates that dissipation of vorticity must be important somewhere in the boundary-current region to achieve a steady balance, but the purely inertial theories are useful in determining the scales of strongly nonlinear currents and in clearly isolating the dynamical problem of closing the circulation. The inertial theories also give a fair model of the western boundary current in its formation stage where fluid from the interior enters the boundary layer before dissipation has an opportunity to act.

We may think of the inertial theory as the other extreme limit of the boundary layer problem. Whereas the Munk model, for example, is the limit $\mathrm{Re} \to 0$, the purely inertial problem would appear to be the limit $\mathrm{Re} \to \infty$. However, care must be taken because the formal limit of infinite Reynolds number, in which dissipation is ignored completely, is a singular limit of the governing equations.

### *Purely Inertial Limit*

In the purely inertial limit the boundary layer equation (2.6.2) can be written as:

$$J(\psi, \zeta + \beta y) = 0 \quad \text{where} \quad \zeta = \frac{\partial^2 \psi}{\partial x^2}. \tag{2.8.1}$$

The vorticity is conserved on streamlines in the boundary layer in the inertial limit, and (2.8.1) is the condition that isolines of the total vorticity coincide with streamlines of the flow. The first integral of (2.8.1) is:

$$\frac{\partial^2 \psi}{\partial x^2} + \beta y = Q(\psi) \tag{2.8.2}$$

where $Q(\psi)$ is an arbitrary function of $\psi$. The determination of $Q$ is generally difficult in practice, if easy to understand conceptually. Each streamline entering the boundary-layer possesses at its entry point, where the relative vorticity is negligible, both a known value of $\psi$ and a value of $\beta y$. The latter is the vorticity, aside from the constant $f_0$, that is assigned to the fluid element as it enters the boundary layer. One can therefore associate $\psi$ with the vorticity at the entry point in terms of the position of the entry point. Each entry point $y$ corresponds to an interior value of $\psi$. At the same time $\beta y$ yields the vorticity, and the vorticity can therefore be linked to the value of $\psi$ at the entry point of the streamline into the western bounary layer. That relationship between stream function and total vorticity is then maintained throughout the dissipation-free portion of the fluid element's path in the boundary layer. Thus for large $x$ (compared to $\delta_I$) the sought after relation is determined by inverting $\psi = \psi(y)$ i.e.:

$$y = y(\psi) \tag{2.8.3}$$

so that:

$$\beta y = \beta y(\psi) = Q(\psi). \tag{2.8.4}$$

This determines the function $Q(\psi)$, and (2.8.2) becomes, generally, a nonlinear ordinary differential equation for $\psi$.

A simple often quoted example (see Pedlosky 1987) is for the case where the interior streamfunction near the western boundary (at $x = 0$) can be locally approximated by $\psi_I = Uy$, representing a uniform *westward* flow. In this case the inversion (2.8.4) yields the linear relation:

$$Q(\psi) = \frac{\beta}{U}\psi \tag{2.8.5}$$

so that the solution of (2.8.2) which matches to the interior for large $x$ and satisfies the no normal flow condition at $x = 0$ becomes in this case:

$$\psi = \psi_I\left(1 - e^{-x/\delta_I}\right) \tag{2.8.6}$$

where

$$\delta_I = \sqrt{\frac{U}{\beta}} \tag{2.8.7}$$

is the inertial boundary layer thickness.

The inability of the purely inertial model to reattach itself smoothly to the Sverdrup interior is signaled by the inability to find solutions which tend asymptotically to the interior solution in the case where $U < 0$, i.e., where the interior flow outside the boundary layer is *eastward.*

We are left then in the inertial limit with only a partial, incomplete solution. Furthermore, the solution is incomplete in several senses. First of all, the inertial solution is obtainable only in the region where the interior flow is westward, and the purely inertial solution can therefore hold only in the southern portion of the subtropical gyre where the interior flow is westward and into the boundary layer. We have demonstrated this in the simple example of (2.8.5) and (2.8.6). The general proof for arbitrary interior flows external to the boundary layer is due to Greenspan (1962). The inertial theory simply cannot be used where the flow must leave the boundary layer and hook on to the region of eastward interior flow. The property of vorticity conservation which we used to integrate (2.8.1) and obtain (2.8.2) is exactly the property which physically disallows the inertial model to hold along the whole path of the parcel's transit of the boundary layer for the physical reasons associated with the necessary dissipation of vorticity that we describe above, and which the purely inertial model ignores.

The second inadequacy of the purely inertial model is its inability to satisfy the second boundary condition on $x = 0$, either (2.4.5), (2.4.6), or (2.4.10). These two limitations are of course connected since the purely inertial limit represents a singular perturbation of the vorticity equation by neglecting the diffusion of vorticity. Note that retaining bottom friction while ignoring horizontal vorticity diffusion is also a singular perturbation of the vorticity equation. Although it retains dissipation, it lowers the mathematical order of the differential equation.

### *The Sublayer*

In order to satisfy either the no-slip, no-stress, or no-flux (of vorticity) condition, friction must enter the problem. If $\delta_I \gg \delta_M$, a viscous sublayer must exist within the inertial layer, next to the wall, in which viscous forces are important due to the locally strong gradient in the meridional velocity. Consider the situation shown schematically in Fig. 2.8.1. The interior flow enters the inertial boundary layer, which has a width $\delta_I$, with a westward velocity $O(U_I)$. In the inertial boundary layer the fluid is turned northward and has a characteristic speed, by simple mass balance, $v_i = U_I(L/\delta_I)$, where the subscript $i$ refers to quantities in the inertial boundary layer.

For the lateral friction terms to enter in the boundary-layer equation the scale of the sublayer must be sufficiently small so that the relatively large number of derivatives in the frictional terms compensate for the relative smallness of $A_H$. That is, if the scale of the sublayer is $\delta_*$, to obtain a balance between the nonlinear terms and the linear viscous terms we must have:

$$v_i \frac{\partial^2 v_i}{\partial x \partial y} + u_i \frac{\partial^2 v_i}{\partial x^2} = O\left(\frac{v_i^2}{\delta_* L}\right) = O\left(A_H \frac{\partial^3 v_i}{\partial x^3}\right) = O\left(\frac{A_H v_i}{\delta_*^3}\right). \tag{2.8.8}$$

We have used the fact in making the above estimates that in the sublayer the velocity parallel to the boundary is of order $v_i$. If the estimates in (2.8.8) are exploited and the relation between $U_I$ and $v_i$ is used, we finally obtain:

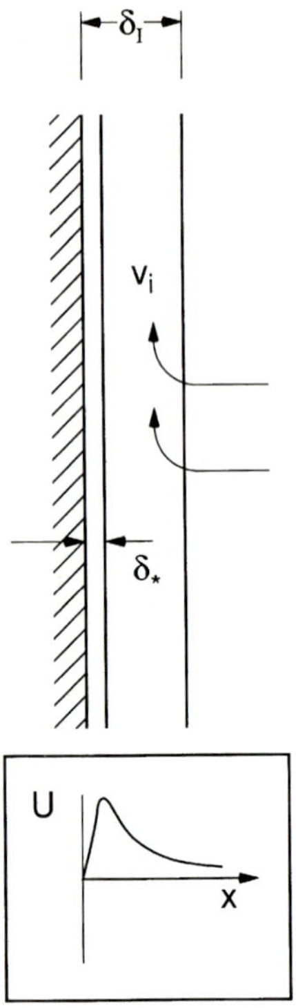


**Fig. 2.8.1.** Hypothesized sublayer of width $\delta_*$ lies within the inertial boundary layer, whose width is $\delta_I$. The oncoming fluid in the interior is turned northward in the inertial layer. The sublayer serves to satisfy the no-slip, no-stress, or no-vorticity flux condition

$$\delta_* = \left(\frac{A_H \delta_I}{U_I}\right)^{1/2} = \delta_I \left(\frac{\delta_M}{\delta_I}\right)^{3/2} = \frac{\delta_I}{\sqrt{R_\delta}}. \tag{2.8.9}$$

Since the boundary-layer Reynolds number, $R_\delta$, is much greater than 1 in the inertial limit (i.e., when $\delta_I \gg \delta_M$), we are guaranteed that $\delta_*$ is much less than the inertial width $\delta_I$. It does seem possible then to place a sublayer tucked within the inertial boundary layer next to the wall in order to satisfy the second boundary condition. We have not actually shown that this sublayer solution always exists but it does seem plausible. In the sublayer (2.6.5) applies without the term proportional to $\beta\psi$ since the scale is too small for the effect of the planetary vorticity gradient to enter directly. The equivalent pressure gradient in the boundary layer is then given by (2.6.9). According to classical boundary-layer theory for nonrotating fluids (Schlichting 1979), for which:

$$u\frac{\partial v}{\partial x} + v\frac{\partial v}{\partial y} = \beta\psi_I(x_w, y) + A_H \frac{\partial^2 v}{\partial x^2} \tag{2.8.10}$$

would be the governing equation, solutions for northward flow in the boundary layer exist as long as $\beta\psi_I$ is positive, as it is in the subtropical gyre. That is, in the sublayer the effective pressure gradient tends to accelerate the fluid in the direction of flow. Were it otherwise, the sublayer would tend to stall next to the wall and separate from the boundary, and the structure hypothesized in Fig. 2.8.1 would break down.

It is important to note that most of the streamlines in the western boundary layer do *not* go through the sublayer in this picture. They remain in the outer inertial layer. This means that for fluid elements on those streamlines dissipation still remains unimportant. However, as we have already noted, dissipation must be important on *every* streamline. It therefore appears unlikely that in the limit of large boundary layer Reynolds number the structure of the motion can remain split between an inertial region and a dissipative region throughout the passage of fluid through the boundary layer. The resolution of this perplexing difficulty forms the crux of the remaining discussion of this chapter.

To confront this question we must go beyond the heuristic discussion of this section and consider a detailed analysis of the boundary layer dynamics for arbitrary Reynolds number. First, however, we describe an additional important feature of the purely inertial theory.

### *The Fofonoff Mode*

The purely inertial model contains a free, nonlinear mode of motion first discovered by Fofonoff (1954). Fofonoff looked for a solution of (2.2.9) in the absence of both forcing and dissipation. The mode is therefore a free, unforced mode of motion for steady flow. It must be nonlinear, i.e., relative vorticity must be important in the solution; otherwise the total vorticity would be simply $\beta y$. For inertial motion isolines of total vorticity and streamlines coincide. If the total vorticity were just $\beta y$, the isolines of total vorticity would be latitude circles and intersect the eastern boundary where $\psi = 0$. There can be no flow emanating from the eastern boundary so that the only way lines of constant $y$ and lines of constant $\psi$ could coincide is if $\psi$ were 0 everywhere. This would imply no motion.

In the purely inertial limit (2.2.9) reduces to:

$$J(\psi, \nabla^2\psi + \beta y) = 0 \tag{2.8.11}$$

which is the generalization of (2.8.1) to flow in the entire basin. In analogy with (2.8.2), a first integral of (2.8.11) is just the statement that the total vorticity is constant on streamlines, that is:

$$\nabla^2\psi + \beta y = Q(\psi). \tag{2.8.12}$$

Whereas the function $Q(\psi)$ could be deduced from the structure of the interior field for the boundary layer problem, here $Q(\psi)$ is completely arbitrary. Fofonoff *arbitrarily* chose the linear form:

$$Q(\psi) = \frac{\beta}{U}\psi + y_0. \tag{2.8.13}$$

Clearly this form is taken for its analytical simplicity, and in principle any more complicated nonlinear form could serve equally well. However, as we see below, this choice turns out to be particularly fortuitous. The constants $y_0$ and $U$ in (2.8.13) are completely arbitrary, but we clearly identify $U/\beta$ with the square of the inertial boundary-layer thickness.

The vorticity integral (2.8.12) with the specification of (2.8.13) can be solved in general, but it is more illuminating to consider the case where:

$$\beta L^2/U \gg 1$$

so that boundary-layer techniques can be used. In this limit the asymptotic solution of (2.8.12) which satisfies the conditions that $\psi = 0$ on the boundaries of the basin at $x = 0$ and $x_e$ (the eastern and western boundaries) and at $y = y_n$ and 0 (the northern and southern boundaries) is:

$$\begin{aligned}\psi = U(y - y_0)\left(1 - e^{-x/\delta_I} - e^{-(x_e - x)/\delta_I}\right) \\ - U(y_n - y_0)e^{-(y_n - y)/\delta_I} + Uy_0e^{-y/\delta_I}\end{aligned} \tag{2.8.14}$$

where $\delta_I = \sqrt{U/\beta}$.

The solution consists of an interior, constant, westward flow of magnitude $U$. It impinges on the western boundary where inertial boundary layers of constant width $\delta_I$ turn the flow along the boundary. The interior flow must be westward to have an inertial boundary layer according to Greenspan's theorem. We can see from (2.8.12) and (2.8.13) that if the interior flow were eastward, a solution could still exist, but the solutions required to satisfy the boundary conditions on the walls of the basin would be spatially wavelike with a *wavelength* $\delta_I$. These waves would fill the interior. When $U < 0$, on the other hand, the effect of the boundaries is limited to boundary layers *on all four boundaries.*

The flow impinging on the western boundary turns northward north of the line $y = y_0$ and turns southward for streamlines impinging on the wall south of $y_0$. The flow in the western boundary layer then flows to the northern and southern boundaries and subsequently flows eastward on both those boundaries. The amount of transport in each of these eastward jets depends on the position of the bifurcation point $y_0$ which, we remember, is arbitrary. The eastward transport in the northern boundary layer is $U(y_n - y_0)$ while the eastward transport in the southern boundary layer is $Uy_0$. These boundary-layer flows impinging on the eastern boundary are turned along the eastern boundary to flow meridionally. The flow decelerates along its path in the eastern boundary layer and gradually emerges from the eastern boundary layer and joins the interior. Figure 2.8.2 shows the Fofonoff mode for the two choices $y_0 = y_n$ and 0. In the first case there is no northern boundary layer, and the circulation is counterclockwise (cyclonic) in the gyre while in the second

**Fig. 2.8.2.** Streamlines in the Fofonoff mode. In the *Upper panel,* the constant $y_0$ (see text for definition) is $y_n$ and there is no northern boundary-layer flow. In the *Lower panel,* $y_0$ is chosen to be zero, and so there is therefore no southern boundary layer

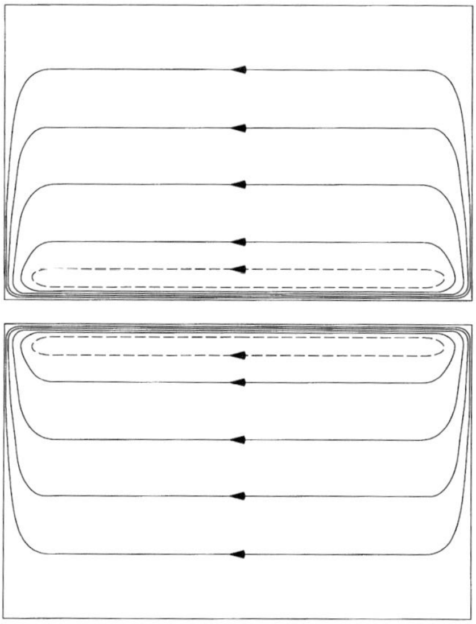

case the flow has no southern boundary layer and is anticyclonic. Note that in each case there are eastern and western boundary layers, and the interior flow is westward.

A purely inertial theory then can allow eastern as well as western boundary layers, and this emphasizes the point originally made by Stewart (1964) that it is really friction which decides whether the intensification of the oceanic circulation can occur on the east or west. We note in this regard that it would not be possible to place a viscous sublayer within an inertial layer on the eastern boundary, for then the pressure gradient within the sublayer would be in the direction opposing the motion.

The purely inertial Fofonoff mode of course satisfies only the no normal flow condition. Further, note that since the total vorticity is constant on streamlines, the isolines of total vorticity coincide with the streamlines of Fig. 2.8.2. Each isoline is therefore significantly distorted from latitude circles by the nonlinear advection of vorticity. Instead of striking the eastern boundary and thus negating the possibility of flow, the isolines of vorticity are dragged by the flow to close on themselves. The key feature of the Fofonoff mode can be considered to be the structure of the vorticity isolines. The contours of vorticity form loops instead of open lines, and the fluid in the Fofonoff mode flows effortlessly around these closed loops so that (2.8.12) is satisfied automatically. We see the emergence of similar nonlinear modes in later chapters in our discussion of the circulation of a stratified ocean. In each case the wrapping around of vorticity contours allows the existence of free modes of steady motion. Whether those modes can be excited by forcing in the presence of dissipation turns out, as we see below, to depend on the nature of the dissipation mechanism.

## 2.9 The Nonlinear Munk Problem

We have already seen that a purely inertial model of the western boundary current is an inadequate model for the closure of the circulation. Dissipation must be important to close the circulation, and yet in any realistic model of the western boundary current the magnitude of the velocity is such that nonlinearity cannot be ignored as in the Munk theory. Thus nonlinearity and dissipation must be considered simultaneously. As might be imagined, this requires numerical rather than analytical approaches to the full circulation problem. However, numerical methods have their limitations as well. The degree of affordable resolution and the fact that the complex results are numerical rather than relational are both severe impediments to understanding. Before discussing the numerical work that has been performed on the circulation problem, and especially with the problem of the closure of the circulation in mind, we discuss in the next few sections some theoretical results which are helpful in understanding and anticipating the numerical work.

We start here by considering the Munk problem for the western boundary layer in the presence of nonnegligible nonlinearity. As in the linear Munk problem, we ignore the effect of bottom friction. The problem was studied originally by Il'in and Kamenkovich (1964) and by Ierley and Ruehr (1986). We shall follow the discussion of Ierley and Ruehr and also Ierley (1987).

We consider the boundary layer equation in the form (2.6.2). It is useful for the analysis that follows to consider the equation in nondimensional form. We choose as the $x$ length scale the Munk layer thickness, $\delta_M$, and for the $y$ scale we choose $L$, the basin or gyre scale. For the stream function scale we take a characteristic value of the interior streamfunction at the western boundary, $\Psi_I$, and then we obtain as our *nondimensional* boundary-layer equation:

$$\frac{\partial^4\psi}{\partial x^4}=\lambda\left[\frac{\partial\psi}{\partial x}\frac{\partial^3\psi}{\partial^2 x\,\partial y}-\frac{\partial\psi}{\partial y}\frac{\partial^3\psi}{\partial x^3}\right]+\frac{\partial\psi}{\partial x}. \tag{2.9.1}$$

Here $\lambda$ is the ratio $(\delta_I/\delta_M)^2$ where the inertial boundary-layer thickness is based on $U=\Psi_I/L$ (see 2.3.6) . Note that Ierley defines his inertial boundary current thickness somewhat differently so that he introduces a factor of $\pi$ in his relation between $\lambda$ and the ratio $(\delta_I/\delta_M)^2$. All the variables in (2.9.1) are nondimensional.

Suppose we consider a Sverdrup interior whose (nondimensional) streamfunction near the western boundary is:

$$\psi=\frac{1}{\pi}\sin\pi y \tag{2.9.2}$$

so that the zonal velocity impinging on the western boundary is of the form $-\cos\pi y$, for $0\le y\le 1$. Near $y=0$ this velocity is $-1$ while it is $+1$ near $y=1$.

The boundary-layer equation (2.9.1) is a nonlinear *partial* differential equation, and this combination renders the problem very difficult. Progress can be made if we consider instead the oncoming flow as being constant and either $-1$ or $+1$. If the actual oncoming flow is as in (2.9.2), this can be thought of as a crude attempt at a Taylor series solution in $y$ around either $y = 0$ or $y = 1$ in which we keep only the first term. This is an important point to keep in mind for it means that our results do not apply all along the boundary layer but are focused instead on these particular regions near $y = \pm 1$. On the other hand, we may simply momentarily abandon trying to consider the whole gyre and instead consider special impinging flows which are constant with latitude in order to find boundary layer solutions just as we did in the inertial boundary layer with our choice (2.8.5).

In either case this implies that in the case where the oncoming zonal flow has velocity $-1$, we search for solutions of the form:

$$\psi = \phi(x)y \tag{2.9.3}$$

which when inserted in (2.9.1) yields the ordinary differential equation:

$$\phi'''' = \lambda(\phi'\phi'' - \phi\phi''') + \phi' \tag{2.9.4}$$

where a prime denotes a differentiation in $x$. The problem is now reduced to an ordinary differential equation in $x$. It is left for the reader to verify that for the case where the interior flow is eastward and equal to $+1$ the same equation for the $x$ structure obtains with $\lambda$ replaced with $-\lambda$.

The vorticity equation (2.9.4) can be integrated once in $x$ to obtain:

$$\phi''' = \lambda(\phi'^2 - \phi\phi'') + \phi - 1 \tag{2.9.5}$$

where we have used the condition that $\phi \to 1$ as $x$ goes to $\infty$ while the derivatives of $\phi$ on the boundary-layer scale go to zero for large $x$. The boundary conditions for (2.9.5) are, for no slip:

$$\phi(0) = \phi'(0) = 0 \tag{2.9.6a}$$

and for slip:

$$\phi(0) = \phi''(0) = 0. \tag{2.9.6b}$$

The third boundary condition is set at $x = \infty$, namely that $\phi$ approaches 1.

Ierley and Ruehr (1986) solved this nonlinear problem numerically by a "shooting method". That is, if three conditions were set at the origin, the problem would be an initial value problem which would be very easy to integrate numerically. Instead, however, we have a nonlinear boundary value problem with two conditions set at the origin and one condition to be satisfied at $\infty$, and this is much more difficult. The shooting method guesses the third "missing" initial condition on $x = 0$ and integrates the equation forward to $x = \infty$. If, improbably, the guess were correct, and the boundary condition at infinity were satisfied, the calculation would be over. In the more likely case that the initial guess is not correct, the value of the initial guess is iterated until the boundary condition at $x = \infty$ is satisfied. In the case of the no-slip condition the missing initial value that must be guessed is the second derivative, while in the case of slip boundary conditions it is the first derivative at the origin that must be guessed. In each case Ierley and Ruehr called the guessed, missing derivative at the origin $\beta$. This is an unfortunate terminology from the oceanographic point of view for it has nothing to do with the planetary vorticity gradient, but we retain their terminology for the discussion of this section (in which the dimensional planetary vorticity gradient does not explicitly appear).

The results of careful numerical calculations are rather surprising. Figure 2.9.1 shows the relation between $\lambda$ and $\beta$ for which solutions to the boundary-layer problem could be found in the case of no-slip boundary conditions. The solid line is the numerical result (which agrees closely with the earlier numerical work of Il'in and Kamenkovich 1964), and the dotted curve is the result of an analytical approximation of Ierley and Ruehr based on an expansion and continued fraction solution for small $\lambda$. For $\lambda > 0$ there is a single solution represented by a single value of $\beta$ for each $\lambda$. This is the branch labeled II in the figure. For $\lambda < 0$ (eastward interior flow) there is a range of $\lambda$ for which two solutions are found, i.e., two values of $\beta$ for each negative $\lambda$ greater than a critical value $\lambda_c$. In the region where two solutions are possible Ierley and Ruehr showed that the solutions on the upper branch are unstable, and therefore only the solutions on the lower branch which continues smoothly from $\lambda > 0$ are relevant. More important from the point of view of the circulation problem is that for $\lambda < \lambda_c$ there are *no* boundary-layer solutions at all. For the no-slip case this critical value of $\lambda = -0.79130$.

If this threshold were to occur for large negative values of $\lambda$ where inertial effects are dominant, the absence of solutions would call to mind Greenspan's theorem on the impossibility of a purely inertial boundary layer for outflow conditions at $x = \infty$. However, the value of the critical value found by Ierley and Ruehr and Il'in and Kamenkovich corresponds to a ratio of $\delta_I/\delta_M$ of about 0.89. That is, even when the boundary layer is largely dissipative, so that the inertial and viscous terms are of the same order, no boundary-layer solution can be found for outflow conditions. (Note that with Ierley and Ruehr's slightly different definition of the inertial layer thickness the critical $\lambda$ would correspond to an even smaller value of $\delta_I/\delta_M$.)

Figure 2.9.2 shows the profiles of $\phi(x)$ (whose derivative yields $v$) for the realizable solutions. The solid line shows the solution for $\lambda > 0$ $(\lambda = 2)$. The solution is not essentially different than the inertial solution (2.8.6) except that $v$ vanishes at $x = 0$. The dotted line shows the solution near $\lambda = 0$ for which nonlinearity is relatively unimportant. Here, the solution is similar to the Munk solution. Note the slight overshoot of $\phi$ before it asymptotes to its final value of unity as $x$ tends to infinity. This overshoot describes the weak recirculation present in the Munk solution that we mentioned above. The dashed

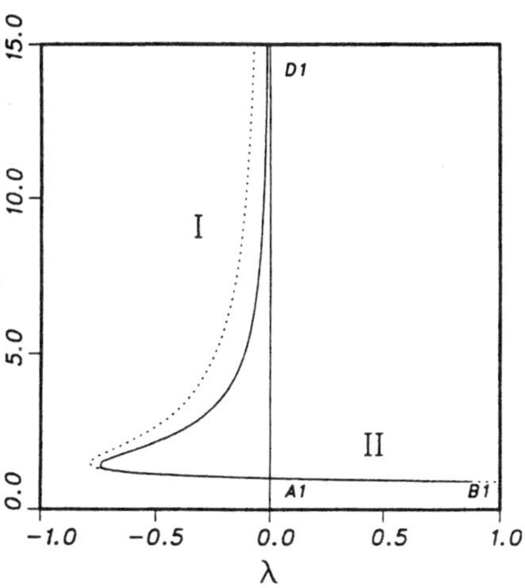


**Fig. 2.9.1.** Relation between the boundary-layer Reynolds number $\lambda$ and the value of the shooting method parameter $\beta$ (*not to be* confused with the planetary vorticity gradient) for which solutions to the nonlinear Munk boundary layer problem can be found for no slip conditions. Only the branch labeled II is stable. (From Ierley and Ruehr 1986)

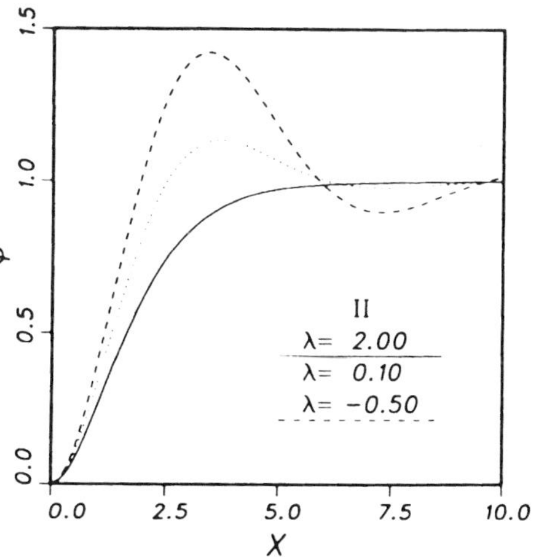


**Fig. 2.9.2.** Profiles of boundary layer streamfunction versus $x$ (scaled by $\delta_M$) for solutions on branch II of Fig. 2.9.1. *Solid line*, a solution for large positive $\lambda$; *dotted curve* solution for a nearly linear solution ($\lambda \approx 0$); *dashed curve* solution for $\lambda$ near (but greater than) the critical value near the nose of the curve in Fig. 2.9.1. (From Ierley and Ruehr 1986)

curve shows the solution for a value of $\lambda$ just short of the critical value $\lambda_c$. The overshoot is greater signifying a much larger recirculation (nearly a third of the transport of the current is now in the recirculation). *Below $\lambda_c$ there are no solutions at all.*

Figure 2.9.3 shows the ($\lambda$, $\beta$) relation for realizable solutions for the case of slip conditions on the wall. For both positive and negative values of $\lambda$ two branches of the solution are found. Again, the only stable branch is the branch which is continuous with the Munk solution at $\lambda \approx 0$. This branch is labeled II in the figure. Again, there is a critical value of $\lambda$ below which no solutions for outflow conditions are possible. In the slip case the range of negative $\lambda$ for which solutions can be found is even more limited. The critical value of $\lambda$ is now $\lambda_c = -0.27$, corresponding to a value of 0.519 for $\delta_I/\delta_M$. In this case the boundary layer analysis fails while the inertial layer thickness is only a bit larger than half the Munk layer thickness. Figure 2.9.4 shows the profiles of $\phi$ for values of $\lambda$ corresponding to a strongly inertial inflow, $\lambda = 2.00$, for a nearly linear boundary layer $\lambda = 0.25$, and for an outflow condition about half the critical value($\lambda = -1.00$). The profiles are similar to the no-slip case.

If we apply the results of this model to the general boundary-layer problem, the consequences are very striking. The inability, for even moderate nonlinearity, to construct a purely boundary-layer solution in the region where

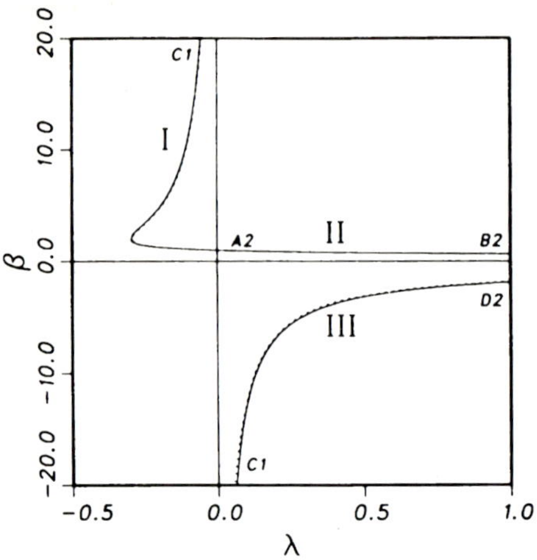


Fig. 2.9.3. As in Fig. 2.9.1 but for slip boundary conditions. The only stable branch is the one labeled *II*. (From Ierley and Ruehr 1986)

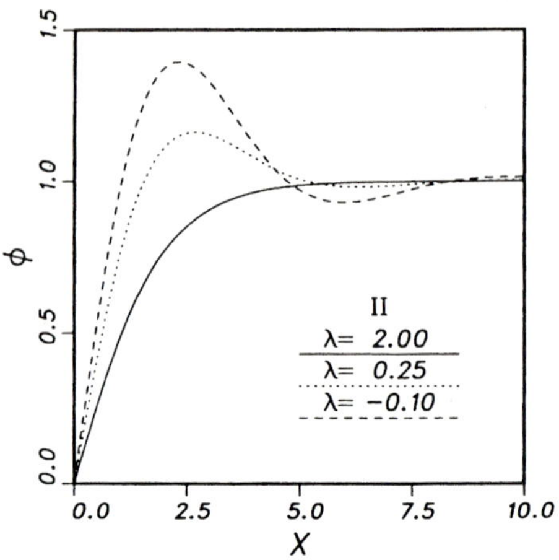


**Fig. 2.9.4.** As in Fig. 2.9.2 but for solutions on branch II of Fig. 2.9.3. *Solid curve*, for $\lambda = 2.00$; *dotted curve*, is for $\lambda = 0.25$; *dashed curve*, for $\lambda = -0.10$. (From Ierley and Ruehr 1986)

the Sverdrup solution requires outflow from the boundary layer strikes at the heart of a simple conceptual understanding of the complete circulation as the sum of a Sverdrup interior flow and a structurally straight forward boundary layer limited to a narrow region near the western boundary. We must keep this in mind as we examine the numerical models of the circulation problem in later sections. This raises the question of how the outflow actually takes place for values of the nonlinearity that are significant.

Figure 2.9.5, from Ierley (1987), shows the result of a calculation of a single layer general circulation model for a value of $\delta_I/\delta_M$ of about unity for slip boundary conditions. That is, rather than making the boundary-layer approximation for the western region of the basin the full vorticity equation is numerically integrated. The solution in the interior consists of a Sverdrup flow, and the western region has what looks like a simple boundary-layer structure in the south. In the northwest corner of the basin the situation is much more complex. A tight recirculation gyre has appeared, and the flow which joins the interior must circumnavigate the rim of the recirculation region before entering the Sverdrup region. Clearly, in this calculation in the region of outflow and entrance into the interior the classical boundary-layer approximation $\partial/\partial x \gg \partial/\partial y$ no longer applies. Recall that it is precisely in this region that the boundary-layer approximation, pivoted about $y = +1$, failed in the treatment by Ierley and Ruehr (1986).

It seems possible then that the dynamics of the full problem, governed by the full vorticity equation, differs qualitatively from the situation described by

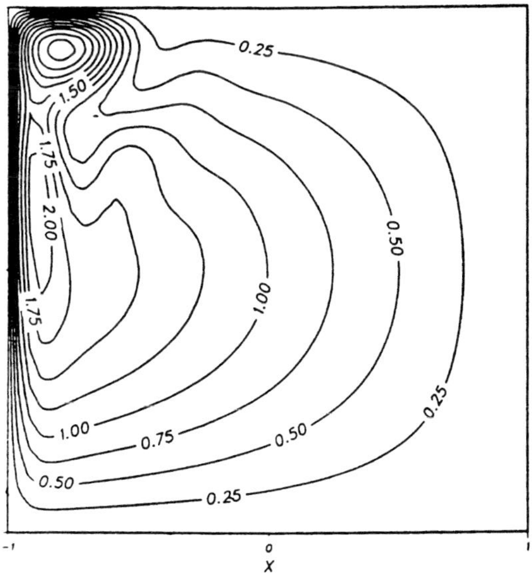


**Fig. 2.9.5.** Calculation of the wind-driven circulation driven by an Ekman pumping of a single-gyre form for the case of no-slip for a value of $\delta_I/\delta M$ of $O(1)$. (From Ierely 1987)

the boundary-layer approximation , $\partial/\partial x \gg \partial/\partial y$. To examine the connection between the full vorticity equation, governed by the partial differential equation (2.3.2), and the boundary-layer problem solved by Ierley and Ruehr, Ierley (1987) examined a "regional" model. That is, he considered a model in which the Sverdrup theory is solved for first and is supposed to be valid over the greater part of the basin. The solution in the vicinity of the western boundary is then determined by solving the vorticity equation in a domain$(L_x, L_y)$ in which $L_x$ is small compared to the full basin scale but large compared to the expected scale of the boundary layer. At the eastern edge of the region the outflow velocity and its vorticity are chosen to exactly match the Sverdrup solution. There is no other forcing for the flow in that region. At the western, northern, and southern boundaries the conditions of no normal flow and, in this case, slip conditions were employed. The scale in $x$ was chosen to be 10 $\delta_M$, and the scale in $y$ was allowed to vary so that $\alpha \equiv L_y/L_x = L_y/10\delta_M$ becomes another parameter of the calculation in addition to $\delta_I/\delta_M$. We might expect that as $\alpha$ becomes large, the results of the regional model would qualitatively approach that of the boundary-layer problem.

Ierley found that for $\alpha$ of $O(1)$ boundary-layer type solutions of the regional model exist, for which $\partial/\partial x \gg \partial/\partial y$, for values of $\lambda$ (evaluated in terms of $\delta_I/\delta_M$) that are supercritical, i.e., for $\lambda < \lambda_c < 0$. Figure 2.9.6a shows a solution for the regional problem with slip conditions for a value of $\lambda = -0.7$ whereas the critical value of $\lambda$ for the boundary layer problem is $-0.297$. In the outflow region of the flow in the figure a boundary layer solution where $x$ derivatives exceed $y$ derivatives in size still exists. For a fixed $\lambda$ as $\alpha$ is increased (i.e., as the $y$ length of the region is expanded) the $y$ scale in the *outflow* region

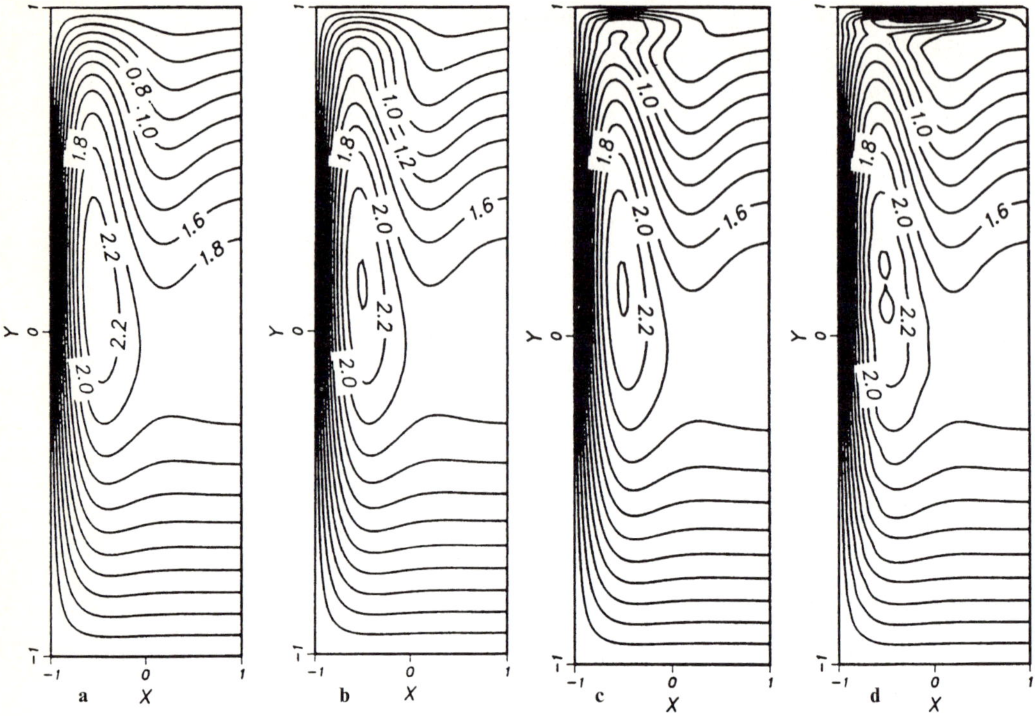


**Fig. 2.9.6a–d.** Sequence of calculations for the regional model described in the text. Each calculation is for the same value of $\lambda$ in the exit region, namely, $\lambda = -0.7$ while the aspect ratio a of the region is increased. **a** $\alpha = 2.5$. **b** $\alpha = 5$. **c** $\alpha = 10$. Here the boundary layer approximation fails in the outflow region. **d** $\alpha = 20$. A strong recirculation region appears in the northwest corner. (From Ierely1987)

decreases (Fig. 2.9.6 b,c,d) until finally, at $\alpha = 10$, the $x$ and $y$ scales become comparable. If $\alpha$ is increased further a cell of *recirculation* appears in the northwest corner of the region, and the outflow must travel around the rim of the recirculation before rejoining the interior.

If we associate the critical value of $\lambda$ with the breakdown of the boundary-layer approximation at the northern boundary [recall that equation (2.9.5) is supposed to apply either in the region near the southern boundary ($\lambda > 0$) or near the northern boundary ($\lambda < 0$)] we can infer, as did Ierley, that the critical value of $\lambda$ is really a function of $\alpha$ and approaches the critical value of the boundary-layer problem only as $\alpha$ tends to infinity. On the basis of extensive numerical calculations Ierley suggested a rough dependence of the form:

$$\lambda_c(\alpha) = \lambda_c(\infty) + 7/\alpha. \tag{2.9.7}$$

More importantly, the analysis suggests that the breakdown of the boundary-layer approximation is intimately connected with the appearance of a tight recirculating gyre in the northwest corner of the gyre which mediates the return of the boundary-layer flow to the interior. The problem of understanding the relation between the interior Sverdrup flow and the western boundary region can therefore be expected to involve regions of flow much more complex than standard boundary-layer regions matched to a Sverdrup interior. The need for the fluid to sufficiently dissipate vorticity along its path after entering the western region of nonlinear flow evidently requires the path of the fluid to become substantially distorted into a permanent recirculating eddy before rejoining the interior. This makes the problem of understanding the relationship between the interior and the boundary layer unexpectedly complex. It further emphasizes the important role of dissipation in determining the structure of the circulation pattern although for the parameter settings studied here the effects are limited to the region near the western boundary.

Given the sensitivity of the circulation to the mechanism of dissipation, it must also be anticipated that the pattern of the circulation depends on the type of dissipation (lateral mixing or bottom friction) and the magnitude of the dissipation compared to the inertial effects and to the boundary conditions (as in the problem studied by Ierley and Ruehr where the no slip boundary layer solutions persist to higher Reynolds numbers than do the slip solutions in the outflow region). Determining exactly what the sensitivity to these conditions is requires us to examine a range of numerical experiments that have been carried out by several investigators over the past few decades. Before describing these experiments in detail we continue with our preparation by discussing several general integral constraints on the circulation and the implications that they have for different types of dissipation and different boundary conditions.

## 2.10 Integral Conditions

### *Integral Statements of Energy and Vorticity*

Consider the vorticity equation (2.2.9) and imagine first that the solution has reached a steady state. Each streamline must be closed, and of course there is no flow across streamlines. We integrate over the area $S$ enclosed by a streamline (see Fig. 2.10.1). Note that the boundary of the basin is also a streamline to which the following analysis must also apply.

When the advective term is integrated over the area $S$ we obtain:

$$\begin{aligned}\iint_A J(\psi, \nabla^2\psi + \beta y)dx\,dy &= \iint_A \vec{u}\cdot\nabla(\zeta+\beta y)dx\,dy\\ &= \iint_A \nabla\cdot(\vec{u}\{\zeta+\beta y\})dx\,dy = \oint_{C_\psi} \vec{u}\cdot\vec{n}\{\zeta+\beta y\}\,ds\\ &= 0.\end{aligned} \tag{2.10.1}$$

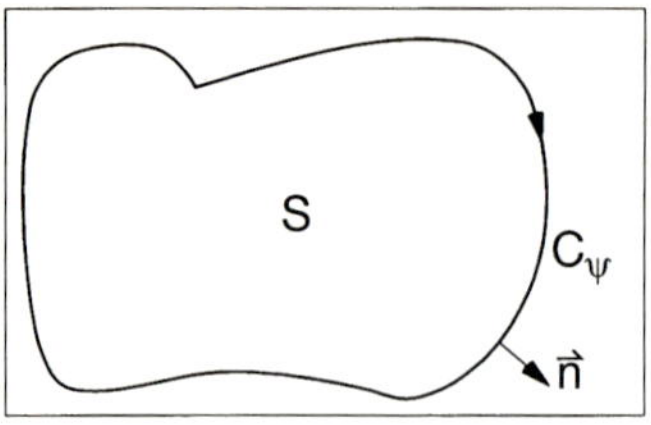


**Fig. 2.10.1.** Streamline traces out the curve $C_\psi$ which bounds an area $S$ over which the vorticity equation is integrated

In (2.10.1) $\vec{n}$ is a unit vector perpendicular to the streamline and $ds$ is a line element along the curve $C_\psi$. In deriving (2.10.1) we use the nondivergence of the velocity and also the divergence theorem to show that within a streamline the advective terms have no net effect. Since there is no flow across the boundary of the region, the advective terms can only redistribute vorticity within the region without changing the total amount contained in $S$. If the remainder of the vorticity equation is similarly integrated we obtain:

$$0 = \iint_S \frac{f_0}{D} w_E \, dA + A_H \oint_{C_\psi} \nabla\zeta \cdot \vec{n} \, ds - r \oint_{C_\psi} \vec{u} \cdot \vec{t} \, ds \tag{2.10.2}$$

where $\vec{t}$ is a unit vector tangent to the curve $C_\psi$.

Equation (2.10.2) tells us that the net vorticity put into the region bounded by $C_\psi$ by the wind-forced Ekman pumping (the first term in the equation), must be balanced by a combination of dissipative effects. Either vorticity must be diffused across the curve by lateral friction at a rate that depends on the horizontal mixing coefficient and the *gradient* of the vorticity on the curve, or it must be destroyed by the action of bottom friction which is proportional to the fluid velocity along the bounding streamline. No matter how small dissipation might be locally with respect to inertia effects, it is dissipation alone that must act to achieve a steady-state balance between vorticity input and its elimination for every streamline. Thus every streamline must experience an amount of dissipation which depends only on the amount of vorticity input and is independent of the size of the dissipation parameters $A_H$ and $r$. This is what renders the problem so difficult in the nonlinear regime. Locally the dynamics may be dominated by inertial effects, but globally there must be a significant role for dissipation.

If the flow is time dependent, the constraint (2.10.2) must be modified. In the presence of eddies and smaller scale fluctuations we can still define a time average circulation by averaging over times long compared to the characteristic eddy fluctuation time. We can then consider $C_\psi$ to be the position of an average streamline. However, in this case the integral of the advection term must be reconsidered. Each dynamical quantity is written in terms of its time average plus a fluctuation, for example:

$$\zeta = \bar{\zeta} + \zeta', \qquad \vec{u} = \bar{u} + \vec{u}\,' \tag{2.10.3}$$

where an overbar represents the time average quantity, and the prime represents the departure from the average. Note that the time average of a linear primed quantity is zero.

Then the time average of the flux of vorticity can be written:

$$\overline{\vec{u}\zeta} = \overline{\vec{u}}\overline{\zeta} + \overline{\vec{u}'\zeta'} \tag{2.10.4}$$

so that the integral of the advection term becomes, instead of (2.10.1):

$$\iint_S \nabla \cdot \left[\overline{\vec{u}}(\overline{\zeta} + \beta y) + \overline{\vec{u}'\zeta'}\right] = \oint_{C_\psi} \overline{\vec{u}}\,(\overline{\zeta} + \beta y) \cdot \vec{n}\, ds + \oint_{C_\psi} \overline{\vec{u}'\zeta'} \cdot \vec{n}\, ds$$

$$= \oint_{C_\psi} \overline{\vec{u}'\zeta'} \cdot \vec{n}\, ds \tag{2.10.5}$$

since only the time-averaged velocity vector is tangent to the mean streamline, and only for this flow is the dot product with the normal to the streamline zero. The fluctuations may then cooperate to carry vorticity across the mean streamline. Hence, in the case of time variable flow we must add another flux term on the right side of (2.10.2) which becomes then, for the time-averaged flow:

$$\begin{aligned} 0 = &\iint_S \frac{f_0}{D} w_E\, dx\, dy + A_H \oint_{C_\psi} \vec{n} \cdot \nabla\zeta\, ds \\ &- r \oint_{C_\psi} \vec{u} \cdot \vec{t}\, ds - \oint_{C_\psi} \vec{n} \cdot \overline{\vec{u}'\zeta'}\, ds \end{aligned} \tag{2.10.6}$$

so that now the flux of vorticity by the eddies can contribute to the vorticity balance. In (2.10.6) we drop the overbar notation for the time-averaged quantities.

If we consider the vorticity balance for the basin as a whole, the curve $C_\psi$ coincides with the boundary of the basin, and both the mean and fluctuating velocities must satisfy the no normal flow condition. In this case (2.10.2) remains valid for the vorticity balance for the time average flow.

An additional informative constraint can be obtained by considering the equation for the energy (see Cessi et al. 1990).

If the vorticity equation is multiplied by $\psi$ and then integrated over the entire basin, we obtain, after frequent use is made of the divergence theorem and the vanishing of the stream function on the boundary:

$$\begin{aligned} \frac{\partial E}{\partial t} = &-\iint_S \frac{f_0}{D} \psi w_E\, dx\, dy - 2rE \\ &- A_H \left[\iint_S \zeta^2 dx\, dy - \oint_C \zeta\, \vec{u} \cdot \vec{t}\, ds\right] \end{aligned} \tag{2.10.7}$$

where now $S$ is the area of the basin and $C$ is its perimeter. $E$ is the total (kinetic) energy in the basin i.e.:

$$E = \iint_S \frac{1}{2}(\nabla\psi \cdot \nabla\psi)dx\, dy. \tag{2.10.8}$$

The energy of the geostrophic motion is increased by the pressure work on the upper surface. The pressure is proportional to $\psi f_0$, and its product with $-w_E$ is the rate at which the the wind works on the ocean beneath the Ekman layer. Downward Ekman pumping ($w_E < 0$) in the subtropical gyre ($\psi > 0$) increases the energy of the gyre. The energy must be balanced by either bottom friction (the term proportional to $r$) or lateral friction (the last term in 2.10.7). Note that the condition of no normal flow does not guarantee the sign of the effect of horizontal friction since the boundary integral in the last bracket is not sign determinant. If the flow satisfies either the no-slip condition, for which $\vec{u} \cdot \vec{t}$ vanishes, or the slip condition, for which $\zeta$ vanishes, the boundary term in the energy equation would also vanish, and the effect of horizontal friction would always be a drain of energy. It certainly is a desirable feature of the dynamics for the lateral mixing definitely to lead to a decrease in energy. However, if the superslip condition or hyperslip condition is used (2.4.10), the boundary term does not vanish, and the effects of lateral friction could actually add energy to the circulation. This is certainly an undesirable feature of the use of that boundary condition (among others, as we saw above in discussion of the Munk problem) even though it springs from an entirely reasonable association of the diffusion term with the action of unresolved turbulent eddies. This once more stresses the important role thrust on the dissipation to balance the inputs of vorticity and energy.

### *Role of Boundary Conditions*

We can use (2.10.2) when applied to the basin as a whole to see how the application of different boundary conditions and different choices of the dissipation mechanism affect the way in which the fluid can contrive to balance the input of vorticity by the wind. The discussion that follows is much influenced by an unpublished manuscript by Ierley and Young.

We can see, according to (2.10.2) that, a priori, the vorticity input can be drained out of the basin by either bottom friction or lateral friction. If the boundary condition on the side wall is that the tangential velocity vanishes there, (2.10.2) shows that the effect of bottom friction, on the basin as a whole, is eliminated and is unable to affect the overall vorticity balance. In this case lateral friction must be solely responsible for the vorticity balance for the entire basin. (This need not be the case for streamlines other than the boundary streamline.) On the other hand, if superslip boundary conditions $\nabla\zeta \cdot \vec{n} = 0$ are used, the lateral friction is unable to flux any vorticity out of the basin, and the vorticity balance requires that bottom friction be the mechanism that the fluid uses to reach equilibrium. Finally, if the slip condition $\zeta = 0$ is used, both mechanisms could enter a priori in the vorticity balance. The choice of boundary condition therefore determines a priori the mechanism of global vorticity balance.

Some models use one or the other, but not both of the dissipation mechanisms. For example, the study of Veronis (1966) employed only bottom friction. In this case there is no second boundary condition because in the absence of lateral friction the order of the differential equation is reduced. This implies that the dissipation of vorticity depends entirely on the magnitude of the velocity along the boundary, and that if $r$ is small, the boundary velocity must be corresponding large.

## 2.11 Integral Balances for the Boundary Layer

Over most of the basin the role of dissipation is locally negligible. This is after all one of the principal ingredients of the Sverdrup balance. It follows that we expect that the western boundary layer is a principal location for dissipation to take place although the results of Ierley and Ruehr (1986) and Ierley (1987) suggest that in the inertial regime the representation of this domain as a simple boundary-layer structure may not be sufficient. It is illuminating in this regard to examine the integral balance for the vorticity, integrating over the open region that stretches eastward from the western boundary and covers the boundary-layer region, as shown in Fig. 2.11.1. The right edge of the region $R$ in the figure is assumed to lie outside the western boundary layer so that at this edge all of the boundary-layer processes involving nonlinearity and friction are negligible.

Consider the boundary layer vorticity equation (2.6.4) with the addition of a term representing the time rate of change of vorticity, i.e.:

$$\frac{\partial \zeta}{\partial t} + \nabla \cdot (\vec{u}\zeta) + \beta v = A_H \frac{\partial^2 \zeta}{\partial x^2} - r\zeta, \quad \zeta = \frac{\partial v}{\partial x}. \qquad (2.11.1a,b)$$

We integrate (2.11.1) over the region $R$ using the condition that $\psi$ vanishes at the western boundary and approaches $\psi_I$ for large $x$. The advection term, by the divergence theorem, can be written in terms of fluxes across the boundary of $R$. At the western and eastern edges of $R$ the fluxes either vanish exactly (at the western boundary where the zonal velocity is zero) or are negligible (at the eastern edge). Thus we have:

$$\frac{\partial}{\partial t}\iint_R \zeta \, dx \, dy = \int_0^\infty dx \left[ v\frac{\partial v}{\partial x}(y_1) - v\frac{\partial v}{\partial x}(y_2) \right] - \beta \int_{y_1}^{y_2} \psi_I(0, y) dy$$
$$- A_H \int_{y_1}^{y_2} \frac{\partial \zeta}{\partial x}(0, y) dy + r \int_{y_1}^{y_2} v(0, y) dy \qquad (2.11.2)$$

or:

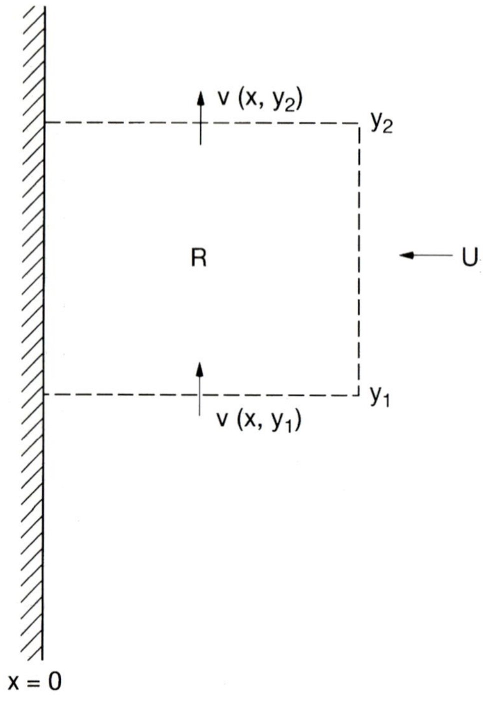


**Fig. 2.11.1.** Open region $R$ covering a segment of the western boundary-layer region over which the vorticity equation is integrated

$$\frac{\partial}{\partial t}\iint_R \zeta\, dx\, dy = \frac{1}{2}\left[v^2(0,y_2) - v^2(0,y_1)\right] - \beta\int_{y_1}^{y_2} \psi_I(0,y)dy$$
$$- A_H\int_{y_1}^{y_2} \frac{\partial\zeta}{\partial x}(0,y)dy + r\int_{y_1}^{y_2} v(0,y)dy. \tag{2.11.3}$$

Thus, in the region $R$ the vorticity balance depends locally on a competition between advection and dissipation. The flux of planetary vorticity, which is the second term on the right side of (2.11.3), brings negative (anticyclonic) vorticity into the boundary layer in the subtropical gyre since $\psi_I$ is positive there and zero on the eastern boundary. Since the velocity is nondivergent:

$$\iint_R \nabla\cdot(\vec{u}f)dx\, dy = \iint_R v\beta\, dx\, dy = \iint_R \beta\frac{\partial\psi}{\partial x}dx\, dy$$
$$= \int_{y_1}^{y_2} \beta\psi_I(0,y)dy \tag{2.11.4}$$

which shows relation between the $\beta$ term in (2.11.3) and the total flux of planetary vorticity into $R$.

This flux into the layer is balanced in a steady state by a frictional flux through the side wall and consumption of vorticity through the bottom and by an inertial flux out of $R$ [the first term on the right side of (2.11.3)]. It is important to note that this inertial advective term, when integrated over $R$, involves only the tangential velocity in the boundary layer *at the wall*. Thus, for no-slip conditions, for which that velocity must be zero, the planetary vorticity imported into the region $R$ from the interior must be locally completely dissipated in that region. Furthermore, it must be dissipated locally only by the lateral friction since the integrated effect of the bottom friction also vanishes for no-slip conditions. This is because the total relative vorticity in the boundary layer (at least in the boundary-layer approximation) vanishes when no-slip conditions apply, i.e.:

$$\int_0^{\infty} \zeta \, dx = -v(0, y) = 0. \tag{2.11.5}$$

If the Sverdrup balance (2.5.3) is used, we can write (2.11.3) in the steady state *for no-slip conditions* as:

$$\int_{y_l}^{y_2} \int_0^{x_e} \frac{f_0}{D} w_E \, dx \, dy = A_H \int_{y_1}^{y_2} \frac{\partial \zeta}{\partial x}(0, y) dy. \tag{2.11.6}$$

This important result was first illustrated by Stewart (1964) who integrated the full-vorticity equation across the entire basin to show that in the steady state *the vorticity put into the latitude strip* $(y_1, y_2)$ *by the wind must be locally dissipated in the same latitude band by a horizontal flux of vorticity out of the basin in that same strip*. On the other hand, for slip conditions *both* bottom friction and local advection can also contribute to the balance of a latitude strip.

The local effect of the boundary conditions can be further emphasized by considering the fate of the total vorticity isolines, i.e., the isolines of $\zeta + \beta y$. Near the eastern boundary of the basin where the relative vorticity is unimportant, only the planetary vorticity is important in evaluating the total vorticity, and thus the isolines are coincident with latitude circles. Each isoline strikes the eastern boundary at a point $y = y_e$, which varies of course from one isoline to another. For each such isoline emanating from the eastern boundary:

$$\zeta + \beta y = \beta y_e. \tag{2.11.7}$$

In the Sverdrup interior where advection of relative vorticity remains negligible the total vorticity isolines remain latitude circles $y = y_e$, and the flow can cross those isolines only because of the presence of a vorticity input from the wind. In the western boundary current, on the other hand, the strong northward flow tends to drag the isolines northward (except in the linear limit where fluid oozes across the vorticity isolines, quickly diffusing away the vorticity anomaly laterally or expunging it through bottom friction).

However, the boundary conditions on the western boundary place a strong constraint on just how easily the isolines can be dragged northward. If, for example, the no-stress condition is used, $\zeta = 0$ on the western boundary and the isoline of total vorticity must attach to the western boundary at the same position, $y = y_e$, that it had on the eastern boundary. If no-slip conditions are applied on the western wall, $\zeta$ is positive ($v$ increasing with $x$) right at the wall. This means that the intersection with the western wall of the isoline in the no-slip case is at:

$$y_w = y_e - \frac{1}{\beta}\frac{\partial v}{\partial x}(0, y_w) \leq y_e. \tag{2.11.8}$$

Thus in both cases of slip and no-slip there is a strong stretching process operating on the vorticity isolines. The total vorticity isolines are anchored at the western boundary at a position equal to or south of the position that they had when they entered the boundary layer. Yet in the nonlinear limit they are being advectively strongly stretched northward within the layer. This should make for a very contorted geometry of the vorticity field in the vicinity of the western boundary. Only in the case of superslip conditions or in the entire absence of lateral friction and its associated boundary conditions are the isolines of vorticity free to slide along the boundaries and be easily wrapped up by the motion.

This has important consequences for the possibility of resonance with the Fofonoff mode. In the inertial, strongly nonlinear parameter range the possibility exists, as we noted above, that the free nonlinear mode found by Fofonoff (discussed in Section 2.8) might be excited by the steady wind forcing. If the mode were to resonate and to dominate the solution, we would have as a first approximation the relation between total vorticity and streamfunction given by (2.8.12). If this is used to evaluate the dissipation balance (2.10.2) we obtain:

$$0 = \iint_S \frac{f_0}{D} w_E \, dx \, dy + A_H \oint_{C_\psi} \nabla Q(\psi) \cdot \vec{n} \, ds - r \oint_{C_\psi} \vec{u} \cdot \vec{t} \, ds. \tag{2.11.9}$$

Since:

$$\nabla Q(\psi) = \frac{dQ}{d\psi} \nabla \psi \tag{2.11.10}$$

and since $dQ/d\psi$ is also a function only of $\psi$ and hence constant on the streamline curve $C_\psi$, it follows that (2.11.9) can be written as:

$$0 = \iint_S \frac{f_0}{D} w_E \, dx \, dy + A_H \frac{dQ}{d\psi} \oint_{C_\psi} \vec{u} \cdot \vec{t} \, ds - r \oint_{C_\psi} \vec{u} \cdot \vec{t} \, ds. \tag{2.11.11}$$

In the absence of bottom friction the first two terms in (2.11.11) would have to balance in the steady state. Since the first term is $< 0$, the second must be positive. However the line integral of the velocity must have the same sign as the integral of the Ekman pumping. This follows from the energy balance (2.10.7) since if there is to be a net energy input by the wind to match the dissipation, the stream function and the Ekman pumping must on average be negatively correlated. If $w_E < 0$, this means $\psi > 0$ or an anticyclonic circulation. This in turn means that the velocity circulation integral in (2.11.11) has the same sign as the Ekman pumping. Therefore in the absence of bottom friction we must have $dQ/d\psi < 0$. This is the opposite sign than that required for the Fofonoff mode, and we conclude that resonance with the Fofonoff mode, is inhibited by the presence of lateral mixing, *as long as the flux of vorticity through the side walls is significant*. If the boundary conditions are superslip conditions, only bottom friction enters to equilibrate the vorticity balance; the second term in (2.1.11) is absent altogether, and there is no prohibition of resonance with the Fofonoff mode. This is consistent with our discussion of the geometry of total vorticity isolines.

## 2.12 Numerical Examples

We now turn our attention to the results of numerical calculations of the wind driven circulation problem. In particular, we concentrate on the effect of the magnitude and type of dissipation and the role of the boundary conditions in shaping the structure of the circulation. On the basis of our discussion up to now we can expect the role of dissipation to be profound. We are particularly interested in the limiting case where the boundary-layer Reynolds number is large, i.e., where $\delta_I > \max(\delta_M, \delta_S)$, and where we would expect from naive scaling arguments that the effects of dissipation would be unimportant. Because of the numerical difficulty of reaching this limit in the circulation problem most calculations have tended at most to be in the regime where inertia and dissipation are of the same importance in determining the boundary-layer structure. Nevertheless the effects of different dissipation parameterizations and boundary conditions are still very striking. We later take up the more difficult question of the limit $\delta_I \gg (\delta_M, \delta_S)$.

Blandford (1971) was the first to take up the systematic study of the role of boundary conditions and dissipation mechanisms in determining the circulation structure. Previous to his calculations there were two pioneering calculations by Bryan (1963) and Veronis (1966). Bryan used only lateral friction with a no-slip condition while Veronis employed only bottom friction and could therefore apply only the no normal flow condition.

A sequence of calculations from the Veronis study are shown in Fig. 2.12.1 for increasing values of $\delta_I/\delta_S$. Veronis chose for his forcing an Ekman pumping (or equivalently a wind stress curl) of the form:

$$w_E = -W_0 \sin\frac{\pi x}{L}\sin\frac{\pi y}{L}. \tag{2.12.1}$$

At low values of the ratio $\delta_I/\delta_S$ the solution is indistinguishable from the linear Stommel solution. Even when the ratio is unity, as in panel a of the

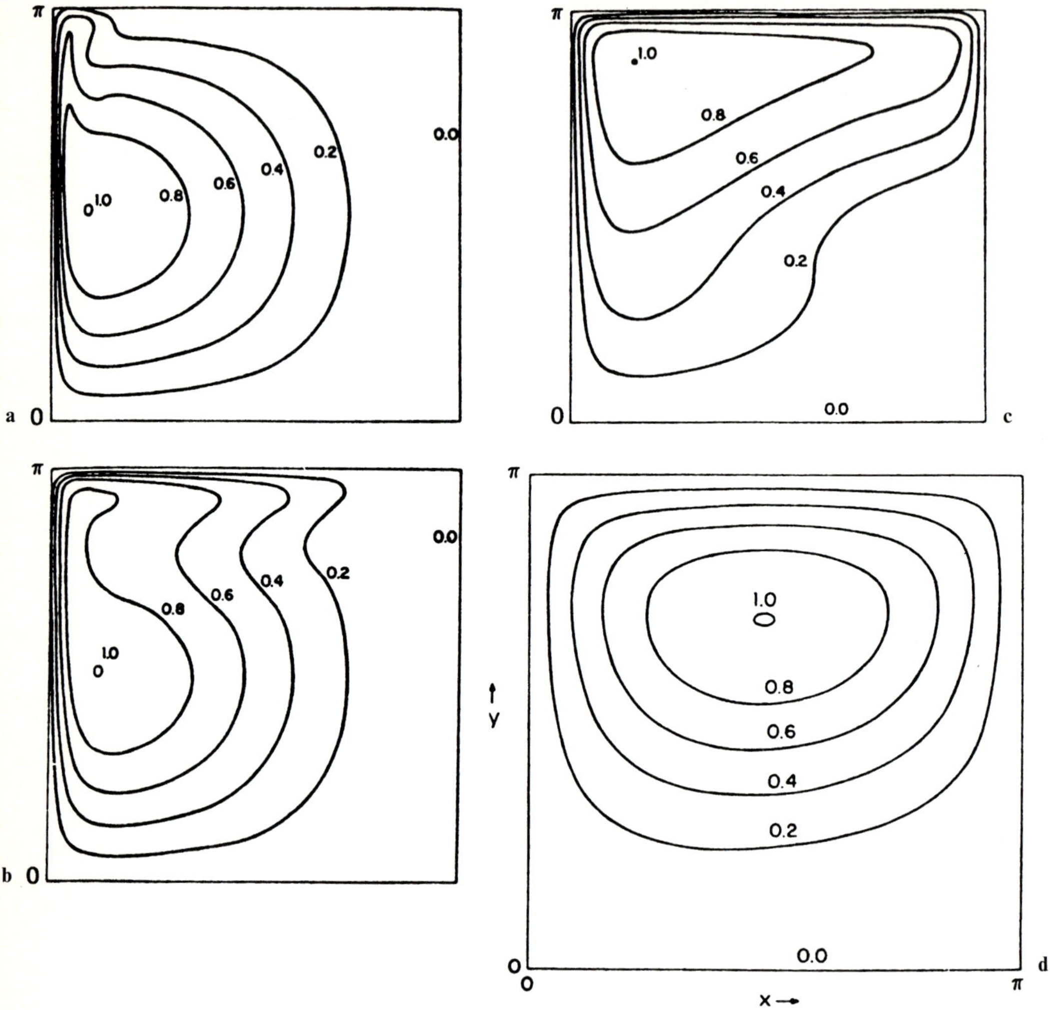


**Fig. 2.12.1a–d.** Solutions of the wind-driven circulation problem for the case where only bottom friction balances the wind input of vorticity. Each panel corresponds to an increasing value of the ratio $\delta_I/\delta_S$ **a** $\delta_I/\delta_S = 1$. **b** $\delta_I/\delta_S = 2$. **c** $\delta_I/\delta_S = 4$. **d** $\delta_I/\delta_S) = 8$. (From Veronis 1966 reprinted from Pedlosky 1987)

figure, there is only a slight north-south asymmetry produced by the northward advection of vorticity. As the inertial effects increase in magnitude, the structure of the flow changes significantly over the entire basin. A strong jet along the northern wall appears for $\delta_I/\delta_S$ of order 2 (panel b). When the ratio reaches 4 as in panel c, there is a strong boundary jet flowing southward on the eastern boundary. Note that in the absence of lateral friction Stewart's (1964) argument about the impossibility of eastern boundary currents becomes irrelevant. Finally, panel d shows the solution for $\delta_I/\delta_S = 8$. In this limit the east-west asymmetry of the circulation is completely lost. As the nonlinearity increases, the transport in the circulation also increases until, in the last panel, the maximum transport of the anticyclonic cell is of the order of ten times larger than would be anticipated by the Sverdrup theory! Indeed, in this limit the Sverdrup theory for the interior is no longer relevant. The solution has resonated with a Fofonoff mode which, as we have seen, is consistent with a circulation balanced only by bottom friction. Niiler (1966) exploited this fact and used the circulation integral and the form of the Fofonoff mode to estimate the large amplitude limit of the Veronis calculation with considerable success.

On the other hand, Bryan's (1963) calculations showed a completely different transformation as the degree of nonlinearity increases. In his model he used no-slip conditions on the eastern and western boundaries and no-stress conditions on the northern and southern boundaries. Figure 2.12.2 shows a sequence of increasing $\delta_I/\delta_M$ from 0.4 to about 0.9, after which the calculated solution became unsteady due to instabilities in the western boundary layer. The similarity to Veronis' calculation for weak nonlinearity is evident. The advection of vorticity northward produces a similar north-south asymmetry. However, for larger values $\delta_I/\delta_M$ of the circulation develops a gyre of *recirculation* the northwest corner and a damped wave train of one or two oscillations in the region of eastward interior flow.

Blandford (1971) then constructed a model with both bottom friction and lateral friction and tested the effects of using either no-slip or no stress boundary conditions (or neither, in the case of $A_H = 0$). In Blandford's calculation $\delta_I/\delta_S$ is 4, while is $\delta_I/\delta_M$ approximately 3.7. Figure 2.12.3 shows the results of three calculations. Each of the three solutions very nearly achieves a steady state although the no-slip case possesses a small residual fluctuation in the energy level of a small percentage due to eddies produced by instabilities in the western boundary layer. In the first panel is the calculation with only bottom friction; that is, in this calculation alone, $A_H = 0$. The result is qualitatively similar to Veronis' calculation with the same value $\delta_I/\delta_S$. A strong northern and eastern boundary current feeds a large interior flow which departs substantially from the Sverdrup solution. In the second panel , in which the slip condition is applied, the effect of the horizontal diffusion eliminates the tendency to form a basin wide Fofonoff mode as our previous discussion anticipated. Instead, a Sverdrup interior is reached by the fluid after it circumvents a large zone of recirculation near the northern boundary, reminiscent of the above calculations of Ierley (1987). The third panel shows the no-slip case. The Sverdrup interior is again maintained by the presence of lateral friction, and now the recirculation shrinks to a relatively small region near the northwestern corner of the basin.

Recall that all three models have about the same overall dissipation time scales, i.e., the Reynolds number and bottom friction level are the same in each case. Only the type of dissipation or the boundary conditions differ, and yet the

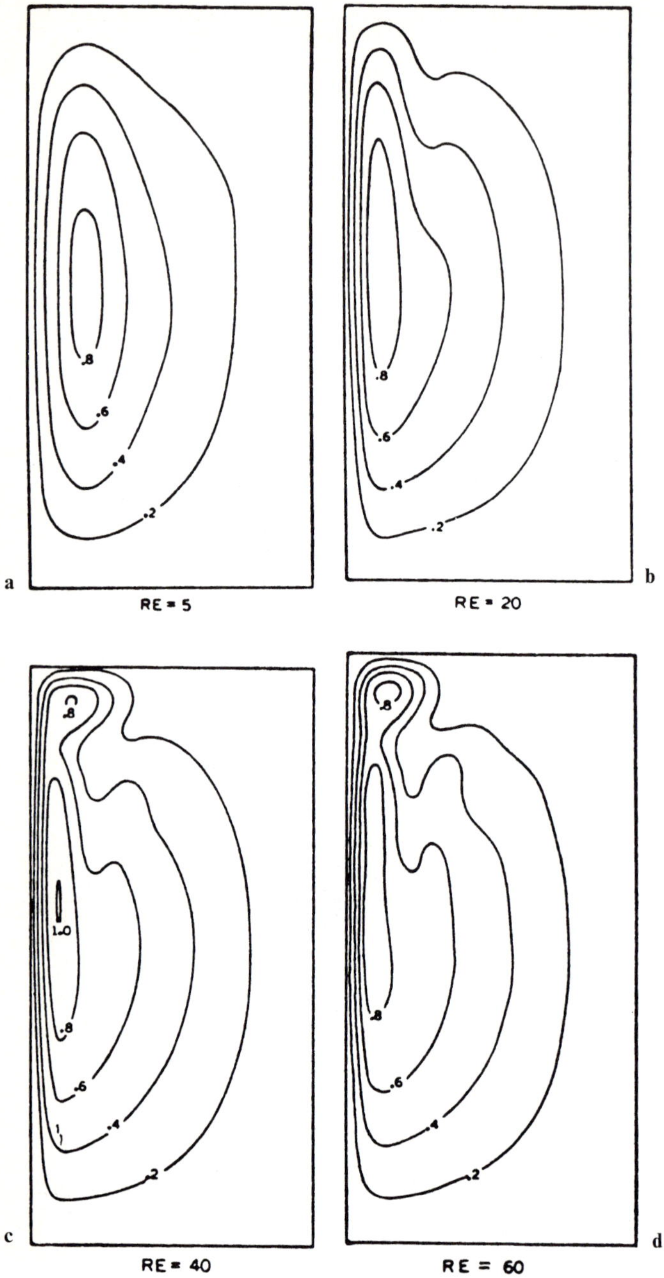


**Fig. 2.12.2a-d.** Sequence of calculations by Bryan (1963) with lateral friction but with no bottom friction and with boundary conditions of no slip. **a** $\delta_I/\delta_S = 1$. **b** $\delta_I/\delta_S = 2$. **c** $\delta_I/\delta_S = 4$. **d** $\delta_I/\delta_S = 8$

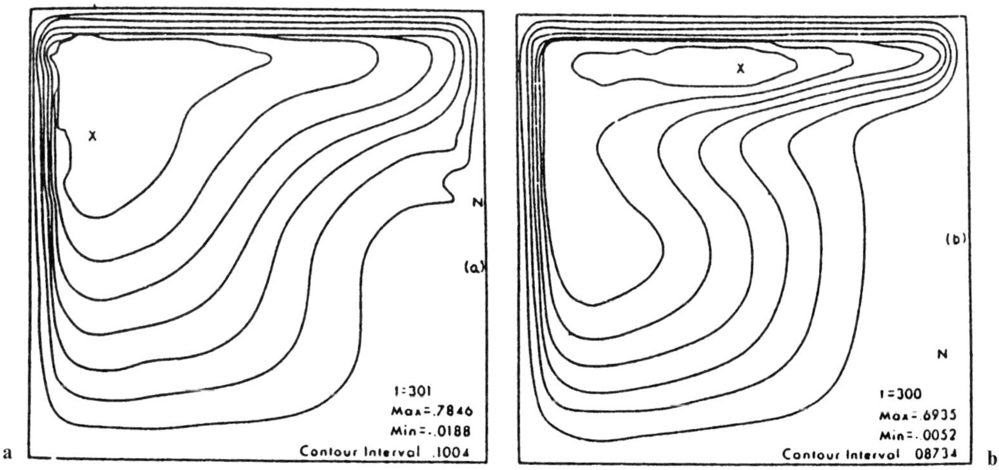


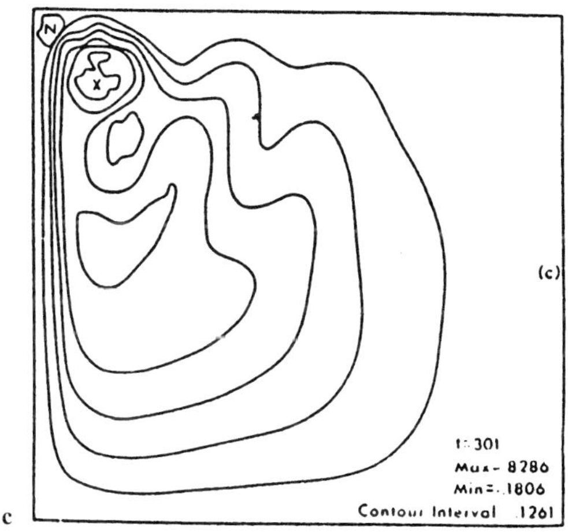


**Fig. 2.12.3a–c.** Calculations by Blandford (1971) of the wind-driven circulation problem. **a** no lateral friction, bottom friction only. **b** lateral friction added with a slip boundary condition on the side walls. **c** As in **b** but with no-slip conditions

global pattern of the circulation differs profoundly from case to case. Of course the discussion of previous sections has prepared us for such differences, but it may be helpful if we probe a bit deeper into the boundary layer balances at this point.

Consider first the Veronis model. In the case of the calculation of Fig. 2.12.1c strong boundary layers on the walls still exist. (In the last panel there is no evidence of a strong intensification toward any boundary.) The nonlinearity is strong enough that the boundary layers are of inertial type with a thickness $\delta_I = \sqrt{U_I/\beta}$ where $U_I$ is the magnitude of the interior velocity. In the absence of lateral friction the dissipation integral for the vorticity equation requires a balance between the vorticity input by the wind and the dissipation in the inertial boundary layers. It is important to remember that without lateral friction there is no sublayer, so that when $\delta_I > \delta_S$, there is only the inertial layer. By mass continuity the order of magnitude of the velocity of the tangential velocity in the boundary layer is:

$$v_i = U_I \frac{L}{\delta_I} = [U_I\beta]^{1/2} \cdot L \tag{2.12.2}$$

where $L$ is a characteristic length of the basin and hence of the extent along the boundary of the boundary layer. From the dissipation integral (2.10.2), the balance requires that, in order of magnitude:

$$\frac{f_0}{D} w_E = \frac{rU_I}{\delta_I} = r[U_I\beta]^{1/2} \tag{2.12.3}$$

or that:

$$U_I = U_S\left(\frac{\delta_I}{\delta_S}\right) \tag{2.12.4}$$

where $U_S$ is the characteristic Sverdrup velocity, $f_0 w_E/D\beta$, and $\delta_I$ is the inertial boundary layer thickness based on that velocity (which is the a priori scaling that we would use to estimate the inertial layer thickness). When the inertial layer thickness, based on the Sverdrup estimate, exceeds the Stommel bottom friction thickness, the interior velocity must rise to a level which *exceeds* the Sverdrup value by the ratio of the inertial to bottom friction thicknesses. This is an example of how the Sverdrup solution can be vitiated even though the wind stress remains small enough that the Sverdrup theory appears to be self-consistent in terms of its predictions of the size of the interior velocity and the smallness, on that basis, of the nonlinear and frictional effects in the interior. The point is that the Sverdrup solution is consistent in every way except in its inability, in this limit, to close the circulation in a western boundary current with enough dissipation to satisfy the global constraints of vorticity balance. That inconsistency is then fatal to the Sverdrup solution's validity.

As $r$ is decreased even further, the solution loses its boundary layer character completely. In this limit the interior velocity is so large that the $\beta$ term in the vorticity equation becomes unimportant, i.e., $U_I \gg \beta L^2$. If the $\beta$ term is ignored:

$$\psi = \left(\frac{1}{2r\pi^2}\right) W_0 \frac{f_0}{D} \sin\frac{\pi x}{L} \sin\frac{\pi y}{L} \tag{2.12.5}$$

becomes an exact solution since the nonlinear terms cancel exactly and the balance of the bottom friction term and the forcing leads directly to (2.12.5). A more complete analysis has been given by Barcilon (unpubl, see also Zimmerman 1993).

Note that in this limit $dQ/d\psi = \nabla^2\psi/\psi < 0$, in distinction to the Fofonoff mode.

On the other hand, for the no-slip boundary condition we can examine the overall vorticity balance with quite different results. Consider the vorticity balance (2.10.2) applied to a streamline which is coincident with the basin boundary.

The flux of vorticity through the side wall in the western boundary layer is proportional to the eastward derivative of the vorticity. The flux occurs through the sublayer with a thickness given by (2.8.9) while the tangential velocity in the sublayer is the same order as that in the inertial layer. Thus in the sublayer our estimate for the vorticity flux over a meridional length $L$ is,

$$\begin{aligned} A_H \frac{\partial \zeta}{\partial x} L &= O\left[\frac{A_H L v_i}{\delta_*^2}\right] = O\left[\frac{A_H u_I L^2}{\delta_*^2 \delta_I}\right] \\ &= \left[\beta L^2 U_I\right] = O\left(\frac{f_0}{D} w_E L^2\right) \end{aligned} \tag{2.12.6}$$

where use has again been made of mass conservation to relate the boundary layer velocity, $v_i$, to the interior velocity and of the definition of the inertial layer thickness. The last step in the estimate chain of (2.12.6) uses the Sverdrup relation. We see from (2.12.6) that *for the basin as a whole* there appears always to be a possible balance between the vorticity input by the wind and the frictional flux through the western boundary sublayer. Recall that in the no-slip case the dissipation balance takes place latitude by latitude, and that there can be no advective export of vorticity anomaly which would produce a large recirculation as in the slip case. This makes at least plausible the relatively small region of flow in Fig. 2.12.3c that departs from the pattern of a western boundary layer/Sverdrup interior picture of the total solution. The perceptive reader will have noted, however, that this still leaves a problem for those streamlines that do not pass through the sublayer. In the limit $\delta_I/\delta_M \gg 1$ this involves most of the Sverdrup streamlines. On these streamlines the vorticity gradient is much smaller than in the sublayer because the length scale is now $\delta_I \gg \delta_*$, and the vorticity flux is therefore much less. How the vorticity balance is achieved for such streamlines is left for Sections 2.13 and 2.14.

For the case of slip boundary conditions the sublayer has the same length scale as in the no-slip case, but the vorticity gradient is weaker in the sublayer by a factor of $\delta_*/\delta_I$ than in the no-slip case because now the sublayer acts only to bring the velocity *gradient* to zero at the wall rather than the velocity. This reduces the flux of vorticity through the side wall by the same factor, and the local balance (2.12.6) is unattainable. Instead the balance is locally met in the western boundary layer by the northward export of the anomalous vorticity where it produces a large region of recirculation. The flow which rejoins the interior recirculates around this small-scale gyre near the northern boundary. This augments the dissipation of vorticity along the path of the streamline until the balance is achieved.

A very interesting series of experiments with slip boundary conditions were carried out by Böning (1986). Both lateral mixing and bottom friction were used. In these experiments both the ratio $\delta_I/\delta_M$ and the ratio $\delta_S/\delta_M$ were varied.

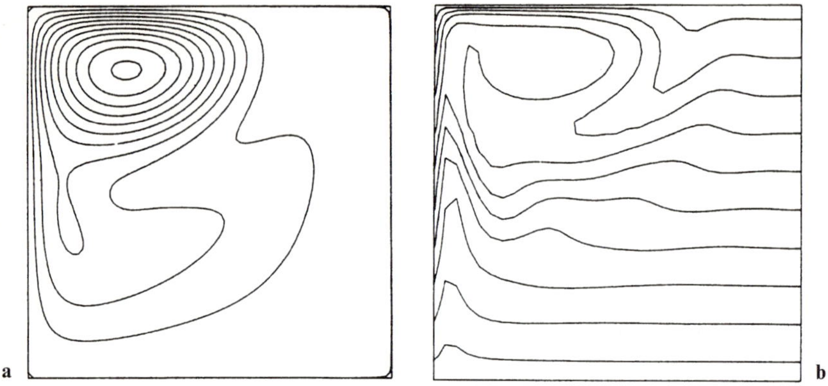

**Fig. 2.12.4a,b.** Calculation of the circulation with slip boundary conditions and no bottom friction for. $\delta_I/\delta_M = 1.0$. **a** the streamline pattern. **b** the isolines of total vorticity $\zeta + \beta y$. Note the relatively uniform value of total vorticity in the northwest zone of recirculation. (From Böning 1986)

In the absence of bottom friction, $\delta_S/\delta_M = 0$, Böning showed the familiar development of a strong recirculation in the northwest corner of the basin. Figure 2.12.4 shows the circulation driven by an Ekman pumping of the form $w_E = -W_0 \sin \pi y/L$ for the case where $\delta_I/\delta_M = 1.0$ Thus the boundary layer dynamics are of mixed dissipative/inertial character. Panel a shows the stream field and panel b the field of total vorticity, $\zeta + \beta y$. Note that in b we see the isolines of total vorticity anchored on the western boundary at the same latitude as on the eastern boundary. The isolines are dragged northward in the boundary layer and are greatly contorted by the recirculation although they can always be traced back to their point of origin on the western boundary. A tongue of high total vorticity is seen being drawn around the recirculation gyre in the northwest. The recirculating gyre itself, however, is a zone of relatively uniform total vorticity.

This can be rationalized by reference to the dissipation integral for the vorticity. The speed in the recirculation region is so much larger than the Sverdrup interior that, excluding those streamlines that lie next to the boundary and rejoin the interior, i.e., concentrating on the recirculating gyre itself, the nonlinearity tends to dominate, and to a first approximation we could write:

$$\zeta + \beta y = Q(\psi) \tag{2.12.7}$$

in the recirculation. In the absence of bottom friction (2.11.11) would apply without the last term. In this parameter range the recirculation has been produced by the vorticity anomaly exported from the western boundary layer and gathered in the northwest corner of the basin. In this region, however, the Ekman pumping is very small since it vanishes exactly on the northern boundary of the gyre. Thus in the absence of bottom friction, and with weak forcing over the area of the recirculation gyre to which (2.11.11) is applied, we must have $dQ/d\psi \approx 0$ to achieve a balance in (2.11.11). That is, the total vorticity tends to become uniform in this limit. At this level of moderate nonlinearity the maximum transport in the basin, i.e., the maximum value of the streamfunction exceeds the maximum of linear theory by a factor of 2.33.

As the ratio $\delta_I/\delta_M$ is increased, the zone of recirculation extends eastward until it touches the eastern boundary and then extends southward for further increases in $\delta_I/\delta_M$, filling an ever larger part of the basin.

A very interesting heuristic model of the recirculation in this parameter regime has been described by Cessi et al. (1987). They suggest that the recirculation is driven by the *anomaly* of total vorticity produced at the northern boundary of the gyre by the nonlinear advection of vorticity in the western boundary current. If this anomaly is *specified,* the recirculation can be calculated and is driven directly by the anomaly. The wind forcing enters only indirectly into the problem to the extent that it produced the inertial boundary current which has transported the anomalous vorticity to the northern boundary. In the case where the recirculation is limited in the north-south direction to less than the full extent of the basin, Cessi et al. provide an estimate of the southward extent of the recirculation. If $Q_n$ is the total vorticity of the field on the northern boundary, and $f_n$ is the planetary vorticity there, the southward extent, $\Lambda$, of the recirculation gyre from the northern boundary is given by:

$$\Lambda = 3(f_n - Q_n)/2\beta. \tag{2.12.8}$$

Note that the total vorticity on the northern boundary, $Q_n$, must be *less* than the planetary value $f_n$. This is expected since the inertial boundary layer carries fluid which has entered the boundary layer with total vorticity equal to $f_s$, which is the planetary vorticity that it has when it leaves the Sverdrup interior at the southern latitude where $f = f_s < f_n$. As the circulation intensifies, and the anomaly increases in strength, the north-south scale increases. (see also Ierley and Young 1988). Unfortunately, there is no simple deductive way to link the anomaly which drives the recirculation to the external forcing and external parameters, but the discussion of Cessi et al. is illuminating because it describes a clear physical mechanism for the production of the recirculation related to the nonlinear production of anomalous vorticity near the northern boundary.

Figure 2.12.5 shows another calculation in the sequence by Böning, this time for $\delta_I/\delta_M = 1.5$. The recirculation zone now fills most of the basin. The Sverdrup interior and the western intensification have both disappeared. The transport is an order of magnitude greater than would be found in linear theory. This is quantitatively similar to the Veronis model with only bottom friction when a resonance with the Fofonoff mode occurs. In the present case, in the absence of bottom friction, the same use of (2.11.11) for a gyre that fills the basin and is therefore in balance with an $O(1)$ forcing, tells us that contrary to the Fofonoff mode, $dQ/d\psi < 0$. We can in fact see from Fig. 2.12.5b the coincidence of streamlines and isolines of total vorticity in the gyre.

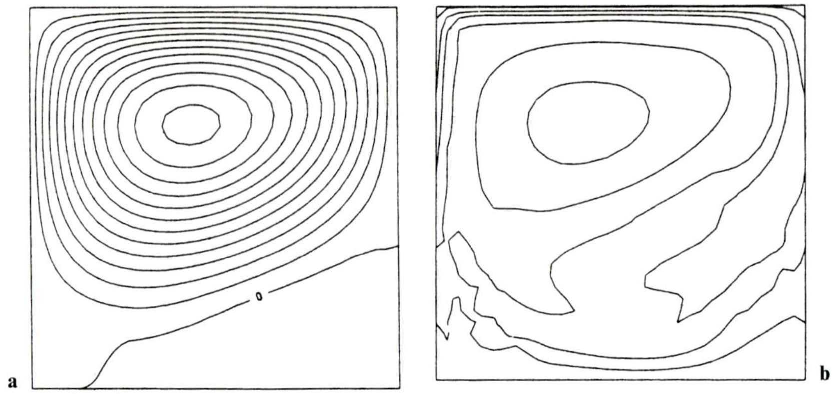


**Fig. 2.12.5a,b.** As in Fig. 2.12.4 but with $\delta_I/\delta_M = 1.5$. (From Böning 1986)

If we fix the ratio $\delta_I/\delta_M = 1$ and increase the bottom friction we find (Böning 1986) that the recirculation shrinks in both size and magnitude. The bottom friction is very efficient in dissipating the vorticity put in by the wind, reducing the need for the strong recirculating gyre. Figure 2.12.6 shows the circulation for $\delta_I/\delta_M = 1.0$ and $\delta_S/\delta_M = 0.25$ (panels a and b) and $\delta_S/\delta_M = 0.5$ (panels c and d). In the first case Böning estimates that roughly 50% of the energy put in by the wind is dissipated by bottom friction even when $\delta_S/\delta_M$ is only 0.25. When $\delta_S/\delta_M$ is 0.5, the percentage of the energy dissipated by the bottom friction is 65%.

A very illuminating series of calculations have been carried out by Ierley and Young (pers. comm., unpubl manuscript) in which the western boundary layer is quite nonlinear, $\delta_I/\delta_M = 1.08$ and $\delta_S/\delta_M = 0.17$. i.e., similar to the parameter setting of Böning's calculation in Fig. 2.12.6. Their results are shown in Fig. 2.12.7.

Ierley and Young compared the solutions for all four boundary conditions. Panel a is shows the solution for so-called *hyperslip* conditions in which the normal gradient of the *total* vorticity vanishes on each boundary, panel b shows the superslip condition solution in which the normal gradient of the relative vorticity vanishes on each boundary, while panel c shows the slip solution and panel d the no-slip solution. Figure 2.12.8 shows the total vorticity fields for each solution. Note that the conditions of hyperslip and superslip are identical on the western boundary and differ only on the northern boundary. Indeed, the two solutions are very similar. Bottom friction is largely responsible for the equilibration of both of the super- and hyperslip conditions. The recirculation, even at this moderate nonlinearity stretches across the entire basin, and the potential vorticity isolines are wrapped around with the streamlines to form a Fofonoff gyre in each case. The slip and no slip cases yield circulations that differ strikingly from each other as well as from the super-/hyperslip cases. The recirculation is more limited in the slip case than in

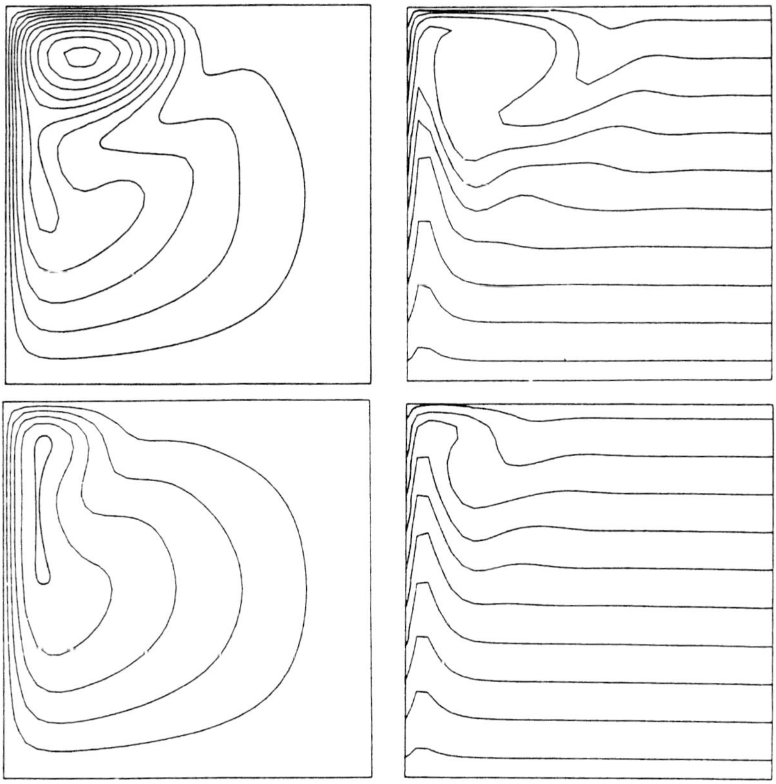

**Fig. 2.12.6a-d.** Pair of calculations with $\delta_I/\delta_M = 1$ but with $\delta_S/\delta_M = 0.25$ **a**, the streamfield and **b** the total vorticity and **(c,d)** for $\delta_S/\delta_M = 0.5$. Note the reduction in the recirculation region as the bottom friction increases. Also compare with Fig. 2.12.4 where $\delta_S/\delta_M = 0$. (From Böning 1986)

the previous two examples, as lateral friction is largely responsible for the equilibration, and a large region of Sverdrup flow south of the recirculation exists. In the recirculation region, as Fig. 2.12.8c clearly shows, the total vorticity is completely homogenized. The no-slip calculation in panel d of both figures shows an even smaller region of recirculation and the development of a damped Rossby wave train in the northwest portion of the basin. The solutions are steady, in large part due to the stabilizing effect of the bottom friction.

## 2.13 Moore's Hypothesis

The stationary, damped Rossby waves in the no-slip case seen in the numerical solutions of the previous section have been suggested by Moore (1963) as an approximate solution of the western boundary-layer equation in the region

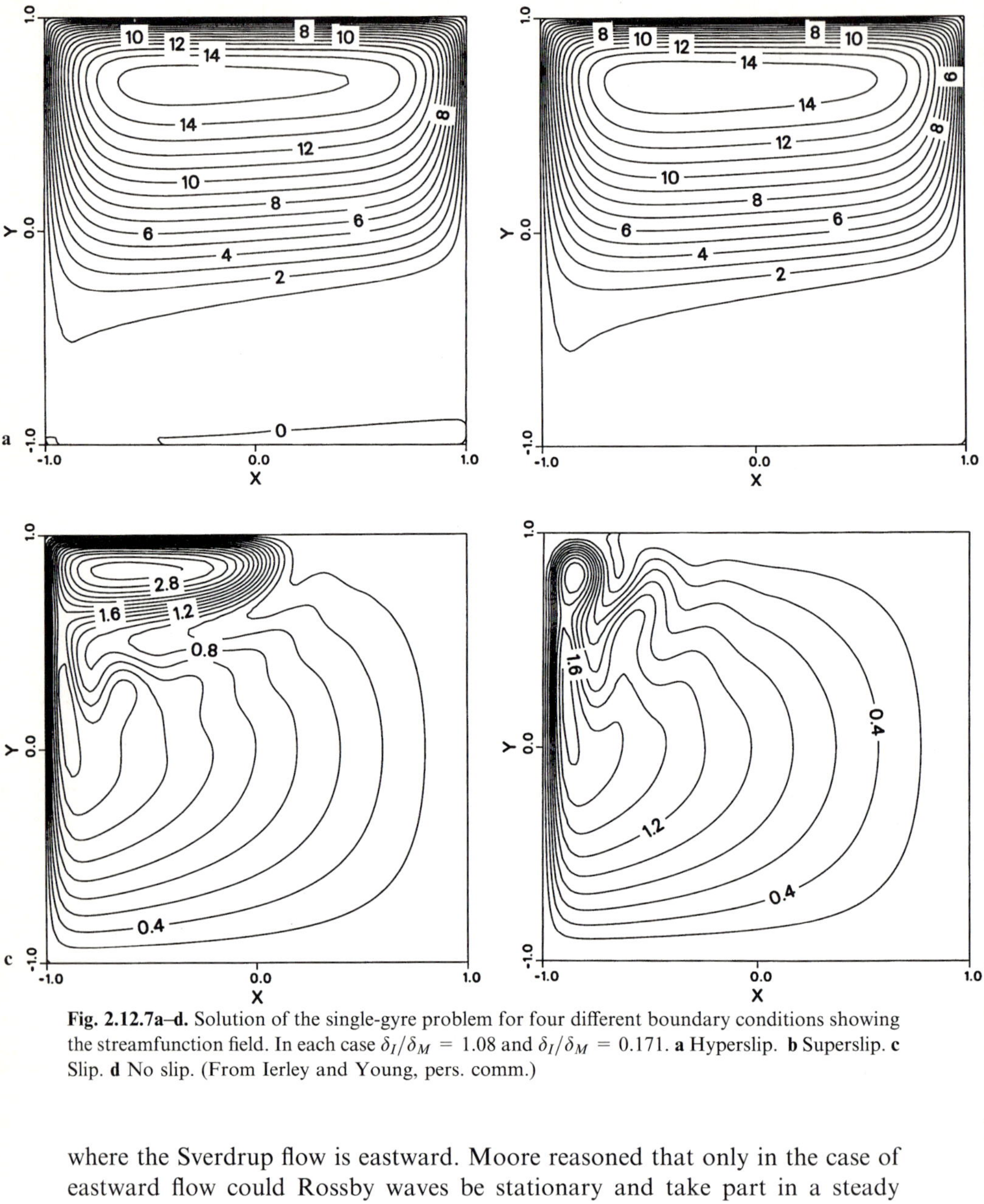


**Fig. 2.12.7a–d.** Solution of the single-gyre problem for four different boundary conditions showing the streamfunction field. In each case $\delta_I/\delta_M = 1.08$ and $\delta_I/\delta_M = 0.171$. **a** Hyperslip. **b** Superslip. **c** Slip. **d** No slip. (From Ierley and Young, pers. comm.)

where the Sverdrup flow is eastward. Moore reasoned that only in the case of eastward flow could Rossby waves be stationary and take part in a steady solution of the circulation problem, and he suggested the damped Rossby wave as a model for the boundary-layer structure in the northern part of the gyre. Moore did not examine the full boundary-layer problem but looked only at the asymptotic region where the boundary layer is supposed to merge smoothly to the interior. For simplicity he considered the case with no bottom friction.

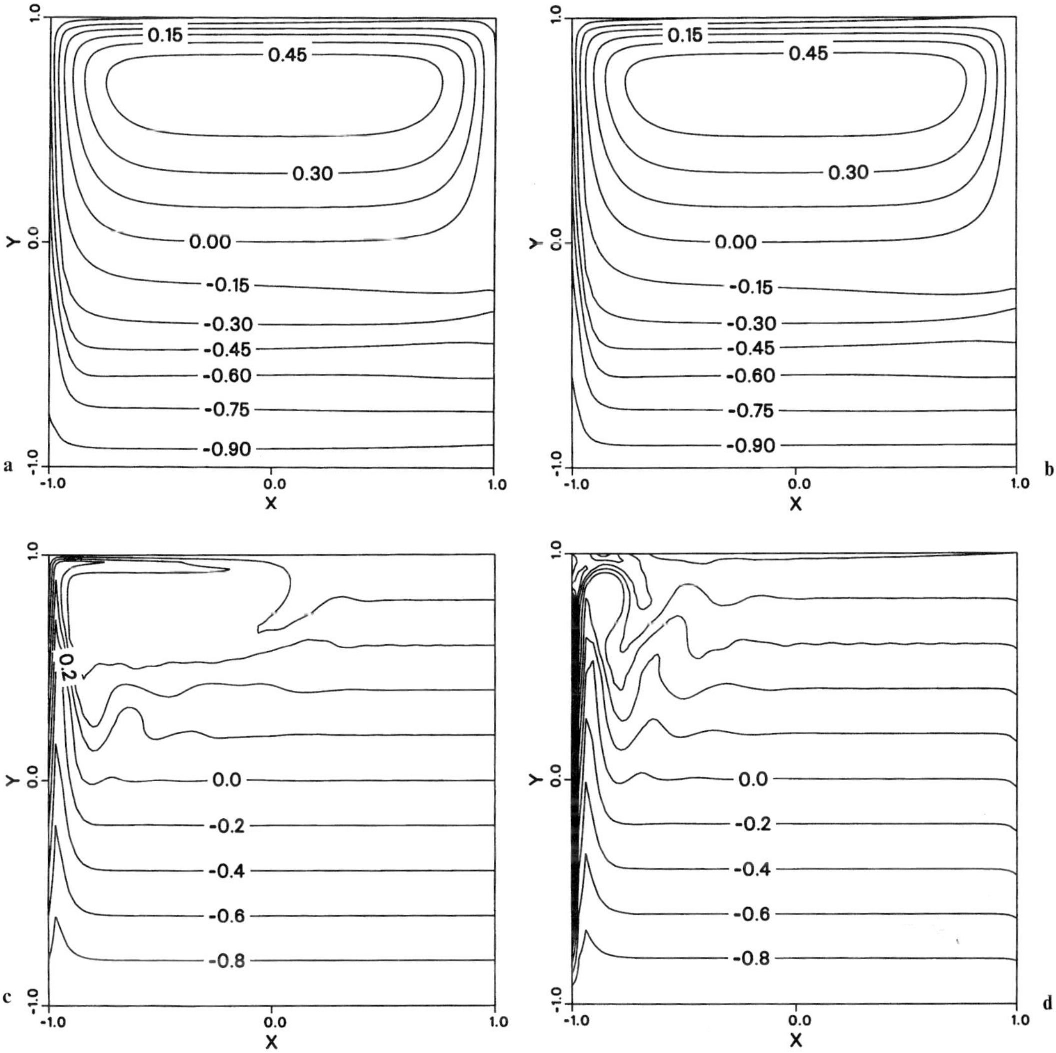


**Fig. 2.12.8a–d.** As in Fig. 2.12.7, but the total vorticity field is shown instead of the streamfunction

In such a transition region the boundary-layer flow, Moore reasoned, could be thought of as a small perturbation on the Sverdrup flow, $\psi_I$, Thus, if we write the full boundary-layer solution as the sum:

$$\psi = \psi_I(x, y) + \phi(x, y) \tag{2.13.1}$$

where:

$$\phi \ll \psi_I$$

in the transition region between the boundary layer and the interior the linearized boundary-layer equation becomes:

$$U_I \frac{\partial^3 \phi}{\partial x^3} + \beta \frac{\partial \phi}{\partial x} = A_H \frac{\partial^4 \phi}{\partial x^4} \tag{2.13.2}$$

where use has been made of the inequality $\partial\phi/\partial x \gg \partial\phi/\partial y$ to simplify the equation. $U_I$ is the Sverdrup zonal flow. In the limit of interest, $\delta_I > \delta_M$, approximate wavelike solutions of (2.13.2) can be found in the form:

$$\phi = \Phi(y)\cos(kx)\exp(-x/\ell)$$

where the wavelength, $\lambda$, is given by the standard formula (Pedlosky 1987) for the stationary Rossby wave:

$$\lambda = \frac{2\pi}{k} = 2\pi\left(\frac{U_I}{\beta}\right)^{1/2} \tag{2.13.3}$$

and the frictional decay scale is given by:

$$\ell = \frac{U_I^2}{\beta A_H} = O\left(\delta_I \frac{\delta_I^3}{\delta_M^3}\right) \gg \delta_I. \tag{2.13.4}$$

Moore and later Pedlosky (1987) suggested this as an approximation to the western boundary layer structure in the region of eastward Sverdrup flow, $U_I > 0$. However, there are severe deficiencies with this picture as a description of the western boundary layer. Most clearly, the Moore solution is only a partial approximation to the full problem. Even were it valid, it would apply only in the transition zone for the boundary layer as it merges into the interior. Nothing in the Moore solution recognizes either the dynamics deep within the layer or the boundary condition that must be satisfied at the western boundary. Why should such a wavelike solution be manifest in the experiments only in the case of no-slip conditions if the Moore argument has general validity as a picture of the boundary layer? More devastating are the results of Ierley and Ruehr (described in Section 2.9) that showed that no boundary layer can exist in an outflow region of a flow this nonlinear. No boundary-layer structure matching the interior with a damped wavelike structure (or any other) was found to be possible. Thus, the Moore picture should probably be regarded less as a model of the boundary layer in the northern part of the gyre and more in the nature of a description of a dynamical regime of its own, existing in the transition zone between the small recirculation gyre in the northwest corner of the flow (which does not satisfy the fundamental boundary-layer approximation, $(\partial\phi/\partial x \gg \partial\phi/\partial y)$ and the region of the Sverdrup interior. We can speculate that it shows up only in the no-slip calculation because only in this case is the recirculation limited enough, for these parameter values, to allow a large enough region of Sverdrup interior zonal flow to support the stationary Rossby wave. The strong recirculations evident in the three other examples of Fig. 2.12.7 leave little room for a Rossby wave field in the northern part of the gyre.

## 2.14 Inertial Runaway

The original hypothesis in our approach to constructing the dynamic picture of the ocean circulation started with the notion of a Sverdrup interior. The incomplete character of the Sverdrup theory means that its validity depends on the existence of a western boundary current to accept the Sverdrup transport in the south and return it in the north after having made sufficient alterations of the total vorticity of the fluid in its travel through the boundary layer so it can smoothly rejoin the interior. Furthermore, our interest is in a western boundary layer, whose existence, basic structure, and ability to perform the tasks just described are not sensitive to the strength or type of the explicit dissipative mixing or the boundary condition applied at the western boundary. These hopes, really, are based on an understanding that neither the parameterization of the mixing or the applied boundary condition are known or deducible from first principles. We would prefer that the global structure of the ocean circulation did not depend on our arbitrary representations of small-scale mixing.

If dissipation of the large-scale motion *is* important, we might hope, as has often been expressed, that with sufficiently powerful computers and sufficiently fine spatial and temporal resolution the naturally occurring eddy field due to instabilities of the large-scale flow would provide the necessary brake on the circulation and allow us, as the resolution is continuously improved, to simultaneously reduce the level of explicit dissipation, $A_H$, say, until the calculated solutions become independent of the explicit mixing parameter. The eddies would take over the role of the explicit dissipation and produce a deductive relation between the large-scale motion and its own dissipation. Finally, in such a case our circulation would saturate at a level and structure independent of $A_H$ in a realistic circulation structure independent of arbitrary parameterizations.

There are several reasons why, after some thought, we should anticipate difficulty with such an asymptotic scenario. First of all, we have seen that the boundary conditions associated with the explicit dissipation, no matter how small, are vital in determining the ability of the flow to transfer vorticity into or out of the basin. Regardless of the advection of vorticity by the eddies within the basin, the flux of vorticity through either the side wall or the bottom depends on the level of the explicit dissipation. As we have seen, whether this flux occurs through the side walls or the bottom depends on the boundary condition at the side wall.

The motion of the fluid, even on the eddy scale of oceanic turbulence, is fundamentally two-dimensional. Oceanic eddies that we might hope to resolve have horizontal scales of tens to hundreds of kilometers and vertical scales of only 1–2 km. A fundamental attribute of such two-dimensional turbulence is the transfer of energy from small *to* large scales (Batchelor 1969). The effect of the eddies in two-dimensional motion is to energize larger scales rather than draining away the energy of the large-scale motion. This would leave the explicit parameterization of small-scale dissipation, acting directly on the large-scale, as the only way to balance the large-scale energy and vorticity input by the wind. It would be futile in this case to expect the resolved eddy field to inhibit the amplitude of the large-scale motion.

In the presence of the $\beta$ effect Rhines (1975) showed that there is a scale of the order of $\sqrt{U/\beta}$ *above which* the eddy energy is unable to cascade. However, the value of $U$ is determined by the motion field itself. If the circulation is limited to the order of the Sverdrup velocity, we know the above parameter would be of the order of the inertial boundary-layer width, which is a modest 100 km. On the other hand, if the explicit dissipation is unable to put a brake on the large-scale field, $U$ might be much larger, and the Rhines arrest scale would be larger as well. Hence the barrier to the cascade to larger scales is part of the answer to the problem and cannot be deduced a priori.

We have already seen in the calculations of Böning (1986) the tendency for the circulation, in the case with slip boundary conditions, to become much stronger than the Sverdrup theory would predict as the boundary-layer Reynolds number becomes large. This is not entirely surprising either, given our global estimate in Section 2.12, of the inadequate flux of vorticity in the sublayer in the slip case when $\delta_I/\delta_M \gg 1$. We discuss the no-slip case below; however, the numerical demands placed on calculations with the no-slip condition (because of the rapid variation of the velocity in the sublayer) make it difficult to examine the limit $\delta_I/\delta_M \gg 1$ for the no-slip condition.

### *Slip Conditions*

The most extensive and careful calculations of the steady, slip condition problem have been carried out recently by Ierley and Sheremet (1995). Calculations by Böning previously showed that for large $\delta_I/\delta_M$ allowing time dependence nevertheless naturally yields a steady solution for reasons that we elaborate below. Hence, indeed especially, for large $\delta_I/\delta_M$ steady solutions are the relevant solutions to the full problem.

Figure 2.14.1 shows a schematic diagram of the solution curve of the steady solutions obtained numerically by Ierley and Sheremet. On the abscissa is plotted the boundary-layer Reynolds number $R = (\delta_I/\delta_M)^3$. On the ordinate is the maximum value of the streamfunction. For small values of $R$ solutions resembling the Munk solution are found. As $R$ increases, the circulation becomes more nonlinear, as we saw in Böning's solutions. The remarkable feature discovered by Ierley and Sheremet, and shown in the figure, is the existence in a range of $R$ of *multiple* solutions of the vorticity equation for the same forcing and dissipation. Furthermore, this multiplicity of solutions occurs in a very moderate range of $R$. For example, for $\delta_M/L = 0.04$, the multiplicity occurs in the range $R_H = 1.0377 \leq R \leq 1.3203 = R_L$. As might be expected, solutions in the middle branch are found to be unstable with respect to the

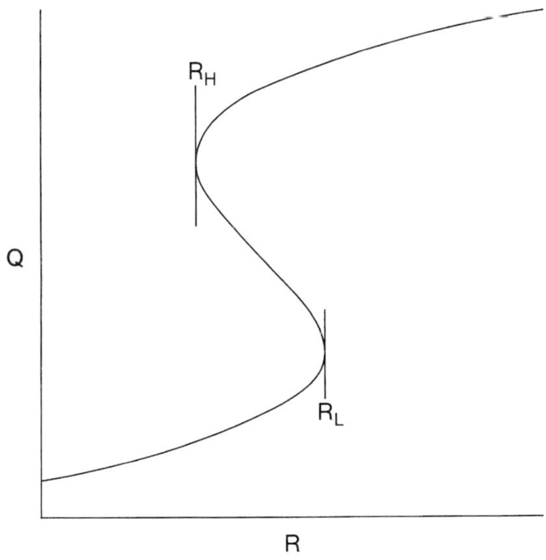


**Fig. 2.14.1.** Representation of the solution curve for slip boundary conditions. *Abscissa,* is the boundary-layer Reynolds number, R = $(\delta_I/\delta_M)^3$; *ordinate, Q* a measure of the maximum transport in the circulation, i.e., the maximum value of the streamfunction. (From Ierley and Sheremet 1994)

solutions on the high or low branch. Thus as $R$ slowly increases from zero on the lower branch, the solution jumps to the higher branch when $R = R_L$, *i.e.*, when $\delta_I/\delta_M$ only slightly exceeds unity (1.097 in the example described above). The upper branch solutions are stable and possess enormous circulation transports, with interior velocities of the order of 50 m/s! Ierley and Sheremet call this behavior "inertial runaway". For $R > R_L$ the upper branch solution is the *only* steady solution.

Note that at this characteristic velocity the Rhines arrest scale is of the order of 5000 km and therefore does not inhibit the transfer of energy to the large scales in any way. Indeed, Ierley and Sheremet are able to argue that the upper branch is an attractor for even the time-dependent solutions of the vorticity equation for $R > R_L$. The fundamental reason for this is the global stability of the steady motion in this limit. The solution to the lowest order for large $\delta_I/\delta_M$ is given in the form of (2.12.5), which is the lowest eigenmode of the equation:

$$\nabla^2\psi = -\lambda\psi \tag{2.14.1}$$

subject to conditions that the slip condition be satisfied at the boundary. This satisfies the condition that $J(\psi, \nabla^2\psi) = 0$, which is the limit of the vorticity equation for large $\delta_I/\delta_M$. The solution of this form thus satisfies the approximate vorticity equation in the very large $R$ limit, it satisfies the boundary condition, and it is the eigenfunction of (2.14.1) of largest scale that "fits" in the basin. For a two-dimensional flow to be unstable, energy must be passed to larger as well as smaller scales by the disturbance (see Pedlosky 1987). Since the motion described here has already locked onto the largest possible scale consistent with the dynamics and the boundary conditions, it is stable to other eigenmode perturbations that have smaller scales. The natural solutions in this limit are therefore steady. This is quite different than the no-slip case where the strong shears in the sublayer maintain a region of small spatial scale which is the source of continuous instability (Ierley and Young 1991).

Figure 2.14.2 shows the sequence of solutions for $\psi$ that evolve as $R$ is increased. In the range of $R$ of $O(1)$, the solution jumps to the upper branch, loses its east-west asymmetry, and produces a very rapid flow with transports of order 15 times the Sverdrup theory prediction.

Figure 2.14.3 shows the three possible solutions in the region of $R$ where the solution is not unique. The first panel is the lower branch solution. This solution is representative of the concentrated recirculation found by Böning and appears as the Munk solution evolves in structure along the lower solution branch with increasing nonlinearity. The third panel is the solution on the upper branch. In this solution $Q$ is, to a good approximation, a function of $\psi$ with $dQ/d\psi < 0$ consistent with (2.14.1). The unstable solution of the middle branch shown in the middle panel would not be realized in direct computation of the time-dependent problem.

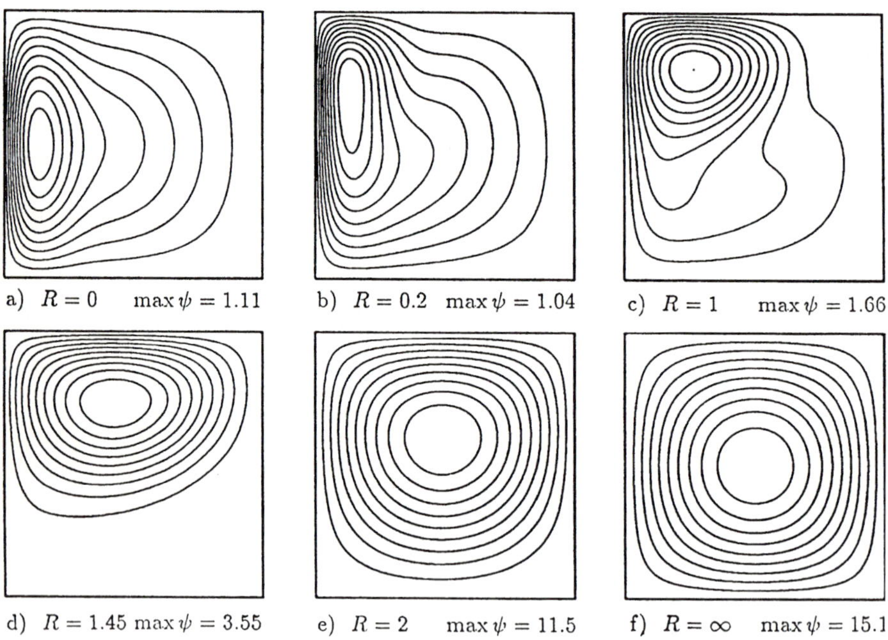


**Fig. 2.14.2a–f.** Streamfunction as calculated by Ireley and Sheremet (1994) showing the monotonic transition from the Munk solution to the upper branch, large amplitude flow. In these calculations $\delta_M/L = 0.06$. **a** $R = 0$ **b** $R = 0.02$ **c** $R = 1$ **d** $R = 1.45$ **e** $R = 2$ and **f** $R = \infty$

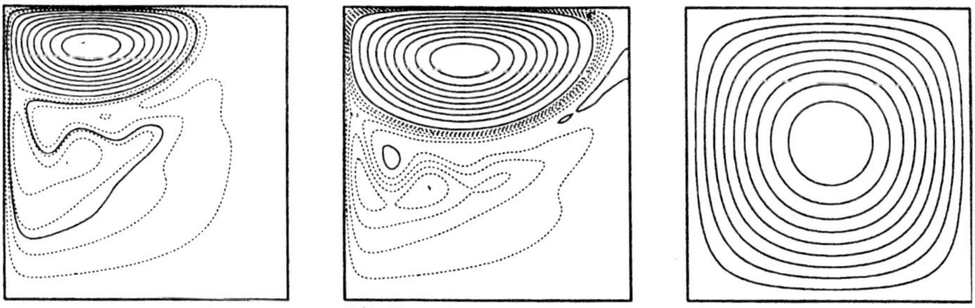


**Fig. 2.14.3a–c.** Solutions on the lower, middle, and upper branch of the solution curve in the region of multiple solutions for the case $R = 1$ and $\delta_M/L = 0.02$. (From Ierley and Sheremet 1994)

These are extraordinary results. At least for these boundary conditions the solution to the general circulation problem is neither unique nor necessarily in Sverdrup balance, even when it is unique, if $R$ exceeds $R_L$. The fact that these critical values of $R$ are order 1 and therefore in a range that cannot be dismissed as artificially high and unrealistic emphasizes the problematic character, from a theoretical point of view of simple Sverdrup theory. If we believe that Sverdrup theory is observationally plausible for the interior, the implication is (*if the present models and the slip condition are relevant*) that the small-scale mixing in the western boundary current is strong enough to keep the boundary-layer Reynolds number sufficiently low to prevent inertial runaway.

## *No-Slip Conditions*

The numerical task of finding solutions for large $\delta_I/\delta_M$ in the case of no-slip conditions is made particularly difficult by the existence of the strong viscous sublayer in which the tangential velocity is brought to rest. This is a source of time-dependent motions, and the solution for Reynolds numbers $\gg 1$ is typically time dependent. In the case of no-slip boundary conditions we saw above in (2.12.6) that an apparent vorticity flux balance for the basin *as a whole* can exist even for large boundary-layer Reynolds numbers. We might therefore imagine that at least in this case the Sverdrup theory for the interior remains valid as the value of the explicit dissipation parameter $A_H$ is continuously reduced.

However, this consideration is relevant only to the streamline which coincides with the basin boundary. In the limit $\delta_I/\delta_M \gg 1$; the boundary layer for the *steady* problem splits into an outer inertial layer and the viscous sublayer. Most of the streamlines flow in the outer layer and only a small fraction, $(\delta_I/\delta_M)^{-1}$, are in the sublayer. For the streamlines that traverse the western boundary-layer region in the outer layer there is an insufficiency crisis for the flux of vorticity across the boundary of the area encircled by the streamline. If the length scale used to calculate the flux is the inertial-scale, which would be appropriate for streamlines in the outer inertial layer rather than the viscous sublayer scale, a calculation similar to (2.12.6) yields a flux of vorticity which is too small by a factor $(\delta_M/\delta_I)^3$. This all assumes, of course, that the region of parameter space where the boundary layer splits its structure between an inertial outer layer and a viscous sublayer is actually realized in calculations which allow time dependence and eddies. Let us consider this possibility first, i.e., that the asymptotic separation of scales actually occurs.

Several suggestions have been made for the way in which a possible balance could occur on these streamlines for *steady* solutions in the nonlinear limit that would simultaneously retain the Sverdrup interior. Pedlosky (1987) argued that the damped Rossby waves found in the Moore solution act as a sort of baffle whose wiggly character increases the length of each streamline sufficiently that the viscous flux of vorticity across the streamlines outside the sublayer would balance the vorticity input by the wind and pass the vorticity flux to the sublayer at the boundary where it would be diffused out of the basin. Pedlosky presented a scaling argument attempting to demonstrate the plausibility of such a balance. Cessi et al. (1990) criticized that scenario and carried out careful and detailed numerical calculations with a regional model. Cessi et al. demonstrated that the Moore scenario, in which a Rossby wave field mediates the transition between an inertial western boundary layer and the interior, would occur for $\delta_I/\delta_M \gg 1$, only if the north-south dimension of the domain is allowed to increase in the same ratio. Under this frankly unrealistic condition the necessary dissipation occurs, not in the Rossby wave field but in a long viscous loop region pressed up against the boundary with a Munk layer thickness. Note that in the limit where $y$ derivatives tend to zero the nonlinearity vanishes, and the boundary-layer equation reduces to the Munk problem for arbitrarily large $\delta_I/\delta_M$. The dissipation in the Rossby wave field, in comparison to the loop current, is found to be negligible. The flaw in Pedlosky's argument lies in attributing to the current in the Rossby wave field a fluid velocity of the order of the western boundary layer velocity. He assumes the Rossby wave field is the meandering of a coherent *narrow* jet with an inertial layer width. Instead, if the wave field mediates the transition of the entire northern part of the gyre as in Moore's scenario, the outflow is broader and is of basin scale in width, and the velocities and vorticity flux are correspondingly smaller by a factor $\delta_I/L$. The flux then fails to balance the input of vorticity by that amount.

An alternative scenario was suggested by Cessi et al. (1990). They pointed out that the more realistic configuration of the flow for no-slip boundary conditions and finite and fixed north-south basin extent is the pattern shown in Fig. 2.12.7d. An intense, localized recirculation exists in the northwest corner of the basin. Streamlines which feed the interior Sverdrup flow navigate around the edge of the recirculation gyre, dissipating their anomalous vorticity before rejoining the interior. In the calculation shown in the figure $\delta_I/\delta_M$ is of order unity and a substantial amount of vorticity is actually dissipated in the western

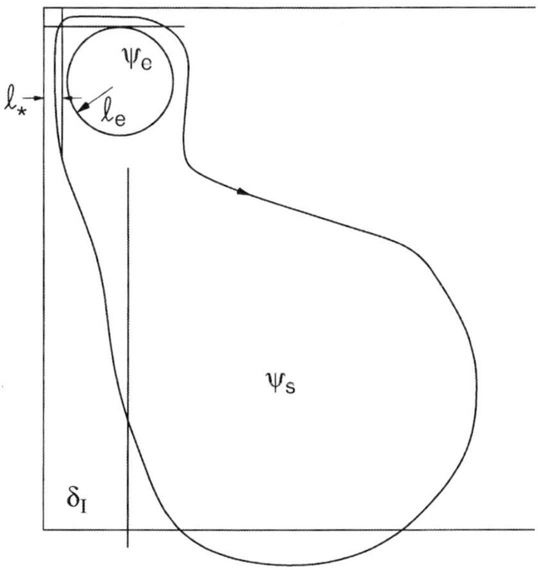


**Fig. 2.14.4.** Schematic presentation of the hypothesized circuit of a streamline in the limit $\delta_I/\delta_M \gg 1$. The Sverdrup flow enters the inertial boundary layer where little dissipation occurs. It flows around the rim of the recirculation gyre in that gyre's viscous sublayer dissipating vorticity and then rejoins the Sverdrup interior. The hypothesis fails in the limit $\delta_I/\delta_M \gg 1$

boundary current on all streamlines. The boundary current has yet to reach the asymptotic state for large $\delta_I/\delta_M$ where it would split into the outer inertial layer and inner viscous layer. Could the recirculation gyre satisfy the dissipation balance (2.10.2) and deliver enough mass flux to the interior to feed an interior flow in Sverdrup balance in that asymptotic limit? To accomplish this the Sverdrup flux would have to circulate around the perimeter of the recirculation eddy or gyre, and in order to dissipate enough vorticity this flow would have to occur in the viscous sublayer of the recirculation gyre. A schematic rendering of this hypothesized flow is shown in Fig. 2.14.4.

Suppose the strength of the circulation in the recirculation eddy is $\psi_e$ and its characteristic horizontal scale $\ell_e$. For simplicity we assume the eddy is nearly circular, but it is straightforward to consider elliptic gyres of any eccentricity without changing the basic result of the following calculation. Thus the characteristic tangential velocity of the gyre is $v_e = \psi_e/\ell_e$.To estimate the scale of the recirculation eddy's sublayer we balance the advection of vorticity (or momentum) with dissipation. If $\ell_*$ is the sublayer scale, the diffusion of momentum in the sublayer, is:

$$A_H \nabla^2 v_e = O\left(A_H \frac{\psi_e/\ell_e}{\ell_*^2}\right). \tag{2.14.2}$$

The advection of momentum can be estimated as:

$$v_e \frac{\partial v_e}{\partial y} = O\left(\frac{\psi_e^2}{\ell_e^2}\right) \tag{2.14.3}$$

and equating them yields:

$$\ell_* = \left(\frac{A_H}{\psi_e}\right)^{1/2} \ell_e. \tag{2.14.4}$$

The vorticity input over the area embraced by the streamline in the interior (which we assume is of the *order* of the basin area, $L^2$) is of the order:

$$\frac{f_0}{D} w_E L^2 = \beta \psi_s L \tag{2.14.5}$$

where the last equality follows from Sverdrup theory. The flux of vorticity across the streamline threading around the recirculation eddy over the distance of order $\ell_e$ can be estimated as, using (2.14.4):

$$A_H \frac{\partial \zeta}{\partial x} \ell_e = O\left(A_H \frac{\psi_e/\ell_e}{\ell_*^2} \ell_e\right) = \frac{\psi_e^2}{\ell_e^2}. \tag{2.14.6}$$

The order of the transport through the sublayer of the recirculation gyre is the fraction $\ell_*/\ell_e$ of the recirculation gyre's transport $\psi_e$. If this transport is equated to the Sverdrup transport:

$$\frac{\ell_*}{\ell_e} \psi_e = \psi_s \tag{2.14.7}$$

and if the dissipation of vorticity on the rim of the sublayer is set equal to the input of vorticity within the streamline area, i.e., if (2.14.6) is equated to (2.14.5), we obtain for our estimate of $\ell_e$:

$$\ell_e = \left(\frac{\delta_I}{\delta_M}\right)^3 L \tag{2.14.8}$$

where:

$$\delta_I = \sqrt{\frac{U_I}{\beta}} = \sqrt{\frac{\psi_s/L}{\beta}}. \tag{2.14.9}$$

Thus the attempt to maintain a Sverdrup interior by imposing on the recirculation gyre the task of dissipating the vorticity input and delivering, around its rim, the flow that rejoins the interior Sverdrup flow forces a recirculation gyre that is larger than the basin dimension! This contradiction indicates that it is unlikely even with no-slip conditions, that inertial runaway can be prevented in the limit of large $\delta_I/\delta_M$, at least for steady solutions. Thus, for all the boundary conditions that we have so far considered the limit of large boundary-layer Reynolds number reacts back on the interior, destroying the interior Sverdrup balance.

A remaining possibility is that the time-dependent eddies generated in the no-slip calculations formulated as an initial value problem and marched forward in time can provide the missing vorticity flux in the manner indicated in (2.10.6). This would be equivalent, conceptually, to having an internal governor on the magnitude of the boundary-layer Reynolds number. If the steady-solution Reynolds number is too large, the eddies which naturally develop would produce a resolved turbulent flux (equivalent to a larger viscous flux) and allow the inertial flow to deliver a sufficient flux of vorticity across mean streamlines to the sublayer on the whole western boundary. The sublayer could then flux that vorticity across the boundary in the manner that we have noted above and allow a balance to be struck for each time-mean streamline. The evidence for this conceptually attractive scenario is still problematic. Very few numerical, full-basin calculations come close to the asymptotic limit of large $\delta_I/\delta_M$ in the no-slip case.

A few years ago Ierley, Young, and this author initiated a series of high-resolution, low-dissipation computations of two and five-layer geostrophic models which are dynamically similar, insofar as the recirculation is concerned, to the single-layer model discussed in this chapter. The goal was to establish a baseline circulation whose naturally resolved eddy field would saturate and render the solutions independent of further changes in the explicit dissipation $A_H$. We were dismayed to find that every reduction in $A_H$ led to further gross changes in the circulation pattern, and that convergence or saturation as a function of Reynolds number was never achieved. Figure 2.14.5 shows two images of the calculated circulation. Panel a shows the time-averaged upper-layer circulation of the two-layer model, which has a boundary-layer Reynolds number equal to 10, corresponding to a value of $\delta_I/\delta_M = 2.2$. The calculation employs no-slip boundary conditions on the western and eastern boundaries and slip conditions on the northern and southern boundaries. In this case we see a large recirculation reaching nearly three-quarters of the way across the basin. Further reduction in $A_H$, leading to a value of $\delta_I/\delta_M = 4.7$, yields the circulation shown in Fig. 2.14.5b. The recirculation reaches right across the basin and the interior is increasingly disturbed by the recirculation. This is the circulation with the largest $\delta_I/\delta_M$ that we were able to achieve numerically. However, the development sequence is similar to that which we saw above for the slip case in which the recirculation first extends across the northern boundary until the eastern boundary is struck and then extends southward, increasingly filling the basin. That last development has so far not been observed in no-slip calculations but may plausibly be expected.

If so, this would imply that for every boundary condition and for every form of dissipation the very existence of the Sverdrup interior depends vitally on the existence of a sufficiently large *explicit* frictional dissipation in the western boundary current. If the explicit dissipation, representing very small-scale turbulent motions is not sufficient, the solution achieves inertial runaway and circulates ever faster, reaching speeds far in excess of those observed.

### *Internal Compensation in a Two-Gyre Model*

There is one rather singular case in which the inertial runaway can be avoided. This occurs when no net vorticity is put into the basin. This can be achieved in

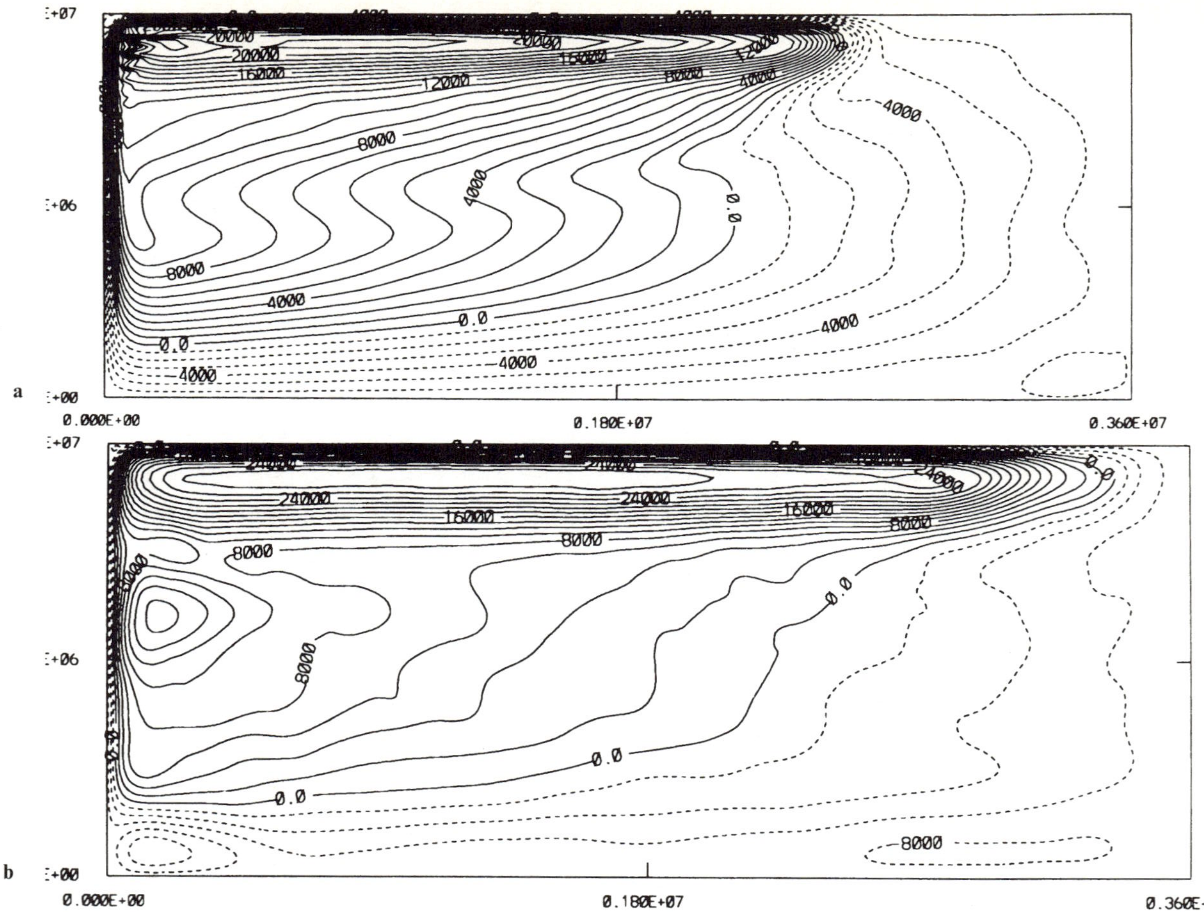


**Fig 2.14.5a,b.** Calculation of a two-layer model with no-slip conditions on the eastern and western boundaries and slip conditions on the northern and southern boundaries. The circulation pattern of the upper layer is shown. In the region of the recirculation the pattern in the lower layer is similar, i.e. the flow in this region is essentially barotropic. **a** a calculation with $\delta_I/\delta_M = 2.2$. **b** as in **a** but $\delta_I/\delta_M = 4.7$. Note the growth of the recirculation gyre

several ways, but the one of greatest oceanographic relevance is the case in which a southern region of downward Ekman pumping is balanced by a northern region of the same size in which Ekman suction is equal and opposite to that of the southern region. This balances the input of anticyclonic vorticity in the southern (subtropical) gyre with an input of cyclonic vorticity in the northern (subpolar) gyre. Thus, no net flux of vorticity is required to leave the basin through side walls or bottom. This completely eliminates the need for the boundary layers to do more than close the mass flux. The vorticity balance can be achieved internally. Marshall (1984) carried out a very interesting numerical study in which the Ekman pumping had the form:

$$w_E = -W \sin\left(\frac{2\pi y}{L}\right), \quad 0 \le y \le L \tag{2.14.10}$$

which naturally produces two gyres with an intergyre boundary, according to Sverdrup theory, at $y = L/2$.

The dissipation in Marshall's model is a combination of bottom friction and lateral friction. The latter is of an artificial form designed to be inconsequential on the scale of the circulation and introduced only to damp out very small-scale fluctuations. The bottom friction was chosen small enough so that the ratio $\delta_I/\delta_S$ is roughly 4.0, i.e., highly inertial. In the single-gyre model such a parameter setting would lead, as we have seen, to the development of a steady Fofonoff mode. In the two-gyre calculations reported by Marshall the circulation rapidly becomes unstable. A jet, which is identified with the eastwardly flowing Gulf Stream after separation from the eastern boundary near $y = L/2$, becomes an unstable, meandering current spawning a vigorous eddy field. The eddies flux anticyclonic vorticity in an amount equal to that put in by the wind, from the southern gyre to the northern gyre, where the wind-forced Ekman pumping of the opposite sign removes it from the basin, establishing an equilibrium. The time-averaged circulation is shown in Fig. 2.14.6. Panel a shows the stream function for the time-averaged flow. Panel b shows the isolines of total vorticity, and panel c shows the superposition of the isolines of $\psi$ and $\zeta + \beta y$. We see that in the time mean a small recirculation does exist, forced by the advective transport of vorticity anomaly to the intergyre boundaries. However, the region of recirculation is quite limited. The isolines of total vorticity are distorted primarily in the region of the western boundary current and the recirculation. Elsewhere they remain latitude circles, a sign of the survival of the Sverdrup interior. The superposition of the isolines of the two fields shows that in the western boundary layer, which is largely inertial, and in which forcing and explicit dissipation are negligible, the mean (time-averaged) flow nevertheless crosses isolines of mean total vorticity. This occurs only because the eddy field provides sufficient flux of vorticity in this region to carry away the potential vorticity anomaly that would otherwise be generated by the advection of mean vorticity.

The singular character of this interesting calculation must be emphasized. If the Ekman pumping is not exactly antisymmetric (as it is not in the natural

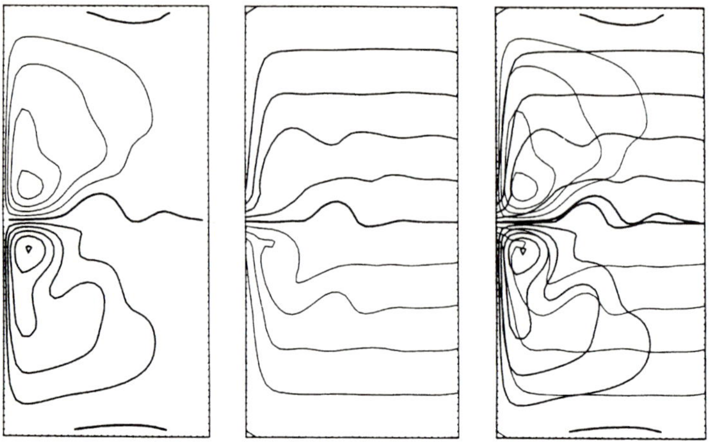

a, b c

Fig. 2.14.6a-c. Mean flow calculated for a two-gyre circulation in which the Ekman pumping is antisymmetric about the midline of the basin. There is no net vorticity input and $\delta_I/\delta_M = 4$. **a** Time-averaged streamlines. **b** Time-averaged field of total vorticity. **c** Superposition of the isolines of the two fields. (From Marshall 1984)

ocean) there is a net input of vorticity that must be expelled through the boundaries of the basin. In this case we return to the situation that we previously described.

## 2.15 Discussion

The homogeneous ocean model has been introduced less as a model for the actual ocean circulation, or even a model for its vertical average, than as a tractable vehicle for the discussion of general and fundamental physical problems. We have concentrated attention in this chapter on one of these problems, perhaps the central one, of the relationship between the attractive (because simple) but incomplete Sverdrup solution for the vertical average of the interior flow and the dynamics of the western boundary current. Because the flow is recirculates endlessly, the complete physics of the western boundary current has been seen to react back on the Sverdrup solution. As the Reynolds number for the suggested western boundary layer becomes greater than 1, the fluid (anthropomorphically speaking) has great difficulty in dissipatively shedding the vorticity put in by the wind so that it can rejoin the Sverdrup interior. Of course, it is really we and not the fluid that experience the difficulty. The complexity of the circulation in the realistic limit of large Reynolds number strains both our conceptual understanding of the physics of the flow and our ability analytically or numerically to model it. Moreover, certain rather arbitrary aspects of our models, such as the parameterization of mixing by small-scale turbulence and the concomitant condition on the tangential velocity on the side wall, have turned out to have an influence on the total solution far beyond our expectations. This implies that the solutions which we do achieve depend on the least reliable part of our physical theory. It is common in present-day numerical modeling to choose a parameterization of turbulent mixing and the associated boundary condition in order to achieve a numerical "solution" that is judged to be realistic or at least pleasing to the investigator. There is nothing wrong with such a strategy as a device to obtain flows for analysis of their dynamics, but we must distinguish such solutions from a completely deductive theory for the circulation, which is after all our ultimate goal.

The homogeneous ocean model has also been used to investigate other important dynamical issues connected to a more detailed description of the circulation. Among the most important of these is the issue of determining the latitude of separation of the western boundary current from the coast and its intrusion into the interior. Such interesting issues are beyond the scope of the present book, and the interested reader is referred to recent work by Cessi (1991), Deng (1993), and Haidvogel et al. (1992). Suffice it to say that in each of these treatments, small-scale dissipation once more appears as a determining factor in the gross dynamics of the large-scale flow.

The following chapters concentrate on theories for the vertical structure of the oceanic general circulation. In almost all cases these theories are incomplete in the sense that their dynamics is limited to the region outside of the western boundary current region. They therefore suffer all the uncertainties and dynamic ambiguities of the Sverdrup theory. Indeed, except for the discussion in Chapter 6 of equatorial dynamics, they all take the Sverdrup theory as a starting point for determining of the vertical average of the flow field in the interior. The reader must therefore bear in mind that from a purely deductive point of view these theories are fundamentally incomplete and depend on strong assumptions about the possibility of completing the solution in the western regions without fundamentally disturbing the interior. We may believe that the Sverdrup theory is justified from an observational point of view, although, as we saw in Chapter 1, this is still itself a controversial point. We also use the theory freely as a building block for more complicated theories of the stratified ocean circulation. We must admit, however, that we have so far fallen short of our goal in deciding the issue on first principles.

## References


Barcilon, V. Large Rossby number wind-driven barotropic ocean circulation.

Batchelor, G.K., 1969: Computation of the energy spectrum in homogeneous two- dimensional turbulence. *Phys. Fluids.(Supplement II)*, **12**, 233–238.

Blandford, R. R., 1971: Boundary conditions in homogeneous ocean models. *Deep Sea Res.*,**18**, 739–751.

Böning, C.W., 1986: On the influence of frictional parameterization in wind-driven ocean circulation models. *Dyn. of Atm. and Oceans*. **10**, 63–92.

Bryan, K, 1963: A numerical investigation of a nonlinear model of a wind-driven ocean. *J. Atmos. Sci.*, **20**, 594–606.

Carrier, G.F. and Robinson, A. R., 1962: On the theory of the wind-driven ocean circulation, *J. Fluid Mech.*, **12**, 49–80.

Cessi, P.R., 1991: Laminar separation of colliding western boundary currents. *J. Marine Res.*, **49**, 697–717.

Cessi, P.R., Condie R.V. and Young, W. R., 1990: Dissipative dynamics of western boundary currents, *J. Mar. Res.* **48**, 677–700.

Cessi, P.R., Ierley G. and Young, W., 1987: A model of the inertial recirculation driven by potential vorticity anomalies. *J. Phys. Ocean.*, **17**, 1640–1652.

Charney, J.C., 1955: The Gulf Stream as an inertial boundary layer. *Proc. Natl. Acad. Sci, USA*, **41**, 731–740.

Deng, J., 1993: The problem of Gulf Stream separation. *J. Phys. Ocean.*, **23**, 2182–2200.

Fofonoff, N.P., 1954: Steady flow in a frictionless homogeneous ocean. *J. Mar. Res.*, **13**, 254–262.

Greenspan, H.P., 1962: A criterion for the existence of inertial boundary layers in the oceanic circulation. *Proc. Nat. Acad. Sci.* , **48**, 2034–2039.

Haidvogel, D.B., McWilliams J. C. and Gent, P.R., 1992: Boundary current separation in a quasi-geostrophic, eddy resolving ocean circulation model. *J. Phys.Ocean.*, **22**, 882–902.

Hellerman, S. and Rosenstein, M., 1983: Normal monthly wind stress over the world ocean with error estimates. *J. Phys. Ocean.*, **13**, 1093–1104.

Hendershott, M.C., 1987: Single layer models of the general circulation. In: *General circulation of the ocean* . Eds. Abarbanel, D.I, and Young, W.R. *Springer-Verlag*, Berlin, Heiderlberg, New York, pp 291.

Ierley, G.R., 1987: On the onset of recirculation in barotropic general circulation models. *J. Phys. Ocean.*, **17**, 2366–2374.

Ierley, G.R., and Ruehr, O.G. 1986: Analytic and numerical solutions of a nonlinear boundary-layer problem. *Studies in App. Math.*, **75**, 1–36.

Ierley, G.R., and Sheremet, V. 1995: Multiple solutions and inertial runaway of the wind driven circulation. (Submitted for publication) *J. Mar. Res.*, **53**, 703–737.

Ierley, G.R., and Young, W.R. 1988: Inertial recirculation in a $\beta$-plane corner. *J. Phys. Ocean.*, **18**, 683–689.

Ierley, G.R., and Young W.R.: A discussion of boundary conditions and potential vorticity balances in quasigeostrophic general circulation model. (*unpublished manuscript.*)

Ierley, G.R., and Young W. R., 1991: Viscous instabilities in the western boundary layer. *J. Phys. Ocean.*, **21**, 1323–1332.

Il'in, A.M. and Kamenkovich, V.M. 1964: The structure of the boundary layer in the two-dimensional theory of ocean currents. (in Russian), *Okeanologiya*, **4**, 756–769.

Marshall, J.C., 1984: Eddy-mean flow interaction in a barotropic ocean model. *Q.J. R. Met. Soc.*, **110**, 573–590.

Moore, D.W., 1963: Rossby waves in ocean circulation. *Deep-Sea Res.*, **10**, 735–747.

Morgan, G.W., 1956: On the wind-driven ocean circulation. *Tellus*, **8**, 301–320.

Munk, W.H., 1950: On the wind-driven ocean circulation. *J. Meteor.*, **7**, 79–93.

Niiler, P. P. 1966: On the theory of the wind-driven ocean circulation. *Deep Sea Res.*, **13**, 597–606.

Pedlosky, J., 1987: Geophysical fluid dynamics, Chapter 5, *Springer-Verlag*, Berlin, Heidelberg, New York, pp 70.

Rhines, P.B., 1975: Waves and turbulence on a beta plane. *J. Fluid Mech.*, **69**, 417–4443.

Schlichting, H. 1979: Boundary layer theory, 7th edn. McGraw-Hill, New York, pp 817.

Stewart, R.W. 1964: The influence of friction on inertial models of oceanic circulation. In: *Studies on oceanography*: Papers dedicated to Professor Hidaka in commemoration of his 60th birthday. Yoshida, K. ed., Tokyo Univ., Geophys. Inst. 3–9.

Stommel, H, 1948: The westward intensification of wind-driven ocean currents. *Trans. Am. Geophys. Union*, **29**, 202–206.

Veronis, G. 1966: Wind-driven ocean circulation Part II. *Deep-Sea Res.*, **13**, 30–55.

Zimmerman, J.T.F. 1993: A simple model for the symmetry properties of nonlinear wind driven ocean circulation. *Geophys. Astrophys. Fluid Dyn.*, **71**, 1–15.

# 3 Vertical Structure: Baroclinic Quasi-Geostrophic Models

## 3.1 Introduction

Observations of the general circulation of the oceans, described briefly in Chapter 1, show that the most vigorous motion, driven primarily by the wind, takes place in the upper 1–2 km of the ocean and then diminishes dramatically with depth. Although it is difficult to obtain reliable measurements of time-averaged velocity in the midocean because of the presence of fluctuations on time scales of months, the signature of this intensification of the velocity field near the upper boundary of the ocean is evident in the more easily observed density field of the oceans. The thermal wind relation (see below) relates the vertical shear of the horizontal velocity to the horizontal density gradient. Therefore regions of sharp density variations horizontally can be identified as regions of strong currents. Figure 3.1.1, taken from the Levitus *Atlas* (1982), shows a zonal average of the density field in the Atlantic Ocean. The region of strong density gradient is seen to be limited to the upper ocean. The horizontal and vertical density gradients are captured largely in the upper 1km and diminish sharply with depth. This phenomenon of the rapid drop-off with depth of the motion and the associated variations of the density field presents us with a theoretical problem which is fundamental to our understanding of the dynamics of the ocean, while being at the same time of considerable difficulty. The link between the density field and the velocity means that the problem of explaining the observed oceanic density field is nothing less than the problem of explaining the full three-dimensional structure of the oceanic circulation.

In Chapter 2 we saw that the theory of even the homogeneous model of the ocean circulation still presents theoretical challenges and difficulties that have not been completely overcome.The homogeneous model tells us nothing about the vertical structure of the ocean's circulation, and we would therefore expect a theory for the baroclinic circulation to be even more complex. Nevertheless, noteworthy progress has been made on the problem of the vertical structure of the ocean circulation in recent years. In most cases the discussion has concentrated on theories for the structure of the flow in the midocean, at distances removed from the regions of strong western boundary currents. In all cases a starting point of the theoretical development has been acceptance of the Sverdrup balance for the vertical average of the circulation in the interior. We

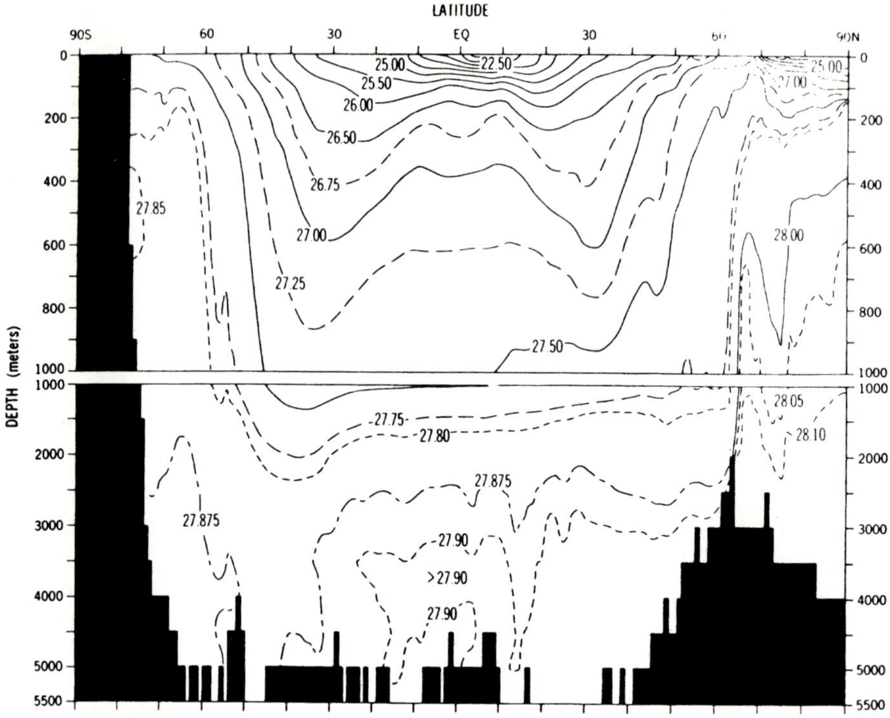


**Fig. 3.1.1.** Meridional cross-section of the zonally averaged density field in the Atlantic ocean. Notice the change of scale beneath 2000 m. The region of strong density gradient lies above this depth. (From Levitus 1982)

employ this approximation throughout our discussion as well, and the reader has been already alerted to the theoretical and observational qualifications associated with the theory whose acceptance we take as conditioned by the discussion of the first two chapters.

Several obvious questions arise almost immediately about the vertical structure. If the ocean is heated or cooled continuously at the surface, why should the resulting density variation not diffuse down into the ocean to form a smooth, nearly uniform density decay from the surface value to the observed frigid temperatures of the abyss? Current estimates of the magnitude of small-scale turbulent mixing of temperature (Ledwell et al. 1993) yield values for the coefficient of vertical mixing, $\kappa$, of about 0.1 cm$^2$/s. For an ocean depth of 5 km this would yield a characteristic diffusion time of less than 100 000 years, which is very short compared to the age of the ocean. Enough time would therefore seem to have elapsed to establish the conductive gradient in a resting fluid. However, although the diffusion time is short compared to geological time scales, it is still very long compared to the characteristic advective time $L/U$ of the ocean gyre, which is of the order of one or two decades, and therefore advection must dominate vertical diffusion. Thus it must be advective, nonlinear dynamics that primarily determines the vertical and horizontal structure of the density field which in turn, through the effects of buoyancy and rotation, shapes the motion field. The problem posed for explaining the density and velocity structure with depth is fundamentally nonlinear, and the theoretical challenge is keen.

The problem as it naturally presents itself to us involves the simultaneous explanation of the vertical and horizontal structures of the density field and renders the physics very difficult. It has been possible to effect great simplifications in the theoretical treatment of the problem by employing quasi-geostrophic models of the stratified, wind-driven ocean circulation. These models essentially split the problem in two. Since the vertical structure of the horizontal velocity field is related to the *horizontal* density gradient, we can with these models concentrate first on describing the structure of the horizontally varying part of the density field accepting, as given, the vertical density gradient. This considerably simplifies the problem and, as we see below, renders it *nearly* linear, although with important and essential nonlinear aspects. This simplification comes naturally with quasi-geostrophic theory (Pedlosky 1987) since in this theory the density field is linearized about a known and horizontally and temporally constant vertical stratification. Thus the density field is written:

$$\rho_{total} = \rho_s(z) + \rho(x, y, z, t) \tag{3.1.1}$$

where, $\rho(x, y, z, t)$ is the density variation around the laterally averaged density field, $\rho_s(z)$, which is a function only of depth. A principal feature and requirement of quasi-geostrophic theory is that:

$$\frac{\partial \rho}{\partial z} \ll \frac{\partial \rho_s}{\partial z}. \tag{3.1.2}$$

Actually the total horizontal variation of the density over the scale of the gyre is certainly as large as the vertical variation of the mean density, as can be seen from Fig. 3.1.1. Therefore, the quasi-geostrophic model cannot be expected to describe well the full density structure. Rather, since the velocity depends only on the horizontal gradient of $\rho$, we can ask it only for a theory for the vertical structure of the velocity field. A theory for the full vertical and horizontal structure of the density field is termed *thermocline theory,* which we take up in Chapter 4. We see below, however, that quasi-geostrophic theory fares unexpectedly well in anticipating many of the results of the more complex theory, especially for those aspects of the theory, such as the vertical variation of the velocity field, that depend less on the overall variation of the density field itself and pivot more on the local horizontal gradient of the density field.

Another simplification that has been of great value in illuminating the basic physical ideas of the dynamics has been the use of layer models of the ocean circulation. These models eliminate the continuous depth variable from

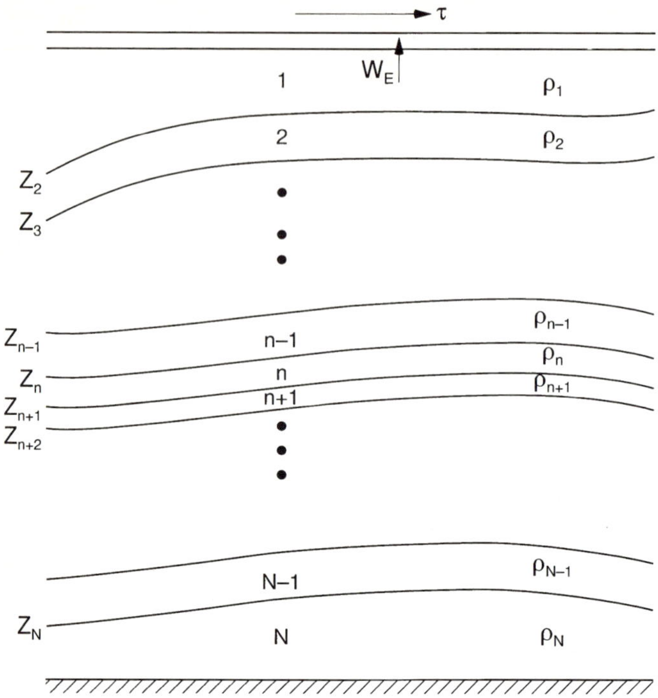


**Fig. 3.1.2.** Schematic presentation of an $N$-layer quasi-geostrophic model. The density is constant in each layer. The top of the $n$th layer is at $z = z_n$ and the thickness of each layer is $h_n$

the problem and replace it with a discrete index which specifies the layer. Often the physics can be successfully illuminated with only a few layers. This reduction in the number of continuous variables is the essential mathematical simplification offered by the layer model. Figure 3.1.2 schematically presents the layer model that we employ heavily in this chapter. $N$ layers of fluid, each of constant but different density, $\rho_n$, lie beneath a very thin Ekman layer whose divergence of transport yields an Ekman pumping velocity, $w_E$, that interacts directly with only the uppermost layer.

We discuss the model in detail in subsequent sections, but the question that we raised earlier about the depth of penetration of the motion can now be asked again from a completely different perspective. Instead of wondering why the velocity field is so restricted in its depth of penetration we may instead ask how any layer below the first layer is ever set into motion at all. In the absence of strong frictional coupling in the vertical direction it is now hard to see how the velocity can penetrate any further than the first layer. Since the number of layers is arbitrary, and we would expect increasing realism to require greater vertical resolution by employing more and thinner layers, it is at first sight hard to see why the circulation is not squeezed into only the uppermost layer. Since the total ensemble of the layers must carry the Sverdrup transport at each geographical location, this vertical squeezing would result, in the limiting case of an infinite number of layers, in the Sverdrup transport being carried in the first layer of infinitesimal thickness, i.e., with a delta function distribution of velocity with depth. One way of characterizing different theories for the vertical structure of the velocity field is by the mechanism used in each theory to resolve the perplexity of this absurd limit for the velocity field as the vertical resolution is improved. The important theoretical task ahead is in clarifying what has been overlooked in the oversimplified view of the dynamics that has led to this apparent paradox for the vertical structure of the velocity. The present chapter describes quasi-geostrophic theories that deal with this issue.

## 3.2 The Quasi-Geostrophic Model

The formulation of the quasi-geostrophic model, both in its layer form and for a continuously stratified fluid, can be found in many texts. A careful development of the theory can be found in Pedlosky (1987), in which the equations are developed in an asymptotic series in the small parameter:

$$R_0 = \frac{U}{fL} \tag{3.2.1}$$

$R_0$ is the Rossby number and $U$, $f$, and $L$ are characteristic values for the horizontal velocity, the Coriolis parameter, and the horizontal length scale of the motion, respectively. This section develops the equations in a more informal manner concentrating on the physical character of the derivation.

Along with the Rossby number the aspect ratio of the motion, $\delta$, is a small parameter. If $D$ is the vertical scale of the motion, then:

$$\delta = \frac{D}{L} \ll 1. \tag{3.2.2}$$

The smallness of the Rossby number, $R_0$, implies that to the lowest order the horizontal velocity field is in geostrophic balance, while the smallness of the aspect ratio implies that the vertical equation of motion can be approximated by the hydrostatic balance, i.e.:

$$\begin{aligned} fv &= \frac{1}{\rho}\frac{\partial p}{\partial x} \\ fu &= -\frac{1}{\rho}\frac{\partial p}{\partial y} \\ \rho g &= -\frac{\partial p}{\partial z} \end{aligned} \tag{3.2.3a,b,c}$$

where $u$ and $v$ are the velocities in the eastward and northward directions. The parameter $f$ is the Coriolis parameter:

$$f = 2\Omega \sin \theta \tag{3.2.4}$$

where $\theta$ is latitude.

The ability to use a Cartesian coordinate frame is granted by the application of the $\beta$ plane approximation. That is, we assume that the north-south scale of the motion, of order $L$, is small enough that:

$$\frac{\beta L}{f} = \frac{L}{R} \cot \theta \ll 1 \tag{3.2.5}$$

where $R$ is the earth's radius. This allows $f$ to be replaced by a constant value $f_0$, typical of the region of the motion, in the geostrophic relations (3.2.3 a,b), with an error of $O(\beta L/f_0)$.

The scale of the lateral variations of the density field is related to the scale of the velocity field through the application of both the geostrophic and the hydrostatic approximation. If $\Delta\rho_V$ is the scale of the vertical variation of density associated with $\rho_s$ and $\Delta\rho_H$ is the scale of the horizontal variation of the density field, geostrophy and hydrostatic balance imply that $g\Delta\rho_H/L = O(fU/D)$, or:

$$\begin{aligned} \frac{\Delta\rho_H}{\Delta\rho_V} &= O\left(\frac{f_0 UL}{g'D}\right) \\ &= R_0 F \ll 1 \end{aligned} \tag{3.2.6}$$

where:

$$F = \frac{f_0{}^2 L^2}{g'D} \tag{3.2.7}$$

and where $g'$ is the reduced gravity, $g\ (\Delta\rho_V/\rho_0)$. We also assume that the vertical variation of density, although much larger than the horizontal variation (in quasi-geostrophic theory), is still small compared with the average value of the density, $\rho_0$. This set of approximations allows $\rho$ to be replaced with $\rho_0$ in the geostrophic balance (the Boussinesq approximation), and eliminating the pressure between the hydrostatic and the geostrophic balances thus, yields the quasi-geostrophic form of the thermal wind relation:

$$\begin{aligned} f_0 \frac{\partial u}{\partial z} &= g \frac{1}{\rho_0} \frac{\partial \rho}{\partial y} \\ f_0 \frac{\partial v}{\partial z} &= -g \frac{1}{\rho_0} \frac{\partial \rho}{\partial x} \end{aligned} \tag{3.2.8}$$

If the density and Coriolis parameters are approximated as constants in (3.2.3a,b), it follows that to the lowest order the horizontal velocity can be written in terms of the geostrophic stream function, $\psi$, i.e.:

$$\vec{u} = \hat{k} \times \nabla\psi, \qquad \psi = \frac{p}{\rho_0 f_0} \tag{3.2.9}$$

so that the horizontally velocity is nondivergent to the lowest order. The unit vector $\hat{k}$ is directed along the local vertical direction. The first divergent contribution to the horizontal velocity field comes from either departures from geostrophy, $O(R_0)$, from horizontal variations in the density field, $O(fUL/gD)$, or from variations in the Coriolis parameter, $O(L/R)$. For gyre-scale motions it is the last effect that is the most significant. Thus to order $L/R$, i.e., to $O(\beta L/f_0)$ the horizontal motion is nondivergent, so that to lowest order, [i.e. to $O(L/R)$]:

$$\frac{\partial w}{\partial z} = 0. \tag{3.2.10}$$

For a flat-bottom ocean, or for one in which the bottom relief is restricted to order $L/R$ or less, the vertical velocity must vanish on the lower boundary. This, with (3.2.10) requires that $w$ vanish *to lowest order*, throughout the region of geostrophic flow. This eliminates w as an agent for the vertical advection of momentum and vorticity, so that the equation for the vertical component of the vorticity is:

$$\frac{\partial \zeta}{\partial t} + u\frac{\partial \zeta}{\partial x} + v\frac{\partial \zeta}{\partial y} + \beta v = -f\nabla \cdot \vec{u} + \operatorname{curl} \Im \tag{3.2.11}$$

where $\Im$ is the horizontal frictional force per unit mass and *curl* is the vertical component of the vector curl operator. The horizontal divergence in (3.2.11) involves velocities of higher order in the parameters $R_0$ or $\beta L/f_0$ than the velocities in geostrophic balance (which are obviously nondivergent) and thus cannot be calculated from the geostrophic streamfunction.

In the layer model shown in Fig. 3.1.2, equations (3.2.9) and (3.2.11) apply within each layer, and we label the variables within each layer with a subscript $n$. The density in each layer is a constant, for example $\rho_n$. The thermal wind relation (3.2.8) shows that therefore the horizontal velocity *within* each layer is independent of $z$. However, normally it varies from layer to layer. The vertical velocity is a linear function of $z$ within each layer.

We can write the pressure within each layer, say, for layers $n$ and $n-1$, as:

$$p_n = -\rho_n g z + \rho_0 \pi_n(x, y, t)$$

$$p_{n-1} = -\rho_{n-1} g z + \rho_0 \pi_{n-1}(x, y, t) \tag{3.2.12}$$

which automatically satisfies the hydrostatic equation. It is the hydrostatic approximation which requires that $\pi_n$ be independent of $z$. At the interface $z = z_n(x, y, t)$ between layers $n$ and $n-1$, the pressure must be continuous, i.e., $p_n$ must equal $p_{n-1}$. Thus:

$$z_n = \frac{\pi_n - \pi_{n-1}}{\gamma_{n-1}} \tag{3.2.13a}$$

where:

$$\gamma_n \equiv \frac{\rho_{n+1} - \rho_n}{\rho_0} g \tag{3.2.13b}$$

is a useful measure of the stratification in the layer model. The geostrophic streamfunction within each layer can be written conveniently in terms of $\pi_n$, namely:

$$\psi_n = \pi_n / f_0 \tag{3.2.14}$$

The velocity difference, or shear, between layer $n-1$ and layer $n$, is then proportional to the slope of the interface between them, i.e.:

$$\vec{u}_{n-1} - \vec{u}_n = \hat{k} \times \nabla(\psi_{n-1} - \psi_n) = -\frac{\gamma_{n-1}}{f_0} \hat{k} \times \nabla z_n\,. \tag{3.2.15}$$

Consider now the full three-dimensional continuity equation, integrated across the $n$th layer. The horizontal velocities are independent of $z$ within the layer so we obtain:

$$h_n \nabla \cdot \vec{u}_n = -w_n(z_n) + w_n(z_{n+1}) \tag{3.2.16}$$

where it is understood that the divergence operator and the velocity on the left side of (3.2.6) refer to the two-dimensional, horizontal component of the velocity alone. The thickness of the $n$th layer is $h_n = z_{n-1} - z_n$.

If we add $dh_n/dt$ to both sides of (3.2.16) we obtain a very useful form of the equation for mass conservation, i.e.:

$$\frac{dh_n}{dt} + h_n \nabla \cdot \vec{u}_n = -\left[w*(z_n) - w*(z_{n+1})\right] \tag{3.2.17}$$

where $w*(z_n)$, defined as:

$$w*(z_n) = w_n(z_n) - \frac{dz_n}{dt} = w_n(z_n) - \frac{\partial z_n}{\partial t} - \vec{u}_n \cdot \nabla z_n \tag{3.2.18}$$

is the flux of fluid, per unit horizontal area in the $x$,$y$ plane, across the $n$th interface. It is obvious from mass conservation that the flux leaving the $n$th layer across $z_n$ enters at the same horizontal location into the $n-1$st layer so that $w*$ is continuous across $z_n$ and could be equally well evaluated on either side of the interface. Thus we could write (3.2.18) equally well as:

$$w*(z_n) = w_{n-1}(z_n) - \frac{\partial z_n}{\partial t} - \vec{u}_{n-1} \cdot \nabla z_n,$$

or (3.2.19)

$$w*(z_{n+1}) = w_n(z_{n+1}) - \frac{\partial z_{n+1}}{\partial t} - \vec{u}_n \cdot \nabla z_{n+1}.$$

The velocity $w*(z_n)$ is positive whenever fluid from layer n crosses the interface into layer $n-1$. This does not require that the vertical velocity, $w_n$, itself be positive if there is strong motion across the interface. If fluid crosses the interface between the layers, a sudden density transformation must take place at the interface as fluid is changed, say at $z = z_n$, from a density $\rho_n$ to a density of $\rho_{n-1}$. This implies that nonadiabatic processes must be active at the interface if $w* \neq 0$. If $w*(z_n)$ is positive, this does not, however, imply that the vertical velocity at the interface $w_n(z_n)$ must be positive, only that some process allows fluid to transform its density nonadiabatically at the interface. Figure 3.2.1 shows three helpful examples. In the first, the interface is flat and motionless. The vertical velocity in the fluid is positive, and fluid crosses the interface, changing its density as it does so. In this case the vertical velocity and the cross-interface velocity are both positive and in fact equal to each other. In Fig. 3.2.1b, the fluid is absolutely stationary, and the interface descends. The vertical velocity is zero, but the cross-interface velocity is again positive as the upper layer absorbs lower layer fluid whose density changes in the process. In the last example (Fig. 3.2.1c) the interface slopes downwards to the right more steeply than the three-dimensional velocity vector. In this case the vertical velocity is *negative,* and yet the cross-interface velocity is again positive.

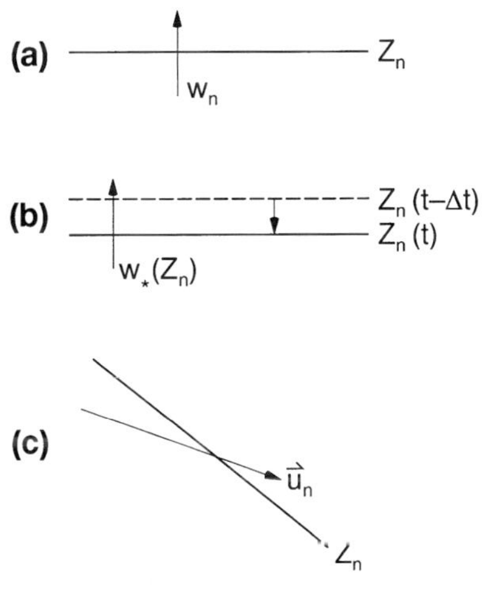


**Fig. 3.2.1a–c.** Schematic presentation of the relationship between the vertical velocity and the cross-interface velocity $w*(z_n)$ in three extreme cases. **a** The interface is flat and stationary. The cross-isopycnal velocity equals the vertical velocity. **b** The fluid is stationary, and the interface descends. The vertical velocity is zero and the cross-interface velocity is *positive*. **c** The interface slopes downward to the right more steeply than the three-dimensional velocity in the $n$th layer. The cross interface velocity is positive and the vertical velocity is *negative*

If the motion is adiabatic, the fluid must conserve its density and remain in its original layer. In this case the cross-interface velocity is zero, and (3.2.18) reduces to the adiabatic boundary condition at the interface $w_n(z_n) = dz_n/dt$ . Note that although the cross-isopycnal velocity is zero, the vertical velocity at the interface is not, in general, zero. In general, for both adiabatic and nonadiabatic motion, unless the interface is flat, or the horizontal velocity is very weak and associated with weak slopes of the interfaces, the cross-isopycnal velocity differs substantially from the vertical velocity.

It is conceptually helpful momentarily to consider the same process of cross-isopycnal motion in a continuously stratified fluid. If $Q$ is the heat added locally to a nearly incompressible fluid, the density equation can be written (Pedlosky 1987):

$$\frac{d\rho}{dt} = \frac{\partial \rho}{\partial t} + u\frac{\partial \rho}{\partial x} + v\frac{\partial \rho}{\partial y} + w\frac{\partial \rho}{\partial z} = -\frac{\alpha Q}{c_p} \tag{3.2.20}$$

where $\alpha$ is the coefficient of thermal expansion and $c_p$ is the specific heat of constant pressure, and a linear equation of state-relating temperature and density is used. Since

$$\left.\frac{\partial z}{\partial \chi}\right|_{\rho} = -\frac{\partial \rho}{\partial \chi}\Big/\frac{\partial \rho}{\partial z} \tag{3.2.21}$$

for $\chi = t$, $x$, or $y$ and where the derivative on the left side of (3.2.21) is taken at constant density. Thus, dividing (3.2.20) by $\partial\rho/\partial z$ yields:

$$w* \equiv w - \left[\frac{\partial z}{\partial t} + \vec{u} \cdot \nabla z\right] = \frac{\alpha Q}{c_p(-\partial \rho/\partial z)}, \tag{3.2.22}$$

where $z$ is the height of a surface of constant density, i.e. an isopycnal. The cross-isopycnal flux is kinematically exactly as we defined it for the cross-interface flux in the layer model, and we see in (3.2.22) that it is explicitly related to the presence of internal heating and cooling where this cross-isopycnal flux occurs.

The cross-interface velocity at the $n$th interface is continuous, which implies that at that interface:

$$w_{n-1}(z_n) - w_n(z_n) = (\vec{u}_{n-1} - \vec{u}_n) \cdot \nabla z_n \tag{3.2.23}$$

so that the jump in the vertical velocity from the $n$th layer to the $n-1$st layer is proportional to the projection along the interface slope of the shear in the horizontal velocity. However, since the horizontal velocity is in geostrophic balance, it follows from (3.2.15) that the shear across the interface is *perpendicular* to the direction of interface slope. Thus, the scalar product on the right side of (3.2.23) is identically zero. The vertical velocity across the interface is continuous (to lowest order in Rossby number) even though the horizontal velocity is discontinuous.

If we apply the vorticity equation (3.2.11) to each layer, and use the mass conservation equation (3.2.17) to eliminate the divergence (which cannot be calculated from the geostrophic velocity), we obtain:

$$h_n\left[\frac{d\zeta_n}{dt}+\beta v_n\right]=-f_0\frac{dh_n}{dt}+f_0\left[w*(z_n)-w*(z_{n+1})\right]+h_n\,\mathrm{curl}\,\Im_n. \tag{3.2.24}$$

The layer thickness can be written in terms of its horizontally averaged value $H_n$ and the spatially variable part $\delta h_n$, i.e.:

$$h_n=H_n+\delta h_n(x,y,t) \tag{3.2.25}$$

and the geostrophic, hydrostatic relation between the interface slope and the geostrophic pressure anomaly allows us to estimate the ratio of the two terms in (3.2.25) as:

$$\frac{\delta h_n}{H_n}-O\left(\frac{f_0UL}{\gamma_{n-1}H_n}\right)=O(R_0F)\ll 1 \tag{3.2.26}$$

which is the layer model equivalent of (3.2.6). Note that $F$ can be larger than unity as long as its product with the Rossby number is small enough for the ratio in (3.2.26) to be small. If these conditions are satisfied, the integrated vorticity equation for each layer can be written as:

$$\frac{d}{dt}\left[\zeta_n+\beta y-\frac{f_0\delta h_n}{H_n}\right]=\frac{f_0}{H_n}\left\{w*(z_n)-w*(z_{n+1})\right\}+\mathrm{curl}\,\Im_n \tag{3.2.27}$$

where the quantity on the left side in brackets is the quasi-geostrophic approximation to the full form of the layer potential vorticity $q_n=(f+\zeta_n)/h_n$, (Pedlosky 1987). That is, in quasi-geostrophic theory:

$$q_n=\zeta_n+\beta y-\frac{f_0\delta h_n}{H_n}. \tag{3.2.28}$$

Both the vorticity and the thickness anomaly can be related to the geostrophic streamfunction through (3.2.13) and (3.2.14), so that the vorticity equation can be written entirely in terms of the geostrophic streamfunction, $\psi_n$ as:

$$\begin{aligned}\frac{dq_n}{dt}&\equiv\frac{d}{dt}\left[\nabla^2\psi_n+\beta y-\frac{f_0^2}{H_n}\left\{\frac{\psi_n-\psi_{n-1}}{\gamma_{n-1}}-\frac{\psi_{n+1}-\psi_n}{\gamma_n}\right\}\right]\\&=\frac{f_0}{H_n}\left\{w*(z_n)-w*(z_{n+1})\right\}+\mathrm{curl}\,\Im_n\end{aligned} \tag{3.2.29}$$

where the quantity in the curly brackets on the left side of (3.2.29) is the representation of the variable part of the thickness of layer $n$ in terms of the geostrophic streamfunction of layer $n$ and of the streamfunction of the two adjacent layers. In the absence of cross-interface velocity and in the absence of friction, i.e., when the fluid motion in a layer is inviscid and adiabatic, the potential vorticity of that layer is conserved, i.e., its total time derivative vanishes.

There are two layers that require special attention. The first is layer 1, the uppermost layer under the Ekman layer. Its upper surface, assuming the Ekman layer is very thin, is actually the interface in pressure contact with the overlying atmosphere. The density jump at the interface with the atmosphere is so large, i.e., $O(1)$, that the strong stratification at that surface renders the motion of the upper surface negligible compared to that of its lower interface by a factor $\Delta\rho/\rho_0$, where $\Delta\rho$ is the density jump from the first to the second layer. Compared to $z_2$ the upper interface is *kinematically* rigid although it adjusts slightly to match the atmospheric pressure at the upper surface. At the same time the mass flux across the upper interface with the Ekman layer is the known Ekman pumping velocity, $w_E$, so that (3.2.27) becomes for layer 1:

$$\frac{d}{dt}\left[\nabla^2\psi_1 + \beta y + \frac{f_0}{H_1} z_2\right] = \frac{f_0}{H_1}\left[w_E - w*(z_2)\right] + \operatorname{curl}\Im_1 \tag{3.2.30}$$

or:

$$\begin{aligned}\frac{dq_1}{dt} &\equiv \frac{d}{dt}\left[\nabla^2\psi_1 + \beta y - \frac{{f_0}^2}{\gamma_1 H_1}(\psi_1 - \psi_2)\right] \\ &= \frac{f_0}{H_1}\left[w_E - w*(z_2)\right] + \operatorname{curl}\Im_1 \ .\end{aligned} \tag{3.2.31}$$

Layer $N$ is the lowermost layer and is in contact with the rigid lower boundary at $z = Z_b(x, y)$. The rate of change of the thickness of the lower layer is simply:

$$\frac{dh_N}{dt} = \frac{dz_N}{dt} - \frac{dZ_b}{dt} \tag{3.2.32}$$

while the vertical velocity pumped out of the Ekman layer at the bottom of layer $N$ and which is equal to $w*(Z_b)$ is given by (Pedlosky 1987):

$$w*(Z_b) = \frac{\delta_E}{2}\zeta_N \tag{3.2.33}$$

where $\delta_E = (2A_v/f_0)^{1/2}$ and $A_v$ is the turbulent vertical mixing coefficient of momentum. When these relations are used in the vorticity equation for layer $N$ we obtain:

$$\begin{aligned}\frac{dq_N}{dt} &\equiv \frac{d}{dt}\left[\nabla^2\psi_N + \beta y - \frac{{f_0}^2}{\gamma_{(N-1)} H_N}\{\psi_N - \psi_{N-1}\} + \frac{f_0}{H_N} Z_b\right] \\ &= \frac{f_0}{H_N} w*(z_{(N-1)} + \operatorname{curl}\Im_N - r_N\nabla^2\psi_N\end{aligned} \tag{3.2.34}$$

where the bottom friction parameter, $r_N$, is given by:

$$r_N = \frac{\sqrt{A_v f_0/2}}{H_N} \tag{3.2.35}$$

The system of equations consists of (3.2.29) for the general $n$th intermediate layer and (3.2.31) and (3.2.34) for the upper- and lower most layers. Note that only the upper layer is driven by the Ekman pumping, and that in the absence of interfacial coupling by either cross-interface velocities or by frictional coupling between layers (which would be included in $\Im_n$) there would be apparently nothing to maintain the lower layers in motion. This is the paradox to which we referred above, and to which we return shortly as the principal topic of this chapter.

In each layer the total derivative of any variable, and in particular the potential vorticity, $q_n$, for this layer can be written in terms of the streamfunction and its Jacobian with the derived variable as:

$$\frac{dq_n}{dt} = \frac{\partial q_n}{\partial t} + \frac{\partial \psi_n}{\partial x}\frac{\partial q_n}{\partial y} - \frac{\partial \psi_n}{\partial y}\frac{\partial q_n}{\partial x} \equiv \frac{\partial q_n}{\partial t} + J(\psi_n, q_n). \tag{3.2.36}$$

## 3.3 Special Cases of the Potential Vorticity Equation

There are two special examples of the quasi-geostrophic layer model of particular interest for the discussion of this chapter. In the first case the entire ocean, beneath the upper Ekman layer, consists of only two layers, the lowermost being in contact with the ocean floor, as shown in Fig. 3.3.1a. In this

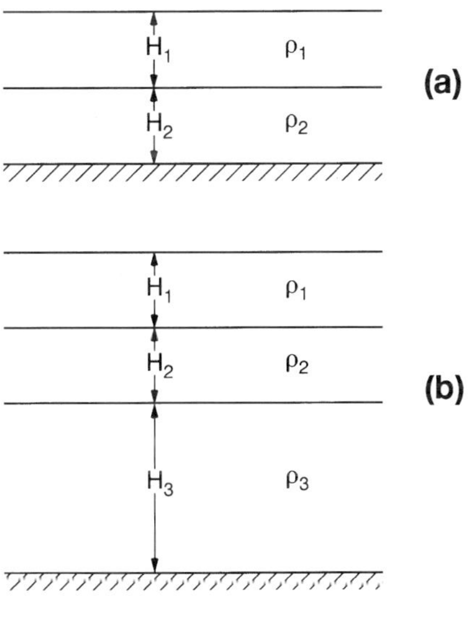


**Fig. 3.3.1a,b.** Two-and three-layer models. **a** The two layers fill the entire depth of the ocean. **b** There are three layers, the upper two of which are in motion for steady flows while the third, deep layer is at rest. If the third layer is deep enough its motion can be neglected for time-dependent motions as well

case our governing equations are (3.2.31) which we rewrite here:

$$\frac{d}{dt}\left[\nabla^2\psi_1 + \beta y - \frac{f_0{}^2}{\gamma_1 H_1}(\psi_1 - \psi_2)\right] = \frac{f_0}{H_1}\left[w_E - w*(z_2)\right] + \operatorname{curl}\Im_1 \tag{3.3.1}$$

and the lower layer equation (3.2.34) for the case where $N = 2$. The equation which closes the system is then:

$$\frac{d}{dt}\left[\nabla^2\psi_2 + \beta y - \frac{f_0{}^2}{\gamma_1 H_2}\{\psi_2 - \psi_1\} + \frac{f_0}{H_2}Z_b\right]$$
$$= \frac{f_0}{H_2}w*(z_2) + \operatorname{curl}\Im_2 - r_2\nabla^2\psi_2. \tag{3.3.2}$$

In the second case the model consists of two layers resting over a deep third layer, which is supposed to represent a resting abyss. The equations for the three-layer model are (3.3.1) for the upper layer while the second layer, from (3.2.29) with $n = 2$, satisfies:

$$\frac{d}{dt}\left[\nabla^2\psi_2 + \beta y - \frac{f_0{}^2}{\gamma_1 H_2}\left\{\psi_1 - \psi_2 - \frac{\gamma_1}{\gamma_2}(\psi_2 - \psi_3)\right\}\right]$$
$$= \frac{f_0}{H_2}\left[w*(z_2) - w*(z_3)\right] + \operatorname{curl}\Im_2 \tag{3.3.3}$$

while layer 3 satisfies (3.2.34) with $N = 3$. For a flat-bottom ocean this is:

$$\frac{d}{dt}\left[\nabla^2\psi_3 + \beta y - \frac{f_0{}^2}{\gamma_2 H_3}(\psi_3 - \psi_2)\right]$$
$$= \frac{f_0}{H_2}w*(z_3) - r_3\nabla^2\psi_3 + \operatorname{curl}\Im_3\ . \tag{3.3.4}$$

Now consider the class of motions in which layer 3 represents a resting abyss and the velocity in layer 3 is therefore supposed to be zero. This obviously can only occur if $w*(z_3)$ is zero; otherwise motion in layer 3 would be forced by the cross-isopycnal velocity from layer 3 to layer 2. If $w*(z_3)$ *is* zero, and $\psi_3$ is set equal to zero, (3.3.3) becomes:

$$\frac{d}{dt}\left[\nabla^2\psi_2 + \beta y - \frac{f_0{}^2}{\gamma_1 H_2}\left\{\psi_1 - \psi_2 - \frac{\gamma_1}{\gamma_2}\psi_2\right\}\right]$$
$$= \frac{f_0}{H_2}w*(z_2) + \operatorname{curl}\Im_2. \tag{3.3.5}$$

Such a model is called a "two-and-a-half-layer model" in recognition of the putative dynamic triviality of the third layer in this limit.

To check the consistency of the assertion that layer 3 is at rest, i.e., that $\psi_3$ is zero, we should examine (3.3.4). If $\Im_3$ also vanishes when $\psi_3$ is zero, the equation for layer 3 (3.3.4) becomes, assuming that $\psi_3$ is zero:

$$\frac{f_0{}^2}{\gamma_2 H_3}\frac{\partial \psi_2}{\partial t} = 0. \tag{3.3.6}$$

For steady-flow problems this is trivially satisfied and the $2\frac{1}{2}$ layer model is consistent. For time-dependent problems this approximation requires that either the thickness of the third layer becomes very great, or that the density jump across the interface between layers 2 and 3 becomes very large so that this remaining term is negligible.

If the density jump across the interface $z = z_2$ is very much larger than the density jump across the interface $z = z_3$, the two models (3.3.2) and (3.3.3) coincide for a flat bottom ocean.

It is important to note that when the motion is time dependent, the fluid in layer 3 is generally put into motion by the motion of the interface separating it from layer 2 so that the model of a resting abyss is valid only for steady motion (in the absence of dissipative coupling). This has important consequences for the interpretation of the Rossby wave modes in the system. These waves are important messengers of information, and they must propagate, in general, within the circulation. The propagation of the waves is affected by the circulation, and the properties of the propagation in turn dynamically shape the circulation itself. The properties of the Rossby wave dynamics in the presence of the gyre flow prove to be vital in understanding the structure of the resulting steady circulation. The difference between steady wave motion in the presence of the circulation and simple propagating waves in the absence of a background steady flow is significant. There is more than a Galilean transformation involved in this difference, for the steady motion is usually vertically sheared, which means that the interfaces between the layers are sloping. This variation of the layer thicknesses alters the potential vorticity gradient which supports the Rossby waves and changes their propagation characteristics.

## 3.4 Rossby Waves

Although studying problems of the steady circulation of the wind driven gyres, some of the principal results can be understood in a particularly helpful way in terms of the wave properties of the system. Anderson and Gill (1975) studied the oceanic spin-up problem and its relation to the Rossby wave modes of the system to demonstrate the way in which the Sverdrup interior is established and comes into a steady state after an initially impulsive imposition of the wind stress. We do not repeat their analysis here, to which the reader is referred for a detailed discussion, but the broad outlines of their results may be understood from an examination of the linearized quasi-geostrophic equations.

Consider the three-layer model for a flat-bottom ocean in which dissipation due to frictional and cross-interface motion are ignored. Suppose the motion is linearized *about a state of rest* and is set into motion by the application at time $t = 0$ by the imposition of an Ekman pumping. The equations of motion are the linearized versions of (3.3.1), (3.3.3), and (3.3.4). We are also particularly interested in the 2 1/2 layer limit described in the previous section where the depth of layer 3 becomes very large.

The linear form of the layer equations is:

$$\begin{aligned}
&\frac{\partial}{\partial t}\left[\nabla^2\psi_1 - F_1(\psi_1 - \psi_2)\right] + \beta\frac{\partial\psi_1}{\partial x} = \frac{f_0}{H_1}w_E \\
&\frac{\partial}{\partial t}\left[\nabla^2\psi_2 - F_2(\psi_2 - \psi_1) - G_2(\psi_2 - \psi_3)\right] + \beta\frac{\partial\psi_2}{\partial x} = 0 \\
&\frac{\partial}{\partial t}\left[\nabla^2\psi_3 - G_3(\psi_3 - \psi_2)\right] + \beta\frac{\partial\psi_3}{\partial x} = 0
\end{aligned} \tag{3.4.1a,b,c}$$

where:

$$\begin{aligned}
F_n &= \frac{f_0{}^2}{\gamma_1 H_n} \qquad n = 1, 2 \\
G_n &= \frac{f_0{}^2}{\gamma_2 H_n} \qquad n = 2, 3.
\end{aligned} \tag{3.4.2a,b}$$

The analysis of Anderson and Gill can be summarized as follows. At the initial instant, all of the three linear, vertical modes of the system are excited by the Ekman pumping, which for simplicity is taken as independent of $x$. The solution can be approximated during this initial period, short compared to a characteristic Rossby wave period, by considering only the time derivative terms in (3.4.1) and ignoring the $\beta$ effect. If $w_E$ is turned on impulsively and then held steady, this yields solutions for each of the $\psi_n$, or more precisely for each of the vertical modes, which increase linearly with time. The solutions are valid in the interior of the basin away from the eastern boundary since this approximate solution, independent of $x$, does not satisfy the boundary condition of no normal flow on the eastern wall.

This phase of linear growth is halted at a position $(x, y)$, for each vertical mode when the free Rossby wave corresponding to the mode propagates from the eastern boundary and passes the point under consideration. This wave carries the information about the boundary condition at the eastern boundary and puts the solution at the point $(x, y)$ in conformity with the boundary condition at $x_e$.

The three free, vertical modes of the system carrying this information from the eastern boundary are the Rossby modes of the layer model. They consist of the barotropic mode and two baroclinic modes. The speed of each mode depends on the wave length of the mode, and for the baroclinic modes the speeds also depend on the stratification. The limiting case of interest is when the length scale of the forcing and the modes are long compared to the Rossby deformation radius for each of the baroclinic modes, i.e., when the relative vorticity in the above equations can be neglected with respect to the terms involving $G_n$ and $F_n$. The speeds are given below but for now we need to observe only that the group velocity of the baroclinic waves in this limit is always westward (Pedlosky 1987). After all the wave modes pass the observation point $(x, y)$ on their westward journey, the solution becomes steady and under the influence of the steady Ekman pumping. Hence after the passage of the waves the solutions must satisfy:

$$\begin{aligned} \beta\frac{\partial\psi_1}{\partial x} &= \frac{f_0}{H_1}w_E \\ \beta\frac{\partial\psi_2}{\partial x} &= 0 \\ \beta\frac{\partial\psi_2}{\partial x} &= 0. \end{aligned} \tag{3.4.3a,b,c}$$

Each of the $\psi_n$ must satisfy:

$$\psi_n = 0, \quad x = x_e \quad n = 1, 2, 3. \tag{3.4.4}$$

Hence in the steady state only $\psi_1$ differs from zero. The upper layer carries all the Sverdrup transport while the lower two layers are at rest. The baroclinic Rossby waves arriving from the eastern boundary "turn off" the initial surge of motion in the lower layers. The long baroclinic Rossby waves are the messengers which, arriving from the east, signal the fluid of the presence of the blocking action of the eastern boundary. The planetary potential vorticity contours in this linear limit are just the lines of constant latitude along which in linear, steady-state theory the streamfunction must be constant, and the lower, unforced layers must therefore be at rest in the steady state since these contours strike the eastern boundary where $\psi_2$ and $\psi_3$ are zero. Thus, the motion in the lower two layers is "blocked" by the intersection of the potential vorticity contours in those layers with the eastern boundary.

The baroclinic Rossby modes are relatively slow (see below), and their speeds differ little from the speeds associated with the fluid motion in the gyre circulation. To consider the qualitative effect on the propagation of the waves by the gyre's mean flow, we can linearize the motion about a mean zonal flow rather than a state of rest. For simplicity, and in preparation for the results of our investigation below of the nonlinear steady flow problem, we choose a mean flow consisting of an eastward velocity $U$, which is the same for the upper two layers. The third layer has no mean velocity. That is, we now consider the streamfunction written as:

$$\begin{aligned} \psi_1 &= -Uy + \varphi_1(x, y, t) \\ \psi_2 &= -Uy + \varphi_2(x, y, t) \\ \psi_3 &= \varphi_3(x, y, t). \end{aligned} \tag{3.4.5}$$

Now the general perturbation problem in the presence of this basic shear flow is complex. On small scales, of the order of the Rossby deformation radius, i.e., of the order of:

$$L_d = \left( \frac{\gamma_1 H_2 H_1}{f_0{}^2 (H_1 + H_2)} \right)^{1/2} \tag{3.4.6}$$

baroclinic instabilities occur (Pedlosky 1987). Even stable waves have complex dispersion relations. We can ask a simpler, and for our purposes, more decisive question. Can the mean shear flow arrest the westward propagating Rossby waves and render them stationary? If so, this could shield a portion of the deeper layers of fluid from the "turn - off" message carried from the eastern boundary. To investigate this we need to consider only the steady version of (3.3.1) ,(3.3.3), and (3.3.4). Since there is no mean flow in layer 3, we anticipate that the steady solution in layer three is one in which layer 3 is entirely at rest. As we noted above, this is consistent for steady flows. With $\varphi_3 = 0$, we obtain for the steady, nondissipative, unforced versions of (3.3.1) and (3.3.2):

$$\begin{aligned} U \frac{\partial}{\partial x} \left[ \nabla^2 \varphi_1 - F_1(\varphi_1 - \varphi_2) \right] + \beta \frac{\partial \varphi_1}{\partial x} &= 0 \\ U \frac{\partial}{\partial x} \left[ \nabla^2 \varphi_2 + F_2(\varphi_1 - \varphi_2) \right] + \beta \frac{\partial \varphi_2}{\partial x} &= 0. \end{aligned} \tag{3.4.7}$$

If steady wave motions of the form:

$$\varphi_n = A_n \exp i(kx + ly), \quad n = 1, 2 \tag{3.4.8}$$

are sought for (3.4.7) it follows that nontrivial solutions for the amplitudes $A_1$ and $A_2$ exist only if either:

$$\begin{aligned} U &= \beta / K^2 \quad \text{or} \\ U &= \beta / \left[ K^2 + F_1 + F_2 \right] \end{aligned} \tag{3.4.9a,b}$$

where $K^2 = k^2 + l^2$.

In the first case (3.4.9a) the amplitudes of the steady waves in the upper two layers are equal. This represents a barotropic motion *of the upper two layers* in which the interface between the upper two layers is not deformed, although since the lowest layer, layer 3, is at rest, the total motion is actually baroclinic. Note that the speed required to halt the wave's westward propagation is independent of the stratification.

The second possibility (3.4.9b) represents a steady wave in which the interface between the upper two layers is deformed by the motion and in which the motion of the two layers is purely baroclinic, i.e., in which $H_1\varphi_1 = -H_2\varphi_2$.

For waves with scales of the order of the gyre, i.e., with wavelengths, $\lambda$, of the order 2000 km, the speed of the zonal flow required to arrest the "barotropic" mode is of the order of:

$$U_{barot.} = \frac{\beta\lambda^2}{4\pi^2} = O(200\,\mathrm{cm/s}) \tag{3.4.10}$$

for $\beta = 2 \times 10^{-13}\mathrm{cm}^{-1}\mathrm{s}^{-1}$. This is far greater than any realistic gyre scale velocity, and we conclude that such quasi-barotropic waves are not arrested by the mean flow. In fact, the barotropic mode swiftly traverses a typical ocean basin in a matter of weeks. For a basin of width 4000 km it would take only 3 weeks to cross the basin. In that time the steady barotropic mode carrying the average Sverdrup transport is established. On the scale of the slow advective motion of the gyre, of the order 1–10 cm/s, this is essentially an instantaneous response.

On the other hand, for the same scales, for which $\lambda \gg L_d$, the required speed for the second, baroclinic mode is given by (3.4.9b), which in this limit, and using the definition of $L_d$ given by (3.4.6), is:

$$U_{baroc.} = \beta L_d^2 = \frac{\beta}{f_0{}^2}\frac{\gamma_1 H_1 H_2}{H_1 + H_2}. \tag{3.4.11}$$

Characteristic values of $L_d$ are of the order of 50 km, so that the required arresting velocity is of the order of 5 cm/s. This is a speed much more characteristic of the gyre scale mean flow, and it is more plausible that this criterion could be met.

Suppose that the condition *is* met in the gyre: what would be its implications? Consider a latitude line along which the Rossby wave propagates westward. If we consider the mean gyre flow as a zonal flow, we can apply the above results directly. Suppose the Ekman pumping is a function only of latitude. Then on the latitude line where the Ekman pumping vanishes, between the subpolar and subtropical gyres, the meridional velocity is zero, and the flow is in fact strictly zonal there. This zonal velocity is of course zero on the eastern boundary and increases in magnitude further west. Now let the baroclinic Rossby wave start its journey westward from the eastern boundary. Near the eastern boundary its speed exceeds that of the zonal flow, and the Rossby waves propagate westward, turning off the lower layer flow. At some more western longitude, $x = x_r$, the zonal speed, which increases with distance from the eastern boundary, may become large enough to satisfy (3.4.11). In this case the further westward progress of the Rossby wave is arrested, and the information that it carries about the presence of the eastern boundary is not delivered to fluid further west than the longitude $x_r$. Fluid west of this point is then sheltered from the Rossby wave that can make no further headway against the current. In regions west of this point there is then no requirement that layer 2 be at rest. The position of this critical point depends on the value of the arrest speed given by (3.4.11) and the strength of the gyre's zonal flow. The critical speed is usually written as:

$$c_R = \frac{\beta\gamma_1 H_1 H_2}{f_0^2(H_1 + H_2)} \tag{3.4.12}$$

and the notation $c_R$ is used to recall the fact that it is the ability of the current to arrest the westward propagating Rossby wave that (at least in linear theory) allows layer 2 to fend off the signal from the eastern boundary and remain in motion.

However, it is important to realize that the critical speed in (3.4.12) is *not* the speed of a freely propagating Rossby wave in a resting fluid brought to rest by a Galilean shift of the wave speed to zero by a barotropic flow. Recall that the speed $U$ in the preceding discussion is actually baroclinic and alters the potential vorticity gradient of the layers by tilting the interface between layers 2 and 3.

If we were to solve the linear problem in the absence of a mean zonal flow posed by (3.4.1a,b,c), we would find three modes of oscillation corresponding to the barotropic and baroclinic modes of the system. The barotropic solution has a westward phase speed:

$$c_0 = -\frac{\beta}{K^2} \tag{3.4.13}$$

which does correspond to the first steady arrest speed (3.4.9a).

It is a straightforward matter, and is left to the reader, to show that the speeds of the two baroclinic modes in the long wave limit are:

$$c_{1,2} = -\frac{\beta}{2}\frac{F_1 + F_2 + G_2 + G_3}{F_1G_2 + (F_1 + F_2)G_3} \pm \frac{\beta}{2}\frac{\sqrt{(F_1 + F_2 + G_2 + G_3)^2 - 4(F_1G_2 + (F_1 + F_2)G_3)}}{F_1G_2 + (F_1 + F_2)G_3}. \tag{3.4.14}$$

Only in the limit $G_3 \to 0$, $G_2 \to 0$, do these speeds approach the arrest speeds given by (3.4.9a,b). The first of these becomes infinite in the long wave limit, as does (3.4.9a). The second root, corresponding to the plus sign in (3.4.14), remains finite and approaches the baroclinic arrest speed of (3.4.9b). The critical speed that we have found, in which the lower layer is sheltered from the eastern boundary condition by the presence of the eastward Sverdrup flow, equals the speed required to arrest the slow, baroclinic, Rossby wave traveling on the interface between the upper two moving layers. The arrest speed generally differs from the speed of the free Rossby modes in a resting ocean of the same stratification due to the differing background potential vorticity gradients in the two cases. In our further discussion we refer to $c_R$ as the baroclinic Rossby wave speed, but the reader should keep in mind that this is only a short-hand term for the more complex description of the speed required to arrest the baroclinic wave on the interface by a flow limited to the upper layers of the thermocline.The arresting flow is barotropic only with respect within the upper two layers. It is baroclinic with respect to the full three-layer model, and this is why the wave speed does not satisfy a simple Galilean transformation.

## 3.5 The Quasi-Geostrophic Circulation Problem

### *Formulation*

The heuristic analysis of the previous section shows that the nature of the circulation, i.e., the distribution of the Sverdrup circulation among the density layers of the ocean, may depend on the nature of the circulation itself. The inability of the Rossby waves to breast the Sverdrup current reopens the possibility that deeper layers, not directly forced by the Ekman pumping, are in motion in regions shielded from Rossby wave arrivals from the eastern boundary. These qualitative remarks, while illuminating, do not tell us with any precision what these regions of deeper motion are and do not describe the motion in those regions. The fact that the existence of such regions of deep motion depends on the motion itself is a reflection of the fundamental nonlinearity of the circulation problem, even in the interior where the Sverdrup balance for the vertically averaged circulation is assumed to be valid. Indeed, we discuss in the introduction to this chapter some of the perplexing and very fundamental issues associated with developing a theory for the vertical structure of the ocean circulation even if the Sverdrup balance is accepted for the vertical average of the motion.

The attempts to develop a theory of the circulation of the upper ocean and its vertical structure took a great step forward with the publication by Rhines and Young of several important papers in the early years of the past decade (Rhines and Young 1982a,b; Young and Rhines 1982). Employing the quasi-geostrophic model, Rhines and Young were able to separate the problem for the full vertical structure of the density field from the problem of determining the vertical structure of the velocity field. Although the specification of the basic density field means that the theory is only a partial theory for the full vertical structure of the circulation, it illuminates very clearly the way in which the circulation itself can shape its own structure through the nonlinear interaction of the motion with the field of potential vorticity. As in the case of the Fofonoff mode described in Chapter 2, the ability of the flow to reshape its own potential vorticity isolines from latitude circles into curves which close on themselves opens up new paths of allowable motion. For a baroclinic ocean this process allows penetration of the wind-driven circulation below the upper most layer. In Chapter 4 we discuss an alternative mechanism, ventilation, that also produces deep motion. The synthesis of these two ideas forms the present basis of our conceptual understanding of the stratified wind-driven circulation.

We employ a layer model for the circulation to begin with and ignore cross-interface motion. This means that in every layer the fluid of a given density remains in that layer, and there are thus no sources of fluid entering the layer. Except for the uppermost layer no layer receives fluid from the upper Ekman layer. In each layer the fluid *recirculates* endlessly in the steady state with each fluid element repeating its round and round trajectory in the gyre an infinite number of times. This recirculatory aspect of the flow is a crucial feature of the dynamics of the model. The model confronts directly the issue of how layers which are not *directly* forced by Ekman pumping can be set into motion.

Consider the layer equations for either the two-layer [(3.3.1) and (3.3.2)] or the $2\frac{1}{2}$ layer model in the steady state [(3.3.1) and (3.3.3)] in the absence of cross-interface motion. We can write both steady-state models in a unified compact form as:

$$J(\psi_1, q_1) = J\left[\psi_1, \nabla^2\psi_1 + \beta y + F_1(\psi_2 - \psi_1)\right] = \frac{f_0}{H_1} w_E + \operatorname{curl}\Im_1 \tag{3.5.1}$$

and

$$\begin{aligned} J(\psi_2, q_2) &= J\left(\psi_2, \nabla^2\psi_2 + \beta y + F_2(\psi_1 - \psi_2) - G_2\psi_2 + \frac{f_0}{H_2} Z_b\right) \\ &= -r_2\nabla^2\psi_2 + \operatorname{curl}\Im_2. \end{aligned} \tag{3.5.2}$$

If the model is a two-layer model in which the two layers fill the ocean, $G_2$ is set to zero (although the term in $G_2$ is actually trivial since its Jacobian with $\psi_2$ is identically zero; it is retained, however, for conceptual purposes). If the model is considered to be a $2\frac{1}{2}$ layer model in which the two layers lie over a resting abyss, the interaction with the bottom is absent, and the bottom friction parameter $r_2$ and the topographic contribution, $Z_b$, should be set equal to zero instead.

Over the scale of the gyre motion, $L$, the relative vorticity can be estimated as:

$$\nabla^2\psi_n = O(\psi_n/L^2) \tag{3.5.3a}$$

while the contribution to the potential vorticity of the variation of the layer thicknesses is estimated as:

$$(F_n, G_2)\psi_n = O(\psi_n/L_d^2) \tag{3.5.3b}$$

and dominates the role of relative vorticity as long as $L^2 \gg L_d^2$. This condition is easily met, and thus in (3.5.1) and (3.5.2) the relative vorticity can be ignored. Note that in making this approximation we are assuming that the parameter $F$, defined by (3.2.7), is much greater than 1. The quasi-geostrophic theory, in which the layer variations are required to be small with respect to their average values, remains valid as long as $R_0F \ll 1$, which, as we saw in Section 3.2, is met as long as $L$ is of the order of 2000 km or less, which is still much greater than $L_d = O(50\,\text{km})$.

In this limit our equations are:

$$
\begin{aligned}
F_1 J(\psi_1, \psi_2) + \beta \frac{\partial \psi_1}{\partial x} &= \frac{f_0}{H_1} w_E + \operatorname{curl} \Im_1 \\
F_2 J(\psi_2, \psi_1) + \beta \frac{\partial \psi_2}{\partial x} &= -r_2 \nabla^2 \psi_2 + \operatorname{curl} \Im_2 - J\left(\psi_2, \frac{f_0}{H_2} Z_b\right)
\end{aligned}
\tag{3.5.4a,b}
$$

where we have used the fact that the Jacobian of any variable with itself is identically zero.

### *The Barotropic Solution*

If each of (3.5.4 a,b) is multiplied by $H_1$ and $H_2$ respectively, and the resulting equations are added together, the Jacobian terms in (3.5.4 a,b) cancel since $F_1H_1 = F_2H_2$ and $J(\psi_1, \psi_2) = -J(\psi_2, \psi_1)$ This cancellation is not fortuitous but rather a mathematical statement of the physical fact, described above, that the vertical velocity is continuous at the interfaces of the layers. Therefore the integral of the vorticity equations in the vertical, or the layer equivalent of the thickness weighted sum over the layers, eliminates the stretching of the individual layers by the internal interfaces, leaving only the stretching (or compression) at the upper surface by the Ekman pumping and the stretching (or compression) at the lower boundary by the pumping of the lower frictional boundary layer (the bottom friction term) and the topographic stretching term. The Jacobians in (3.5.4 a,b) represent the stretching effects of the internal interfaces and they cancel themselves when the barotropic mode is considered since, if one layer is lengthened by the stretching, the other layer must be shrunk in compensation.

Thus for the vertical average we have:

$$
\begin{aligned}
&\beta \frac{\partial}{\partial x}\{H_1\psi_1 + H_2\psi_2\} + J(\psi_2, f_0 Z_b) \\
&\quad = f_0 w_E + \sum_{n=1}^{2} H_n \operatorname{curl} \Im_n - r_2 H_2 \nabla^2 \psi_2
\end{aligned}
\tag{3.5.5}
$$

We discussed in Chapter 1 certain necessary conditions for the validity of the Sverdrup balance. We see them reappear here in our layer model. We have already assumed that the relative vorticity is negligible in the midocean gyre. We now assume that the direct role of friction within each layer, modeled by the term $\Im_n$, and the effect of bottom friction, proportional to $r_2$, are each negligible to the *lowest order*. At the same time we ignore the interaction with the bottom topography, either because we (unrealistically) assume in our model that the bottom is flat, or that our model is a two-layer model over a resting abyss that shelters the layers from the effect of the topography (as well as bottom friction, of course). With these assumptions the equation for the vertical average of the interior motion once again becomes the Sverdrup balance:

$$\beta \frac{\partial \psi_B}{\partial x} = \frac{f_0}{H} w_E \tag{3.5.6}$$

where:

$$\psi_B \equiv \frac{H_1 \psi_1 + H_2 \psi_2}{H} \tag{3.5.7}$$

and $H = H_1 + H_2$.

The term $\psi_B$ is the barotropic streamfunction which, when multiplied by $H$, yields the total Sverdrup transport at each geographical position. It is independent of the stratification and the layer thicknesses and yields no information about the partitioning of the flow between layers. It is this partitioning that the theory developed here is designed to explain.

Assuming that the Sverdrup flow must satisfy the boundary condition of no normal flow on the eastern boundary (3.5.6) yields $\psi_B$ as:

$$\psi_B = -\frac{f_0}{\beta H} \int_x^{x_e} w_E(x', y) dx' \tag{3.5.8}$$

so that $\psi_B$ can be considered a known function.

### *Geostrophic Contours*

With the total transport determined by the Sverdrup solution (3.5.8) we can rewrite and $\psi_1$ and $\psi_2$ as:

$$\begin{aligned} \psi_1 &= \frac{H\psi_B - H_2\psi_2}{H_1} \\ \psi_2 &= \frac{H\psi_B - H_1\psi_1}{H_2}. \end{aligned} \tag{3.5.9a,b}$$

This allows the layer equations (3.5.4a,b) to be rewritten in turn as:

$$\begin{aligned} J\left(\psi_1, \beta y + \hat{F}\psi_B\right) &= \frac{f_0}{H_1} w_E + \text{curl}\, \Im_1 \\ J\left(\psi_2, \beta y + \hat{F}\psi_B + \frac{f_0}{H_2} Z_b\right) &= -r_2 \nabla^2 \psi_2 + \text{curl}\, \Im_2 \end{aligned} \tag{3.5.10a,b}$$

where:

$$\hat{F} = \frac{f_0{}^2 H}{\gamma_1 H_1 H_2} = 1/L_d^2 \tag{3.5.11}$$

$L_d$ is the relevant deformation radius for the two-layer model.

It is important to note that the use of the known Sverdrup streamfunction renders the problem for $\psi_1$ and $\psi_2$ *linear*. The equations (3.5.10a,b) are linear in $\psi_1$ and $\psi_2$ and the solution for either one of them determines them both through their algebraic relation to $\psi_B$. Although the original problem for the vertical structure is nonlinear, the linearization of the problem comes about due to the independence of the barotropic mode from the baroclinic mode. The opposite is not true and the barotropic mode which yields the Sverdrup transport, as is clear from (3.5.10), shapes the solution for the vertical structure.

For simplicity, we consider in our examples the situation in which the topographic term $Z_b$ does not enter, i.e., either a $2\frac{1}{2}$ layer model or a two-layer model with a flat bottom.

The case of greatest interest to us is the parameter regime where, in the interior, both the bottom friction term and the term curl$\mathfrak{T}_2$ are negligible to the lowest order. They must be negligible in any case if our assumption of the validity of the Sverdrup balance is to hold. If those terms are neglected to lowest order, (3.5.10b) can be approximated as:

$$J(\psi_2, \hat{q}_2) = 0 \tag{3.5.12a}$$

where:

$$\hat{q}_2 = \beta y + \hat{F}\psi_B \tag{3.5.12b}$$

so that $\psi_2$ must be a function only of $\hat{q}_2$, i.e.:

$$\psi_2 = \Psi_2(\hat{q}_2) \tag{3.5.13}$$

where $\Psi_2(\hat{q}_2)$ is an arbitrary function of its argument.

That is, in the nondissipative limit, the streamlines in layer 2 must coincide with the isolines of the known function $\hat{q}_2$. At the same time (3.5.2) tells us that when the bottom friction and curl$\mathfrak{T}_2$ are negligible, the Jacobian of $q_2$ and $_2$ is zero. Thus the potential vorticity isolines in layer 2 must coincide with isolines of $\psi_2$, which in turn are identical to the isolines of the known function $\hat{q}_2$. When topography is ignored, the potential vorticity in layer 2 is thus given by:

$$\begin{aligned} q_2 &= \beta y + F_2(\psi_1 - \psi_2) - G_2\psi_2 \\ &= \beta y + \hat{F}\psi_B - (\hat{F} + G_2)\psi_2 \\ &= \hat{q}_2 - (\hat{F} + G_2)\psi_2 = \hat{q}_2 - (\hat{F} + G_2)\Psi_2(\hat{q}_2). \end{aligned} \tag{3.5.14}$$

Hence in the weak dissipation limit both the streamlines and the isolines of potential vorticity coincide with the isolines of the function $q_2$. These isolines are so fundamental to the dynamics that they are given a special name. They are called *geostrophic contours* because the nondissipative geostrophic flow must travel along the paths traced out by these curves. The geostrophic contours are shaped by the flow itself. In the quasi-geostrophic approximation the curves are completely determined once the Sverdrup transport is calculated. Thus the shape of the contours depends directly on the structure of the wind forcing and its strength. When the wind forcing is weak, the shape of the geostrophic contours is determined by the $\beta$ term, and the geostrophic contours coincide with latitude circles. As the strength of the wind forcing increases, the contours can become increasingly distorted from latitude circles. This effect is also favored by large values of $\hat{F}$, i.e., either by a weak stratification or by small layer thicknesses. The basic physics of this distortion of the geostrophic contours lies in the variation of the interface between layers 1 and 2 forced by the wind-driven circulation. To obtain an intuitive understanding of the shaping of the contours, note that if layer 2 were at rest $q_2$ would be equal to $\hat{q}_2$. In this case the full Sverdrup transport would be carried in the upper layer. In order for layer 2 to be at rest the interface between layer 1 and layer 2 must slope to eliminate any pressure gradient in the second layer according to (3.2.13a). The thickness of layer 2 therefore varies even if the layer is at rest. This alters the field of the potential vorticity in the second layer and distorts the geostrophic contours. Note that the gesotrophic contours obtained in this thought experiment in which layer 2 is at rest are the same contours that would obtain if layer 2 is in motion. In this more general case $q_2$ differs from $\hat{q}_2$ although their isolines coincide by (3.5.14).

The geostrophic contours are also the characteristic curves of the linear partial differential equation (3.5.10). Information is propagated along the characteristic curves. In particular the information concerning the eastern boundary condition of no normal flow is carried along the geostrophic contours to points within the basin.

Since the Sverdrup streamfunction (3.5.8) always vanishes at the eastern boundary, the geostrophic contours near the eastern boundary of the ocean become identical with latitude circles. Hence a geostrophic contour emanates from each point on the eastern boundary and initially trends westward along a latitude circle (see Fig. 3.5.1). If the wind forcing is strong enough, the curves become increasingly distorted and the possibility arises that the curves become bent around certain regions. If the forcing is strong enough, the bending can become so great that the geostrophic contours close on themselves, as shown in the figure, isolating the enclosed regions from the eastern boundary. Into those regions girdled by a geostrophic contour, information about the eastern boundary is unable to penetrate.

Since $\psi_2$ is constant along the geostrophic contours, the lower layer streamfunction must be zero on all the geostrophic contours that strike the eastern boundary where $\psi_2$ is zero. Hence along all contours such as the one labeled A in Fig. 3.5.1 there can be no motion in the nondissipative limit. The flow is *blocked* on all such contours and in such regions the entire Sverdrup transport is carried in the upper layer. In the island regions or *pools*, such as the area marked $P$ in the figure, the contours do not strike the eastern boundary,

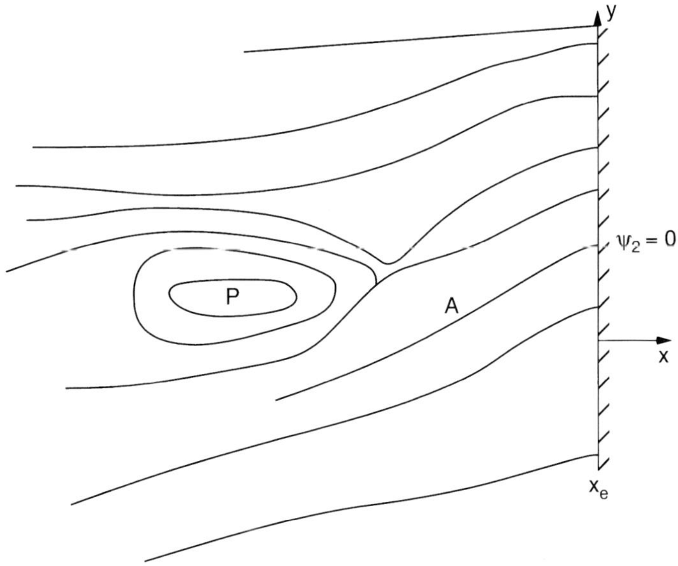


**Fig. 3.5.1.** Schematic presentation of the geostrophic contours emanting from the eastern boundary. Each point on the eastern boundary is a starting point for a geostrophic contour which threads into the basin. Not all regions need be covered by the contours emanating from the eastern boundary. If the contours are sufficiently distorted by the Ekman pumping, islands of closed contours may appear that are not connected to the eastern boundary

and motion in layer 2 is possible. In the low dissipation limit it must consist of a recirculation around the closed geostrophic contours. The only constraint on such motion in the inviscid limit is that the streamlines must coincide with the geostrophic contours, i.e., as described by (3.5.13), and that the potential vorticity in layer 2, $q_2$, must be folded around on itself so that it too coincides with the geostrophic contours. The motion is not otherwise determined by these nondissipative considerations, and were dissipation neglected completely, it would still be *consistent* to have a solution in which the lower layer remains at rest everywhere, i.e., in the pool regions as well as on the blocked contours since a resting fluid satisfies (3.5.12a). As is seen below, it is the weak but persistent effect of dissipation acting inexorably on the regions of recirculation around closed geostrophic contours which determines the $O(1)$ motion in the pool regions.

## 3.6 A Midocean Example

A particularly illuminating example of the ideas discussed in the previous section has been described by Rhines and Young (1982a).

The net Sverdrup transport across the basin at any latitude is proportional to the integral of the Ekman pumping velocity along each latitude circle which spans the basin. If that integral differs from zero, a western boundary current is required to return the Sverdrup transport along the western boundary. We saw in Chapter 2 that there are still unresolved dynamical issues associated with the closure of the circulation in the low dissipation limit. In order to focus attention on the physics by which lower layers of the thermocline can be set into motion when the geostrophic contours close on themselves and avoid the eastern boundary, Rhines and Young considered an Ekman pumping velocity that is antisymmetric in longitude about a midpoint longitude, which we choose to be $x = 0$, so that the integral of the Ekman pumping in $x$ across the basin is zero for all $y$, obviating the need for a western boundary current.

In particular, consider the example in which the region of Ekman pumping is limited to the circular region of radius $r_1$, as shown in Fig. 3.6.1. In particular, the Ekman pumping velocity is chosen to be:

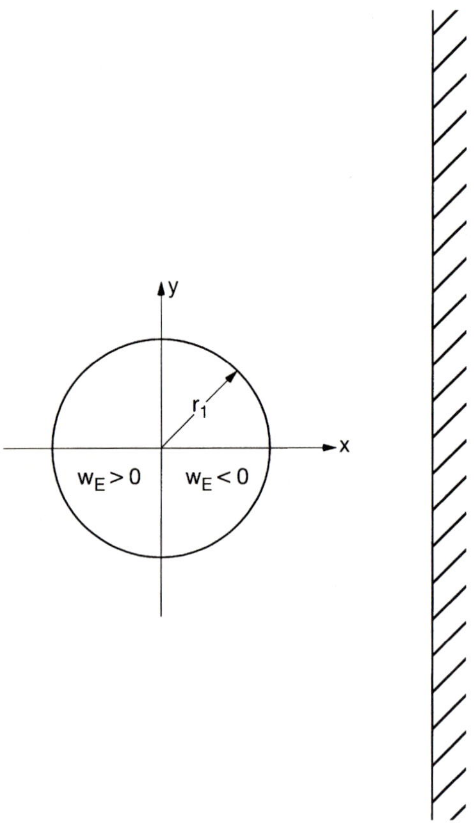


**Fig. 3.6.1.** The circular region or radius $r_1$ in which the Ekman pumping differs from zero in the example described in Section 3.6. The origin of the cordinates is at the center of the circle and $w_E = -\alpha x$ within the circle and is zero outside the circle

$$\begin{aligned} w_E &= -\alpha x \quad r < r_1 \\ &= 0 \qquad r > r_1 \end{aligned} \tag{3.6.1}$$

where $r = \sqrt{x^2 + y^2}$. We can think of $\alpha$ as being the amplitude of the Ekman pumping, $W_0$, divided by the disk's radius, $r_1$, so that within the "disk" of forcing, $w_E = -W_0(x/r_1)$.

The Sverdrup streamfunction is given by (3.5.8). The integral is zero for all $x$ east of the eastern edge of the circle $r = r_1$ and for all values of $|y| > r_1$. Within the latitude limits $-r_1 \le y \le r_1$ the Sverdrup solution is thus given by the integral from the point $x$ to the eastern rim of the circle at:

$$X(y) = \sqrt{r_1^2 - y^2}. \tag{3.6.2}$$

Thus:

$$\psi_B = \frac{f_0 \alpha}{\beta H} \int_x^{X(y)} x' dx' \tag{3.6.3}$$

or:

$$\begin{aligned} \psi_B &= \frac{\alpha f_0}{2\beta H} \left[ X(y)^2 - x^2 \right] \\ &= \frac{\alpha f_0}{2\beta H} \left[ r_1^2 - x^2 - y^2 \right] \quad r < r_1 \\ &= 0 \quad r > r_1. \end{aligned} \tag{3.6.4}$$

It is important to note that because of the antisymmetric form of the Ekman pumping the Sverdrup transport is equal to zero outside the circle of forcing and yields a barotropic flow with circular streamlines within the disk of radius $r_1$.

Now that $\psi_B$ is determined, the geostrophic contours can be easily calculated. Within the disk of radius $r_1$ they are given by the contours of:

$$\begin{aligned} \hat{q}_2 &= \beta y + \frac{\alpha f_0 \hat{F}}{2\beta H} \left[ r_1^2 - x^2 - y^2 \right] \quad r < r_1 \\ &= \beta y \quad r > r_1. \end{aligned} \tag{3.6.5}$$

Inside the disk the geostrophic contours are circles centered on $x = 0$, $y = y_0$, where:

$$y_0 = \frac{\beta^2 H}{\alpha f_0 \hat{F}} \tag{3.6.6}$$

i.e., inside the disk:

$$\hat{q}_2 = \frac{\beta}{2y_0}\left[r_1^2 + y_0^2 - x^2 - (y - y_0)^2\right] \quad r < r_1. \tag{3.6.7}$$

The isolines of $\hat{q}_2$, which within the disk are arcs of circles centered on $y = y_0$, are lines of constant latitude outside the disk. For weak forcing (small $\alpha$) $y_0$ is greater than $r_1$, and the geostrophic contours remain open. Each circular arc within the disk connects to a geostrophic contours outside the disk, which is a line of constant $y$. Each contour then strikes the eastern boundary, and the motion on the contour is therefore blocked, and layer 2 remains at rest. Figure 3.6.2 shows the contours for such a weak forcing case where $y_0/r_1$ is 2. The contours of $\hat{q}_2$, on which $\psi_2$ is constant, are only slightly distorted within the region of forcing, and every point within this region is connected by a geostrophic contour to a point on the eastern boundary where the streamfunction in layer 2 vanishes. Thus in this case, within the whole region of forcing, the Sverdrup transport is carried in the upper layer, and the lower layer is at rest. In the language of Section 3.4, Rossby waves are able to propagate from the eastern boundary and cover the entire lower layer with a signal which turns off the motion in the layer.

When the forcing is increased sufficiently that $y_0 < r_1$, the center of the circles on which $\hat{q}_2$ is constant lies within the disk of forcing. In this case there are isolines of $\hat{q}_2$ which close on themselves within the region of forcing and are disconnected from the geostrophic contours which emanate from the eastern boundary. The largest such circle within the forcing region just brushes the boundary of the disk at $x = 0$, $y = r_1$. From (3.6.5) this implies that $\hat{q}_2$ is equal to $\beta r_1$ on this outer circle of the zone of isolated geostrophic contours. Thus the outer boundary of the pool of detached $\hat{q}_2$ contours is determined by setting $\hat{q}_2 = \beta r_1$ in (3.6.7), so that its boundary is given by:

$$x^2 + (y - y_0)^2 = (r_1 - y_0)^2. \tag{3.6.8}$$

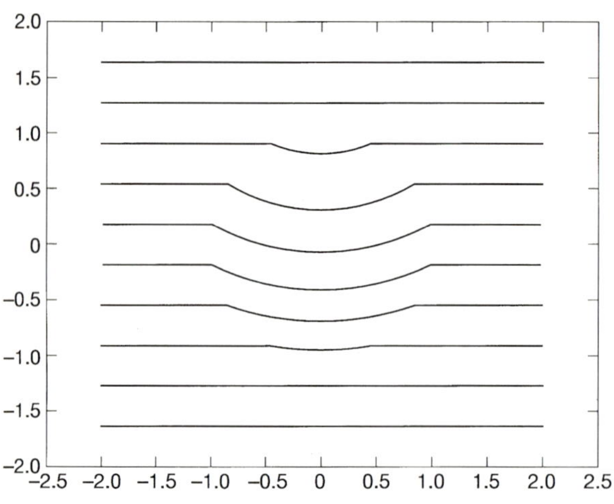


**Fig. 3.6.2.** Isolines of $\hat{q}_2$ which are the geostrophic contours for the forcing described in the text. In the figure $y_0/r_1 = 2$ so that all the geostrophic contours are open and strike the eastern boundary

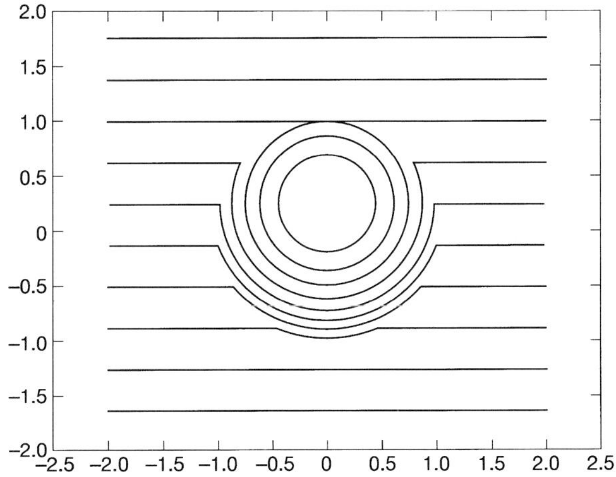


**Fig. 3.6.3.** As in Fig. 3.6.2 but now $y_0/r_1 = 0.25$. The domain of closed geostrophic contours, isolated from the eastern boundary, is clearly seen

Figure 3.6.3 shows the case in which $y_0/r_1 = 1/3$. The region of closed, circular geostrophic contours is clearly observable, and the outermost circle is given by (3.6.8). Outside this circle, which is offset northward from the center of the center of the circular disk of the forcing, the lower layer streamfunction must vanish since in the outer region all geostrophic contours strike the eastern boundary. Inside the enclosed pool the fluid in layer 2 can be in motion. As the forcing increases, the value of $y_0$ decreases until finally, as $y_0$ tends to zero, the region of closed contours coincides with the region of forcing. In this limit the entire lower layer within the forcing region can be set into motion. We have not yet determined what the motion is, only that the forcing has opened up a domain in the lower layer where motion is allowed in the same low dissipation limit that is consistent with the Sverdrup balance for the interior.

The critical value of the forcing which is required in order to have closed geostrophic contours is determined by the condition that $y_0 \le r_1$. Using (3.6.6), this is equivalent to the condition that:

$$\alpha = \frac{W_0}{r_1} \ge \frac{\beta^2 H}{f_0 \hat{F} r_1}$$

or equivalently that:

$$W_0 \ge \frac{\beta^2 L_d^2 H}{f_0}. \tag{3.6.9}$$

Now $W_0$ is the characteristic value of the Ekman vertical velocity. From the Sverdrup balance the characteristic horizontal velocity is related to the Ekman velocity by dividing by the factor $\beta H/f_0$. Thus the condition (3.6.9) is equivalent to the condition that the characteristic horizontal velocity exceeds $\beta L_d^2$, which as we saw in Section 3.4 is required to arrest the progress of the westward propagating baroclinic Rossby wave.

If we choose $f_0 = 10^{-4}\mathrm{s}^{-1}, \beta = 10^{-13}\mathrm{cm}^{-1}\mathrm{s}^{-1}$, H = 1 km, and $L_d = 40$ km, the critical value of $W_0$ is $1.6 \times 10^{-4}$cm/s, well within the range of realistic values of $w_E$. The example thus suggests that the wind forcing present in the Sverdrup interior may be strong enough to wrap the geostrophic contours around on themselves and detach them form the eastern boundary, at least in a part of the region, and allow motion in layers deeper than the one directly forced by the Ekman pumping. What remains to be seen is what that region is for more realistic forcing in which the Sverdrup transport requires a western boundary layer, and how the motion, if it exists within the domain of closed geostrophic contours, can be determined.

## 3.7 Determination of the Recirculation

When the motion is nearly inviscid, the local dynamics is determined by the conservation of potential vorticity in layer 2. When the geostrophic contours are attached to the eastern boundary, free geostrophic motion is blocked, and only weak motion directly forced by the frictional coupling to the upper layer takes place. In the weak dissipation limit this motion is negligible. However, in the region of closed geostrophic contours a free geostrophic, recirculating mode is possible. In the inviscid limit a recirculating motion of arbitrary amplitude is possible as long as the streamlines coincide with the geostrophic contours. As long as:

$$\psi_2 = \Psi_2(\hat{q}_2) \tag{3.7.1}$$

the requirements of inviscid dynamics are consistently satisfied. The paths of the motion have been laid down by the formation of the closed geostrophic contours, and only a weak forcing, acting persistently over time, is required to set the fluid in motion and to determine both its amplitude and detailed form subject, always, to (3.7.1). That weak forcing comes from the weak frictional coupling of layer 2 to layer 1 through viscous stresses. We can think of the motion in layer 2 as the resonance of a free, nonlinear, steady mode and its resonant character therefore requires only a tiny forcing to produce an order 1 motion. We normally think of a resonance producing a large response to an $O(1)$ forcing. In the oceanic recirculation problem it is instead the production of an $O(1)$ motion by infinitesimally small forcing that is relevant.

The mathematical expression of this comes from the integral of the vorticity equation over a *closed* contour of constant $\hat{q}_2$. We may rewrite (3.5.10b) as:

$$J(\psi_2, \hat{q}_2) = -r_2\nabla^2\psi_2 + \mathrm{curl}\,\Im_2. \tag{3.7.2}$$

The dominant term in (3.7.2) is locally the inertial term on the left side. The condition that it must self cancel to within the order of the small dissipative

**Fig. 3.7.1.** Area A enclosed by the geostrophic contour $C$ over which the potential vorticity equation in layer 2 is integrated

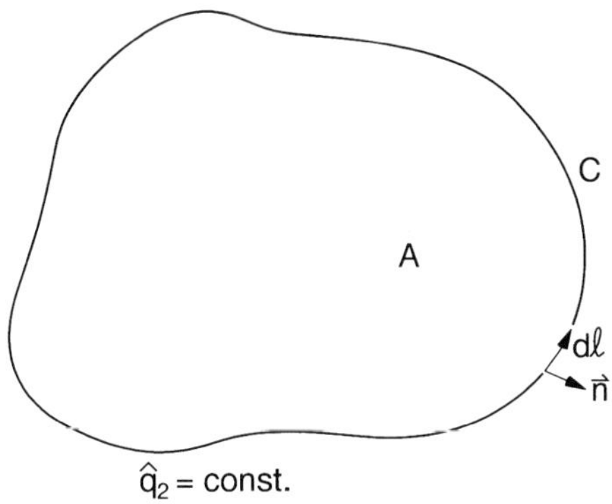


effects on the right side of the equation is what gives rise to (3.7.1) as a local balance. Consider, however, the area $A$, enclosed by the contour $C$ defined by the curve $\hat{q}_2 = \text{constant}$ and shown in Fig. 3.7.1. We integrate (3.7.2) over that area and note that:

$$\begin{aligned}\iint_A J(\psi_2, \hat{q}_2)dx\,dy - \iint_A \vec{u}_2 \cdot \nabla\hat{q}_2\,dx\,dy &= \iint_A \nabla \cdot \vec{u}_2\hat{q}_2\,dx\,dy \\ &= \oint_C \vec{u}_2\hat{q}_2 \cdot \vec{n}\,d\ell = \hat{q}_2 \oint_C \vec{u}_2 \cdot \vec{n}\,d\ell = 0 \end{aligned} \tag{3.7.3}$$

where $\hat{q}_2$ has been brought outside the line integral in the last step in (3.7.3) since by definition it is constant on $C$. The last integral is zero by mass conservation (or equivalently by the use of the geostrophic stream function to evaluate the integral). The locally dominant inertial advective effects integrate to zero in a region bounded by a closed curve across which there is no net transport of potential vorticity. Indeed, in the inviscid limit in which streamlines and geostrophic contours coincide, $C$ is also a streamline, and there is no flow across $C$ to the lowest order. Although inertial effects may dominate locally, the integrated balance over A reduces to a condition on the dissipative sources and sinks of potential vorticity in layer 2. This is exactly analogous to the dissipation integral constraints discussed in Chapter 2.

Thus, the area integral of (3.7.2) becomes:

$$\begin{aligned} 0 &= -r_2 \iint_A \nabla^2\psi_2\,dx\,dy + \iint_A \operatorname{curl}\Im_2\,dx\,dy \qquad \text{or} \\ 0 &= -r_2 \oint_C \vec{u}_2 \cdot \vec{t}\,d\ell + \oint_C \Im_2 \cdot \vec{t}\,d\ell. \end{aligned} \tag{3.7.4}$$

$\hat{t}$ is the unit tangent vector to $C$.

This constraint is valid for arbitrary levels of dissipation. When the dissipation is small, however, (3.7.1) is also valid everywhere *locally* on $C$. It is the integral constraint (3.7.4) that determines the otherwise arbitrary relationship between $\psi_2$ and $\hat{q}_2$.

It is useful to consider a concrete example by specifying the form of $\Im_2$. We can think of $\Im_2$ as representing the effects of unresolved eddies coupling layer 2 with layer 1, for example. One of the simplest representations of this coupling might be written as:

$$\Im_2 = -A_2(\vec{u}_2 - \vec{u}_1). \tag{3.7.5}$$

In this form the coupling is represented as a simple drag law proportional to the velocity difference between the two layers. If this form is used in (3.7.4) we obtain:

$$0 = -r_2 \oint_C \vec{u}_2 \cdot \vec{t}\, d\ell + A_2 \oint_C (\vec{u}_1 - \vec{u}_2) \cdot \vec{t}\, d\ell \tag{3.7.6}$$

or

$$A_2 \oint_C \vec{u}_1 \cdot \vec{t}\, d\ell = (r_2 + A_2) \oint_C \vec{u}_2 \cdot \vec{t}\, d\ell. \tag{3.7.7}$$

We can use (3.5.9a) to write the $\vec{u}_1$ in terms of the barotropic velocity, obtained from $\psi_B$ and $\vec{u}_2$, i.e.:

$$\vec{u}_1 = \frac{H}{H_1}\vec{u}_B - \frac{H_2}{H_1}\vec{u}_2 \tag{3.7.8}$$

where:

$$\vec{u}_B = \hat{k} \times \nabla\psi_B \tag{3.7.9}$$

from which it follows that:

$$\oint_C \vec{u}_2 \cdot \vec{t}\, d\ell = \left[\frac{A_2 H/H_1}{r_2 + A_2(1 + H_2/H_1)}\right] \oint_C \vec{u}_B \cdot \vec{t}\, d\ell. \tag{3.7.10}$$

This relation is valid for arbitrary values of $r_2$ and $A_2$. However, if the dissipative effects are small, (3.7.1) applies and can be used to relate the integral of $\vec{u}_2$ to the integral of the barotropic velocity, i.e.:

$$\begin{aligned}\oint_C \vec{u}_2 \cdot \vec{t}\, d\ell &= \oint_C (\hat{k} \times \nabla\psi_2) \cdot \vec{t}\, d\ell \\ &= \oint_C (\hat{k} \times \nabla\Psi_2) \cdot \vec{t}\, d\ell = \oint_C (\hat{k} \times \nabla\hat{q}_2)\frac{d\Psi_2}{d\hat{q}_2} \cdot \vec{t}\, d\ell.\end{aligned} \tag{3.7.11}$$

Now since $\Psi_2$ is a function only of $\hat{q}_2$, its derivative with respect to its argument is also a function only of $\hat{q}_2$ and is constant on the curve $C$ and can therefore be moved outside the integral. Since $\hat{q}_2 = \hat{F}\psi_B + \beta y$, it follows that the last integral in (3.7.11) can be written:

$$\frac{d\Psi_2}{d\hat{q}_2}\left[\oint_C \left\{\hat{F}\vec{u}_B\cdot\vec{t}\,d\ell\right\} - \oint_C \left\{\beta\hat{i}\cdot\vec{t}\,d\ell\right\}\right] = \frac{d\Psi_2}{d\hat{q}_2}\oint_C \left\{\hat{F}\vec{u}_B\cdot\vec{t}\,d\ell\right\} \tag{3.7.12}$$

since the integral around the closed curve $C$ of the dot product of the unit vector in the $x$ direction, $\hat{i}$, with the tangent vector to $C$ is zero.

When (3.7.11) and (3.7.12) are used in (3.7.10) we obtain, as long as the circulation of the barotropic velocity around $C$ differs from zero (and we note again that this is a known quantity):

$$\frac{d\Psi_2}{d\hat{q}_2} = \frac{A_2}{\hat{F}\left\{r_2\frac{H_1}{H} + A_2\right\}}. \tag{3.7.13}$$

It is important to note that in this case the relation between $\psi_2$ and $\hat{q}_2$ is linear and depends only on the dissipation parameters. Thus integrating (3.7.13), we obtain:

$$\psi_2 = \Psi_2(\hat{q}_2) = \frac{A_2}{r_2\frac{H_1}{H} + A_2}\left\{\frac{\hat{q}_2(x,y)}{\hat{F}} - \frac{\hat{q}_{2_0}}{\hat{F}}\right\} \tag{3.7.14}$$

where $\hat{q}_{2_0}$ is the constant value of $\hat{q}_2$ on the outermost closed geostrophic contour. On this contour the streamfunction in layer 2 must vanish by continuity with the resting region outside the pool of closed $\hat{q}_2$ contours. Thus the lower layer motion is now completely determined. The process has occurred in two steps. First, the inviscid dynamics shapes the geostrophic contours which determines the region in which motion in layer 2 can occur. The paths within that region on which the recirculation can take place are identified as the isolines of $\hat{q}_2$, i.e., the geostrophic contours. Then weak dissipation, acting on the free geostrophic mode, forces it into motion with an amplitude which is determined by a balance between the weak frictional coupling and dissipation.

There are two particularly interesting limits of (3.7.14). If bottom friction is very strong compared to the coupling with layer 1, i.e., if the ratio, $r_2H_1/A_2H$, is very large, the streamfunction in the lower layer is expunged by the bottom friction. We can think of this as a strong coupling with either a solid boundary at the bottom or as a strong frictional coupling with a very deep and nearly motionless deeper layer. Then, even though the geostrophic contours allow motion, it is dissipated by bottom friction faster than it can be generated by the weak coupling with the directly wind-driven layer above.

On the other hand, if the bottom friction is relatively small so that $r_2H_1/A_2H \ll 1$, the amplitude of the streamfunction in layer 2 becomes independent of the value of the mixing coefficient $A_2$.

In the case of the example in Section 3.8 the outermost closed streamline in the lower layer just touches the outer circle of the region of forcing at the point

$(x, y) = (0, r_1)$. At this point the barotropic streamfunction is zero and $\beta y = \beta r_1$ so that on this outer most curve $\hat{q}_2 = \hat{q}_{2_0} = \beta r_1$, and (3.7.14) becomes:

$$\psi_2 = \frac{A_2}{A_2 + r_2 \dfrac{H_1}{H}} \left\{ \psi_B + \beta \frac{y - r_1}{\hat{F}} \right\}. \tag{3.7.15}$$

Once $\psi_2$ is known $\psi_1$ is determined from (3.5.9a).

Figure 3.7.2 shows the streamfunctions in the upper and lower layers for the example of the disk forcing in the limit of negligible bottom friction and thus when the form of the circulation is independent of the magnitude of the frictional parameters. Panel a shows the lower layer streamfunction for the case $r_1/y_0 = 4$. The circulation in layer 2 is limited to the zone of closed geostrophic contours and its streamfunction is given by (3.7.15) with $r_2 = 0$ and with $\psi_B$ given by (3.6.4). Panel b shows the circulation in the upper layer. Outside the region of closed $\hat{q}_2$ contours the upper layer flow coincides with the streamlines of the Sverdrup transport, and therefore the upper layer streamlines are portions of circles centered at $x = y = 0$. Above the region of closed $\hat{q}_2$ upper layer shares the total Sverdrup transport with the layer 2. The meridional velocity in the upper layer is therefore less and the zonal velocity somewhat larger. The streamlines thus "flatten" in this region, as can be seen in the figure. At the boundary of the region there is a kink in the upper layer streamlines. In order that the mass flux in the upper layer across the pool region be continuous, the velocity normal to the pool boundary must be continuous. This implies that the magnitude of the upper layer streamfunction is continuous at the pool boundary. The tangential velocity in the upper layer is not continuous. This follows from the fact that the tangential velocity in the lower layer is discontinuous at the pool boundary, and that the tangential velocity of the upper layer is determined by (3.7.8). That is, the Sverdrup velocity is everywhere continuous so that if any component of the velocity in one layer is discontinuous, the same component in the other layer must have a compensating discontinuity. This implies that the first derivative of $\psi_1$ normal to the pool boundary is discontinuous, and this shows up as a kink in the streamlines in Fig. 3.7.2b.

The potential vorticity in the pool region in layer 2 can now be directly determined from (3.5.14). For the form of the dissipation that we have chosen (3.7.5) it follows that:

$$\begin{aligned} q_2 = {} & \hat{q}_2(x, y) \left[ \frac{r_2 H_1/H - G_2 A_2/\hat{F}}{A_2 + r_2 H_1/H} \right] \\ & + \hat{q}_{2_0} \left( 1 + \frac{G_2}{\hat{F}} \right) \frac{A_2}{A_2 + r_2 H_1/H}. \end{aligned} \tag{3.7.16}$$

In the limit where bottom friction is unimportant ($r_2 \to 0$), and where the density jump across the interface between layers 2 and 3 is much larger than the

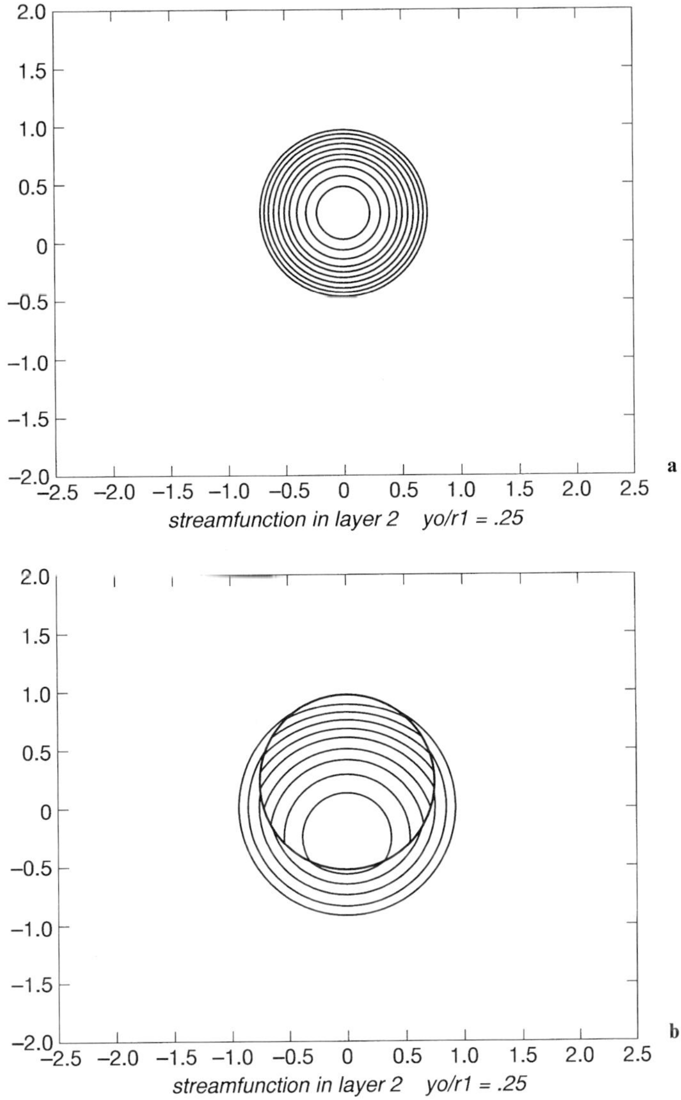


**Fig. 3.7.2a,b.** Streamlines in the $2\frac{1}{2}$ layer model driven by the disk-shaped Ekman pumping discussed in the text. For the case $r_1/y_0 = 4$, **a** streamlines in layer 2. **b** Streamlines in layer 1. Note the kink in the upper layer streamlines at the boundary of the region (*heavy line*) in which the lower layer is in motion

density jump across layers 1 and 2 so that $G_2 \ll \hat{F}$, the potential vorticity in layer 2 becomes equal to a constant everywhere within the pool of motion in the layer, i.e.:

$$q_2 = \hat{q}_{2_0} \tag{3.7.17}$$

which is the potential vorticity of the outermost geostrophic contour. For example, in the disk problem of the last section $q_2$ would be equal to $\beta r_1$ everywhere in the pool. In this limit the potential vorticity in the pool region is rendered uniform by the action of dissipation. As Rhines and Young put it, the potential vorticity in the pool becomes *homogenized*. It is important to note that outside the pool region the potential vorticity in layer 2, where the fluid is motionless, possesses strong gradients on the order of the planetary vorticity gradient.

The sensitive relationship of the potential vorticity in layer 2 to the form of the dissipation is an interesting and important one. In the example discussed we chose a form of dissipation (3.7.5) in a rather arbitrary manner. Other choices are equally, if not more, plausible. For example, in the $2\frac{1}{2}$ layer model we might include a frictional drag law for layer 2 which includes a drag with a deep nearly motionless layer beneath layer 2 as well as a frictional coupling with layer 1. That is, we might write $\Im_2$ as:

$$\Im_2 = -A_2\left[(\vec{u}_2 - \vec{u}_1) + \frac{\gamma_1}{\gamma_2}\vec{u}_2\right] \tag{3.7.18}$$

where the first term represents the interaction with layer 1 and the second the drag with an inert layer 3. The ratio of the density jumps at the upper and lower interfaces has been added and might be argued to weigh the drag law such that the momentum exchange with layer 1 or layer 3 is enhanced by a weaker density difference between the layers across the relevant interface. In fact, the drag law (3.7.18) is chosen since it leads to a dissipative term in the potential vorticity equation for layer 2 in the form:

$$\begin{aligned} \operatorname{curl} \Im_2 &= A_2\left\{\nabla^2(\psi_1 - \psi_2) - \frac{G_2}{F_2}\nabla^2\psi_2\right\} \\ &= A_2\nabla^2 q_2/F_2. \end{aligned} \tag{3.7.19}$$

In this parameterization of the mixing term the dissipation of potential vorticity enters as a lateral diffusion of potential vorticity. It is left to the reader to show that, reproducing the steps leading to (3.7.13), (3.7.14) and (3.7.16) that we have now, with this representation of mixing:

$$\frac{d\Psi_2}{d\hat{q}_2} = \frac{A_2 H/H_1}{\hat{F}\left[r_2 + A_2\left(\frac{H}{H_1} + \frac{G_2}{F_2}\right)\right]} \tag{3.7.20}$$

and,

$$\psi_2 = \frac{A_2}{\left[r_2\frac{H_1}{H} + A_2\left(1 + \frac{G_2}{\hat{F}}\right)\right]}\left(\frac{\hat{q}_2}{\hat{F}} - \frac{\hat{q}_{2_0}}{\hat{F}}\right). \tag{3.7.21}$$

Compared with (3.7.14) this represents only a slight change in determining the streamfunction in layer 2. The potential vorticity in layer 2 is determined by (3.5.14) and (3.7.21), i.e.:

$$\begin{aligned} q_2 = &\left(\hat{F}\psi_B + \beta y\right)\left[\frac{r_2 H_1/H}{r_2\frac{H_1}{H} + A_2\left(1 + \frac{G_2}{\hat{F}}\right)}\right] \\ &+ A_2 \frac{(1 + G_2/\hat{F})}{r_2\frac{H_1}{H} + A_2\left(1 + G_2/\hat{F}\right)}\hat{q}_{2_0} \ . \end{aligned} \tag{3.7.22}$$

For the $2\frac{1}{2}$ layer model, where $r_2$ is set equal to zero, the potential vorticity in the pool region in layer becomes exactly equal to the *constant* $\hat{q}_{2_0}$ which is the potential vorticity on the bounding streamline of the pool in the lower layer. This is a general result which follows from the diffusion law for potential vorticity (3.7.19).

## 3.8 Homogenization of Potential Vorticity

When we consider layers that are shielded from the direct forcing of the Ekman pumping, the conservation of potential vorticity in such layers is upset only by the presence, in the equation for the potential vorticity, of dissipative terms on the right side which include [see (3.2.29) and (3.2.34)] an unspecified sink of potential vorticity curl $\Im_n$. There are, in addition, possible nonadiabatic effects represented by the cross-interface velocities $w*$ and, for a layer in contact with the bottom, a frictional interaction with the bottom. If the motion preserves density to the lowest order and if the layer is not in contact with the bottom, the sink (or conceivably a source) term curl $\Im_n$ is the only agent that acts on the layer's potential vorticity. Although it may be a small effect locally compared with the advective terms in the potential vorticity equation, we saw in section (3.7) that when integrated over the area bounded by a closed streamline it is the only term that survives since the advection of potential vorticity across a streamline is zero.

In the low dissipation limit where the potential vorticity is, to first approximation, conserved along streamlines the functional relation between the potential vorticity and the streamlines is determined, as we have seen in the previous example, by the cumulative, integrated effects of the dissipation. The potential vorticity and streamfunction in these layers, which are without direct wind forcing, are both constant on geostrophic contours, but the determination of the amplitude of each of these functions on each geostrophic contour is fixed by the long-term action of dissipation.

We saw in the previous section that when the dissipation of potential vorticity can be written as a diffusion of the potential vorticity itself, the potential vorticity in the pool region becomes a constant. We can see this directly without detailed calculation. The proof is given by the Prandtl-Batchelor theorem (Batchelor 1956), whose outlines we give here.

Suppose curl $\Im_n$ can be written as:

$$\text{curl}\,\Im_n = \nabla \cdot (\kappa \nabla q_n) \tag{3.8.1}$$

where we let the diffusion coefficient, $\kappa$, be a function of position. We discuss below the motivation for such a law for potential vorticity dissipation, but for now we focus on the consequences of (3.8.1).

Then in any layer $n$, where (3.8.1) applies and is the only source or sink of potential vorticity, the integral of the potential vorticity equation over an area bounded by a closed streamline yields:

$$\iint_{A\psi} \vec{u}_n \cdot \nabla q_n dx\,dy = 0 = \iint_{A\psi} \nabla \cdot k \nabla q_n dx\,dy. \tag{3.8.2}$$

The divergence theorem applied to the right side of (3.8.2) yields:

$$0 = \oint_{C\psi} \kappa \nabla q_n \cdot \vec{n}\,d\ell \tag{3.8.3}$$

where $C\psi$ is the closed contour coincident with a streamline in layer $n$. The result (3.8.3) is general as long as the diffusion of potential vorticity is the only nonadvective influence on the potential vorticity. The net diffusive flux out of the area $A\psi$ must be zero since there are no other sources of $q_n$ in the region. This provides an *integral* constraint on the potential vorticity. If, however, the flow is nearly without dissipation the *local* balance is one in which $q_n$ is conserved and therefore constant on streamlines, i.e.:

$$q_n = Q_n(\psi_n). \tag{3.8.4}$$

Although (3.8.4) is a local condition, it is supposed to be valid everywhere on $C_\psi$ and thus can be used to evaluate (3.8.3). Thus:

$$\frac{dQ_n}{d\psi_n}\oint_{C\psi}\kappa\vec{u}_n\cdot\vec{t}d\ell=0 \tag{3.8.5}$$

Note that we take $dQ_n/d\psi_n$ outside the integral since it is a function only of $\psi_n$ and hence constant on $C\psi$. As long as $\kappa$ is positive, the integral in (3.8.5) differs from zero and so for all closed streamlines in layer $n$ on which there is motion:

$$\frac{dQ_n}{d\psi_n}=0. \tag{3.8.6}$$

Thus, there is no variation of the potential vorticity from streamline to streamline within the region of motion. The potential vorticity is constant and equal to the potential vorticity, $q_{n0}$, on the outermost streamline which bounds the pool of constant potential vorticity.

The sequence of events leading to the homogenization of $q_n$ is shown in Fig. 3.8.1. The streamline bounding the domain has uniform potential vorticity since in the nondissipative limit the potential vorticity must be constant on each streamline. Rhines and Young (1983) estimate the time required to establish this first step of the process, and the reader is referred to their paper for details. As long as the Peclet number, $\text{Pe}=UL/\kappa$, is large, the time required, while longer than the advective scale, $L/U$, is still much less than the diffusion time $L^2/\kappa$. Indeed, Rhines and Young show that the ratio of this establishment time to the diffusion time is $O(\text{Pe}^{-2/3})$. $L$ and $U$ are chosen as characteristic of the Sverdrup interior circulation. Thus in this first step the flow itself encircles a region, coincident with a closed geostrophic contour, and wraps it with an isoline of potential vorticity. Having established that the boundary of this region is a line of constant $q_n$, diffusion then slowly acts to diffuse this constant boundary value into the whole area bounded by the contour. This second process requires the diffusion time, $L^2/\kappa$, to complete.

It is important that the diffusion coefficient be *small* for the potential vorticity to be homogenized by this process. Only if $\kappa$ is small (more precisely,

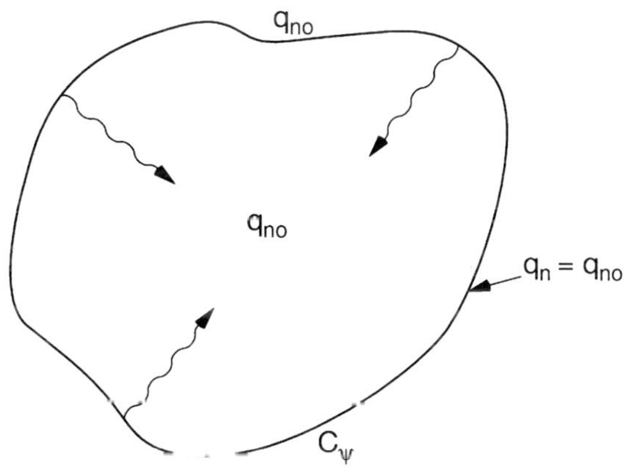


**Fig. 3.8.1.** Schematic presentation of the process of potential vorticity homogenization. A closed streamline on which the potential vorticity is nearly constant as a consequence of potential vorticity conservation slowly diffuses that value into the area that it encircles, eventually rendering the potential vorticity uniform everywhere within the region

only if Pe $\gg$ 1) is the flow able to establish the boundary of the pool as a line of constant $q_n$ which is then diffused inwards establishing a uniform internal value. If the diffusion coefficient is large, and diffusion dominates advection, the generally nonuniform values of $q_n$ on the boundary of the gyre would be diffused inwards. Diffusion would control the potential vorticity, but $q_n$ would not be uniform. If $\kappa$ is a function of position, it is necessary only that it be small on the *bounding* contour so that a uniform value of potential vorticity can be established there by advective wrapping of the $q_n$ contours. Large values of $\kappa$ interior to the pool's boundary would then shorten the homogneization period. Thus we need only require that the mixing of vorticity be small on the boundary established by the outermost of the closed geostrophic contours.

In the $2\frac{1}{2}$ layer model we can use (3.5.14) to write the advection of potential vorticity equation in layer 2 entirely in terms of the known function $\hat{q}_2$ since:

$$\psi_2 = \frac{\hat{q}_2 - q_2}{\left(\hat{F} + G_2\right)}. \tag{3.8.7}$$

The part of $\psi_2$ that is proportional to $q_2$ is incapable of advecting $q_2$ since it is aligned with the $q_2$ field. As Rhines and Schopp (1991) point out, this reduces the potential vorticity equation to the equation for a passive scalar, $q_2$, advected by a known and unchanging field of motion, i.e., with (3.8.1) and (3.8.7) the potential vorticity for layer 2 becomes the *linear* equation:

$$\frac{\partial q_2}{\partial t} + \frac{1}{(\hat{F} + G_2)} J(\hat{q}_2, q_2) = \nabla \cdot \kappa \nabla q_2 \tag{3.8.8}$$

where we have reintroduced the time derivative to emphasize the fact that during the entire baroclinic evolution of the $q_2$ field the potential vorticity of layer 2 is advected and diffused as if it were a colored dye in the fluid over which it exerts no control, and whose structure is unable to further influence its own evolution. It is entirely the structure of the geostrophic contours as determined by $\hat{q}_2$ that determines the evolution of $q_2$.
The principal issue is the identification (3.8.1), i.e., the ability to write the dissipative term in the potential vorticity equation as a diffusion of potential vorticity. In the examples given above we treated the large-scale flow as if it were laminar and steady and used the specification of a particular form of viscous inter layer drag (3.7.18) to arrive at a diffusion law for potential vorticity.

Instead of thinking of the ocean as a completely steady, laminar flow, we should, as we discussed in Chapter 1, think of our equations as applying to the large-scale, time-averaged flow, in which smaller scale, turbulent eddies are embedded. As long as the large-scale flow satisfies $U/L^2 \ll \beta$, the flow is stable to barotropic instabilities which would laterally transport relative vorticity (Pedlosky 1987). This is precisely the parameter condition that the Sverdrup theory itself requires for validity, and which condition is likely to be correct in the midocean. Therefore the Sverdrup vorticity balance for the *barotropic* field is unlikely to be upset by the eddies. On the other hand, the fact that the large-scale field is large compared to the Rossby deformation radius encourages baroclinic instability (Pedlosky 1987) if the vertical shear exceeds a critical value. This critical value for instability is exactly the value that would make the potential vorticity gradient in meridional direction vanish at the northern boundary of the gyre where the velocity is zonal. If the motion in the lower layer is zero, use of (3.5.14) shows that this condition would be identical to the condition that closed geostrophic contours elude the eastern boundary, as is shown in detail in the next section. The instability produces a rectified transport of potential vorticity, as has been noted in many numerical experiments (see, for example, Holland 1978). Indeed, the equivalence of the condition for closed geostrophic contours and the criterion for instability satisfyingly identifies the mechanism for the production of motion with the necessary condition for the existence of that motion. When the motion is allowed, the advent of baroclinic instability is capable of producing it.

The potential vorticity equation for the time averaged mean flow in layer 2, temporarily denoted by an overbar, can be written, if the average effect of the eddies in included:

$$\bar{\vec{u}}_2 \cdot \nabla \bar{q}_2 = \operatorname{curl} \Im_2 - \nabla \cdot \overline{\left(\vec{u}\,'_2 q'_2\right)} \tag{3.8.9}$$

where the primed variables represent the fluctuations in the eddy field. Although the average of any linear fluctuation variable is zero by definition, the average of the quadratic terms is generally not zero, giving rise to non-linear transports of potential vorticity on the large scale.

If the fluctuations were small, linear perturbation theory would relate the fluctuation of the potential vorticity to the Lagrangian displacements of the fluctuation field. If $\xi'_i$ is the fluctuation displacement in the $i$th coordinate direction ($i = 1, 2$), linear theory suggests that:

$$q'_2 = -\xi'_i \frac{\partial \bar{q}_2}{\partial x_i}. \tag{3.8.10}$$

The convention of implied summation over the repeated index $i = 1, 2$, has been used in (3.8.10) and $x_1 = x$ while $x_2 = y$. The content of (3.8.9) is that a parcel arrives at a distance $\xi'_i$ from its initial position carrying with it the potential vorticity of the mean field. This is a small amplitude, conservative dynamics approximation for the perturbation field. If (3.8.10) is used to evaluate the potential vorticity flux by the eddies:

$$-\overline{u'_{2i}q'_2} = \overline{u'_{2i}\xi'_j}\frac{\partial \bar{q}_2}{\partial x_j} = \kappa_{ij}\frac{\partial \bar{q}_2}{\partial x_j} \tag{3.8.11}$$

for $i$ and $j$ running from 1 to 2. The tensor $\kappa_{ij}$ is determined by the dynamics of the eddy field, and there is very little that one can specify about it a priori that is rigorous and useful. Young (1986) discusses the efforts at the establishing the parameterization (3.8.11) in detail and clearly demonstrates the difficulty of reaching (3.8.11) in a completely satisfying manner. If we accept (3.8.11) as a plausible attempt at a parameterization of small-scale mixing we are left with:

$$\frac{dQ_2}{d\psi_2}\oint_{C\psi} n_i\kappa_{ij}\left(\hat{k}\times\vec{u}_2\right)_j d\ell = 0 \tag{3.8.12}$$

as a consequence of integrating (3.8.9) over a closed, mean streamline if the explicit dissipation is ignored as small in comparison with the eddy flux.

Only if $\kappa_{ij}$ is diagonal, does (3.8.12) reduce to (3.8.4) and only in this case can we argue that $q_2$ must be homogeneous to satisfy (3.8.11), for in general the integral in (3.8.12) is not sign definite. If we choose the simplest parameterization for the eddy field, in which $\kappa_{ij} = \kappa\,\delta_{ij}$, the potential vorticity field is homogenized. If the diffusion tensor is not diagonal, the possibility exists that the line integral itself vanishes. This would require a rather detailed balance of eddy fluxes across the streamlines. We recall that in the low dissipation limit this curve is established a priori by the geostrophic contours which are independent of the eddy field. It seems implausible that the eddy field would adjust perfectly to annul the net potential vorticity flux, and it is *more likely* that the field of potential vorticity is rendered homogeneous within the pool (Rhines and Young 1982b).

We must conclude that the parameterization of mixing by a potential vorticity diffusion, which leads to homogenization of potential vorticity, is not rigorously established. Nevertheless, we see in Section 3.11 that there is considerable evidence, both in observations and in numerical experiments, that large regions of uniform potential vorticity emerge. Whether such zones in the natural ocean are due to the particular processes described above is not altogether clear. Before reviewing this evidence we consider applications of the ideas to the dynamics of the oceanic gyres.

## 3.9 Application of the Theory to the Subtropical Gyre

In the example described in Section 3.7 the Ekman pumping is chosen so that its integral across the basin is zero. This obviates the need to close the circulation in a western boundary current. The net input of vorticity by the wind is zero, and we are therefore able to avoid the vexing difficulty of removing in the western boundary current the accession of potential vorticity achieved by the fluid in its journey through the interior. The ideas of recirculation driven by potential voritcity homogenization, developed in the previous section, appear to require a clear understanding of the potential vorticity transformations everywhere on the closed contour of the recirculating cell. Young and Rhines (1982) nevertheless boldly applied the basic ideas developed above to a theory for the structure of the midlatitude subtropical gyres in which western boundary currents play an important role, and the issues associated with the closure of the circulation through the western boudary layer remain unresolved. Their theory as applied to the oceanic subtropical gyre is discussed in this section.

Note first that within quasi-geostrophic theory, for which (3.2.29), (3.2.31), and (3.2.34) apply, there is no dynamical difference between the subtropical gyre where the Ekman velocity, $w_E < 0$, and the subpolar gyre, where $w_E > 0$. This can be seen by the symmetry of the governing equations. The equations are invariant under the transformation:

$$\begin{aligned} w_E &\rightarrow -w_E \\ y &\rightarrow -y \\ \psi_n &\rightarrow -\psi_n \\ \Im_n &\rightarrow -\Im_n \\ w* &\rightarrow -w*. \end{aligned} \tag{3.9.1a – e}$$

The last two of these invariance statements require a basically linear relation between the mixing and the motion field. Thus, if latitude is measured from the line of zero Ekman pumping, there is no way in quasi-geostrophic theory of distinguishing between the subtropical and subpolar gyres. As shown in the next chapter, when we allow $O(1)$ variations in the vertical positions of the density surfaces or interfaces there is a significant dynamical difference between the two gyres, and the considerations of the present discussion are more relevant to the subtropical gyre where $w_E < 0$. The present discussion is restricted to the subtropical gyre. The quasi-geostrophic solution for the subpolar gyre can be obtained by the transformation (3.9.1)

The barotropic streamfunction for the $2\frac{1}{2}$ layer model, in which bottom friction and topography can be ignored, is again given by:

$$\psi_B \equiv \frac{H_1\psi_1 + H_2\psi_2}{H} = -\int_x^{x_e} \frac{f_0}{\beta H} w_E(x', y)dx'. \tag{3.9.2}$$

We consider here the simple case in which $w_E$ is a function only of $y$. Not only is this a simple case algebraically, it also is the archetypal situation in which a western boundary current must close the circulation. In this case:

$$\psi_B = -\frac{f_0}{H}\frac{w_E(y)}{\beta}(x_e - x) \geq 0 \tag{3.9.3}$$

so that the geostrophic contours are given by the isolines of:

$$\begin{aligned}\hat{q}_2 &= \beta y + \hat{F}\psi_B \\ &= \beta y - \frac{f_0 w_E(y)}{\beta H}\hat{F}(x_e - x)\end{aligned} \tag{3.9.4}$$

where $\hat{F} = f_0^2 H/\gamma_1 H_1 H_2 = 1/L_d^2$ and $\gamma_1$ is the reduced gravity associated with the interface between layers 1 and 2.

Consider the geostrophic contour that originates at the eastern boundary at the latitude $y_e$. The value of $\hat{q}_2$ for this contour is $\beta y_e$ and the isoline of this geostrophic contour thus satisfies:

$$(x_e - x) = \frac{\beta^2(y_e - y)H}{f_0\hat{F}[-w_E(y)]}. \tag{3.9.5}$$

Recall from (3.5.14) that $\hat{q}_2$ is equal to $q_2$, the potential vorticity in layer 2, *if layer 2 is motionless*. This gives us a physically immediate way to think about the $\hat{q}_2$ isolines.

If we momentarily imagine then, that the Sverdrup circulation were limited to the upper layer, the anticyclonic swirl in the subtropical gyre would depress the interface between layer 1, which is in motion, and layer 2, which is at rest and whose base must therefore be level. The thickness of layer 2 would thus decrease westward from the eastern boundary. Its upper surface has been depressed downward by the motion of the upper layer, and its lower surface is flat. An isoline of constant $q_2 = f/h_2$ must therefore trend southward as it strikes westward from the eastern boundary so that the decrease in $f$ offsets the decrease in $h_2$. If the isolines are diverted southward strongly enough, this can open up a region in the northwest, shielded from the eastern boundary, in which motion in layer 2 is no longer prohibited. Baroclinic Rossby waves from the eastern boundary are guided around, and prevented from penetrating, a zone in which contours of $\hat{q}_2$ are disconnected from the eastern boundary.

Although each point of the eastern boundary is a point of origin of a $\hat{q}_2$, isoline not every point in the basin is bathed by isolines starting from the eastern boundary if the distortion of the geostrophic contours is strong enough.

As an example, consider the Ekman pumping:

$$w_E = -W_0 \sin \pi \frac{y}{L}. \tag{3.9.6}$$

The geostrophic contours are given by the isolines of:

$$\hat{q}_2 = \beta y + \frac{f_0 W_0}{\beta H} \hat{F}(x_e - x) \sin \pi \frac{y}{L}. \tag{3.9.7}$$

Figure 3.9.1 shows the geostrophic contours for three values of the parameter $\alpha = f_0 W_0 / \beta^2 L_d^2 H$ which measures the relative contribution to the potential vorticity of the thickness variation of layer 2 due to the wind forcing and the planetary vorticity gradient. For small values of this parameter the isolines depart only slightly from latitude circles, and the whole basin is covered with geostrophic contours emanating from the eastern boundary. Panel a shows the case in which $\alpha = 0.1$, and the basin is indeed covered with such blocked contours on which the motion in layer 2 is negligible. In panel b, in which $\alpha = 0.3$, we see the appearance of a region in the northwest corner of the basin in which the geostrophic contours are so distorted that they curve around to hit the western boundary twice and are completely disconnected from the eastern boundary. Information from the eastern boundary cannot enter this region, which is outlined by the critical $\hat{q}_2$ contour indicated by the plus signs. This is the outermost $\hat{q}_2$ contour which avoids the eastern boundary. Within this region flow can circulate within the interior and *through the western boundary current,* and the motion is not turned off by the Rossby wave emanating from the eastern boundary. The Rossby waves are deflected around the pool and swept southwestward. As $f_0 W_0 / \beta^2 L_d^2 H$ increases, the extent of the pool region also increases, as shown in panel c, where $\alpha = 1$.

Whether or not the pool region appears depends on the east-west extent of the basin. For example in panel b, were the basin limited to the region $0 \leq x \leq L$, the closed contours would lie west of the western boundary of the ocean at $x = 0$ and lie outside the basin.

The critical value of the forcing required to produce a pool region in the basin can be determined directly from (3.9.5). As $y$ approaches $y_e$ it appears that $x$ must approach the eastern boundary at $x = x_e$ since the right hand side of (3.9.5) vanishes at $y = y_e$. However, consider the expression for $x_e - x$ if $y_e$ lies asymptotically close to the northern boundary. If $y_e$ is equal to $L$ than the denominator of the right hand side also vanishes as $y$ approaches $y_e$ since $w_E$ vanishes at $y = L$. Thus for the isoline emanating from the corner at $y = L$, $x = x_e$, the ratio on the right side of (3.9.5) becomes indeterminate and must be evaluated by expanding numerator and denominator in a Taylor series around the point $y = y_e = L$. Doing so, we obtain for the zonal position, $x_r$, of the geostrophic contour as it emanates from $y = y_e = L$:

$$x_e - x_r = \frac{\beta^2 H}{f_0 \hat{F} \left( \partial w_E / \partial y \right)_{y=L}}. \tag{3.9.8}$$

This yields the intersection of the boundary of the pool region with the northern latitude of the gyre at $y = L$, as shown in Fig. 3.9.1b,c. This outermost contour starts from the critical point $(x, y) = (x_r, L)$. As long as $x_r$

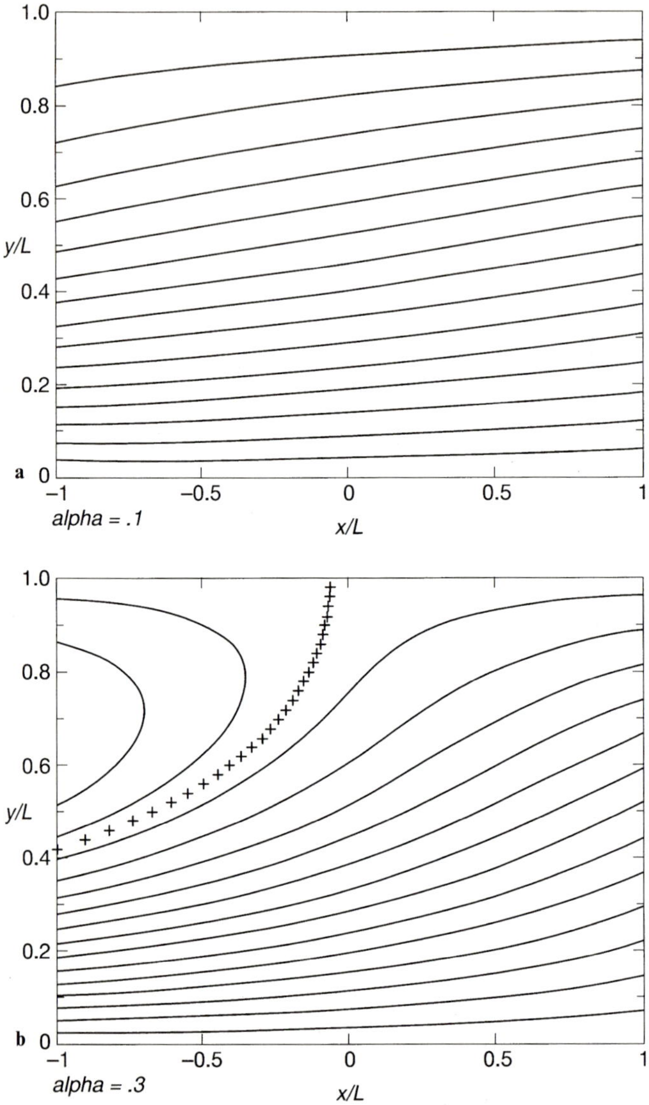


**Fig. 3.9.1a–c.** Geostrophic contours in an rectangular ocean basin $-L \leq X \leq L$, $0 \leq y \leq L$. The forcing is given by $w_E = W_0 sin\pi(y/L)$, and the parameter measuring the forcing is $\alpha = \frac{f_0 W_0}{\beta^2 L_d^2 H}$. **a** Geostrophic contours when $\alpha = 0.1$. The entire basin is covered with geostrophic contours originating at the eastern boundary. **b** $\alpha = 0.3$. A small pool of gestrophic contours emanting from the *western* boundary has arisen. The boundary of the pool is given by the critical curve $Xs(y)$ *(plus signs)*. **c** For a larger value of $\alpha$, here equal to 1.0, the region of the pool has increased

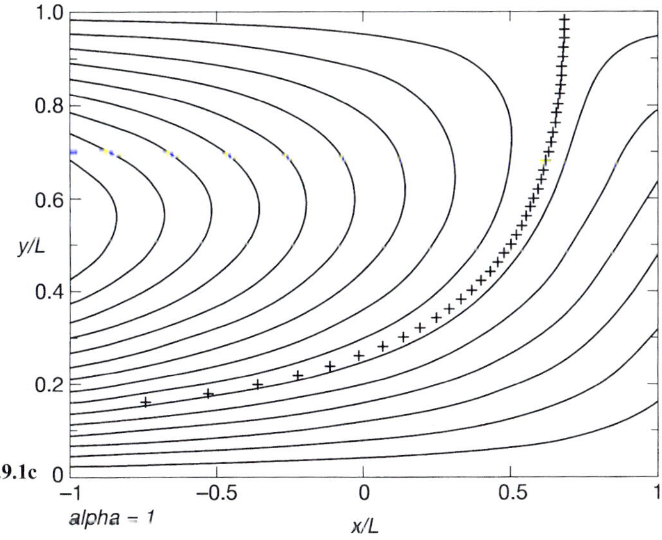


**Fig. 3.9.1c**

lies within the basin, a region in the northwest corner of the gyre in layer 2 can be set into motion. The condition (3.9.8) can be rewritten:

$$\frac{x_e - x_r}{H}\frac{\partial w_E}{\partial y}\frac{f_0}{\beta} > \frac{\beta}{\hat{F}} = c_R \tag{3.9.9}$$

The left side of (3.9.9) can be recognized as the barotropic zonal velocity at the point $(x_r, L)$ if we refer to (3.9.3). The condition that closed geostrophic contours appear in the gyre is equivalent to the statement (3.9.9) that there exists a point $x_r$ on the northern boundary of the gyre (where the eastward barotropic velocity is strongest) such that this barotropic velocity is large enough to arrest the westward propagating Rossby wave that would otherwise turn off the flow in layer 2. This is precisely the heuristic result of (3.4.11). The stronger the forcing, the further eastward this point is. For weak forcing the point is driven westward and outside of the basin, i.e,. west of the western boundary. The critical point is found at the northern boundary because the eastward barotropic flow is strongest there, and the Rossby wave is arrested in its most eastward point there. If there is no position at this latitude where the Rossby wave can be arrested, there can be no other position where it can happen in the basin, and layer 2 is at rest. As we see from Fig. 3.9.1, the pool of closed geostrophic contours is broadest at the gyre's northern boundary.

To find the flow in layer 2 in the pool region which is shielded from the baroclinic Rossby waves requires, a choice for the dissipation parameterization. We follow Young and Rhines (1982) and imagine that the dissipation can be modeled as a diffusion of potential vorticity. We saw in Section 3.8 that under the proper conditions this would lead to homogenization of potential vorticity. Were these conditions met in the present case we could set $q_2$ equal to its value on the outermost geostrophic contour. Since:

$$\psi_2 = \frac{(\hat{q}_2 - q_2)}{(\hat{F} + G_2)} \tag{3.9.10}$$

once $q_2$ is known, $\psi_2$ is determined. The outermost geostrophic contour of the pool is the contour which starts from the critical point $(x_r, L)$. The contour can then be thought of as running westward along $y = L$ to the western boundary (as well as eastward to the eastern boundary; this splitting is what defines $x_r$ as a critical point). Along this contour, by continuity with the region external to the pool, $\psi_2 = 0$. At the critical point $(x_r, L)$ the streamfunction in the upper layer also vanishes because there is no flow across the boundary at $y = L$ where $\psi_n$ is zero. Thus $q_2 = \beta L$ on the outer contour and is equal to this value everywhere within the pool if the potential vorticity in layer 2 becomes homogenized.

For the moment let us accept this scenario of potential vorticity homogenization. This yields, for the flow within the pool region:

$$\psi_2 = \frac{\hat{q}_2 - \beta L}{\hat{F} + G_2} = \frac{\psi_B + \beta/\hat{F}(y - L)}{1 + \dfrac{\gamma_1 H_1}{\gamma_2 H}} \tag{3.9.11}$$

and using the definition (3.9.2):

$$\psi_1 = \psi_B \left( \frac{1 + \gamma_1/\gamma_2}{1 + \dfrac{\gamma_1}{\gamma_2}\dfrac{H_1}{H}} \right) - \frac{\beta}{\hat{F}}(y - L) \left( \frac{H_2/H_1}{1 + \dfrac{\gamma_1}{\gamma_2}\dfrac{H_1}{H}} \right). \tag{3.9.12}$$

Outside the pool region, i.e., east and south of the critical outer contour, the lower layer is at rest, and the upper layer carries the full Sverdrup transport, i.e.:

$$\begin{aligned} \psi_2 &= 0 \\ \psi_1 &= \frac{H}{H_1}\psi_B. \end{aligned} \tag{3.9.13}$$

Figure 3.9.2 shows the circulation in the Sverdrup interior for the Ekman pumping of (3.9.6) . Panel b shows the lower layer flow. It is limited to the pool region and displays a large-scale anticyclonic swirl. The flow that is shown is that in the interior, and a return flow in the western boundary current is not shown, nor of course have we calculated it. Panel a shows the flow in the upper layer. Outside the pool region it carries the full Sverdrup transport. Within the pool region it shares the transport with the lower layer. The upper layer streamfunction is continuous at the pool boundary, but its derivative is discontinuous, reflecting the discontinuity in tangential velocity in both the

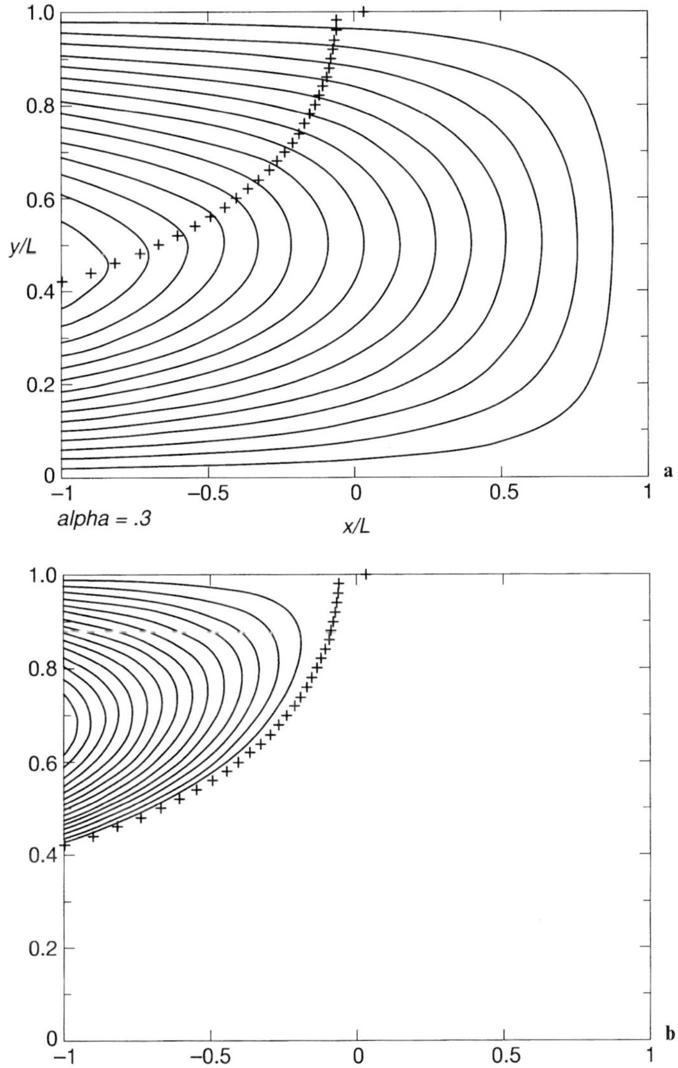


**Fig. 3.9.2a,b.** Streamfunction in the 2.5-layer model with the assumption that the potential vorticity in layer 2 is homogenized within the pool of closed geostrophic contours. In the figure $\alpha = \frac{f_0 W_0}{\beta^2 L_d^2 H} = 0.3$. The layer depths and the density jumps at the two active interfaces are taken to be equal. **a** $\psi_1$. Note the kink in the upper layer streamlines at the boundary of the pool (*plus signs*). **b** $\psi_2$. The motion is a swirling recirculation limited to the pool. An implied return flow through the western boundary current must take place in both layers

upper and lower layer across the pool boundary. This shows up in the figure as a kink in the upper layer streamlines at the pool boundary and is especially evident in the figure towards the southwest terminus of the pool boundary.

In the parameter setting of Fig. 3.9.2, where $H_1 = H_2$ and $\gamma_1 = \gamma_2$, it follows from (3.9.11) and (3.9.12) that the meridional velocity in the upper layer in the pool is exactly twice as large as the meridional velocity in the lower layer. If the density difference between the two layers, $\gamma_1$, should diminish, the meridional flow would become increasingly barotropic. The zonal flow has a vertical shear:

$$u_1 - u_2 = -\left[\frac{\gamma_1/\gamma_2}{1 + (\gamma_1 H_1/\gamma_2 H)}\right]\frac{\partial\psi_B}{\partial y} + \frac{H_2}{H_1}\frac{\beta}{(\hat{F} + G_2)} \tag{3.9.14}$$

large enough to tilt the interface and offset the planetary potential vorticity gradient even as the ratio $\gamma_1/\gamma_2 \to 0$.

The criterion (3.9.9) has another interpretation more closely related to a possible mechanism for setting the lower layers into motion. In the absence of motion in layer 2 (and all lower layers) the potential vorticity in layer 2 would be given by:

$$\begin{aligned} q_2 &= \beta y + F_2\psi_1 \\ &= \beta y + \hat{F}\psi_B. \end{aligned} \tag{3.9.15}$$

Along the northern boundary of the gyre the flow is strictly zonal, and the potential vorticity gradient is in the meridional direction. On $y = L$ the gradient of potential vorticity in layer 2 is given by:

$$\begin{aligned} \frac{\partial q_2}{\partial y} &= \beta + \hat{F}\frac{\partial\psi_B}{\partial y} \\ &= \beta - \hat{F}u_B. \end{aligned} \tag{3.9.16}$$

Therefore the condition that the Rossby wave be arrested by the barotropic zonal flow is equivalent to the condition that the meridional potential vorticity gradient vanishes at the point $x_r$ on the northern boundary. Further westward, where $u_B$ is larger, the potential vorticity gradient in layer 2 is negative, while it is positive in the upper layer, as the reader may verify. Thus along the northern boundary of the pool the conditions for baroclinic instability are satisfied (Pedlosky 1987). The instability produces eddies on scales of the order of the deformation radius, i.e., quite small compared to the gyre scale, so that the application of the theory for the instability of horizontally uniform flows can be applied to the slowly varying (on the eddy scale) gyre scale flow. One of the principal effects of baroclinically produced eddies is vertically to transfer the momentum of the mean flow so as to reduce the vertical shear which is the source of the instability. This sets layer 2 into motion. Thus the emergence of the closed geostrophic contours, isolating a pool region in which motion is allowed, also suggests the simultaneous appearance of a mechanism of small-scale eddy driving of the deeper layers to produce the motion that moves along the geostrophic contours.

For realistic values of the parameters $x_r$ lies within the ocean basin. For example with $f_0 = 10^{-4}\text{s}^{-1}, \beta = 10^{-13}\text{cm}^{-1}\text{s}^{-1}$, H = 800m, and $L_d^2 = 1/\hat{F} = 25\times 10^{12}\text{cm}^2$ and $W_0 = 10^{-4}\text{cm/s}$ with L = 1000 km, then $x_e - x_r$, the distance from the critical point to the eastern boundary is 1270 km for the Ekman pumping of the form (3.9.6). This is close enough to the eastern boundary to fit the critical point into all the major ocean basins.

In the pool region of constant potential vorticity in layer 2, the base of layer 2 is depressed by the anticyclonic motion shown in Fig. 3.9.2. We assumed layer 3 to be at rest, but now that layer 2 is in motion and the isolines of potential vorticity in layer 3 are distorted from latitude circles by the flexing of the base of layer 2, it is possible that *within* the pool region of layer 2 these $q_3$ contours may now become closed, repeating the process that occurred in layer 2. Should they close they would be nested within the pool region of layer 2. Within this subpool the motion can be recalculated by attributing to layer 3 a constant potential vorticity equal to $\beta L$, since it too touches the northern boundary of the gyre. The process can be repeated indefinitely with deeper layers set into motion if the thickness field of each layer, presumed to be at rest, is altered enough by the motion of the layers above. Since the intensity of the motion decreases with depth, this becomes increasingly difficult to achieve, and in most cases a layer is finally reached in which its critical point on the northern boundary of the gyre lies outside of the basin, and this layer and all deeper layers remain at rest. It is important to note here that the pools are nested in the sense that each deeper layer in motion has its pool completely contained within the region of the pool of the layer above it since it is only the motion of the upper layer that can distort a layer's potential vorticity contours away from latitude circles. The region of uniform potential vorticity would then form an inverted bowl or cone with the vertex at depth in the northwest corner and then opening outward to the south and east as one rises in the water column. Note that each layer would have the potential vorticity $\beta L$ within its pool so that the potential vorticity is constant from layer to layer within the moving region as well as being uniform within the moving part of each layer. At each depth the horizontal plan form would resemble the pool region found in the $2\frac{1}{2}$ layer model shown in Fig. 3.9.2.

The vertical resolution of the layer model can be improved by adding more layers and making each layer thinner. In the limit, the layers become infinitesimally thin, and the fluid has a continuous stratification of density. In this limit the layer equations must be changed to equations governing a fluid with a continuous variation of density, $\rho$. The limiting form of this bowl is obtained as each of the layers has an infinitesimal thickness, i.e., when the fluid has a continuous vertical stratification. Although the continuously stratified problem is generally a very difficult one, it is tractable within the quasi-geostrophic approximation.

## 3.10 Quasi-Geostrophic Model with Continuous Stratification

It is relatively straightforward to move from the layer model to a formulation of the continuous model within the quasi-geostrophic approximation.

The equations of geostrophy (3.2.3) and thermal wind apply unchanged, as well as the relation between the pressure and the geostrophic streamfunction (3.2.9). The equation for the relative vorticity (3.2.11) is also unaltered. However, instead of vertically integrated equations for mass conservation for each layer we must use the continuity equation to eliminate the horizontal divergence in the vorticity equation to obtain:

$$\frac{d\zeta}{dt}=\frac{\partial\zeta}{\partial t}+u\frac{\partial\zeta}{\partial x}+v\frac{\partial\zeta}{\partial y}+\beta v=f_0\frac{\partial w}{\partial z}+\operatorname{curl}\Im \tag{3.10.1}$$

The density equation (3.2.20) can be partially linearized since within the quasi-geostrophic approximation the vertical density difference is much greater than the horizontal density variation. Thus (3.2.20) can be written:

$$\frac{\partial\rho}{\partial t}+u\frac{\partial\rho}{\partial x}+v\frac{\partial\rho}{\partial y}-w\frac{\rho_0}{g}N^2=-\alpha Q/c_p. \tag{3.10.2}$$

The buoyancy, or Brunt-Väisälä, frequency is:

$$N=\sqrt{-\frac{g}{\rho_0}\frac{d\rho_s}{dz}} \tag{3.10.3}$$

which is, at most, a function of $z$ in quasi-geostrophic theory.

The vertical velocity can be eliminated between the vorticity equation and the density equation to obtain the quasi-geostrophic potential vorticity equation for the continuously stratified fluid, i.e:

$$\frac{d}{dt}\left[\zeta+\beta y-\frac{gf_0}{\rho_0}\frac{\partial}{\partial z}\frac{\rho}{N^2}\right]=\operatorname{curl}\Im+\frac{\partial}{\partial z}\left(\frac{f_0\alpha gQ}{\rho_0 c_p N^2}\right). \tag{3.10.4}$$

In deriving (3.10.4) use has been made of the thermal wind relations to interchange the $z$ derivative and the total derivative of $\rho/N^2$. Recall also that $\rho$ refers to the density *anomaly*. The geostrophic and hydrostatic approximations allow (3.10.4) to be written entirely in terms of the geostrophic streamfunction, $\psi$, i.e.:

$$\frac{\partial q}{\partial t}+J(\psi,q)=\operatorname{curl}\Im+\frac{\partial}{\partial z}\left(\frac{f_0\alpha gQ}{\rho_0 c_p N^2}\right) \tag{3.10.5}$$

where $q$, the geostrophic potential vorticity, is given as:

$$q = \zeta + \beta y - \frac{g f_0}{\rho_0}\frac{\partial}{\partial z}\frac{\rho}{N^2} = \nabla^2\psi + \frac{\partial}{\partial z}\left(\frac{f_0^2}{N^2}\frac{\partial\psi}{\partial z}\right) + \beta y. \tag{3.10.6}$$

The Laplacian in (3.10.6) is the "horizontal" Laplacian $\partial^2/\partial x^2 + \partial^2/\partial y^2$ and the Jacobian in (3.10.5) is the Jacobian operator with respect to $x$ and $y$.

The vertical velocity is given at the upper boundary, $z = 0$, by the Ekman pumping velocity so that with the aid of the density equation and the geostrophic and hydrostatic approximations this boundary condition can be written as:

$$w = -\frac{d}{dt}\left(\frac{f_0}{N^2}\frac{\partial\psi}{\partial z}\right) + \frac{\alpha g Q}{c_p \rho_0 N^2} = w_E, \quad z = 0. \tag{3.10.7}$$

In this chapter we consider only adiabatic flows for which $Q$ is zero. Our discussion closely follows that of Young and Rhines (1982).

It is illuminating and convenient to introduce nondimensional variables for the continuous problem. This allows us to write the circulation problem in a neater, more compact form, but more importantly it reveals the characteristic scales of the motion from scaling considerations.

We introduce nondimensional variables *temporarily* denoted by primes in the following manner:

$$\begin{aligned} \psi &= UL\psi' \\ (x, y) &= L(x', y') \\ z &= dz' \\ w &= Ww' \\ q &= \beta L q' \end{aligned} \tag{3.10.8}$$

where the vertical scale, $d$, must be determined. The horizontal scale $L$ is imposed by horizontal scale of the wind forcing, and the scale for the vertical velocity, $W$, is determined by the magnitude of the Ekman pumping.

The Sverdrup relation $\beta v = f\,\partial w/\partial z$, determines the relation between the scales for the horizontal, $U$, and vertical velocity, i.e:

$$W = \frac{\beta U d}{f_0}. \tag{3.10.9}$$

In the expression for the potential vorticity, $q$ (3.10.6), the relative vorticity (the first term) is small with respect to the contribution from stretching of planetary vortex lines (the second term), since their ratio can be estimated as $N^2 d^2 / f_0{}^2 L^2 \ll 1$.

On the other hand, in order for there to be motion below the surface the stretching term must be of the same order as the planetary vorticity term so that the $q$ isolines can distort enough to form internal pools of geostrophic contours which are detached from the eastern boundary. Thus, we require that:

$$\frac{\partial}{\partial z}\left(\frac{f_0{}^2}{N^2}\frac{\partial\psi}{\partial z}\right) = O\left(\frac{f_0{}^2 UL}{N_0{}^2 d^2}\right) = O\left(\frac{f_0{}^3 WL}{N_0{}^2 \beta d^3}\right) \approx \beta y = O(\beta L) \qquad (3.10.10a)$$

where $N_0$ is a characteristic value of the buoyancy frequency. The proposed equality in (3.10.10) yields scales for $d$ and $U$ in terms of the external parameters, i.e.:

$$d = \frac{f_0}{(N_0\beta)^{2/3}} W^{1/3}$$

$$U = \left(\frac{WN_0}{\beta^{1/2}}\right)^{2/3} \qquad (3.10.10b,c)$$

If we use the same parameter values as in Section 3.9 to evaluate these scales, namely, $f_0 = 10^{-4}\mathrm{s}^{-1}, \beta = 10^{-13}\mathrm{cm}^{-1}\mathrm{s}^{-1}, W = 10^{-4}\mathrm{cm/s}$ and use $N_0 = 5 \times 10^{-3}\mathrm{s}^{-1}$, we obtain $d = 700$ m and $U = 1\mathrm{cm/s}$, both of which are reasonable scales for the thermocline depth and mid ocean velocity.

In the layer model the motion is contained in a region which shrinks to the northwest with depth. A similar behavior can be expected in the continuous model and below the surface we can expect the motion to be contained in a bowl-shaped region whose containing surface is given by:

$$z = -D(x, y) \qquad (3.10.11)$$

outside of which the fluid is at rest. Note that in (3.10.11) we drop the prime notation from the dimensionless variables. In the remaining portion of this section all variables are nondimensional unless otherwise indicated.

Figure 3.10.1 shows a schematic of the region in which motion takes place. The boundary of the region in motion is $z = -D(x, y)$ at which the streamfunction and density must be continuous with the surrounding, resting fluid. In the resting fluid $\psi$ and the density *anomaly* are both 0. Hence on the boundary:

$$\psi = 0, \quad \frac{\partial\psi}{\partial z} = 0 \qquad (3.10.12)$$

on $z = -D(x, y)$. At the same time the vertical velocity must match the Ekman pumping velocity at $z = 0$. In the absence of heating ($Q = 0$) and in our nondimensional units this yields, from (3.10.7) in the steady state:

$$w_E = -J\left(\psi, \frac{\partial\psi}{\partial z}\right)\frac{[N_0{}^2]}{[N^2]}. \qquad (3.10.13)$$

Within the bowl and below the Ekman layer, we assume that the potential vorticity has been homogenized so that "layer 1" of the layer model shrinks to an infinitesimally thin layer directly below the Ekman layer. In fact, in the continuous limit of the layer model the dynamics of layer 1 yields the boundary

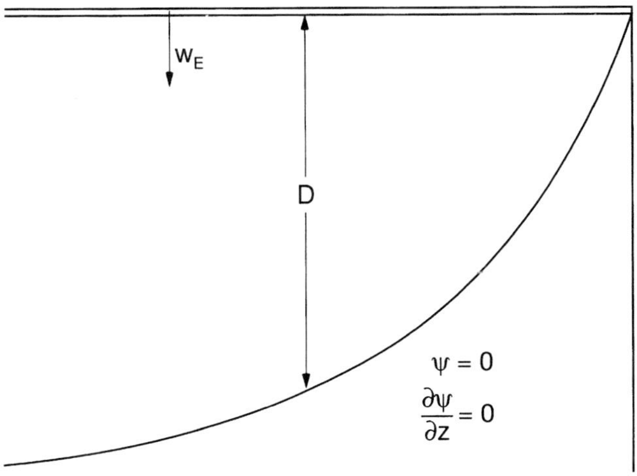


**Fig. 3.10.1.** The continuously stratified model. The motion is confined to a region Z=−D(x,y). On the boundary of the region the streamfunction and density are continuous with the surrounding resting fluid

condition (3.10.13). This then is the situation we discussed in the introduction, in which linear theory compresses the Sverdrup transport into an absurdly thin, delta function zone at the upper boundary. The closing of geostrophic contours and the homogenization of potential vorticity gives the fluid a way to establish deep motion beneath the layers directly influenced by Ekman pumping and to distribute the Sverdrup transport over a nonzero depth, $D(x, y)$. We must find the vertical and lateral extent of the field of motion and the streamfunction within that zone.

For small dissipation the steady form of (3.10.5) implies that:

$$q = Q(\psi, z). \tag{3.10.14}$$

Here $Q$ (not to be confused with the heating function, which is zero in this example) is a function of streamfunction and vertical coordinate. This is similar to the relation (3.8.4) where now $z$ is a continuous counterpart to the layer label n in the layer model. The relation (3.10.14) applies in the bowl region $-D \le z \le 0$. If the potential vorticity is homogenized in this region in the sense that for each $z$ we have $\partial Q/\partial\psi = 0$, then we may write, with the approximation that the relative vorticity is negligible:

$$q = y + \frac{\partial}{\partial z}\left[\frac{{N_0}^2}{N^2}\frac{\partial\psi}{\partial z}\right] = Q(z) \equiv y_0(z) \tag{3.10.15}$$

for $-D \le z \le 0$. Once $y_0$ is specified, we can integrate (3.10.15) and apply the boundary conditions (3.10.12) and (3.10.13) to determine the motion in the bowl-shaped pool. Since (3.10.15) is a second-order ordinary differential equation, and (3.10.12) and (3.10.13) impose three boundary conditions in $z$, there are more conditions than is normal in a second-order problem. However, the problem is actually a free boundary problem since $D(x, y)$ is not known a priori. The boundary of the region in motion is part of the solution and the additional boundary condition is required to determine $D(x, y)$. A very similar situation arises in the more general and difficult continuous problem described in Chapter 4 where the quasi-geostrophic approximation is relaxed, but where the overall structure of the problem is the same as that confronted in the quasi-geostrophic limit.

Our discussion of the layer model allows us to anticipate a useful result. At each level in $z$ the shape of the bowl of moving fluid is tucked up against the northwest corner of the basin such that the outermost closed contour at each depth intersects the northern boundary of the gyre. At that boundary $\psi$ and its $z$ derivatives are zero so that we can determine $y_0$ as the value of the potential vorticity, which is then only the planetary vorticity at the northern boundary, which in nondimensional units is $y = 1$. Thus we eventually choose:

$$y_0 = 1. \tag{3.10.16}$$

We shall see that any other choice of $y_0$ would lead to either a singularity or an unstable solution. For the moment we hold $y_0$ arbitrary.

For simplicity we consider the case of uniform buoyancy frequency for which $N = N_0$. The solution of (3.10.15) subject to (3.10.12) is in that case:

$$\psi = \frac{1}{2}(z + D)^2 (y_0 - y). \tag{3.10.17}$$

The total density field in *dimensional* variables is

$$\rho_{total} = \rho_0 \left[ 1 - N^2 z/g - \frac{f_0}{g}\frac{\partial \psi}{\partial z} \right] \quad \text{(dimensional)} \tag{3.10.18}$$

which in *nondimensional* form, using the scales for $d$ and $U$ of (3.10.11), gives us for the case of constant buoyancy frequency:

$$\frac{\rho_{total}}{\rho_0} - 1 = \frac{N_0{}^2 d}{g}\left\{ z + \frac{\beta L}{f_0}\frac{\partial \psi}{\partial z} \right\}. \tag{3.10.19}$$

The smallness of the parameter $\beta L / f_0$ is required for the validity of the $\beta$ plane approximation and the consequent smallness of the density anomaly, compared to the density distribution in the resting state, is as required by quasi-geostrophic theory.

The remaining boundary condition is the matching of the vertical velocity to the given Ekman pumping velocity at $z = 0$, i.e. (3.10.13). Inserting the solution (3.10.17) into (3.10.13) yields:

$$w_E = \frac{D^2}{2}\frac{\partial D}{\partial x}(y_0 - y). \tag{3.10.20}$$

For the zonal flow to vanish on the eastern boundary we must have:

$$\frac{\partial \psi}{\partial y} = (z + D)(y_0 - y)\frac{\partial D}{\partial y} - \frac{1}{2}(z + D)^2 = 0 \tag{3.10.21}$$

for all $z$ and $y$ on $x = x_e$. The only way (3.10.21) and its $z$ and $y$ derivatives can be satisfied for all $z$ and $y$ is for the bowl to have zero depth on the eastern boundary so that $z = D = 0$ within the bowl on $x = x_e$. Then, the integral of (3.10.20) yields:

$$D^3 = \frac{-6}{y_0 - y}\int_x^{x_e} w_E(x', y)dx'. \tag{3.10.22}$$

In the case in which $w_E$ is a function only of latitude:

$$D = \left[-6(x_e - x)\frac{w_E(y)}{y_0 - y}\right]^{1/3} \tag{3.10.23}$$

which yields the depth of the pool of constant potential vorticity and the shape of the bowl of circulating water. Note that if $y_0$ is chosen to be less than 1, which is the position of the northern boundary of the gyre, an unacceptable singularity develops for $D$ within the gyre. At each level depth the shape in the $x, y$ plane of the pool is given by:

$$(x_e - x) = \frac{(y_0 - y)D^3}{-6w_E(y)} \tag{3.10.24}$$

for a fixed $D$. This should be compared to (3.9.5) which yields the boundary of the pool region in the layer model. The similarity is, in fact, both striking and reassuring. Note that if $y_0$ is greater than 1, the isoline of the pool boundary bends back and strikes the western boundary as $y = 1$ is approached since $w_E$ vanishes at $y = 1$ but does so before the northern boundary is actually reached. The outermost boundary of the pool would then not strike the northern boundary. The choice of $y_0$ greater than 1 is therefore equivalent to choosing an outer boundary for the pool which is coincident with one of the *inner* closed boundaries of Fig. 3.9.2. This would leave an outer strip of the pool with zero velocity although there are closed geostrophic contours threading through this region. Such a choice is unstable since a little forcing from the upper layer sets this outer shell into motion. Thus to fill out the maximum zone of motion allowed and to avoid a singularity within the zone we must choose $y_0 = 1$.

The maximum depth of the pool occurs in the northwest corner of the basin. If $x_w$ is the position of the western boundary (scaled by $L$) the nondimensional maximum depth is:

$$D_{max} = D(x_w, 1) = \left[6(x_e - x_w)\frac{\partial w_E(1)}{\partial y}\right]^{1/3}. \qquad (3.10.25)$$

Figure 3.10.2 shows the isolines of $D$ in the horizontal plane for a square basin and a *nondimensional* Ekman pumping velocity $w_E = -\sin \pi y$. The contours are labeled with the value of $D$ with respect to its maximum value, 2.661. The dimensional value is obtained by multiplication by $d$, which by our earlier estimate for $d$ would yield a maximum depth of about 1860 m. This is somewhat large and is partly connected with our assumption that the Ekman pumping is independent of $x$.

The isolines of $D$ are similar, mathematically, to a set of nested boundaries of the pool regions for a set of layers. The layer version of the continuous bowl would be a finite-difference approximation to the continuous model, and in fact it can be shown (Pedlosky 1987) that the layer models are mathematically equivalent in quasi-geostrophic theory to a finite-difference approximation of the continuous model. Figure 3.10.3 shows the schematic relationship between the continuous model and a multilayer model. Thus the contours in Fig. 3.10.2 can also be thought of as the pool boundaries of a sequence of finite layers.

A three-dimensional view of the bowl is shown in Fig. 3.10.4. The bowl slopes downward to the north and west where it obtains its maximum depth. Beneath the bowl the distortion of the density surfaces is zero, and the fluid is at rest. Note that the bowl slopes up sharply at both the southern and eastern

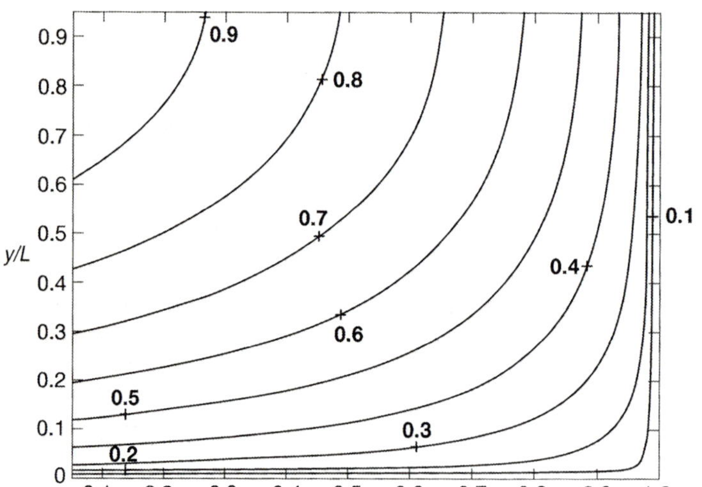


**Fig. 3.10.2.** Contours of $D(x, y)$ showing the shape of the bowl of constant potential vorticity in the continuous model. The contours label the depth scaled by the maximum depth of the bowl which is achieved in the northwest corner

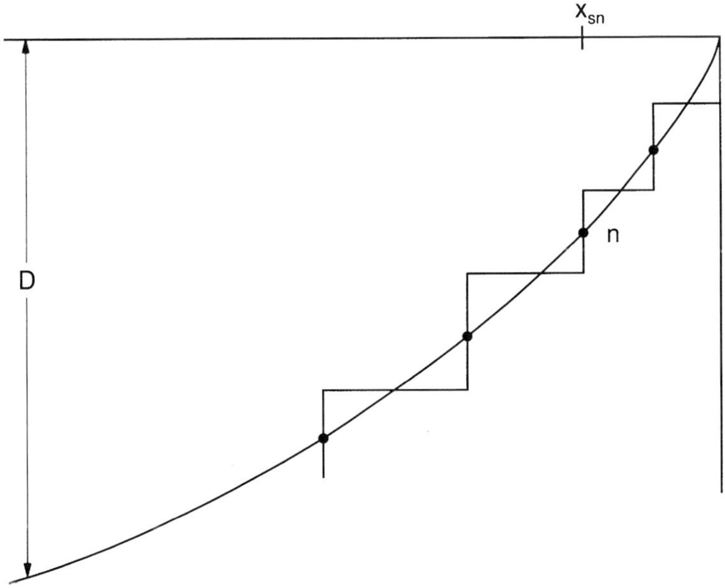


**Fig. 3.10.3.** Relation between the shape of the continuous bowl $D$ and the boundaries of the nested regions in a multilayer model. The eastern edge of the pool in the $n$th layer is found at $xs_n$ which is the intersection of the boundary $D$ with the midpoint of the $n$th layer

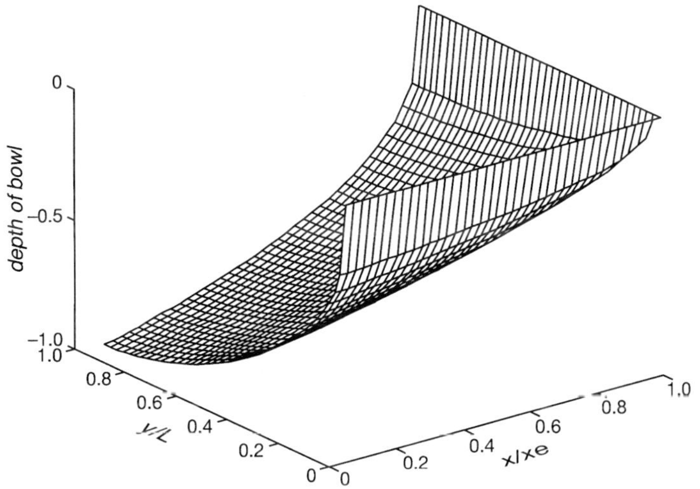


**Fig. 3.10.4.** Three-dimensional perspective view of the bowl looking into the basin from the southwest corner of the gyre

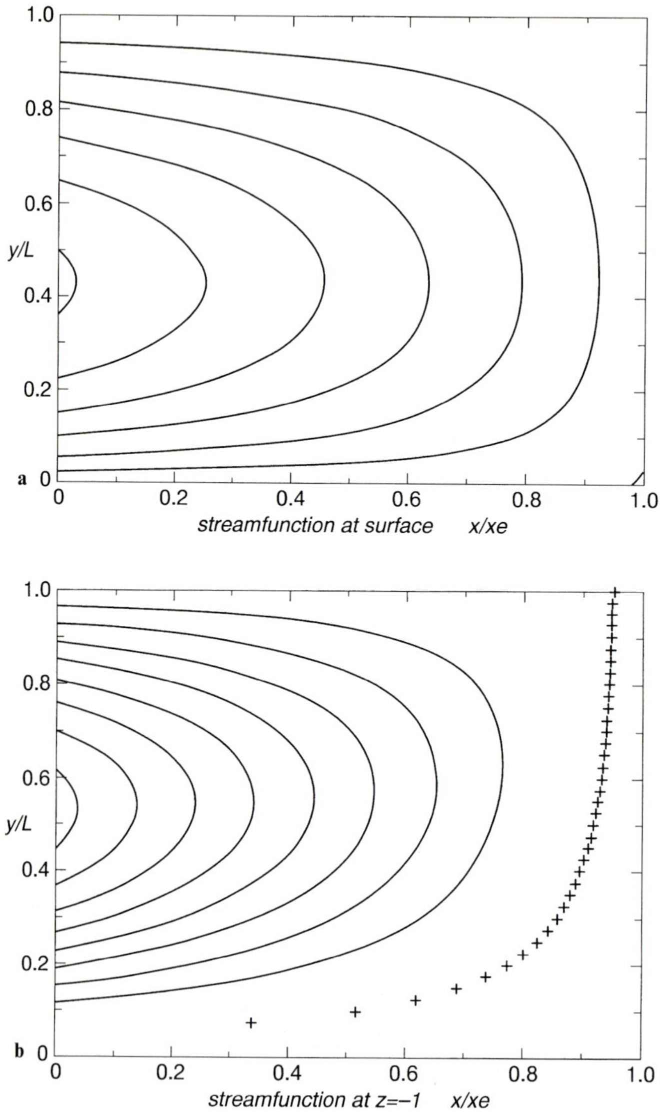


**Fig. 3.10.5a–c.** Streamline pattern in the horizontal plane for the quasi-geostrophic model driven by the Ekman pumping $w_E = -W_0 \text{sinx}(y/L)$, for three selected depths. **a** The flow at the surface. **b** The flow at one scale depth d below the surface. **c** The flow at 2 scale depths below the surface

boundaries of the basin where $D$ has a weak singularity. At $x = x_e$, for example, $D$ goes as $(x_e - x)^{1/3}$ and thus has an (integrable) singularity in its first derivative.

At each level in $z$ the circulation is limited to the bowl and appears as a sequence of recirculations similar to the layer model shown in Fig. 3.9.2b. Now

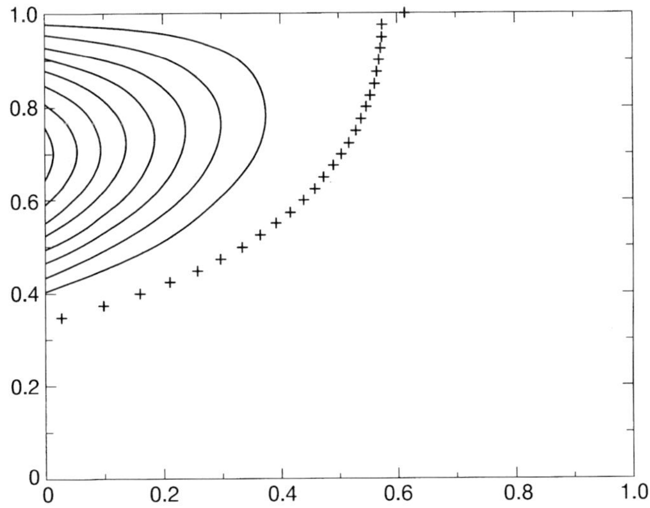


**Fig. 3.10.5c**

at each depth the lateral extent of the pool varies becoming more restricted with depth. Figure 3.10.5 shows a sequence of streamline patterns. Panel a shows the circulation at the surface. Note that it covers the entire basin. This can be anticipated from (3.9.5) in the limit where the layer thicknesses, and thus $H$, go to zero. Although the pool covers the basin, it in infinitesimally thin on the eastern boundary and only skims the surface of the fluid. At greater depth where the bowl has a finite thickness the domain of motion become restricted to the familiar northwest pool, as shown in Fig. 3.10.5 b, which is at $z/d = -1$. At even greater depth the pool shaped zone of moving fluid is even more restricted as in panel c, in which the circulation at $z = -2d$ is shown and is nearly indistinguishable in structure from the predictions of the layer model (Fig. 3.9.2b) even with the very low resolution $2\frac{1}{2}$ layer model.

The disturbed density field can be found from (3.10.19). Panel a of Fig. 3.10.6 shows the density surface in a meridional cross-section and Panel b a zonal cross section. Each case uses the parameter setting $\beta L/f_0 = 0.3$. Careful inspection of panel a shows the slight spreading of the isopycnals for increasing $y$ to offset the increase in $f$ so as to keep the potential vorticity constant. As the bowl rises to the surface at $x = x_e$, the isopycnals outside the bowl become flat on the eastern boundary except for the near surface isopycnals that form a wedge near $x = x_e$ and $z = 0$. In fact at $z = 0$ the density anomaly is proportional to:

$$\frac{\partial \psi}{\partial z}(x, y, 0) = \left[6(x_e - x)\right]^{1/3}\left(-w_E(y)\right)^{1/3}(1 - y)^{2/3}. \tag{3.10.26}$$

Although the density is finite on the eastern boundary, its derivative with respect to $x$ is not. Strong meridional velocities are predicted by the model since the pool must carry the nonzero southward Sverdrup transport at $x = x_e$ in a

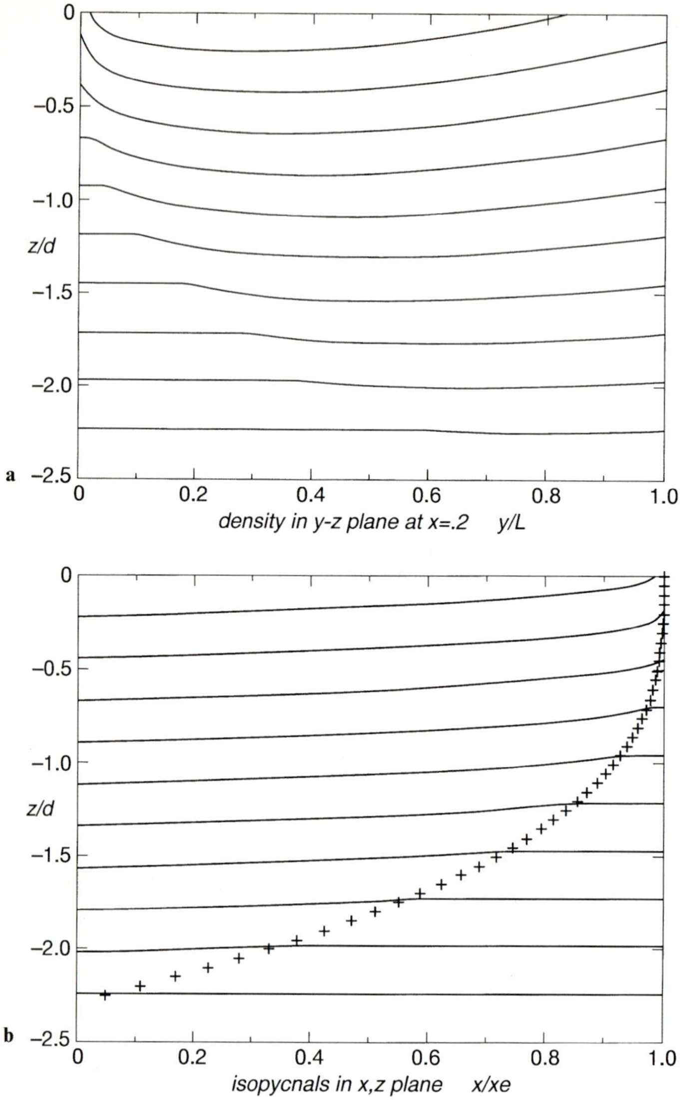


**Fig. 3.10.6a,b.** Isopycnal surfaces for the parameter choice, $\beta L/f_0 = 0.3$. **a** A meridional section of the total density. **b** Zonal section. *Plus signs,* boundary of the pool. Note that the density surfaces are flat outside and below the pool

layer whose depth has shrunk locally to zero. On the northern boundary of the gyre no such singularity occurs as long as $w_E$ goes to zero as rapidly as $1 - y$.

It is important that the continuous model provides qualitatively the same solution as the layer models. This gives us confidence that the generally more tractable layer models are able to capture the basic physical processes of the circulation and represent them adequately. One important reason for this is that although the horizontal velocities may be discontinuous at the layer interfaces, the vertical velocities are continuous, and it is $w$ that drives the fluid by stretching planetary vorticity filaments. One important difference between the layer model and the continuous model is in the nature of the velocity at the lower boundary of the bowl. In both the layer model and the continuous model the boundary of the bowl is a locus of positions where the isopycnal surfaces have a discontinuity in their slope, as seen in Fig. 3.10.6. In the layer model, because of the $O(1)$ jump in the density across each interface, this slope is accompanied by a finite jump in the velocity across the pool boundary. We noted this above in our discussion of the streamline patterns in Sections 3.8 and 3.9. In the continuous model the sloping isopycnals, by the thermal wind equation, support a jump in the *shear* of the velocity rather than the velocity itself. The velocity across the bowl boundary is continuous with the resting fluid around it and is thus zero at $z = -D$. As the number of layers increase, and the density difference decreases in proportion, the velocity jumps of the layer model diminish, and the velocity structure smoothly approaches that of the continuous model.

The principal question that we have avoided so far is the validity of the homogenization hypothesis for situations such as the gyre circulation where the geostrophic contours "close" through the western boundary current.

## 3.11 Numerical and Observational Evidence

We have accepted the hypothesis of the homogenization of potential vorticity and applied it in the previous two sections in the pool regions of the gyre where the geostrophic contours or the isolines of potential vorticity are so bent by the overlying wind-driven flow that they no longer strike the eastern boundary. This escape of the contours from striking the eastern boundary really only makes it plausible that motion occurs in such regions. Whether the potential vorticity should be expected to be homogeneous within those regions depends on a good deal more.

The discussion in Section 3.8 demonstrated that there are several physical conditions that must be satisfied for homogenization. To summarize briefly, these are:

1. The isolines of potential vorticity must close.
2. The dissipation of potential vorticity must be equivalent to a diffusion of potential vorticity.
3. There must be nearly no dissipation of potential vorticity at the lowest order, at least on the bounding contour of the "pool."
4. There must be no other sources of potential vorticity within the region enclosed in the pool.

Only the first condition is clearly satisfied for the gyre circulation problem that we have discussed. The dissipation of potential vorticity by small-scale eddy motion may act as a diffusion, but it is hard to argue that the diffusion coefficient is constant, and in some cases it may even be negative (Holland and Rhines 1980).

Perhaps the most critical difficulty in applying the homogenization argument in the case of the gyre circulation is the fact that the isolines of potential vorticity, including the outermost one, must close in the western boundary current. Normally, the western boundary current is a location of substantial potential vorticity flux, and, as we saw in Chapter 2, it is difficult in general to construct a circulation pattern in which this dissipation is not a controlling factor in the circulation. Moreover, it is precisely the outermost contour that is closest to the western boundary and on which it will be most likely for dissipation to act. In the absence of adequate theories of the western boundary current it is therefore not possible to give a completely convincing a priori argument of whether conditions 3 and 4 are likely to be met.

Young and Rhines (1982) argue that for a quasi geostrophic layer model there are no substantial inputs of vorticity below the first layer, at least in the region outside the western boundary current, and therefore there is no need for the western boundary current at depth to diffuse substantial amounts of potential vorticity through the side wall on the western boundary. They present an interesting heuristic attempt to construct a dissipationless western boundary current for the deeper layers to demonstrate the applicability of the Prandtl-Batchelor theorem to the wind driven gyre. Unfortunately, an essential part of their argument is based on the Moore scenario for the western boundary current in the region of eastward interior flow, which as we saw in Chapter 2, is fundamentally deficient. Indeed, Ierley and Young (1983) later presented a counterexample in which the matching of the western boundary current to the Sverdrup interior places a condition on the interior flow that blocks the process of homogenization. However, the Ierley and Young argument, while rigorous, is limited to a rather artificial parameter regime in which relative vorticity is considered unimportant in the western boundary layer. From a theoretical point of view the issue remains obscure.

The question of the existence of sources of potential vorticity for the flow in the gyre is a complex one which we take up in greater detail in Chapters 4 and 5. However, as we have seen in Fig. 3.1.1, the interfaces between the upper density strata rise to the surface within the latitude band of the subtropical gyre where they are exposed directly to Ekman pumping. The potential vorticity of fluid in the surface mixed layer is determined by a complex interaction with heating and cooling by the atmosphere; then enters the thermocline in these outcrop zones and carries the information of the surface-determined potential vorticity to great depth within the thermocline. For such density layers it seems problematic from the beginning whether we can ignore such sources in determining the potential vorticity structure. It again, appears unlikely a priori, that such *ventilated* layers would satisfy the conditions of the theorem.

### *Numerical Experiments*

With this a priori uncertainty we are left to ask whether observations can decide the issue. Two types of "observations" are relevant. The first of these are the results of numerical experiments. Although numerical calculations cannot reproduce the natural ocean with sufficient fidelity to substitute for direct observations and with which they should never be confused, they can be used to help us test dynamical issues not tractable by analytical methods. Thus, for example, numerical experiments do contain western boundary currents, and within the physics of each experiment, whether it is a faithful reproduction of the natural ocean or not, it is possible to see whether the process of potential vorticity homogenization is disrupted by a western boundary current in which, substantial dissipation is allowed a priori. In particular, quasi-geostrophic experiments are particularly illuminating since they do not allow for isopycnal outcropping, and thus, within the model, the absence of potential vorticity sources during the journey of fluid through the interior in the lower layers allows us to concentrate on the possible role of the western boundary current in upsetting the process of potential vorticity homogenization.

A series of several numerical experiments with the quasi-geostrophic model each has demonstrated the ability of the homogenization of potential vorticity to proceed even with the western boundary current forced to act as a conduit for the closure of the circulation and the potential vorticity isolines. Figure 3.11.1 shows the circulation pattern in the three-layer quasi-geostrophic model as calculated by Rhines and Schopp (1991). In this model layer 1 is directly forced by the Ekman pumping, and layer 3 is too deep (4000 m) to have closed geostrophic contours in the interior, and its circulation is limited to very small weak local gyres driven by strong inertial effects associated with the western boundary current. Layer 2, on the other hand, does have closed geostrophic contours, and in fact we see that layer 2 shows a large region in which $q_2$ is essentially homogenized. The isolines of $q_2$ are expelled from the center of the gyre which is left with a plateau of uniform potential vorticity. Rhines and Schopp show that this zone is determined almost exactly by the expected shape of the analytically determined pool region. A detailed analysis of the calculated western boundary current by the same authors showed it to have a multiple layer structure. Over most of the width of the western boundary current most of the streamlines pass through the current in an outer boundary layer in which friction is unimportant. The potential vorticity of only a small inner region is substantially affected by dissipation. Thus a posteriori the most important requirements of the potential vorticity homogenization process are met on most streamlines. The mixing of potential vorticity *within* the pool turns out not to be small, but as we remarked above, it is necessary only that it be small

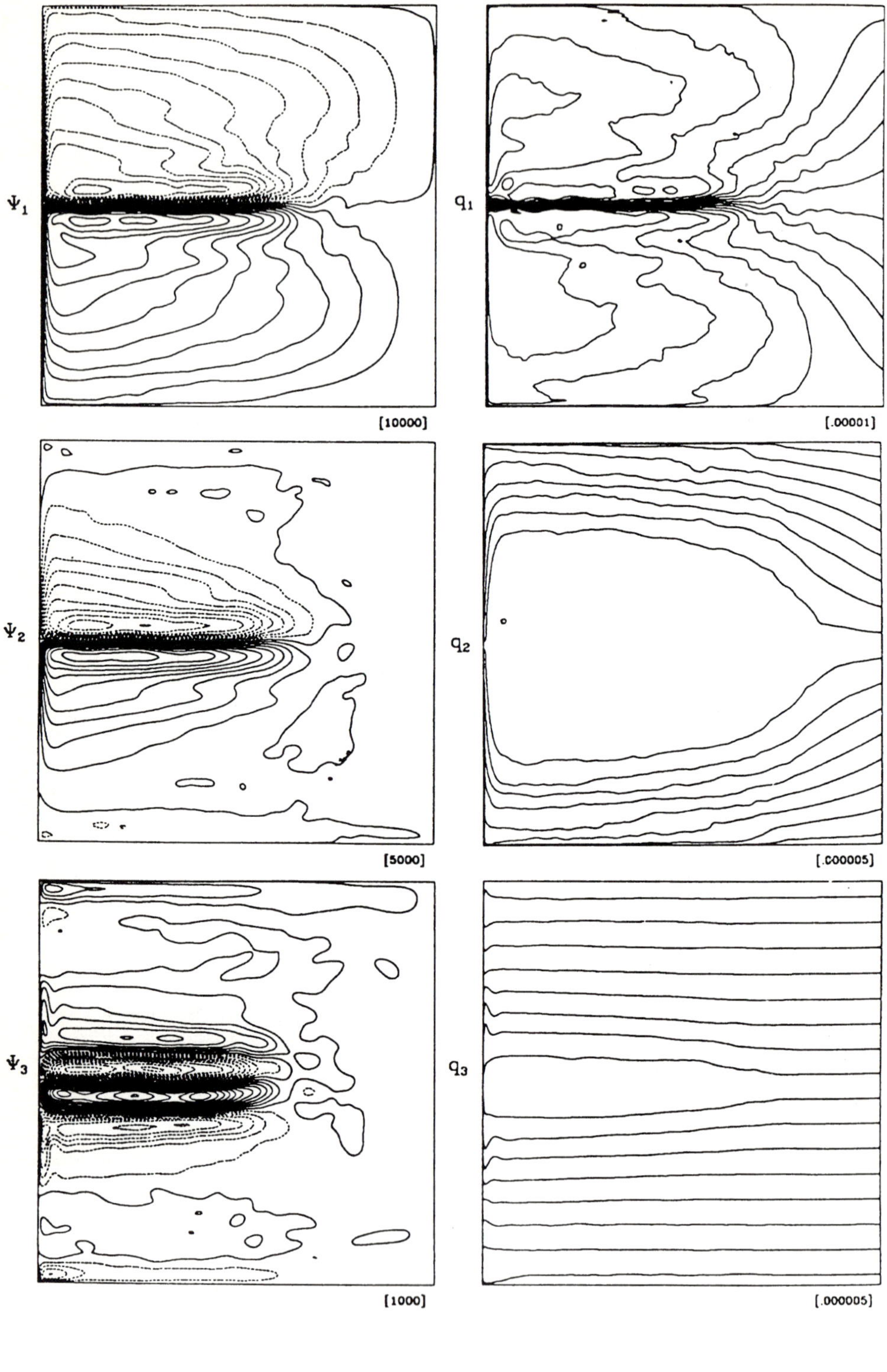
$\Psi_1$
[10000]
$q_1$
[.00001]
$\Psi_2$
[5000]
$q_2$
[.000005]
$\Psi_3$
[1000]
$q_3$
[.000005]

on the edge of the pool where the outer potential vorticity contour must be wrapped around the pool by inertial processes, subsequently enclosing the pool with a constant boundary value for potential vorticity.

It is very important to note that in this experiment, as in most of the detailed numerical investigations of the homogenization process, the wind field that is used is antisymmetric about a midlatitude, yielding a double-gyre circulation. There is no net input of vorticity. As in the Marshall double-gyre calculation described in Chapter 2, this implies that the fluid need not export an order 1 amount of potential vorticity through the western boundary. Eddies can allow the flow to reach equilibration through fluxes of vorticity from the subpolar to subtropical gyres where the vorticity input in one gyre is exactly balanced in the other. This favors the homogenization process by obviating the need for dissipation in the western boundary current.

One of the first numerical demonstrations, of the process of potential vorticity homogenization was that of Rhines and Young (1982) who presented the result of a calculation by Holland shown in Fig. 3.11.2. The figure, again, shows the middle layer of a three-layer model in which the layer evolves to a state in which the central zone possesses uniform potential vorticity. Panel b shows a three-dimensional rendering of $q_2$ over the gyre showing the plateau of constant $q_2$. What is particularly interesting is that in addition to the homogenization of the large-scale, time-averaged potential vorticity, the potential vorticity fluctuations are also expelled from the pool region. Panel c shows the instantaneous field of $q_2$, and it is clear that there are also no fluctuations of $q_2$ in the pool region. This is consistent with a law such as (3.8.9) in which the fluctuations of the potential vorticity are due to the eddy advection of large-scale potential vorticity. If the potential vorticity is uniform on the large scale, the turbulent motion of a fluid element carrying its potential vorticity from one position to another leaves the local value of the potential vorticity unchanged so that the eddying motion is not accompanied by a fluctuation in the potential vorticity.

Thus, in those controlled cases where the physics is otherwise a priori favorable to potential vorticity homogenization the process is seen to proceed to completion even when the fluid must complete its circuit in a western boundary current for which there is no adequate theory, and even when the dissipation of large-scale potential vorticity is accomplished by an eddy field whose effect on the mean flow is not specified a priori as a diffusion process. This is a real conceptual triumph for the theory.

**Fig. 3.11.1.** Streamline and potential vorticity patterns for the time-averaged flow in a three-layer quasi-geostrophic model. Left *panels,* streamfunction in layers 1–3; right panels, potential vorticity in each layer. Layer 1 is in contact with the surface forcing. Layer 3 is very deep (4000 m) and has no closed contours. Layer 2 has geostrophic contours which close through the western boundary layer. Note the homogenization of potential vorticity in this region. (From Rhines and Schopp 1991)

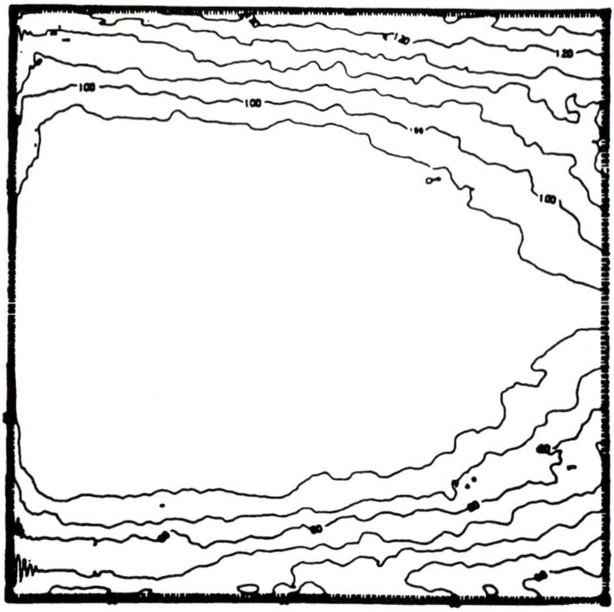

a

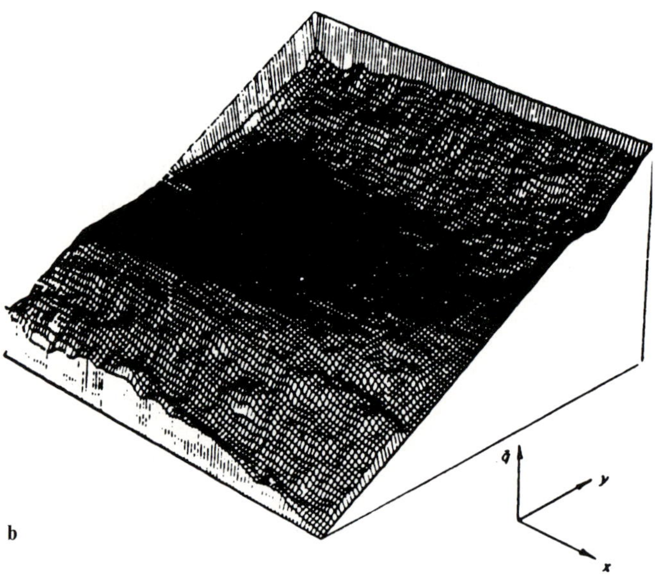

b

**Fig. 3.11.2a–c.** Potential vorticity in the second layer of a three-layer model showing the homogenization of $q_2$. As in Fig. 3.11.1 the flow is driven by an anti symmetric Ekman pumping leading to an antisymmetric double gyre. **a** Contours of the time-averaged $q_2$ showing a zone of uniform $q_2$ straddling the inter gyre boundary. **b** A three-dimensional perspective of the same result. **c** The *instantaneous* contours of $q_2$. Note the expulsion of $q_2$ from the pool even for the instantaneous field of potential vorticity rendering the pool free from potential vorticity fluctuations. (From Rhines and Young 1982b)

Fig. 3.11.2c

*Observational Evidence*

In the midocean, on the large scale, the relative vorticity makes a negligible contribution to the potential vorticity. In general, for an isopycnal layer of thickness $\Delta z$, the potential vorticity on the planetary scale is simply:

$$q = \frac{f}{\Delta z} \tag{3.11.1}$$

which is a finite difference approximation (using density as a vertical coordinate) for the continuous representation of potential vorticity on the planetary scale (Pedlosky 1987):

$$q = f\frac{\partial \rho}{\partial z} \tag{3.11.2}$$

for which (3.10.6) (without the relative vorticity) is the geostrophic approximation.

It follows that in the midocean it is possible to calculate $q$ directly from hydrographic data, that is, entirely from the density field without any reference to the velocity field, which is generally not known. It is therefore possible to prepare maps of the large-scale planetary potential vorticity and to determine whether regions of uniform potential vorticity can be observed. The elegant character of the Rhines and Young theory has inspired several careful investigations of the planetary potential vorticity on the large scale. Principal among these are McDowell et al. (1982), Holland et al. (1984), Keffer (1985), and Talley (1985, 1988).

The results are quite striking. Each density surface can be characterized by whether or not its outcrop line lies in the subtropical gyre in late winter, which is the season when the mixed layer is able to affect the potential vorticity distribution of the permanent thermocline. During the course of the year, as warming of the surface takes place in summer, the surface of the ocean is increasingly covered by warm water, and the outcrop line of colder layers occurs at higher latitudes. When the cooling season starts, the outcrop line of each density surface moves southward, exposing colder water. It is, however, the position of the outcrop line in late winter which is decisive for the ventilation of the time-averaged thermocline. The reason for the choice of this season is presented in more detail in the next chapter, where the physics of the ventilation of the thermocline is discussed. For now we simply accept that season as the appropriate one for deciding whether or not a particular density surface is substantially exposed to Ekman pumping and hence to sources of potential vorticity.

We should expect homogenization to occur only on density surfaces which are unventilated by the mixed layer. That is, they should be deep enough that the positions where the density surface intersects the sea surface, the outcrop line, occur too far north to occur in a region of downward Ekman pumping.

Figure 3.11.3 (McDowell et al. 1982) shows the potential vorticity on selected density surfaces in the North Atlantic. Panel a shows $q$ on a fairly shallow density surface with a mean density between $\sigma_\theta = 26.3$ and 26.5. [The density at a particular depth is a function of temperature, pressure, and salinity. The potential density $\sigma_\theta$ is the density in cgs units after a correction is made to eliminate the effect of the adiabatic increase in temperature that the fluid element suffers as a consequence of the increase in pressure with depth. A more complete discussion can be found in many texts, for example, Pickard and Emery (1982)]. This surface ranges from about 100 m in the eastern part of the subtropical gyre to about 400 m in the west, just outside of the western boundary current. Its winter time outcrop line trends northwest to southeast probably as a result of the movement of the clockwise circulation of the fluid. The winter time outcrop line lies well within the subtropical gyre, and the potential vorticity on this surface is nonuniform and clearly not homogenized as it is subject to communication with the surface Ekman layer. A more recent analysis of the potential vorticity on essentially the same surface has been carried out by Lozier et al. 1996 and is shown in Fig. 3.11.4. This figure shows the depth of the density surface at $\sigma_\theta = 26.5$ and the potential vorticity on that surface. The shape of the gyre in the potential vorticity field is quite apparent. The potential vorticity contours are bent around in the shape of the gyre, but the potential vorticity is not homogenized. Ventilation of the density surface is too strong at this level to allow homogenization to proceed.

A deeper density interval, $\sigma_\theta = 26.5$–$27.0$, is shown in Fig. 3.11.3b. The outcrop line is far enough north that now only weak sources of potential vorticity affect the layer in the subtropical gyre. Its outcrop line lies in a zone of weak Ekman pumping. The potential vorticity is strikingly uniform. Indeed, it seems to extend all the way across the basin to very near the eastern boundary, as we would expect for a layer whose outcrop line is near the edge of the subtropical gyre. However, at the basc of the thermocline, with $\sigma_\theta = 27.0$–$27.3$, the potential vorticity is again nonuniform, with contours which run from the outcrop line and swing around to the western boundary. The interpretation of this layer is subject to some disagreement. McDowell et al. suggest that this

a 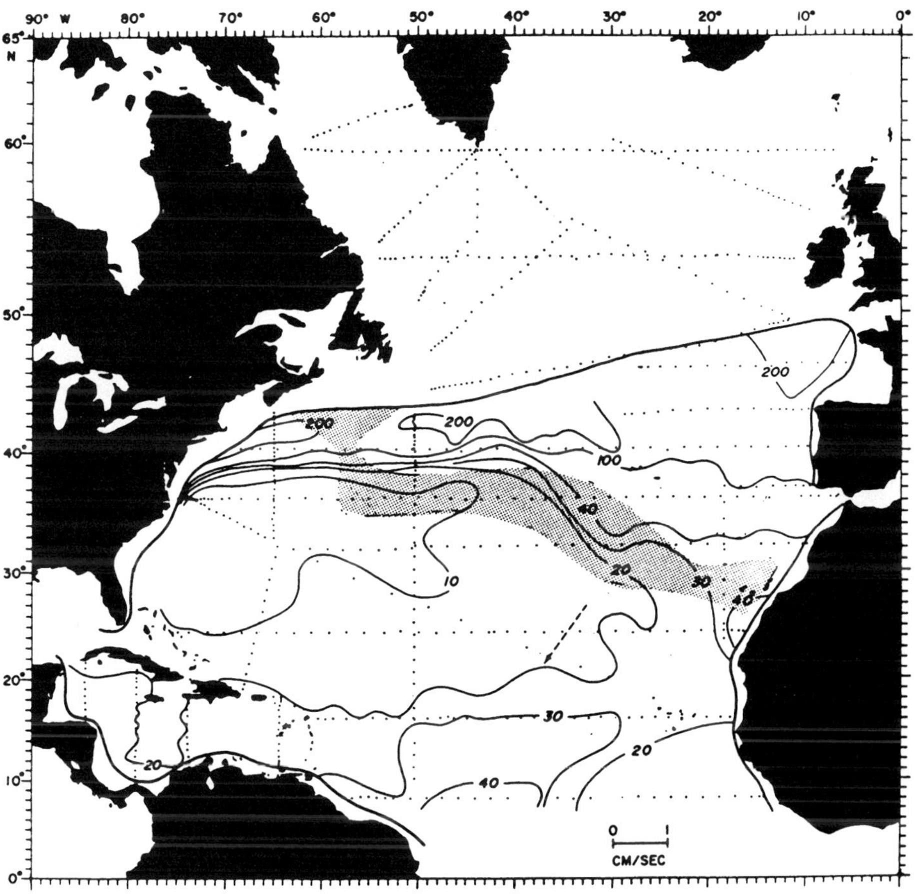


**Fig. 3.11.3a–c.** Planetary potential vorticity from hydrographic data for the North Atlantic at selected density intervals. **a** $\sigma_\theta = 26.3 - 26.5$. **b** $\sigma_\theta = 26.5 - 27$. Note the large region of nearly uniform potential vorticity. **c** $\sigma_\theta = 27.0 - 27.3$. (From McDowell et al. 1982)

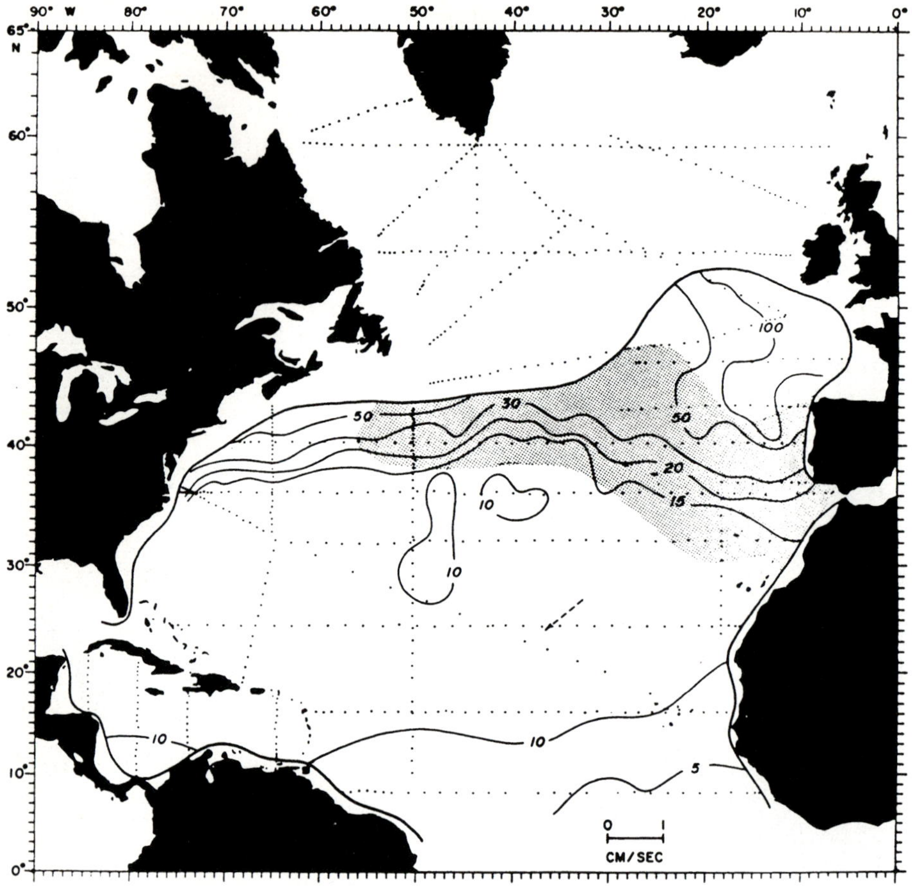


**Fig. 3.11.3b**

layer represents a layer with little or no motion even though the $q$ contours avoid the eastern boundary. Keffer (1985) associates this layer with the subtropical recirculation of Subpolar Mode Water and suggests that it actually forms south of the zero Ekman pumping line so that it is a layer which is ventilated and in motion. Indeed, Keffer has suggested that in the North Atlantic the region of uniform $q$ seen on the $\sigma_\theta = 26.75$ surface may be not connected at all with potential vorticity homogenization. He argues that this surface is in fact far enough south of the inter gyre boundary that it is subject to forcing of its potential vorticity by the mixed layer, and that the uniformity may be due to a fortuitous uniformity of the potential vorticity in the source region for the layer. There is general agreement, however, that the equivalent layer in the North Pacific, $\sigma_\theta = 26.5$ (see also Talley 1988) is absent of sources of potential vorticity and contains a large zone of uniform potential vorticity,

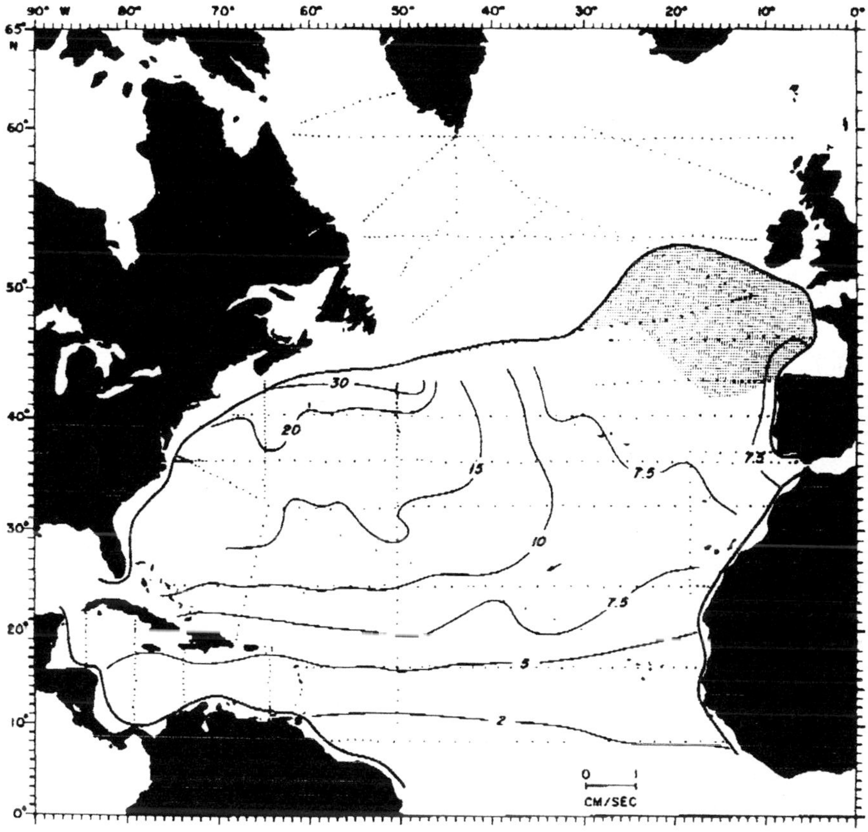


Fig. 3.11.3c

as shown in Fig. 3.11.5, in agreement with our expectations concerning homogenization of unventilated layers.

The observational evidence for the existence of large zones of nearly uniform potential vorticity on *unventilated* density surfaces is strong, circumstantial evidence for the validity of the theory of potential vorticity homogenization. It is not decisive because of our inability conclusively to test the thoretical prediction of the structure of the zones and compare them with observations. There may in fact be several ways to achieve uniformity on isopycnal layers in the subtropical gyre, including layers which outcrop within the gyre. However, the observed penetration of the circulation to density layers well beneath those which are anywhere driven by Ekman pumping in the subtropical gyre (especially evident in the Pacific), and the observation of uniform $q$ on those surfaces is a strong argument in favor of the theory.

The situation is much less clear on density surfaces which outcrop somewhere within the region of downward Ekman velocity. These surfaces

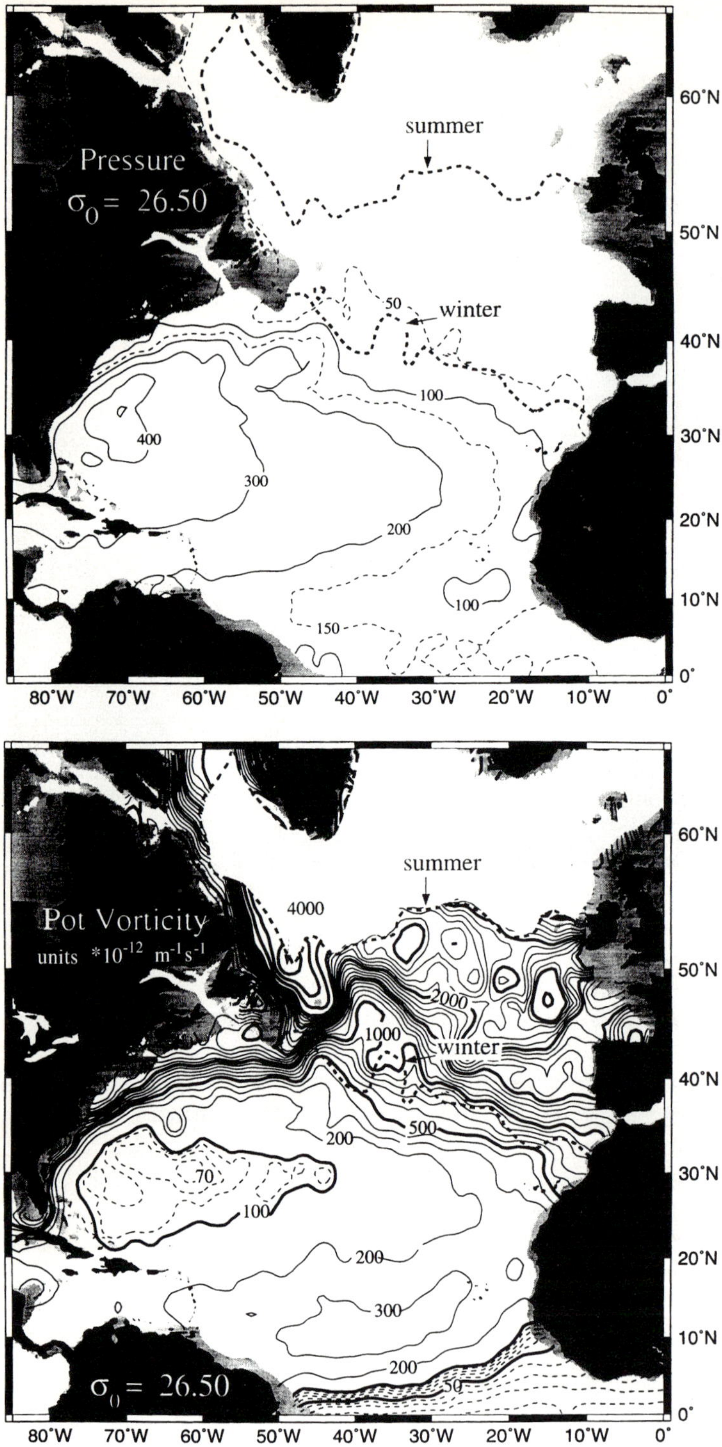


**Fig. 3.11.4a,b.** As in the previous figure for $\sigma_\theta = 26.5$. (From Lozier et al. 1996). **a** Depth of the density surface. **b** Potential vorticity on the same surface

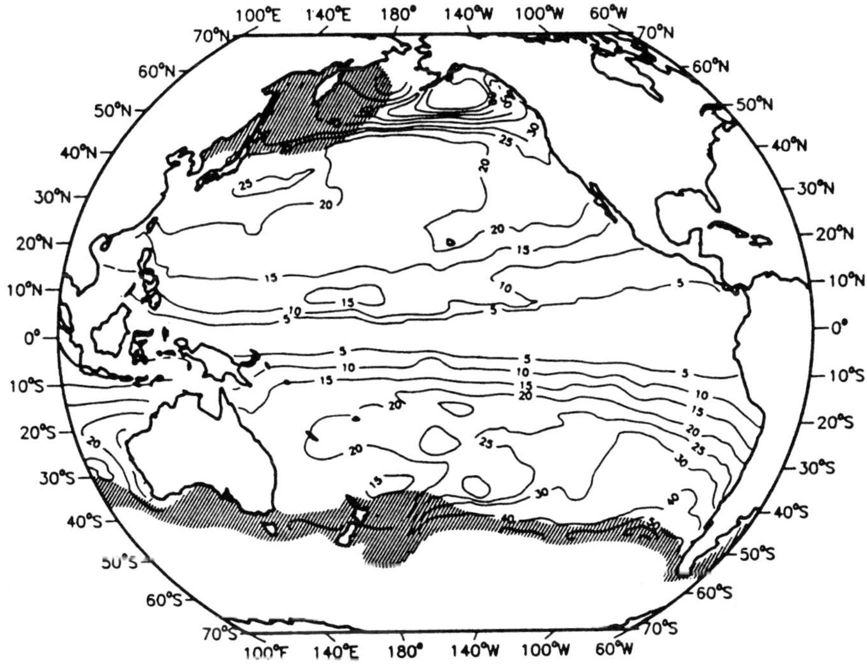


**Fig. 3.11.5.** Potential vorticity on the $\sigma_\theta = 26.25 - 26.75$ surface in the Pacific ocean. (From Keffer 1985)

may be shielded from direct Ekman pumping over most of their horizontal extent. Are these driven by potential vorticity homogenization or by sources of potential vorticity in the zone where they outcrop, or by some combination of both?

Quasi-geostrophic theory is unable to deal with this issue, and the next chapter takes up the question of the dynamics of density layers and interfaces which suffer such large changes in depth that quasi-geostrophic theory is no longer accurate, and whose outcropping introduces an entirely new mechanism to set deeper layers into motion, namely, ventilation of the thermocline by the surface layer of the ocean.

## References

Anderson, D.L.T. and Gill A. E., 1975: Spin-up of a stratified ocean, with application to upwelling. *Deep - Sea Res.*, **22**, 593–596.

Batchelor, G.K. 1956: On steady laminar flow with closed streamlines at large Reynolds numbers. *J. Fluid Mech.*, **1**, 177–190.

Holland, W.R. 1978: The role of mesoscale eddies in the general circulation of the ocean – numerical experiments using a wind-driven quasi-geostrophic model. *J. Phys. Ocean.*, **8**, 363–392.

Holland, W.R. and Rhines P.B., 1980: An example of eddy induced circulation. *J. Phys. Ocean.* , **10**, 1010–1031.

Holland,W.R., Keffer T. and Rhines P.B., 1984: Dynamics of the oceanic circulation: the potential vorticity field. *Nature* , **308**, 698–705.

Ierley G.R. and Young W. R., 1983: Can the western boundary layer affect the potential vorticity distribution in the Sverdrup interior of a wind gyre? *J. Phys. Ocean.*, **13**, 1753–1763.

Keffer, T. 1985: The ventilation of the world's oceans: maps of potential vorticity. *J. Phys. Ocean.*, **15**, 509–523.

Ledwell, J., Watson A.J. and Law C. S., 1993: Evidence for slow mixing across the pycnocline from an open-ocean tracer release experiment. *Nature*, **364**, 701–703.

Levitus, S., 1982: *Climatological atlas of the world ocean*, NOAA Prof. paper **13**, U.S. Govt. Printing Office, Washington, D.C. 173 pp.

Lozier, M.S., Owens W.B., Curry, R.G.,(1996) The climatology of the North Atlantic. Prog. Oceanogr. 36: 1–44

McDowell, S., Rhines P. and Keffer, T., 1982: North Atlantic potential vorticity and its relation to the general circulation. *J. Phys. Ocean.*, **12**, 1417–1436.

Pedlosky, J., 1987: Geophysical fluid dynamics, *Springer,* Berlin Heidelberg NewYork, pp 710.

Pickard, G.L. and Emery, W.J. 1982: Descriptive physical oceanography: an introduction, 4$^{th}$ edn (SI). *Pergamon Press,* Oxford, pp 249.

Rhines, P.B. and Schopp, R., 1991: The wind-driven circulation: Quasi-geostrophic simulations and theory for nonsymmetric winds. *J. Phys. Ocean.*, **21**, 1438–1469.

Rhines, P.B., and Young, W. R., 1982 a: A theory of the wind-driven circulation. I. Mid-ocean gyres. *J. Marine Res.*, **40** (Supplement), 559–596.

Rhines, P.B., and Young, W.R., 1982 b: Homogenization of potential vorticity in planetary gyres. *J. Fluid Mech.*, **122**, 347 –367.

Rhines, P.B., and Young, W.R., 1983: How quickly is a passive scalar mixed within closed streamlines? *J. Fluid Mech.* , **133**, 133–145.

Talley, L.D. 1985: Ventilation of the subtropical North Pacific: the shallow salinity mimimum. *J. Phys. Ocean.*, **15**, 633–649.

Talley, L.D. 1988: Potential vorticity distribution in the North Pacific. *J. Phys. Ocean.*, **18**, 89–106.

Young, W.R. 1986: Baroclinic theories of the wind - driven circulation. In: *General circulation of the ocean.* Eds. Abarbanel, H.D.I. and Young, W.R. *Springer,* Berlin Heidelberg New York, pp 134–200.

Young, W.R., and Rhines P.B., 1982: A theory of the wind-driven circulation. II. Gyres with western boundary layers, *J. Marine Res.*, **40**, 849–872.

# 4 Theory of the Ventilated Thermocline

## 4.1 Introduction

One of the fundamental problems in physical oceanography is explaining the observed density structure of the ocean. The density field varies significantly in all three spatial directions, and perhaps its most striking feature is the limitation of the largest density variations to the upper 1–2 km of the ocean. The density increases with depth almost everywhere in the ocean, but the major part of this increase occurs in the upper kilometer. Figure 4.1.1 from the Levitus *Atlas* (1982) shows basin-averaged density profiles for the major oceans in which this feature is manifest. The density increases rapidly to its abyssal value in a rather short distance compared to the total depth of the ocean. In Section 3.1 we described the link between the motion field and the field of density that this implies. The two are intimately coupled through the advection/diffusion equation for density and the thermal wind relation for the velocity. We also noted in Chapter 3 that the explanation of the density structure requires a dynamical explanation of the circulation.

The density field varies strongly in the horizontal direction. Figure 4.1.2 shows the zonally averaged density field for the Pacific Ocean, also from the Levitus *Atlas*, and its qualitative resemblance to the corresponding figure for the Atlantic, (Fig. 3.1.1) suggests that the problem of understanding the density field is of a general dynamical nature not connected with any peculiar feature of one or another of the major ocean basins. The density field in fact varies so strongly that the overall *horizontal* variation in density is as large as the overall *vertical* variation in density in each of the gyres. This implies that an explanation of the observed density field requires, at a minimum, a relaxation of the quasi-geostrophic approximation in which the horizontal density variation is supposed small with respect to the vertical variation. Isopycnal surfaces experience an $O(1)$ change in depth over the horizontal extent of the gyre (see Fig. 4.1.3; Lozier et al. 1996). At the same time a complete theory for the density field should dynamically predict the vertical as well as the horizontal variation of density and the associated velocity field. Moreover, we anticipate that the horizontal and vertical structures are linked through the motion field. This is the problem of the thermocline. It is nonlinear and difficult.

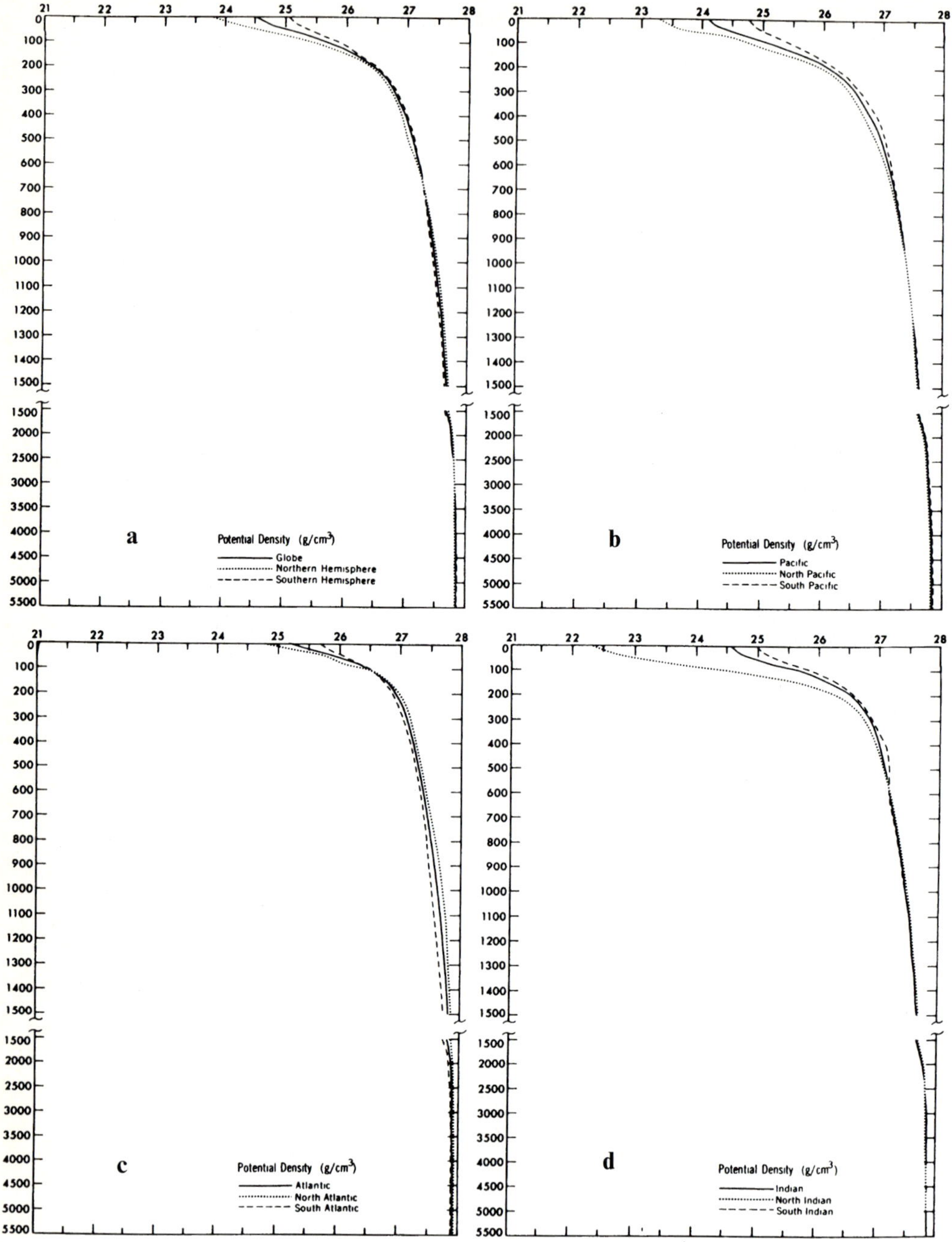


**Fig. 4.1.1a–d.** Basin-averaged potential density versus depth curves for the principal oceans. **a** Global average, **b** Pacific Ocean, **c** Atlantic Ocean, **d** Indian Ocean. (From Levitus 1982)

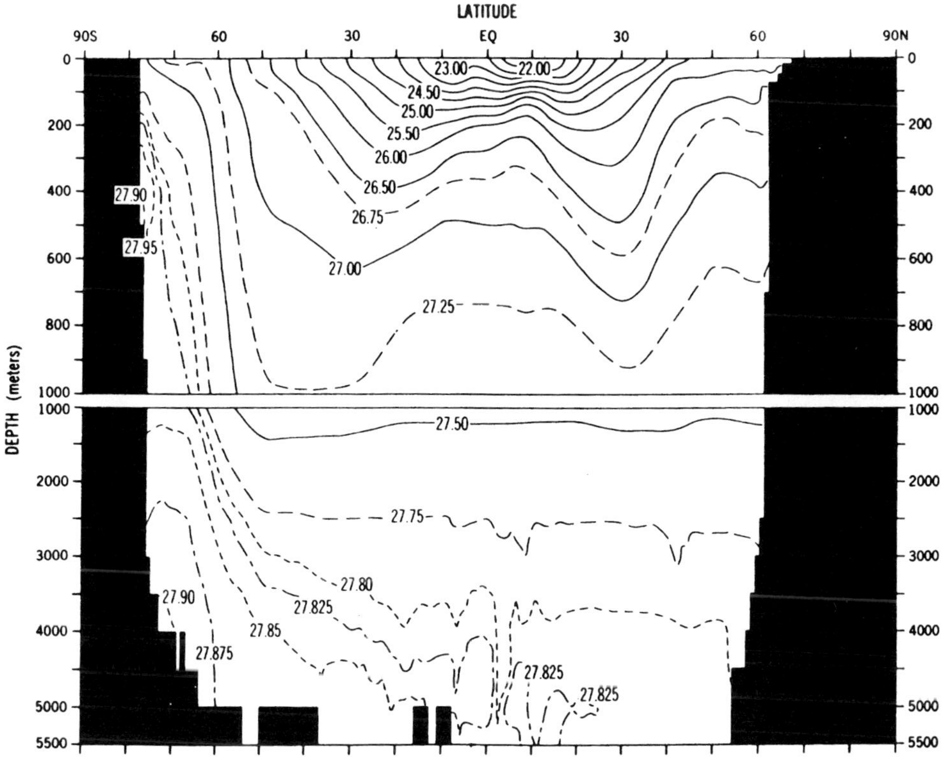


**Fig. 4.1.2.** Zonally averaged potential density field for the Pacific Ocean. Note the change of vertical scale at 1000 m, reflecting the decrease in gradient below this depth. (From Levitus 1982)

The previous chapter described a quasi-geostrophic theory for unventilated density layers. These are layers which are shielded everywhere over the gyre from the effects of Ekman pumping and the alteration of potential vorticity in the oceanic mixed layer. This chapter builds on those results but in addition confronts directly the fact that the density surfaces in a large part of the thermocline outcrop within the domain of the subtropical gyre, and that their dynamics is strongly influenced by this outcropping.

The influence has been long recognized which the interaction of the fluid with the atmosphere-ocean interface has on the structure of the thermocline. Over 50 years ago Iselin (1939) and Montgomery (1938) suggested that the vertical structure of the density field is a mapping of the latitudinal surface variation carried downward into the gyre from the outcrop position by the circulation itself. Montgomery, especially, proposed a clear qualitative picture of the fluid adiabatically sliding down sloping density surfaces and carrying the surface values of tracer properties, such as salinity, into the thermocline. The

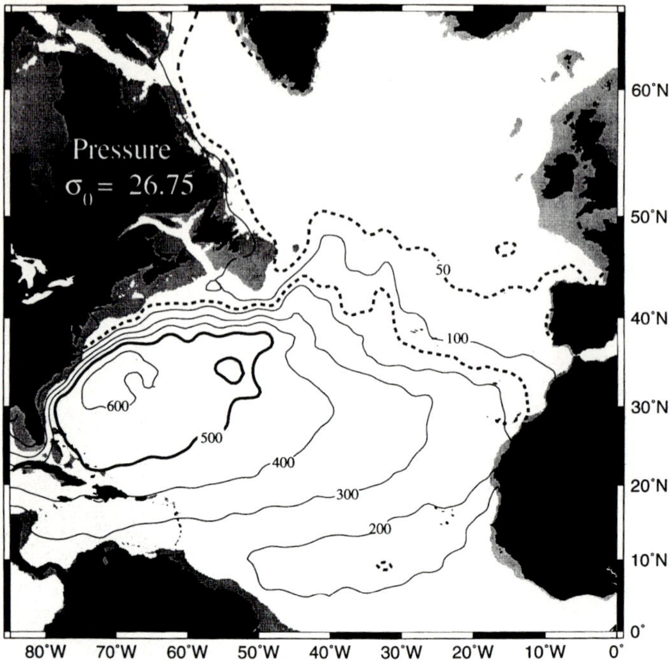


**Fig. 4.1.3.** Depth in decibars (1 decibar very nearly equals 1 m) of the density layer $\sigma_\theta = 26.75$ in the North Atlantic thermocline. (From Lozier et al. 1996)

principal technical difficulty of his explanation from a theoretical viewpoint is of course that the pathways of the flow are determined nonlinearly by the flow itself. The mapping of the surface properties, which are primarily a function of latitude, into a vertical variation of properties at a given geographical location is nonlinear. At each depth the density and other scalar properties, if carried in a quasiconservative manner, originate at different surface positions, and the trajectories connecting each point on a vertical sounding to the point of its origin on the surface is an important and a priori unknown part of the solution.

During the 1950s and 1960s another and quite different conceptual picture of the thermocline was developed. In this picture the thermocline is conceived as a thermal boundary layer trapped near the uppper surface and exists as a balance between the downward diffusion of heat by turbulent mixing countered by the upward advection of cold abyssal water. We now believe that the vertical turbulent diffusion in the thermocline is too small to effect this balance. Recent estimates of the diffusion coefficient, described in Chapter 3, render such a balance unlikely. Nevertheless, this conceptual picture was for a long time extremely influential and led to several very interesting mathematical investigations of the possible thermal structures that would arise from this balance. These efforts have been reviewed by Veronis (1969) and Pedlosky (1987a).

A principal mathematical feature of this early theoretical attempt to deal with the thermocline as a thermal boundary layer was the technical difficulty of the governing nonlinear partial differential equation. The vertical diffusion of density in the theory is balanced by both lateral and vertical advection of density. Both of the latter are of the same order as a consequence of estimates connected with the continuity equation linking the vertical and horizontal velocities. The great mathematical difficulty of including both diffusion and three-dimensional advection led to the search for structurally special solutions of the governing equations of so-called similarity type. That is, it was assumed (see, for example, Robinson and Stommel 1959) that the basic vertical structure of the thermocline is the same everywhere except insofar as the vertical scale is stretched differently at different horizontal locations. Similarity solutions of this type are common in fluid dynamics and are rigorously valid only when there are no significant external length scales to provide a natural scale against which the phenomenon can be measured . This is certainly not the case for the thermocline where the basin scale, the length scale of the wind, and the length scale associated with the surface density field are all important for the physics. It was also hoped that although the similarity solutions were unable to satisfy arbitrary and realistic surface boundary conditions, they would nonetheless be physically revealing. The complexity of the mathematics, however, limited the solutions to a fairly narrow class, and, most disappointingly of all , the physical content of the solutions remained unsatisfactorily sparse. The solutions all had structures that were frozen by the requirement of the similarity form and proved to be largely independent, for example, of the degree of vertical mixing that was assumed. Thus it was impossible to ask the solutions to reveal parametrically interesting structural variations with the degree of mixing. Worst of all, even though the mixing might be assumed small, the mathematical requirement that the similarity form of the solution be consistent with each term in the governing partial differential equation, *regardless of its size*, meant that dissipation played a determining role in selecting the form of the solution everywhere. This is not physically realistic. The search for complete similarity solutions was technically so challenging that the physical connection with the thermocline problem became increasingly tenuous.

An interesting alternative approach was taken by Welander (1959, 1971) in which he proposed a purely adiabatic, advective model for the thermocline more in line with the Iselin and Montgomery ideas discussed above. In Welander's theory the potential vorticity, $q$, is the key dynamic variable. In a nondissipative, steady flow it is constant on both density, $\rho$, and Bernoulli function, $B$, surfaces (Pedlosky 1987b; see also below). Thus in general $q = Q(\rho, B)$ where $Q$ is an arbitrary function of its arguments. Welander *chose* certain forms for $Q$ and found interesting solutions but was unable to describe a relationship between the function $Q$ and the prescribed external boundary conditions for the problem specified in a forward manner. That is, given the surface density field and the Ekman pumping and conditions of adiabatic flow it should be possible to determine the flow and the potential vorticity, i.e.,

$Q$, from first principles. Without such a forward method solutions of the adiabatic problem as well as the similarity solutions of the dissipative problem remained as isolated examples of shrewd analysis whose relationship with the underlying physical problem was obscure.

The Rhines and Young theory discussed in Chapter 3, limited though it is to a quasi-geostrophic theory for the thermocline velocity field, nevertheless restored the physics of the problem to center stage. Shortly after the appearance of the Rhines and Young theory, Luyten, Pedlosky, and Stommel (1983), stimulated by that work, presented a theory for the thermocline which included the physics of outcropping which was lacking in the Rhines and Young theory. The Luyten et al. theory is meant to explain both the vertical and the horizontal variation in the density field for layers which outcrop in the subtropical gyre in terms of the surface density field. The two theories together form a complementary picture of the thermocline circulation, and their union and the subsequent work of others have led to a rich theoretical picture of the midocean thermocline structure. The purpose of this chapter is to describe some of the key aspects of these developments.

As in earlier chapters we assume that the Sverdrup theory is valid in the interior, and that the solutions which we find for the interior structure of the thermocline circulation are consistent with closure of the circulation through a western boundary current. The reader should keep in mind that neither of these two assumptions has been clearly demonstrated.

Before taking up the thermocline theories themselves we review some general dynamical fundamentals. This is useful for placing our approximations in context and allows us later to discuss extensions of the theory to the equatorial regions where the geostrophic approximation fails.

## 4.2 Formulation of the Model

We concentrate here on layer models of the thermocline and defer to Section 4.11 the formulation and solution of the continuously stratified problem. Consider the situation depicted in Fig. 4.2.1. A layer of constant density $\rho_n$ is bounded above by the interface $z = z_n$ where $z_n$ is a function of longitude, $\phi$, latitude $\theta$, and time, $t$. The horizontal velocity in each layer is independent of $z$. This is a modeling decision which becomes rigorously required in the limit when the motion is geostrophic. The thickness of the layer is $h_n(\phi_q, \theta t)$. The local value of the Coriolis parameter is $f = 2\Omega \sin\theta$. At each interface a cross-isopycnal velocity $w*$ is allowed. As the fluid crosses the interface, its density changes instantaneously to adapt to the density of the layer that it enters. As discussed in Chapter 3, this is equivalent to the specification of a local heating (or cooling) to accomplish the density change.

If we consider a small cylindrical control volume in which the area in the horizontal plane is fixed with time such that fluid flows are unimpeded across

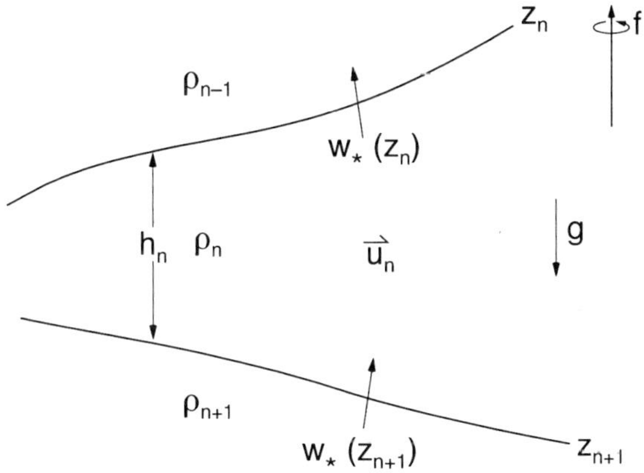


**Fig. 4.2.1.** Layer model. Layer $n$ is bounded above by the interface $z = z_{n+1}(\phi, \theta)$ and from below by $z = z_{n+1}(\phi, \theta)$. At each interface a cross-isopycnal mass flux per unit of area in the horizontal plane, $w*$, is allowed. Within each layer the density is constant, and the horizontal velocity is depth independent

the vertical elements of the cylinder (e.g., Fig. 4.2.2), we can construct a balance for the volume, or, since the layer density is fixed, for the mass of the region. In a unit horizontal area the volume in the control volume is simply $h_n$. The increase of this volume must be due to net volume flux into the control volume. Through the vertical boundaries the *outward* volume flux is $\nabla \cdot (\vec{u}_n h_n)$, where the divergence operator is understood to be the two-dimensional divergence in the horizontal plane. and $\vec{u}_n$ is the horizontal velocity. The net volume flux into the control volume across the two nearly horizontal interfaces bounding the control volume horizontally is $w*(z_{n+1}) - w*(z_n)$. Balancing the net volume flux with the increase in the volume per unit horizontal area leads us, again, to (3.2.17) as the layer form of the mass conservation statement, namely:

$$\frac{\partial h_n}{\partial t} + \nabla \cdot (\vec{u}_n h_n) = w*(z_{n+1}) - w*(z_n). \tag{4.2.1}$$

We have reviewed the argument for the mass balance to prepare for the somewhat more subtle balance required to deal with the momentum. Consider the control volume again, as shown in Fig. 4.2.3. First let us calculate the net pressure force on the volume. For simplicity consider the horizontal $x$ direction. The pressure force on the left edge of the volume is $p_n h_n$ while the

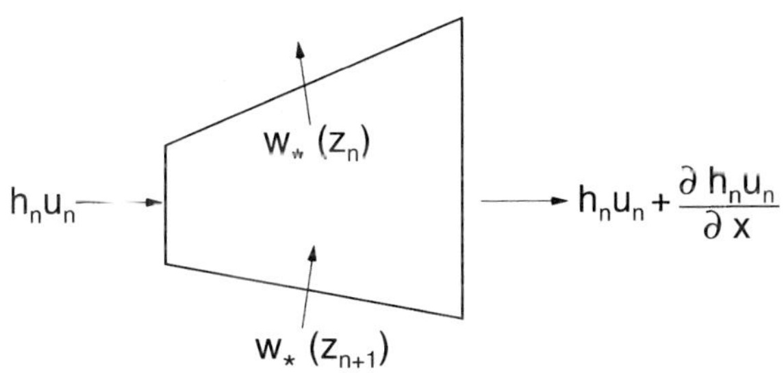


**Fig. 4.2.2.** Elementary control volume constructed for layer $n$. Only the projection along the $x$ axis is shown. The net volume flux into the volume must lead to an increase in the volume of the region

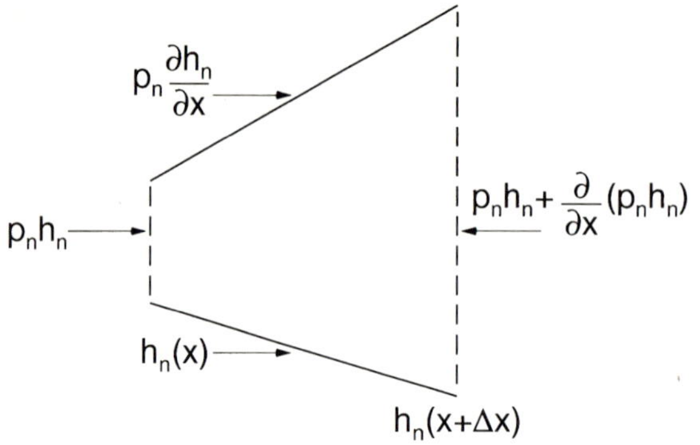


**Fig. 4.2.3.** Pressure forces in the $x$ direction on the control element. Only the projection of the element in the $x - z$ plane is shown

pressure force in the negative $x$ direction on the right face, a distance $\Delta x$ away, is $-p_n h_n - \partial(p_n h_n)/\partial x$. On the sloping interfaces there is a pressure force in the positive $x$ direction equal to $p_n \partial h_n/\partial x$. Thus the net pressure force in the $x$ direction is $-h_n \partial p_n/\partial x$ or in general vector form $-h_n \nabla p_n$.

The volume flux across the density interfaces carries momentum as well as volume. At the surface $z = z_n$, for example, there is a momentum flux per unit of horizontal area equal to $w*(z_n)\vec{u}_n$ *leaving* the layer if $w*(z_n)$ is positive and equal to $-w*(z_n)\vec{u}_{n-1}$ *entering* the layer if $w*(z_n)$ is negative. In each case the momentum flux depends on the direction of the cross-isopycnal flux since this is what determines whether it carries momentum originating in layer $n$ or in layer $n-1$. A similar consideration holds for the $n + 1^{\text{st}}$ interface. If the local rate of change of momentum is set equal to the net forces acting on the control volume plus the net flux of momentum into the volume, we obtain:

$$\begin{aligned}\frac{\partial h_n \vec{u}_n}{\partial t} = &- f\hat{k} \times h_n \vec{u}_n - \nabla \cdot (h_n \vec{u}_n \vec{u}_n) - h_n \nabla p_n / \rho_n \\ &- w*(z_n)[\vec{u}_n \Theta\{w*(z_n)\} + \vec{u}_{n-1}\Theta\{-w*(z_n)\}] \\ &+ w*(z_{n+1})[\vec{u}_{n+1}\Theta\{w*(z_{n+1})\} + \vec{u}_n \Theta\{-w*(z_{n+1})\}] \\ &+ h_n \vec{F}_n \end{aligned} \tag{4.2.2}$$

where we introduce the notation for the Heaviside step function:

$$\Theta(x) = \begin{cases} 1 & x > 0 \\ 0 & x < 0 \end{cases} \tag{4.2.3}$$

in order economically to write the sign-dependent fluxes of momentum across the isopycnal surfaces. The function $h_n \vec{F}_n$, multiplied by the density $\rho_n$, is the frictional force per unit horizontal area acting on the volume. The second term on the right side of (4.2.2) is the dyadic divergence of horizontal momentum and can be rewritten in proper vector form as:

$$\nabla \cdot (h_n \vec{u}_n \vec{u}_n) = (h_n \vec{u}_n) \cdot \nabla \vec{u}_n + \vec{u}_n \nabla \cdot (h_n \vec{u}_n). \tag{4.2.4}$$

If the mass conservation equation is used after expanding the time derivative in the first term in (4.2.2) we obtain for the momentum equation:

$$\begin{aligned} \frac{d\vec{u}_n}{dt} + f\hat{k} \times \vec{u}_n = & -\frac{1}{\rho_n} \nabla p_n + \vec{F}_n \\ & + w*(z_{n+1}) \left\{ \frac{\vec{u}_{n+1} - \vec{u}_n}{h_n} \right\} \Theta\{w*(z_{n+1})\} \\ & + w*(z_n) \left\{ \frac{\vec{u}_n - \vec{u}_{n-1}}{h_n} \right\} \Theta\{-w*(z_n)\}. \end{aligned} \tag{4.2.5}$$

If the cross-isopycnal mass flux at the upper interface $z = z_n$ is positive, there is no change produced in the momentum *per unit volume* of the $n^{\text{th}}$ layer by motion across that interface. Momentum leaves the layer, but as it leaves, the density of the momentum remains unchanged. (Note that the Heaviside function is zero in this case.) On the other hand, if the cross-isopycnal flux across this interface is negative, fluid from the layer above brings in fluid with a different density of momentum proportional to the velocity of the $n - 1^{\text{st}}$ layer. The change in momentum density is then proportional to the rate of cross-isopycnal flux of the difference of the velocities of the two layers. The same occurs at the lower interface. There is a change in momentum density only if fluid enters the layer, in which case the change is proportional to the velocity difference across the layers. If the cross-isopycnal mass flux is outward across both interfaces it leads to no change, by itself, in the velocity of the layer. The layer thickness tends to contract, and there is a consequent change in the total momentum $h_n \vec{u}_n$ per unit area, as described by (4.2.2), but not in the velocity itself (which is the momentum density). The reader is invited now to repeat the argument for a passive tracer (money!) to understand the dependence of the momentum flux on the direction of the cross-isopycnal velocity. Thus, imagine a room full of people who agree to obey the rule that each shares equally the available cash in the room (which is equivalent to attributing the same velocity to every element in the layer under consideration). When a rich or poor person then *enters* the room with an amount of cash which differs from that which each person already carries, the subsequent redistribution of wealth changes the content of each person's pocketbook. If a person, however, *leaves* the room, the cash possessed by each person is *unaltered* although the total wealth of the room, of course, diminishes – in perfect analogy with the change in the total momentum of our layer.

It is convenient to combine the momentum flux associated with the cross-isopycnal mass flux with the friction term and write the sum as a single dissipative term, i.e., we define:

$$\Im_n = \vec{F}_n + w*(z_{n+1})\left\{\frac{\vec{u}_{n+1} - \vec{u}_n}{h_n}\right\}\Theta\{w*(z_{n+1})\}$$
$$+ w*(z_n)\left\{\frac{\vec{u}_n - \vec{u}_{n-1}}{h_n}\right\}\Theta\{-w*(z_n)\} \tag{4.2.6}$$

so that the momentum equation for the $n$th layer can be compactly rewritten:

$$\frac{d\vec{u}_n}{dt} + f\hat{k} \times \vec{u}_n = -\frac{1}{\rho_n}\nabla p_n + \Im_n. \tag{4.2.7}$$

The total derivative:

$$\frac{d\vec{u}_n}{dt} = \frac{\partial \vec{u}_n}{\partial t} + \vec{u}_n \cdot \nabla \vec{u}_n \tag{4.2.8}$$

can be rewritten, using the vector identity $\vec{u}_n \cdot \nabla \vec{u}_n = \frac{1}{2}\nabla(\vec{u}_n \cdot \vec{u}_n) + \hat{k}\zeta_n \times \vec{u}_n$, so that (4.2.7) becomes

$$\frac{\partial \vec{u}_n}{\partial t} + (f + \zeta_n)\hat{k} \times \vec{u}_n = -\nabla\left(\frac{p_n}{\rho_n} + \frac{1}{2}|\vec{u}_n|^2\right) + \Im_n. \tag{4.2.9}$$

The vertical component of the relative vorticity is:

$$\zeta_n = \hat{k} \cdot \nabla \times \vec{u}_n. \tag{4.2.10}$$

We have used throughout the notation that $\hat{k}$ is a unit vector in the *local* vertical direction.

We *define*:

$$q_n = \frac{(f + \zeta_n)}{h_n}, \quad B_n = p_n + \frac{\rho_n}{2}|\vec{u}_n|^2 \tag{4.2.11a, b}$$

as the potential vorticity and the Bernoulli function for layer $n$ respectively. Thus (4.2.9) is:

$$\frac{\partial \vec{u}_n}{\partial t} + \hat{k} \times (\vec{u}_n h_n) q_n = -\frac{1}{\rho_n}\nabla B_n + \Im_n. \tag{4.2.12}$$

If we take the curl of (4.2.9) and examine the $z$ component of the result, we obtain,

$$\frac{\partial}{\partial t}(\zeta_n + f) + \nabla \cdot \left[\vec{u}_n(\zeta_n + f)\right] = \hat{k} \cdot \nabla \times \Im_n$$
$$= -\nabla \cdot \left(\hat{k} \times \Im_n\right). \tag{4.2.13}$$

This implies that the potential vorticity satisfies an equation which can be written entirely in flux form, namely:

$$\frac{\partial}{\partial t}(q_n h_n) + \nabla \cdot \left[\vec{u}_n h_n q_n + \hat{k} \times \Im_n\right] = 0. \tag{4.2.14}$$

The total, vertically integrated potential vorticity per unit horizontal area is $q_n h_n$. According to (4.2.14) its *local* rate of change is given by the divergence of the flux of potential vorticity out of this elementary cylinder. The flux of total potential vorticity is:

$$\vec{S}_q = \vec{u}_n h_n q_n + \hat{k} \times \Im_n. \tag{4.2.15}$$

The flux vector clearly lies in the horizontal plane since both $\vec{u}_n$ and $\hat{k} \times \Im_n$ do. This means that there is *no flux of potential vorticity across the isopycnal surfaces*. This is true even though the cross-isopycnal flux of mass, $w*$, differs from zero. This is a special case of the *impermeability theorem* (Haynes and McIntyre 1987; Marshall and Nurser 1992). This implies that potential vorticity can flux only along density surfaces into the layer from regions where dissipation is important or across boundaries in which the advective flux into the layer is not zero. It is the later case which, as we see below, is relevant to the ventilation process and the production of deep motion in the thermocline.

If the temporal and spatial derivatives in (4.2.14) are expanded, we obtain:

$$\begin{aligned} h_n \frac{\partial q_n}{\partial t} + \vec{u}_n h_n \cdot \nabla q_n + q_n \left[\frac{\partial h_n}{\partial t} + \nabla \cdot (\vec{u}_n h_n)\right] &= \hat{k} \cdot \nabla \times \Im_n \\ &\equiv \operatorname{curl} \Im_n. \end{aligned} \tag{4.2.16}$$

The mass conservation equation (4.2.1) allows (4.2.16) to be written:

$$\frac{dq_n}{dt} = \frac{q_n}{h_n}\left[w*(z_n) - w*(z_{n+1})\right] + \frac{\operatorname{curl} \Im_n}{h_n}. \tag{4.2.17}$$

If (a) the cross-isopycnal flux, $w*$, is zero at both interfaces, and (b) the frictional forces in the layer are zero [note from (4.2.6) that these are not entirely independent],then the potential vorticity:

$$q_n = \frac{f + \zeta_n}{h_n}$$

is conserved following the motion of a fluid column in the layer.

If the flow is also *steady*, so that the local time derivative $\partial/\partial t$ of any fluid property vanishes, then under the condition that there be no cross-isopycnal flux the mass conservation equation reduces to:

$$\nabla \cdot \vec{u}_n h_n = 0. \tag{4.2.18}$$

This allows the horizontal mass flux to be written in terms of a streamfunction, i.e.:

$$\vec{u}_n h_n = \hat{k} \times \nabla \psi_n. \tag{4.2.19}$$

This is valid as long as the flow is steady and adiabatic ($w* = 0$). Under the same conditions the momentum equation, (4.2.12) becomes:

$$q_n \hat{k} \times (\hat{k} \times \nabla\psi_n) = -\frac{1}{\rho_n}\nabla B_n \quad \text{or} \qquad (4.2.20\text{a,b})$$
$$\rho_n q_n \nabla\psi_n = \nabla B_n.$$

Two important and related consequences result from (4.2.20 b). Since, by definition, $\vec{u}_n \cdot \nabla\psi_n = 0$, the dot product of the velocity with the momentum equation yields:

$$\vec{u}_n \cdot \nabla B_n = 0. \qquad (4.2.21)$$

The Bernoulli function is therefore conserved on streamlines of the flow when the flow is steady, adiabatic, and frictionless. The function $B_n$ changes only from streamline to streamline and is therefore only a function of streamfunction $\psi_n$, i.e.:

$$B_n = B_n(\psi_n). \qquad (4.2.22)$$

The potential vorticity is also constant along streamlines for steady, frictionless, and adiabatic flow. (It is be conserved as well following the *trajectories* of fluid elements in unsteady flow as long as it is adiabatic and frictionless.) Thus for steady flow:

$$q_n = Q_n(\psi_n). \qquad (4.2.23)$$

Since $B_n$ is a function only of $\psi_n$ it follows that:

$$\nabla B_n = \frac{dB_n}{d\psi_n}\nabla\psi_n \qquad (4.2.24)$$

thus the momentum equation in the form (4.2.20b) implies that the potential vorticity and Bernoulli function are connected by the differential relation:

$$q_n = \frac{1}{\rho_n}\frac{dB_n}{d\psi_n}. \qquad (4.2.25)$$

Since both $B_n$ and $q_n$ are functions of only $\psi_n$ within the $n^{\text{th}}$ layer for adiabatic and frictionless flow, it follows that at least in principle we can eliminate $\psi_n$ between (4.2.22) and (4.2.23) to obtain:

$$q_n = Q_n(B_n). \qquad (4.2.26)$$

The function $Q_n$ in general is not the same function in (4.2.26) as in (4.2.23). Note also that the potential vorticity is a function not only of $B_n$ but of the layer index $n$, i.e., it depends on the particular isopycnal surface. This is the layer equivalent of the relation for a continuous fluid model in which $q = Q(\rho, B)$ (Pedlosky 1987b) as used by Welander (1971).

## 4.3 Midocean Approximations

We are interested in situations in which the overall variation of the depth of a density interface in midocean varies by order one over the extent of the gyre. However, the midocean slope of the interfaces remains locally small as long as $R_0 = U/fL$ remains small. Estimates for $R_0$ were given in Chapter 3 and we saw that in the midocean $R_0$ is of the order of 0.001 for the velocities of the order 10 cm/s and length scales of the order 1000 km. Of course, as the equator is approached and the Coriolis parameter goes to zero, the local Rossby number becomes large. Thus, the considerations of the present chapter will *exclude the equatorial zone*. We return to the question of equatorial dynamics in Chapter 6.

The ratio of the relative vorticity to the *variable part* of the planetary vorticity (which is what counts in estimating the importance of the relative vorticity in the overall vorticity balance) is of the order, $U/\beta L^2$. For the same scales used to estimate Ro this parameter is of the order $10^{-2}$. Hence to an excellent approximation for the mid ocean the relative vorticity can be ignored with respect to the variable part of the planetary vorticity in the expression for the potential vorticity, i.e., to the order $U/\beta L^2$:

$$q_n = \frac{f}{h_n}. \tag{4.3.1}$$

In distinction to quasi-geostrophic theory we do *not* assume that $h_n$ varies only slightly from a mean value $H_n$, nor do we assume that $f$ can be linearized about a central value. We are examining motions on truly planetary scales for which those two approximations are quantitatively inaccurate. More importantly qualitatively, we examine the case in which the layer thickness vanishes at the sea surface when the layer beneath it outcrops, i.e., when the interface between the two layers reaches the sea surface.

The ratio of the kinetic energy to the pressure in the definition of the Bernoulli function is:

$$\frac{\rho_n|\vec{u}_n|^2}{p_n} = O\left(\frac{\rho U^2}{\rho U f L}\right) = R_0 \ll 1 \tag{4.3.2}$$

so that to a first approximation, $B_n = p_n$.

For small Rossby numbers the first approximation to the momentum equation is the geostrophic balance:

$$f\vec{u}_n = \hat{k} \times \frac{\nabla p_n}{\rho_0} \tag{4.3.3}$$

where we replace the variable density, $\rho_n$, in the horizontal momentum equation with the mean density $\rho_0$, with an error of $O(10^{-3})$.

The smallness of the frictional terms in the vorticity equation and the slow time scale of the large-scale motion allow (4.2.13) to be approximated by:

$$\nabla \cdot (\vec{u}_n f) = 0 \tag{4.3.4}$$

which also follows from the divergence of (4.3.3). Note that in ignoring the contribution of the frictional terms in the interior we are also ignoring the effect of the cross-isopycnal *momentum flux* which is included in $\Im_n$ (4.2.6). Assuming that $w*/h_n$ is of the order of $U/L$ from continuity of mass considerations, the neglected term is of the same order as the horizontal advection of momentum and is therefore negligible in the momentum and vorticity balances as long as $R_0$ and $U/\beta L^2 \ll 1$.

Expanding (4.3.4) and using the continuity equation lead to the more familiar form of the planetary vorticity relation:

$$\beta v_n = f \frac{\partial w_n}{\partial Z}. \tag{4.3.5}$$

As in Chapter 3, the continuity of the cross-isopycnal velocity from one density layer to the next across an isopycnal surface follows from elementary considerations of mass conservation. The continuity of the vertical velocity itself follows from the geostrophic balance for the horizontal velocity, so that the vertical shear of the horizontal velocity across each density interface is *perpendicular* to the direction of slope of the interface. The horizontal velocity in the direction of the interface slope is therefore continuous. Thus, as shown in Section 3.2, the vertical velocity itself is continuous across each density interface. Just as in quasi-geostrophic theory we write the overall integral constraint of the Sverdrup balance for the layer model as the sum:

$$\beta \sum_n v_n h_n = f w_E \tag{4.3.6}$$

where the sum is taken over all the moving layers.

We have discussed above the requirements for the validity of the Sverdrup balance, most fundamental of which is the assumption that there is negligible interaction with the ocean bottom. Indeed, in all the models described in this chapter we assume either that there is a layer below which no motion occurs, or if motion exists down to the bottom, that the bottom is flat and exerts a negligible drag on the *interior* flow.

The hydrostatic relation (3.2.12) of course remains valid on the planetary scale, as also does the condition on the dynamic part of the pressure field that is obtained by matching the total pressure across each density interface. Thus, (3.2.13 a,b) applies and can be written as a difference equation for the dynamic pressure, $\pi_n$, namely:

$$\pi_n - \pi_{n-1} = \gamma_{n-1} z_n, \qquad \gamma_j = \frac{\rho_{j+1} - \rho_j}{\rho_o} g. \tag{4.3.7a,b}$$

Recall that $z_n$ is the interface at the top of the $n^{\text{th}}$ density layer. The solution of (4.3.7) subject to the condition that the dynamic correction to the pressure vanishes below the deepest moving layer for which $n = N$, is

$$\pi_n = -\sum_{i=n}^{N} \gamma_i z_{i+1}. \tag{4.3.8}$$

In the layer model which allows outcropping to occur, the number of active layers differs from location to location, as shown in the schematic in Fig. 4.3.1. If $M$ is the index of the upper most layer at the geographical position under consideration:

$$z_{n+1} = -\sum_{j=M}^{n} h_j. \tag{4.3.9}$$

Therefore, the dynamic pressure can be written, aside from an irrelevant constant, at any location in terms of the layer thicknesses as:

$$\pi_n = \sum_{i=n}^{N} \gamma_i \sum_{j=M}^{i} h_j. \tag{4.3.10}$$

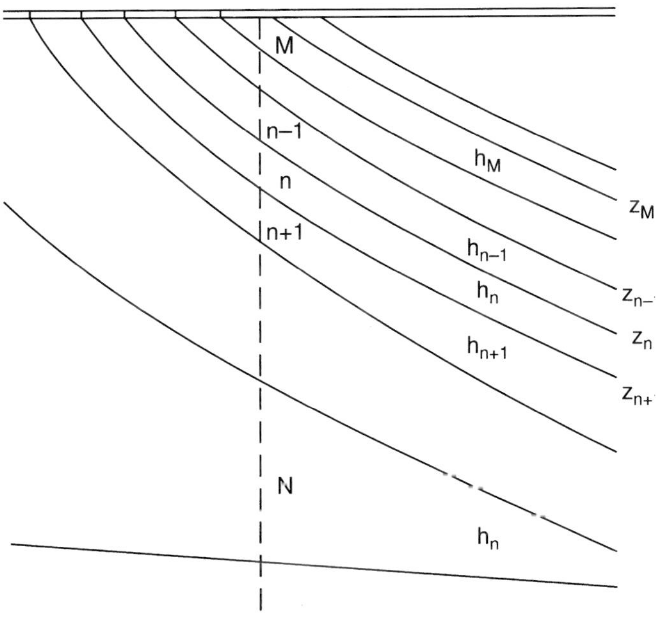


**Fig. 4.3.1.** Schematic presentation of the layer model. At the location indicated (*vertical dotted line*) the uppermost layer is layer $M$. The lowest moving layer is layer $N$. The index $M$ changes from one geographic position to another and in general decreases southward

The geostrophic relation, written in spherical coordinates, and using the dynamic pressure variable $\pi_n$, is:

$$fu_n = -\frac{1}{R}\frac{\partial \pi_n}{\partial \theta}, \quad fv_n = \frac{1}{R\cos\theta}\frac{\partial \pi_n}{\partial \phi} \tag{4.3.11}$$

where $u_n$ and $v_n$ are the eastward and northward velocities, and $\theta$ and $\phi$ are latitude and longitude, respectively. The earth's radius is $R$. The Sverdrup relation is then:

$$\beta \sum_{n=M}^{N} \frac{1}{fR\cos\theta}\frac{\partial \pi_n}{\partial \phi} h_n = fw_E. \tag{4.3.12}$$

With (4.3.8) and the relation:

$$h_n = z_n - z_{n+1} \tag{4.3.13}$$

the Sverdrup relation can be written:

$$\frac{\beta}{fR\cos\theta}\sum_{i=M}^{N} \gamma_j z_{i+1}\frac{\partial z_{i+1}}{\partial \phi} = fw_E. \tag{4.3.14}$$

On the eastern boundary of the basin, at $\phi = \phi_e$, we assume that the geostrophic velocity normal to the boundary is zero which implies that the layer depths, $z_i$, must be constant there. Thus if (4.3.14) is integrated from an arbitrary point in the interior to the eastern boundary, we obtain:

$$\sum_{i=M}^{N} \gamma_i z_{i+1}^2 = -\frac{2f^2}{\beta}\int_{\phi}^{\phi_e} w_E(\phi', \theta) R\cos\theta \, d\phi' + \sum_{i=M}^{N} \gamma_i Z_{i+1}^2 \tag{4.3.15}$$

where $Z_i$ is the constant value of $z_i$ on the eastern boundary.

Deriving (4.3.14) requires a good deal of manipulation of the double sums implied by (4.3.8) and (4.3.12), and it is important to recall that $z_M$ by definition equals zero. The reader may wish to work out some examples involving one or two layers in order to better understand the derivation of (4.3.15) and its consequences.

This is the form of the Sverdrup relation that is most useful in the analysis which follows. It is important to note that the validity of (4.3.15) depends only on the geostrophic and hydrostatic approximation, and that it says nothing about the conservation of potential vorticity in any layer. It is valid in the presence of cross-isopycnal fluxes as long as the momentum balance remains geostrophic.

With the above approximations for the mid ocean, the potential vorticity equation (4.2.17) becomes:

$$\frac{d}{dt}\left(\frac{f}{h_n}\right) = \frac{f}{h_n^2}\left[w*(z_n) - w*(z_{n+1})\right] + \frac{\text{curl}\,\Im_n}{h_n}. \tag{4.3.16}$$

This is a Lagrangian description of the variation of the potential vorticity evolution following an individual fluid element. It is complementary to the Eulerian view which is emphasized in the flux formulation, i.e., when the friction term can be neglected in the midocean:

$$\nabla \cdot [\vec{u}_n h_n q_n] = 0 \tag{4.3.17}$$

where $h_n q_n = f$ in the mid ocean approximation.

Note again that (4.3.17) applies regardless of whether there are cross-isopycnal fluxes. There can be a flux of large-scale potential vorticity into a layer in the midocean only if the layer has a boundary in the horizontal plane across which there can be a normal velocity. One such boundary is an outcropping line in which the edge of the layer, adjacent to the sea surface (or mixed layer), allows potential vorticity to enter the layer and so distort the potential vorticity contours from latitude circles which would be the isolines for a resting fluid.

In steady, frictionless, and adiabatic flow the potential vorticity is a constant on streamlines in all layers not locally forced by the Ekman pumping. For the geostrophic approximation the streamlines coincide with isobars (or isolines of the Bernoulli function). Thus for adiabatic, frictionless flow the potential vorticity satisfies:

$$q_n = \frac{f}{h_n} = Q_n(\pi_n). \tag{4.3.18}$$

Equivalently:

$$\pi_n = \Pi_n(f/h_n) \tag{4.3.19}$$

where $\Pi_n$ is the inverse function to $Q_n$.

If the isolines of $q_n = f/h_n$ strike the eastern boundary along which the pressure anomaly is constant in each layer (to avoid flow through the eastern wall) that constant value is then carried by (4.3.19) throughout the whole region reached by those isolines. The pressure is constant in these regions, and the layer is therefore at rest. Again, in order that the layers beneath the upper layer can be in motion, the potential vorticity isolines in these layers must be sufficiently curved to avoid the eastern boundary. Now, if the layer outcrops in the subtropical gyre, its layer thickness varies from zero at the outcrop line to order 1 farther south, induced by the filling of the layer by Ekman downwelling. This is sufficient, as we see below, to introduce a family of potential vorticity isolines in each layer which start at the outcrop line instead of at the eastern boundary. Thus the process of outcropping, i.e., ventilation, yields another mechanism by which the potential vorticity isolines, which are the highways of the flow in the nondissipative limit, can escape contact with the eastern boundary. In addition, they can make this escape without the necessity for them to close on themselves as in the Rhines and Young mechanism discussed in Chapter 3. The existence of outcropping twists the $q_n$ isolines around to encounter the outcrop line where the mixed layer feeds the layer both fluid and potential vorticity, sending ventilated fluid on its way to depths where it is not directly in contact with Ekman pumping.

## 4.4 The Ventilated Thermocline: The Two-Layer Model

Consider the model shown in Fig. 4.4.1. In the region of the subtropical gyre where the Ekman pumping velocity is downward there are two layers which are somewhere exposed to the Ekman downwelling. In the next section we examine the nature of the solution when more layers are added, in particular when layer 3 is also exposed to Ekman pumping. The exposed layers in the two-layer

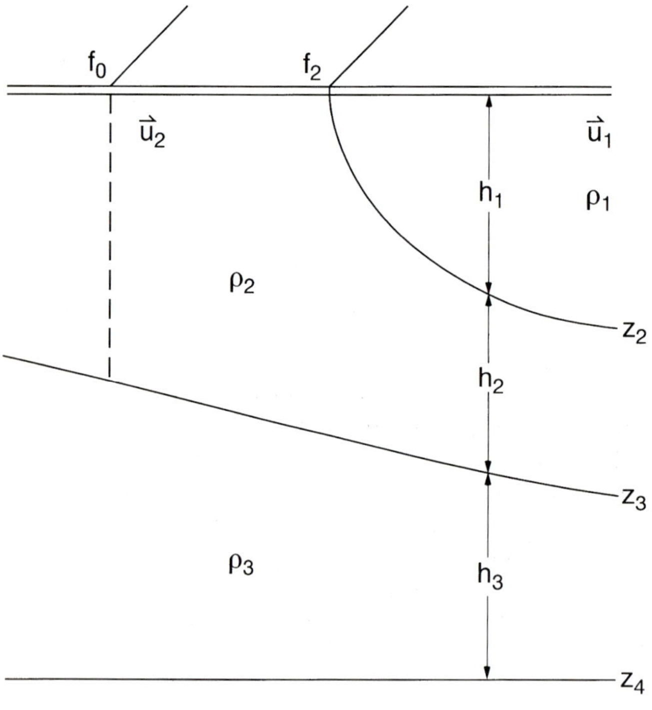


**Fig. 4.4.1.** Schematic presentation of the ventilated-thermocline two-layer model. There are two layers in motion. Layer 3 is assumed to be at rest. The interface between layer 2 and layer 1 rises to meet the mixed layer where $f = f_2$. Ekman pumping is everywhere negative south of the latitude where $f = f_0$

model are labeled 1 and 2. For now, layer 3 and all lower layers do not outcrop, i.e., they are unventilated in the subtropical gyre, and we provisionally assume that they are at rest. The solution so obtained is consistent, but, as we saw in Chapter 3 the possibility that unventilated layers are actually in motion depends on the strength of the forcing and the stratification of these layers. We return to the question of the motion of an unventilated layer 3 and deeper layers later in this chapter. For now, using the results of Chapter 3, we assume simply that layer 3 is so thick that the motion of the overlying layers is unable to distort the potential vorticity in layer 3 sufficiently to close the geostrophic contours in layer 3, which being unventilated must then be at rest. Only layers 1 and 2 are in motion.

The Ekman pumping vanishes on the line, $\theta = \theta_0$ on which $f = f_0$. For now we take this line to be a latitude circle, although this can easily be relaxed.

The specification of the surface density field is equivalent to specifying the position in which layers deeper than the lightest layer are exposed to the interface. These are the lines $\theta = \theta_n$, $n = 1, 2 \ldots N$, on which $f = f_n$. These lines too may in general be functions of longitude, but the theory rapidly becomes technically more complex if this occurs. Fortunately, the surface density and temperature fields in the world oceans are more zonal than not, and this idealization is an acceptable one. In the present case of the two-layer model there is a single outcrop line on which the surface density changes from $\rho_1$ to $\rho_2$ as we move northward beyond the outcrop line. Layers 1 and 2 are therefore, each, exposed somewhere to Ekman pumping, and the principal question of interest is whether a layer that is exposed to Ekman pumping at high latitudes is in motion even at lower latitudes where it is shielded from direct forcing by the Ekman pumping. If so, this provides another mechanism, in addition to the closure of potential vorticity contours and the subsequent process of potential vorticity homogenization, to set thermocline in motion below its uppermost layer.

With the surface density field given, the structure of the thermocline is determined by the geography of the interfaces after they leave the sea surface. This is equivalent to the problem of the mapping of the surface density field onto a vertical line at a fixed latitude and longitude that we referred to in Section 4.7. The subsurface spacing of the density interfaces determines the vertical density variation at each geographical location. Adiabatic flow slides along the density interfaces, bringing fluid of that density to regions below the surface, but we do not know a priori the configuration of these pathways of the flow. In distinction to quasi geostrophic theory where the potential vorticity contours are known ahead of time from the barotropic transport stream function (see Chapter 3), here the flow path is entirely a part of the final solution. It is conceivable, although wrong, to imagine a priori that the flow in the ventilated layers is nonzero only where the layer is exposed directly to Ekman pumping, and that the meridional flow would be halted once an overlying layer interposed itself, shielding the layer from Ekman pumping. However, only the full solution allows us to settle the issue.

We assume, again for simplicity that the eastern boundary is a line of constant longitude on which the *geostrophic* zonal velocity must vanish. Our principal assumption is that the motion is adiabatic and frictionless so that potential vorticity is conserved where the flow is not directly exposed to Ekman pumping. Thus at any location the governing equations are:

(a) the conservation of potential vorticity where the layer is not exposed to Ekman pumping, or, for steady flow:

$$q_n = \frac{f}{h_n} = Q_n(\pi_n) \tag{4.4.1}$$

(b) the Sverdrup balance for layers 1 and 2:

$$\beta \sum_{n=1}^{2} v_n h_n = f w_E \tag{4.4.2}$$

or, equivalently, (4.3.15).

At any location if there are, in the general case, $N - M$ layers in motion, there are $N - M - 1$ of them for which the law of conservation of potential vorticity applies, and the remaining Sverdrup relation fills out our system yielding $N - M$ equations for the $N - M$ unknown interface depths. Once these are known, geostrophy, (4.3.11) and the hydrostatic relation (4.3.10) determine the velocity field completely. The essential problem is determining the potential vorticity functions $Q_n$ which a priori are arbitrary functions. These are determined completely at the outcrop lines by the process of subduction in which southward moving heavier fluid at the surface is forced to slide under the fluid of the lighter layer whose edge is pinned to its surface position by the specification of the surface density field, i.e., by the specification of the position of the outcrop line. It is important to realize that specifying the outcrop line does nothing more than specify the surface density. It does not determine the density structure below the surface.

### *Single-Layer Region*

We start our analysis of the motion in the region $\theta_2 \leq \theta \leq \theta_0$, i.e., in the latitude band between the northern boundary of the gyre and the first outcrop line south of the northern boundary where $f = f_2$. In this region layer 1 is absent. Layer 2 is the only moving layer, and it is directly forced by Ekman pumping. There is no layer in this region in which potential vorticity is conserved and the only nontrivial equation is the Sverdrup relation which reduces to:

$$\beta v_2 h_2 = f w_E \tag{4.4.3}$$

which, with the use of geostrophy, or equivalently with (4.3.15) with $M = N = 2$, yields:

$$z_3^2 = D_0^2(\phi, \theta) + Z_3^2 \tag{4.4.4}$$

The function $D_0^2$ is defined as:

$$D_0^2 = -\frac{2f^2}{\beta\gamma_2}\int_\phi^{\phi_e} w_E(\phi', \theta)R\cos\theta\, d\phi'. \tag{4.4.5}$$

As long as the Ekman pumping velocity is negative, the function $D_0^2$ is positive and an increasing function of distance from the eastern boundary. Recall that $Z_3$ is a constant and equal to the depth of the base of layer 2 at the eastern wall. In the region $\theta_2 \le \theta \le \theta_0$, $h_2 = -z_3$, thus (4.4.4) implies that:

$$h_2^2 = D_0^2 + H_2^2 \tag{4.4.6}$$

Since $D_0^2$ is an increasing function of distance from the eastern boundary the depth of the moving layer increases westward. The layer thickness, $h_2$, is equal to the constant value $H_2$ both on the eastern boundary and on the northern boundary of the gyre where the Ekman pumping and hence $D_0^2$ vanish. The base of layer 2 deepens then as we move south of $\theta = \theta_0$, but the interface rises if the Ekman pumping decreases further southward, and it rises of course to the value $H_2$ all along the eastern boundary . Thus $h_2$ is a single constant along the northern and eastern boundary of the gyre. This constant is arbitrary and unknown. We therefore generate a *family* of solutions each one corresponding to a different depth of the ventilated layer on the intergyre boundary, $\theta = \theta_0$.

The velocity in this region is given by the geostrophic relations. Since:

$$\pi_2 = -\gamma_2 z_3 = \gamma_2 h_2 \tag{4.4.7}$$

it follows that:

$$u_2 = -\frac{\gamma_2}{fR}\frac{\partial h_2}{\partial \theta} = -\frac{\gamma_2}{2fh_2R}\frac{\partial D_0^2}{\partial \theta} = \frac{1}{fh_2}\frac{\partial}{\partial \theta}\int_\phi^{\phi_e}\frac{f^2}{\beta}w_E\cos\theta\, d\phi' \tag{4.4.8a}$$

while the velocity in the meridional direction is:

$$v_2 = \frac{\gamma_2}{fR\cos\theta}\frac{\partial h_2}{\partial \phi} = \frac{f}{\beta}\frac{w_E}{h_2}. \tag{4.4.8b}$$

Note that the meridional velocity is always negative in this region as long as the Ekman pumping is downward. The flow in layer 2 is driven southward by the action of the wind until fluid elements impinge on the outcrop line at $\theta = \theta_2$. At this point the fluid in layer 2 is pushed under the blanket of fluid of density $\rho_1$ which interposes itself between the heavier fluid of layer 2, with density $\rho_2$, and the Ekman pumping. This is the process of subduction, i.e., the fluid in layer 2 *subducts* beneath the fluid in layer 1. What happens next to the fluid in layer 2 which is forced to subduct beneath layer 1?

### *The Process of Subduction*

Consider Fig. 4.4.2 which schematically describes the event. The fluid column in layer 2 which is driven southward by the Ekman pumping is forced to subduct under the edge of layer 1 which then shields it from the Ekman pumping. In the absence of all other sources or sinks of potential vorticity, i.e., for adiabatic, frictionless flow, the potential vorticity in layer 2 is subsequently conserved along a streamline. Thus from (4.3.8) and (4.4.1):

$$q_2 = \frac{f}{h_2} = \frac{f}{(z_2 - z_3)} = Q_2(\pi_2) = Q_2(-\gamma_2 z_3)\,. \tag{4.4.9}$$

We know the solution north of, *and up to*, the outcrop line. Therefore along the outcrop line $h_2$ is known from (4.4.6) as a function of distance (here longitude) along the outcrop line. At the same time $z_3$ is also known along the outcrop line. If the outcrop line slopes with respect to a latitude circle, the equation for the outcrop line gives f as a function of longitude as well so that $q_2$ along the outcrop line is known. Thus to each point along the outcrop line we can assign a value of the potential vorticity $q_2$ and the geostrophic pressure $\pi_2$. The tabular relationship between these yields the function $Q_2$, perhaps in the form of a numerical table of pairs $(q_2, z_3)$. As the streamline of the flow in layer 2 proceeds southward of the outcrop line, the pairing of a particular value of $q_2$ with a particular value of $\pi_2$, or $z_3$, is maintained. This gives us, from (4.4.9),

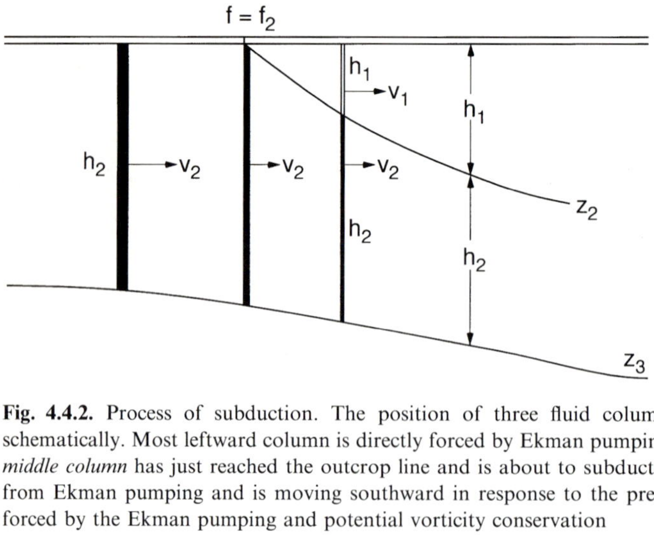


**Fig. 4.4.2.** Process of subduction. The position of three fluid columns in layer 2 is shows schematically. Most leftward column is directly forced by Ekman pumping and moves southward; *middle column* has just reached the outcrop line and is about to subduct; *third column* is shielded from Ekman pumping and is moving southward in response to the pressure gradient in layer 2 forced by the Ekman pumping and potential vorticity conservation

one equation between $z_2$ and $z_3$ while the Sverdrup balance provides the remaining equation required to complete the solution.

When the outcrop line is a latitude circle, $Q_2$ can be determined analytically with great ease. On the outcrop line $z_2$ is *by definition*, zero. If the outcrop line is a latitude circle, on the outcrop line $f = f_2$, which is a constant. Thus on the outcrop line we have the relation:

$$\frac{f}{h_2} = \frac{f_2}{-z_3} = Q_2(\gamma_2 z_3). \tag{4.4.10}$$

This completely determines the functional form of $Q_2$ on every streamline that carries flow from the outcrop line. Although both the potential vorticity and the thickness field may both be complicated functions of longitude along the outcrop line, the functional relationship between them is very simple:

$$Q_2(\gamma_2 z_3) = \frac{f_2}{-z_3}. \tag{4.4.11}$$

Thus everywhere in layer 2 south of the outcrop line, in regions covered by streamlines emanating from the outcrop line, the relationship between potential vorticity and pressure is established once and for all by the relationship already determined during subduction:

$$q_2 \equiv \frac{f}{(z_2 - z_3)} = \frac{f_2}{-z_3}. \tag{4.4.12}$$

Solving for $z_2$ in terms of $z_3$ yields:

$$z_2 = \left(1 - \frac{f}{f_2}\right) z_3. \tag{4.4.13}$$

In terms of the layer thicknesses themselves, we have:

$$\begin{aligned} h_2 &= z_2 - z_3 = \frac{f}{f_2} h \\ h_1 &= -z_2 = \left(1 - \frac{f}{f_2}\right) h \\ q_2 &= \frac{f_2}{h} \end{aligned} \tag{4.4.14a,b,c}$$

where $h$, the total depth of the moving fluid is given by:

$$h = h_1 + h_2 = -z_3. \tag{4.4.15}$$

Note that in the two-layer model the ratio of the layer thicknesses is a function only of latitude and is independent of the stratification, i.e.:

$$\frac{h_1}{h_2} = \left(\frac{f_2}{f} - 1\right). \tag{4.4.16}$$

The further south the fluid moves, the thinner layer 2 becomes in order to conserve potential vorticity, and layer 1 occupies an increasingly greater fraction of the total water column.

The Sverdrup balance for the two layer model is, from (4.3.15) and (4.4.5):

$$z_3^2 + \frac{\gamma_1}{\gamma_2} z_2^2 = D_0^2 + Z_3^2. \tag{4.4.17}$$

Note that we have set $Z_2$, which is the depth of the interface between layers 1 and 2 at the eastern wall, equal to zero. Since there is no zonal flow along the eastern wall all the layer thicknesses, $Z_i$, must be constant along the wall. However, the interface between layers 1 and 2 vanishes along the outcrop line which intersects the eastern wall at $\theta = \theta_2$. Thus $Z_2$ is zero there and must remain zero southward all along the eastern boundary.

Using (4.4.13) in the Sverdrup balance (4.4.17) yields:

$$h = -z_3 = \sqrt{\frac{D_0^2 + Z_3^2}{1 + \gamma_1/\gamma_2(1 - f/f_2)^2}} \tag{4.4.18}$$

from which, by (4.4.14), the individual layer thicknesses and the interface depths (and hence the pressure in both layers) can easily be obtained. Before examining the structure of the solution thus obtained, we must first examine whether our solution satisfies the boundary condition of no normal flow on the eastern boundary.

### *The Shadow Zone*

On the eastern boundary of the ocean the function $D_0^2$ vanishes by its definition (4.4.5). If our solution for $h$ were valid on the eastern boundary, we would have there, from (4.4.18) (note that $Z_3 = -H_2$);

$$h = \frac{H_2}{\sqrt{1 + \gamma_1/\gamma_2(1 - f/f_2)^2}}, \qquad \phi = \phi_e \tag{4.4.19}$$

which is clearly not constant as a function of latitude because of the variation of $f$ with $\theta$. Apparently our solution does not satisfy the boundary condition. It appears that h is constant (and zero) only if $H_2$ were zero. This would be a rather restrictive condition on the solution, and we would prefer to allow the thickness of the layer of ventilated fluid to be nonzero on the eastern boundary and on the intergyre boundary, i.e., to have $H_2$ not equal to zero. However, we then face the conundrum presented by (4.4.19): what is happening?

The difficulty is easy to understand. If fluid is flowing in layer 2 *along* the eastern boundary southward, it must satisfy two conditions. First, each of the layer thicknesses must be constant along the boundary so that the geostrophic zonal velocity vanishes at the eastern boundary. This implies that both $h_1$ and

$h_2$ are constant on the eastern wall. At the same time, however, the fluid column in layer 2 that we can imagine moving along the eastern boundary must keep its potential vorticity, $f/h_2$, constant, which it clearly cannot, and also keep $h_2$ constant since $f$ varies with latitude. This contradiction is the source of the difficulty in extending the solution in its present form to the eastern boundary.

Consider a streamline in layer 2 in the region of subduction, as shown schematically in Fig. 4.4.3. Each point on the outcrop line is the starting point of a subducting streamline. Each streamline is a line of constant $\pi_2$, or equivalently of constant $h$. We can use (4.4.18) to obtain a parametric equation for the streamline emanating from the arbitrary point $(\phi', \theta_2)$ on the outcrop line. Keeping $h$ constant we have:

$$h^2 = \frac{D_0^2(\phi, \theta) + H_2^2}{(1 + \gamma_1/\gamma_2)(1 - f/f_2)^2} = \frac{D_0^2(\phi', \theta_2) + H_2^2}{1 + 0} \tag{4.4.20}$$

so that the equation for the streamline becomes:

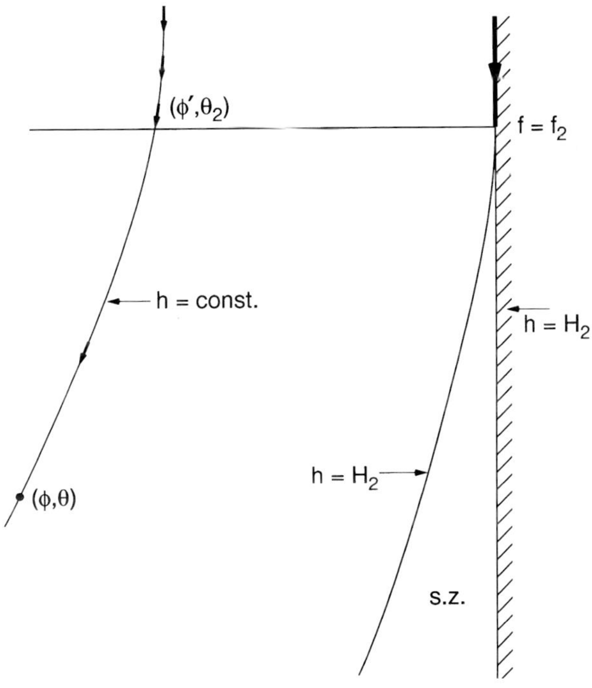


**Fig. 4.4.3.** Schematic streamlines in layer 2 as they emanate from the outcrop line. Through the point $(\phi', \theta_2)$ on the outcrop line a geostrophic streamline passes to the point $(\phi, \theta)$ along a line of constant $z_3$ or $h$. The streamline, which north of the outcrop line is contiguous with the eastern wall, must leave the wall at the outcrop line and trend westward exposing a zone near the eastern wall that is not reached by any streamline originating from the outcrop line. This is the shadow zone

$$D_0^2(\phi,\theta) = D_0^2(\phi',\theta_2)\left[1+\frac{\gamma_1}{\gamma_2}\left(1-\frac{f}{f_2}\right)^2\right] + H_2^2\frac{\gamma_1}{\gamma_2}\left(1-\frac{f}{f_2}\right)^2 \tag{4.4.21}$$

which can be thought of as yielding an equation $\phi = \phi(\theta)$ for the shape of the streamline emanating from the outcrop line at the point $(\phi', \theta_2)$. The function $D_0^2$ is an increasing function of distance from the eastern boundary and vanishes at $\phi = \phi_e$. To examine the path of the streamline emanating from the intersection of the outcrop line and the eastern boundary we set $\phi' = \phi_e$ in (4.4.21) to obtain, for the critical curve $\Phi_s(\theta)$, the equation:

$$D_0^2(\Phi_s,\theta) = H_2^2\frac{\gamma_1}{\gamma_2}\left(1-\frac{f}{f_2}\right)^2. \tag{4.4.22}$$

When $f = f_2$ the right side of (4.4.22) is zero and the left side is also zero since then $\Phi_s = \phi_e$ where $D_0^2$ vanishes. As f decreases and becomes less than $f_2$, the right side becomes positive and the left side must increase from zero, and this implies that $\Phi_s < \phi_e$ for $\theta < \theta_2$. Thus the streamline emanating from the intersection of the outcrop line and the eastern boundary must pull away from the eastern wall as long as $H_2 \neq 0$, in conformity with our earlier more heuristic discussion. A zone near the eastern boundary is opened east of the critical streamline whose path is given by $\phi = \Phi_s(\theta)$. This zone is not reached by any streamline emanating from the outcrop line. All the streamlines from the outcrop line are shifted westward and leave this eastern region unventilated. In analogy with the field of optics, we call this region the *shadow zone*. All points on the outcrop line emit a streamline (analogous to a light ray), but not all points in layer 2 are illuminated by streamlines emanating from those points since the rays bend away from the eastern boundary. In the shadow zone the fluid in layer 2 lies outside of the zone of ventilation. It is for this reason that the solution (4.4.18) does not satisfy the boundary condition on the eastern wall. This solution has been determined using the potential vorticity relation determined on the outcrop line and is valid only in regions covered by ventilating streamlines. Thus east of the critical curve $\phi = \Phi_s(\theta)$ (4.4.18) is not valid and certainly cannot be applied on the eastern boundary. Our solution is therefore a partial solution, valid only in the ventilated region. Note that the extent of the region is a function of the strength of the forcing. The larger the forcing the smaller the distance from the eastern boundary the point $\phi$ must be to achieve the magnitude of $D_0^2$ required to satisfy (4.4.22). Thus the larger the forcing by the Ekman pumping, the smaller is the longitudinal extent of the shadow zone. The thicker the layer of moving fluid just south of the northern boundary of the gyre, the larger is the extent of the shadow zone.

What is the solution in the shadow zone? Note that on the eastern boundary of the shadow zone, which is the eastern boundary of the basin, the depth of the fluid is $H_2$. The same is true on the western boundary of the shadow zone which is the critical streamline given by $h = H_2$. Thus no flow enters the shadow zone across either the eastern or western edges of the wedge-shaped region that forms the zone. In the absence of mixing and other non adiabatic sources of potential vorticity anomalies it is plausible that the layer remains at rest. Since we are dealing, at the lowest order, with an ideal fluid, the motion of this isolated region is actually arbitrary, but it seems reasonable to set the velocity in layer 2 to zero. This is what Luyten et al. (1983) did.

If the velocity in layer 2 is zero in the shadow zone, all the Sverdrup transport in this region must be carried by the upper layer. Thus, east of the line $\phi = \Phi_s(\theta)$ the flow in layer 2 is zero, and therefore everywhere in this region $h = H_2$ or equivalently $z_3 = Z_3$. Thus (4.4.17) reduces to:

$$z_2^2 = h_1^2 = \frac{\gamma_2}{\gamma_1} D_0^2 \tag{4.4.23}$$

which is equivalent to $\beta v_1 h_1 = f w_E$, $v_2 = 0$.

Thus we have one solution (4.4.18) that is valid west of $\phi = \Phi_s(\theta)$ and another solution (4.4.23) that is valid east of $\phi = \Phi_s(\theta)$. On the critical curve which forms the western boundary of the shadow zone and the eastern boundary of the ventilated region the solution must be continuous in the pressure field in each layer, or, equivalently, the layer interface depths must be continuous on $\phi = \Phi_s(\theta)$.

The depth of the lower interface $z = z_3 = -h$, is already continuous on the boundary of the two regions since by construction the eastern boundary of the ventilated region corresponds to the streamline on which $h = H_2$, and this is the value of h *everywhere* in the shadow zone. In order for the interface $z = z_2$ to be continuous we must have $h_1$ continuous at $\phi = \Phi_s(\theta)$. Now, for any streamline in the ventilated region $h_1 = (1 - f/f_2)h$ and for the easternmost ventilated streamline this means that $h_1 = (1 - f/f_2)H_2$ along the eastern boundary of the ventilated region. However, within the shadow zone (4.4.23) applies. For $h_1$ to be continuous we must have:

$$\left(1 - \frac{f}{f_2}\right)^2 H_2^2 = \frac{\gamma_2}{\gamma_1} D_0^2(\Phi_s, \theta) \tag{4.4.24}$$

which is the same as the equation for the shadow zone boundary (4.4.22). Therefore both $z_2$ and $z_3$ are continuous across the shadow zone boundary. Their slopes *normal* to the shadow zone boundary are *not* continuous. The velocity tangential to the shadow zone boundary is discontinuous in the lower layer, passing from zero in the shadow zone to the ventilated value west of $\phi = \Phi_s(\theta)$. Since the total Sverdrup transport is a continuous function, the tangential velocity in the upper layer must also be discontinuous across the shadow zone boundary.

### *Properties of the Two-Layer Solution*

The solution that we have obtained so far is composed of flow in two regions, the shadow zone and the ventilated region in layer 2. There is a strong variation of structure from one region to the other, and this immediately demonstrates the *non-self-similar* nature of the flow. The ideal fluid flow has critical curves, in the two-layer model for example, this is the shadow zone boundary across which the solution possesses weak discontinuities, for example, the jump in the slope of the interface surfaces.

Figure 4.4.4 shows various images of the solution. In this example the Ekman pumping has been chosen in the form:

$$w_E = -W_0 \frac{f_0{}^2}{f^2} \sin(\pi f / f_0) \tag{4.4.25}$$

which diverges linearly at the equator but leaves $D_0^2$ finite, in fact zero, at the equator. In the figures shown the depth is scaled with $H_2$, the depth of the thermocline on the eastern boundary. $W_0$ has been chosen to be $1.5 \times 10^{-4}$ cm/s, the ocean width is 3000 km and $\gamma_2$ has been taken as $1\,\text{cm/s}^{-2}$. $H_2$ is taken to be 670 m. This yields:

$$\frac{D_0^2}{H_2^2} = 2(1 - \phi/\phi_e) \sin(\pi f / f_0). \tag{4.4.26}$$

Figure 4.4.4a shows the streamlines in the lower layer. As anticipated by the previous discussion, the region east of the shadow zone is at rest. The flow in the lower layer consists of a giant anticyclonic circulation exiting from the western boundary in the north and reentering it in the south. We recall that this interior solution assumes that the western boundary layer accepts the interior flow and returns it as required by the solution but our discussion of the simpler one-layer model in Chapter 2 reminds us that the physics of the western boundary current is not that simple and accommodating. We must expect at least a portion of the present solution, in the western region, to be sharply altered by the presence of a western current that may detach from the boundary and impinge on the interior. Or, for small enough dissipation, the

**Fig. 4.4.4a–f.** Circulation pattern for the two-layer model. The outcrop line is purely zonal and is indicated by the *crosses* extending the width of the basin. The shadow zone boundary is delineated by *asterisks* which initiate at the eastern boundary and pull away from the boundary at the intersection of the outcrop line with the eastern wall. **a** Streamlines (and *h*) in the lower layer. The shaded region is the "pool" region of constant potential vorticity described in Section 4.6. **b** Streamlines of the flow in the upper layer. Note the kink in the upper layer streamlines as they cross the shadow zone boundary. **c** Potential vorticity of the lower layer. Note the decrease in the variation of $q_2$ in the ventilated region. **d** Meridional cross section showing the layer thicknesses. **e** Zonal cross-section of the layer thicknesses. **f** Perspective view of the base of the thermocline. Note the region of the shadow zone in which the base of the thermocline is flat

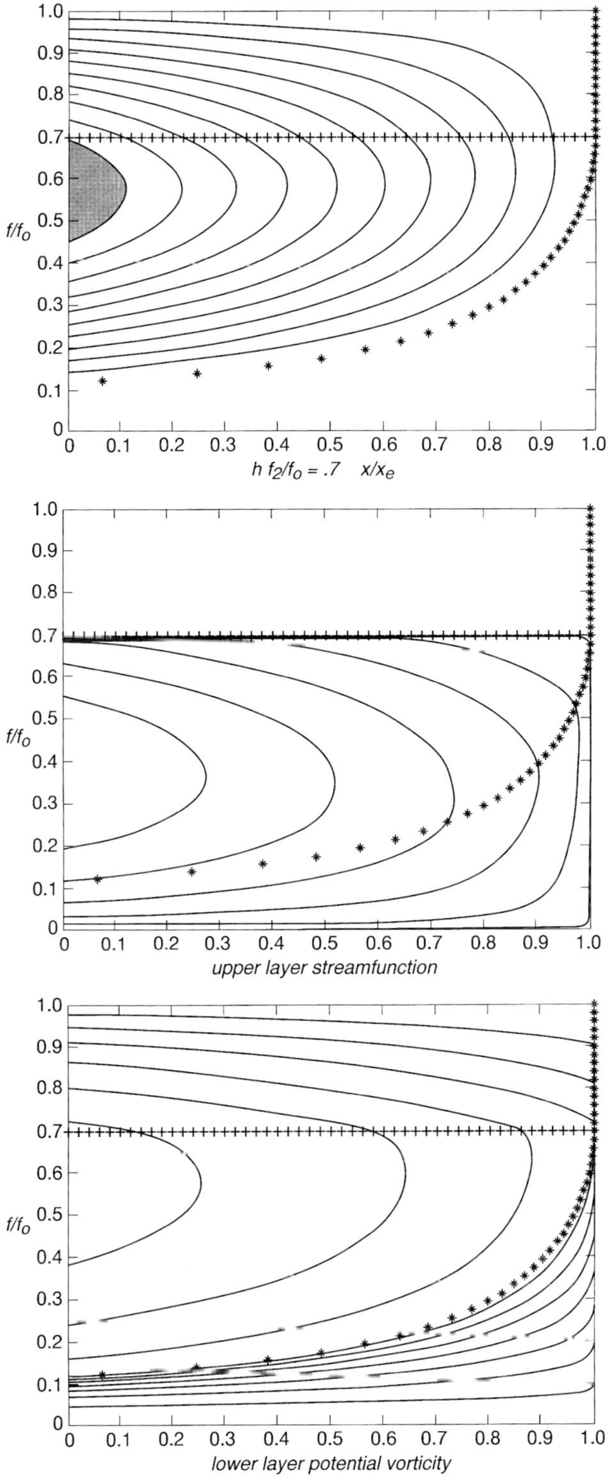
f/fo
h f2/fo = .7 x/xe
upper layer streamfunction
lower layer potential vorticity

a

b

c

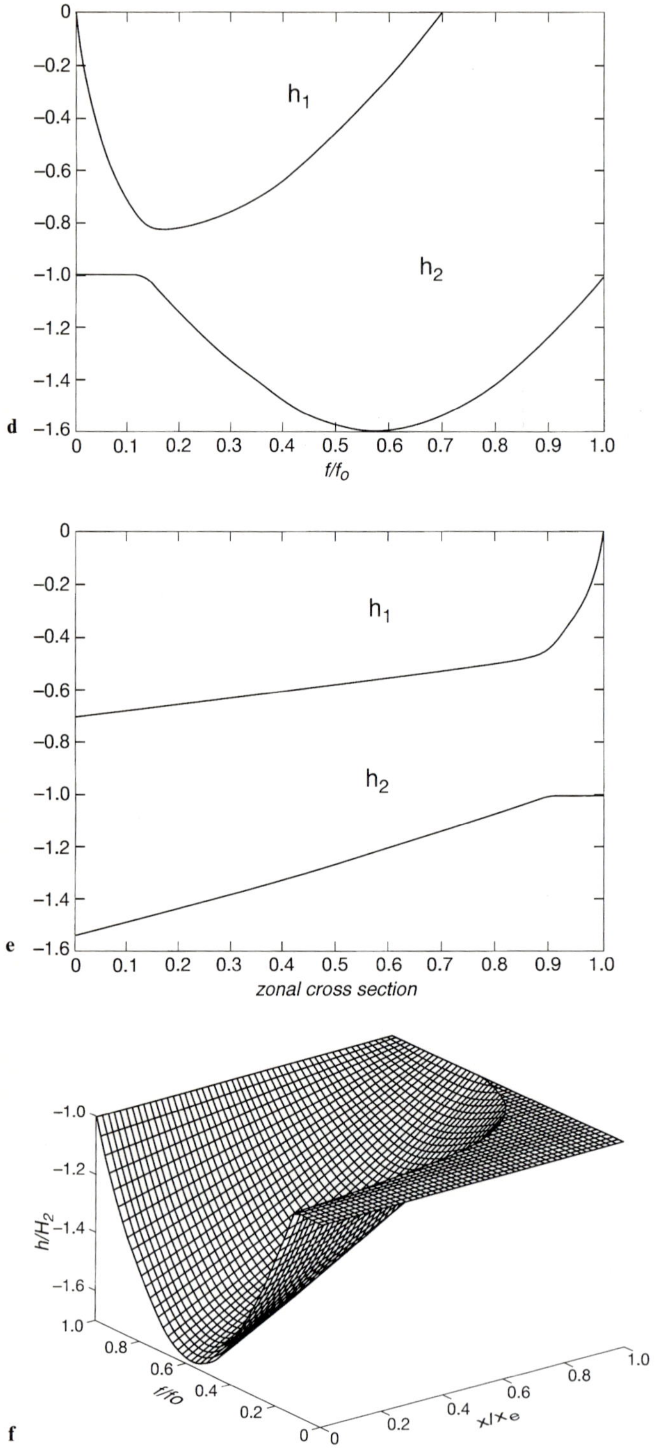


Fig. 4.4.4d–f

whole notion of an interior in Sverdrup balance may become invalid. For now we accept the provisional character of the solution.

In Fig. 4.4.4b we see the streamlines in the upper layer. The shadow zone boundary is also shown. The streamlines in the upper layer cross the shadow zone boundary. West of this boundary the upper layer shares the Sverdrup transport with the lower layer. East of the shadow zone boundary the upper layer carries the full Sverdrup transport. On the boundary, note the kink in the upper layer streamfunction. The tangential velocity in *both* layers is discontinuous at the shadow zone boundary. It is obviously true for the lower layer flow. Since the total Sverdrup transport is continuous everywhere, the upper layer flow must experience a compensating discontinuity to render the total transport continuous.

Panel c of Fig. 4.4.4 shows the potential vorticity, $q_2$, in the lower layer where it is conserved south of the outcrop line. Note the weakening of the variation in $q_2$ in the region of the ventilation south of the outcrop line. The thickness of layer 2 varies much more strongly in regions where either layer 2 carries the full Sverdrup transport, as in the region north of the outcrop line, or in the shadow zone where the upper layer carries the full Sverdrup transport, and thus where the interface between layers 1 and 2 must vary strongly. This is demonstrated in panels d and e, which show meridional and zonal cross sections of the layer thicknesses. In panel d note the flattening of the base of the thermocline south of the intersection with the shadow zone and the simultaneous increase in the slope of the interface between the two layers. The same behavior is evident in panel e, which shows a zonal cross section. Note in particular that the layer thickness in the upper layer vanishes at the eastern boundary. The strong east-west gradient of $q_2$ in the shadow zone is due to the rapid variation in $h_2$ in the shadow zone region where $z_2$ surfaces on the eastern boundary. The overall shape of the base of the thermocline is shown in the perspective representation of $h$ in panel f. The bowl of the thermocline and the ledge of the shadow zone are apparent.

## 4.5 The $\beta$-Spiral

A careful examination of the streamline patterns in Fig. 4.4.4 shows that the streamlines in layer 2 are always twisted clockwise with respect to streamlines in layer 1 at the same geographical location. This can easily be observed where the streamlines in layer 1 cross the shadow zone boundary, which is after all a streamline in layer 2. The velocity vector spirals clockwise with increasing depth in the subtropical gyre. This is a general feature of adiabatic flow in the models of the subtropical gyre and can also be noted in the flow patterns predicted by the Rhines and Young model discussed in Chapter 3. Stommel and Schott (1977) pointed out this fact in an attempt to describe a method to proceed from a knowledge of the density field, and hence the vertical shear of the velocity, to the absolute velocity itself.

Consider the integration of the Sverdrup vorticity equation (4.3.5) over the thickness of layer $n$. Since the horizontal velocity is independent of depth in the layer:

$$\beta v_n h_n = f\left[w_n(z_n) - w_n(z_{n+1})\right]. \tag{4.5.1}$$

Consider now a layer in which the flow is adiabatic, i.e., in which there is no flow across either the upper or lower interface of the layer. Then from (3.2.18) we have, for steady flow:

$$w_n(z_n) = \vec{u}_n \cdot \nabla z_n, \quad w_n(z_{n+1}) = \vec{u}_n \cdot \nabla z_{n+1}. \tag{4.5.2}$$

The hydrostatic and geostrophic relation yields:

$$\frac{\gamma_{n-1}}{f}\nabla z_n = \hat{k} \times (\vec{u}_{n-1} - \vec{u}_n) \tag{4.5.3}$$

with a similar equation for the lower interface. When (4.5.3) and (4.5.2) are used in (4.5.1) we obtain:

$$\beta v_n h_n = -f^2 \hat{k} \cdot \left\{ \vec{u}_n \times \left[ \frac{\vec{u}_{n-1} - \vec{u}_n}{\gamma_{n-1}} - \frac{\vec{u}_{n+1} - \vec{u}_n}{\gamma_n} \right] \right\}. \tag{4.5.4}$$

If the density jumps across each interface are equal (this still allows us to represent an arbitrary stratification on the northern edge of the gyre by choosing the thicknesses of the resting layers there) the right side of (4.5.4) involves the cross-product of the velocity in layer n with the difference of the vertical shears of the velocity across that layer's two interfaces. This is the layer equivalent of the cross-product of the velocity with the *second* derivative of the velocity vector with depth. A stronger condition is obtained by considering the vertical velocity only at the $n^{\text{th}}$ interface.

Using (4.5.2) and (4.5.3) we obtain:

$$\begin{aligned} w_n(z_n) &= \vec{u}_n \cdot \nabla z_n \\ &= -\frac{f}{\gamma_{n-1}}\vec{u}_n \cdot \left[\hat{k} \times (\vec{u}_n - \vec{u}_{n-1})\right] \\ &= -\frac{f}{\gamma_{n-1}}\hat{k} \cdot \left[\vec{u}_n \times \vec{u}_{n-1}\right]. \end{aligned} \tag{4.5.5}$$

If the vertical velocity is negative at interface $z = z_n$, the velocity vector must swing clockwise as we go from layer $n - 1$ down to layer $n$. That is, the velocity vector must spiral clockwise with depth. To derive (4.5.5) we use only geostrophy and the thermal wind relation to express the slope of the interface in terms of the shear across the interface. Fundamentally, it is the sign of the vertical velocity which determines the direction of the spiral.

It is the use of the Sverdrup vorticity equation that allows us to relate the vertical velocity, which is usually too small to be directly observed, with the meridional velocity, $v$. Suppose, as in the model, that there is some layer, $n = N$, below which the total velocity field is zero. Then if we integrate the vertical velocity from that great depth, the Sverdrup vorticity equation implies that:

$$w_n(z_n) = \frac{\beta}{f} \sum_{i=n}^{N} v_i h_i. \tag{4.5.6}$$

Thus, with (4.5.5):

$$\frac{\beta \gamma_{n-1}}{f^2} \sum_{i=n}^{N} v_i h_i = \hat{k} \cdot [\vec{u}_{n-1} \times \vec{u}_n]. \tag{4.5.7}$$

If the total meridional *transport* of all the layers beneath layer $n$ is negative the velocity vector must swing clockwise with depth when moving from layer $n-1$ to layer $n$. Furthermore, from (4.5.4) this vector cross-product becomes larger with increasing depth if $v_n$ is negative.

Using (4.5.7) in the case of the two-layer model, we see directly that the velocities in the two layers must describe the clockwise spiral with depth as shown schematically in Fig. 4.5.1 and observed in the results of the previous section.

Stommel (1984) provided a very illuminating derivation of the $\beta$-spiral. Consider, the two-layer model discussed in the previous section and again imagine that the motion in layer 2 preserves potential vorticity. Consider a column in layer 2 moving southward. Along its trajectory, which is also a streamline, the total depth $h$, which forms the base of layer 2, remains constant, as shown in Fig. 4.5.2. We display the path *along* the streamline in the figure, and it is important to note that we do not assume that the flow is purely southward. The path is only been unwrapped and laid flat on the figure. As it

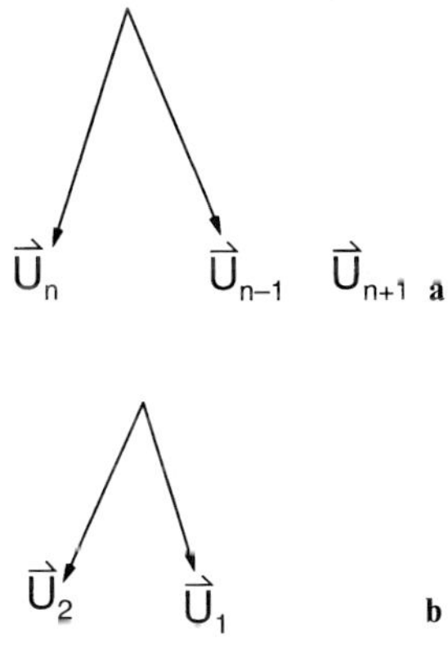


**Fig. 4.5.1.** **a** Velocity vector in layer n must be oriented clockwise with respect to the shear across the layer. The velocity vector must swing clockwise with depth. **b** Velocities in the two-layer model must consequently be arranged so that the velocity in the lower layer flows more south westward than the flow in the upper layer

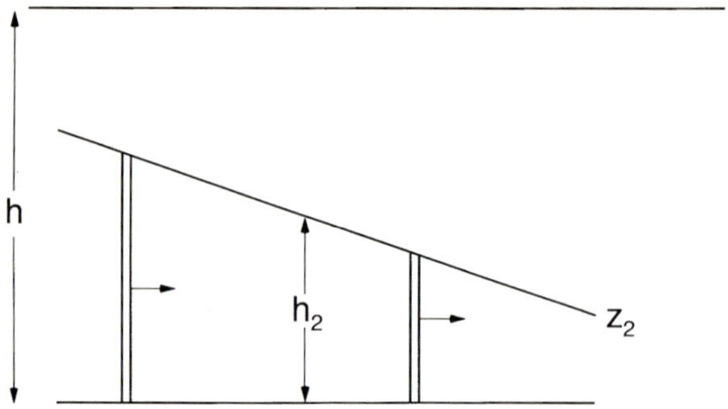


**Fig. 4.5.2.** Fluid column in layer 2 moves along a path of constant h. Thus its base remains at a fixed level. Conservation of potential vorticity implies that the upper surface must slope downwards along the direction of flow. The thermal wind equation then implies that layer 1 has a component of flow into the paper. The velocity therefore spirals clockwise with increasing depth

proceeds southward conserving potential vorticity, $f/h_2$ must be constant for the fluid column. Since f decreases southward, $h_2$ must also decrease. Since the base of the layer is at constant depth *along that stream line* it follows that the upper surface of layer 2 at $z = z_2$ must slope downward in the direction of the flow in the lower layer, as shown in the figure. From the thermal wind relation this slope implies that there must be a shear across the interface perpendicular to, and into the plane of, the paper. Since the lower layer velocity lies in the plane of the paper, this means the upper layer must have a component of velocity into the paper. The velocity must spiral counter-clockwise with increasing $z$ or clockwise with increasing depth. This is the $\beta$-spiral, and it is clear that it is fundamentally related to the conservation of potential vorticity.

If we repeat the derivation leading to (4.5.7) for the two-layer model when there is cross-isopycnal flow across the interface between layers 1 and 2, it is easy to show, starting again from (3.2.18), that the $\beta$-spiral equation is modified and becomes:

$$\frac{\beta}{f} v_2 h_2 = \hat{k} \cdot [\vec{u}_1 \times \vec{u}_2] \frac{f}{\gamma_1} + w*(z_2). \tag{4.5.8}$$

In the presence of heating or cooling, when the cross-isopycnal velocities are no longer zero, the spiral of the velocity vector with depth is no longer assured by the presence of southward flow. Potential vorticity is no longer conserved, and the velocity vector may even be parallel from one layer to the next if the heating or cooling is sufficient. We examine this in more detail in Chapter 5.

## 4.6 The Pool of Constant Potential Vorticity

Examining Fig. 4.4.4a carefully, we see that there is a streamline which issues from the intersection of the outcrop line with the *western* boundary of the basin. This streamline carves out a region on the western flank of the ventilated layer in which all the streamlines within that region emanate from the western boundary south of the outcrop line. Thus the fluid in layer 2 in this region does not originate from the outcrop line. If we follow a streamline of the solution given by (4.4.18), its origin on the outcrop line lies west of the oceanic western boundary for fluid in the shaded region. Hence our solution, which depends on the process of subduction to determine the potential vorticity of the fluid, is inapplicable in the shaded zone. This is the same situation as above in discussing the solution in the shadow zone on the eastern edge of the ventilated region. There we argued that the flow in layer 2 should be at rest since the shadow zone region is bounded on both its eastern and western edges by curves across which there is no flow. The shaded region on the western flank of the ventilated zone, on the other hand, receives fluid from the western boundary, and in fact the fluid clearly recirculates through the western boundary current. Luyten et al. (1983), following the ideas of Young and Rhines (1982), assumed that the recirculation in this region, absent of the influence of any source of potential vorticity, would lead to a final state of homogenized potential vorticity in layer 2, and they called this region the *pool of constant potential vorticity*.

The extent of the pool depends on the position of the outcrop line. Any of the streamlines in layer 2 in Fig. 4.4.4a could be the boundary of the pool region if the outcrop line coincides with the intersection of the streamline with the western boundary. Thus the farther north the outcrop line is, the larger is the area of layer 2 covered by the pool. For outcrop lines that are far enough south there are no streamlines emanating from the western boundary, and there is no pool for such outcrop lines. To find the boundary of the pool consider the trajectory:

$$\phi = \Phi_w(\theta) \tag{4.6.1}$$

which emanates from the intersection of the outcrop line and the western boundary of the basin where $\phi = \Phi_w$. If we again use the streamline equation (4.4.20), we obtain for the trajectory (4.6.1) the implicit relation:

$$\begin{aligned} D_0^2(\Phi_w, \theta) = D_0^2(\phi_w, \theta_2)\left\{1 + \frac{\gamma_1}{\gamma_2}\left(1 - \frac{f}{f_2}\right)^2\right\} \\ + H_2^2 \frac{\gamma_1}{\gamma_2}\left(1 - \frac{f}{f_2}\right)^2. \end{aligned} \tag{4.6.2}$$

Suppose that south of $\theta_2$ the function $D_0^2$ increases southward at a fixed longitude, perhaps due to a southward increase of the Ekman pumping. Near the outcrop line the terms that are quadratic in $1 - f/f_2$ are negligible and near the intersection of the outcrop line with the western boundary (4.6.2) reduces to:

$$D_0^2(\Phi_w, \theta) \approx D_0^2(\phi_w, \theta_2). \tag{4.6.3}$$

If $D_0^2$ increases southward at a fixed longitude, the only way for (4.6.3) to be satisfied is for $\Phi_w$ to become greater than $\phi_w$. By reducing the east-west range of integration in $D_0^2$ we can compensate for the tendency for $D_0^2$ to increase with decreasing latitude. That is, the trajectory must initially swing eastward from the outcrop line. Eventually the terms that are quadratic in $1 - f/f_2$ become important, causing the trajectory to swing westward and thus carve out a closed pool. However, as long as the intersection of the outcrop line with the western boundary lies north of the maximum of $D_0^2$, there is such a pool. Only for outcrop lines south of the maximum of $D_0^2$ is there no pool.

If the potential vorticity is homogenized in the pool, its value must be that of the potential vorticity on the boundary given by (4.6.2). That the potential vorticity is constant on this curve means that everywhere in the pool the potential vorticity in layer 2 is given as:

$$q_2 = \frac{f}{h_2} = \frac{f_2}{h(\phi_w, \theta_2)} = \frac{f_2}{\sqrt{D_0^2(\phi_w, \theta_2) + H_2^2}} \tag{4.6.4}$$

so that in the pool region:

$$h_2 = \frac{f}{f_2}\left(D_0^2(\phi_w, \theta_2) + H_2^2\right)^{1/2} \equiv \frac{f}{f_2} h_w \tag{4.6.5}$$

Since $h_1 = h - h_2$, when (4.6.5) is used with the Sverdrup relation (4.4.17), we obtain a single equation for h whose solution is:

$$h = \Gamma_{12}\frac{f}{f_2}\frac{h_w}{(1+\Gamma_{12})} + \frac{\left[\left(D_0^2 + H_2^2\right)(1+\Gamma_{12}) - \Gamma_{12}(f/f_2)^2 h_w^2\right]^{1/2}}{(1+\Gamma_{12})}. \tag{4.6.6}$$

In (4.6.6) the notation $\Gamma_{12} = \gamma_1/\gamma_2$ is used. This completes the solution in the pool region and completes the full domain of the problem.

The solution, even within the subtropical gyre, divides itself naturally into three distinct regions. From east to west these are (a) the region of the shadow zone in which the lower layer fluid is unventilated and at rest, (b) the ventilated region in which the subducted fluid conserves potential vorticity as set at the outcrop line, and (c) the pool region which is again unventilated but in motion, and in which it is *assumed* that the potential vorticity is homogenized.

There is an interesting interplay between the depth of layer 2 on the northern (and eastern) boundary of the gyre and the relative extents of the various regions. As $H_2$ increases, the streamlines in the ventilated region swing westward, as can be seen by examining (4.4.21). Thus if $H_2$ increases, the shadow zone increases in longitudinal extent, and the zonal extent of the pool region correspondingly shrinks.

## 4.7 The Three-Layer Model

In their original paper Luyten et al. (1983) described a three-layer model of the thermocline. The three-layer model is of interest for several reasons. First of all, in a quantitative way it leads to a better vertical resolution of the velocity and density structure of the circulation. Of course, a three-layer model is still a rather crude model of the continuous thermocline, and the principal motivation for the present discussion of the three-layer version of the model described in Sections 4.4–4.6 is to present certain new qualitative features of the solution. We emphasize these new features and allow the reader to consult the paper of Luyten et al. for details of the solution.

The physical model is presented in Fig. 4.7.1. Three layers are assumed to be in motion. The new element of the model is the addition of another outcrop

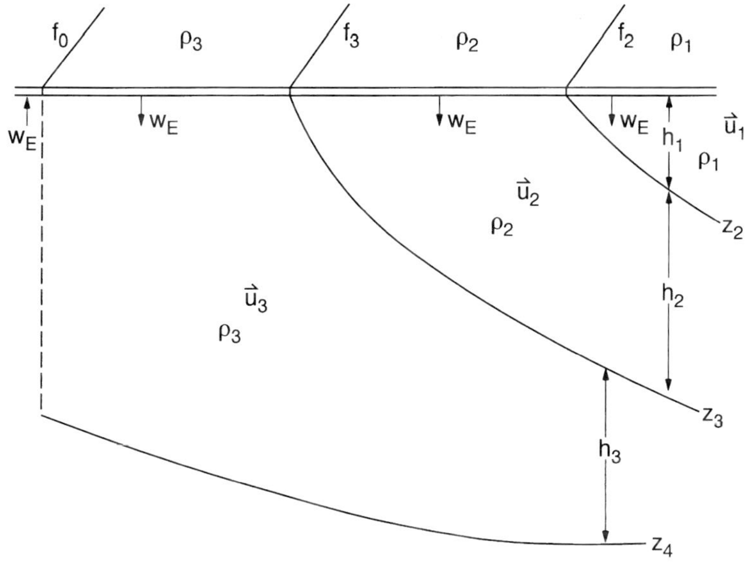


**Fig. 4.7.1.** Three-layer ventilated-thermocline model. Layers 1 through 3 are forced directly by Ekman pumping in the subtropical gyre. The outcrop lines are latitude circles as shown. The surface density field now has three values $\rho_n$, $n = 1, 2, 3$

line south of the first outcrop line. The first outcrop line is at $\theta = \theta_3$ where $f = f_3$. The addition of another layer has required a slight alteration of notation and the new, more southern outcrop line is a latitude circle along which $f = f_2$. This is the latitude in which the layer 2, which was the upper layer of the two-layer model subducts. South of this latitude there are now two ventilated layers each of which is assumed to conserve its potential vorticity. The solution north of this outcrop line is *exactly the same as that obtained for the two-layer model.* This is an important point for it means that as we increase the resolution by adding more layers, the previous solution still holds north of the new outcrop lines and remains undisturbed by the addition of new layers. Information in the ventilated region travels southward along the streamlines as long as potential vorticity is conserved. We need not iterate the solution already found as new layers are added. Thus merely letting $2 \to 3$ and $1 \to 2$, we can rewrite the solution in the region north of $\theta = \theta_2$. That is, for $\theta_3 \leq \theta \leq \theta_0$ the solution in which only layer 3 is moving is given by (4.4.4) or:

$$h = \sqrt{D_0^2 + H_3^2}. \tag{4.7.1}$$

where $h$ is the total depth of the moving fluid (equal to $-z_4$ in this region) while $D_0^2$ is now defined as:

$$D_0^2 = -\frac{2f^2}{\beta\gamma_3} \int_{\phi}^{\phi_e} w_E(\phi', \theta) R \cos\theta \, d\phi' \tag{4.7.2}$$

and $H_3$ is the constant depth of the lowest layer on the eastern boundary.

In the region $\theta_2 \leq \theta \leq \theta_3$ layers 2 and 3 are in motion and contain all the structure described above. In particular, the boundary between the shadow zone and the ventilated region in layer 3 follows the curve given by:

$$D_0^2(\Phi_s, \theta) = H_3^2 \frac{\gamma_2}{\gamma_3} \left(1 - \frac{f}{f_3}\right)^2 \tag{4.7.3}$$

while the relation in the ventilated zone between the thickness of the lowest layer and the total depth is again given by the potential vorticity relation determined at the outcrop line, (4.4.14) i.e.:

$$q_3 = \frac{f}{h_3} = \frac{f_3}{h}$$

so that

$$h_3 = \frac{f}{f_3} h, \quad h_2 = \left(1 - \frac{f}{f_3}\right) h. \tag{4.7.4}$$

The solution itself in each of the ventilated, shadow zone and pool regions is given by the formulae in Sections 4.4–4.6, with an obvious change in

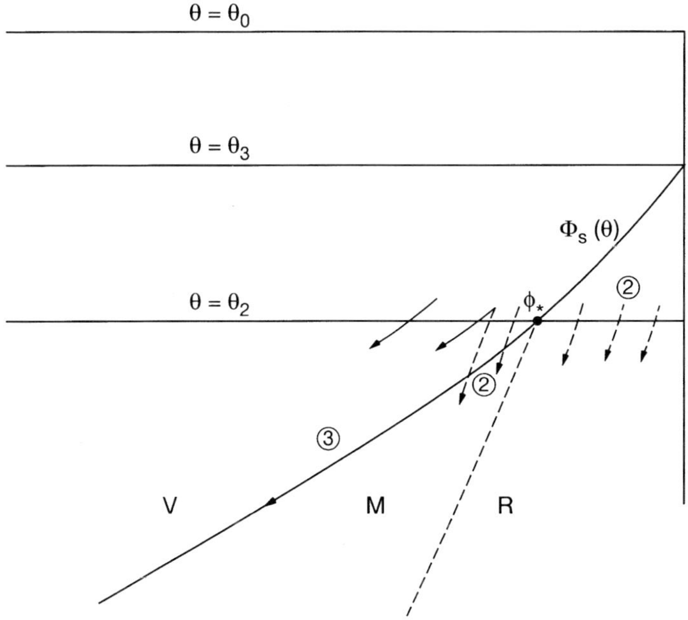


**Fig. 4.7.2.** Schematic plan view of the eastern portion of the domain of flow showing the various dynamical subdomains of motion. South of the southern outcrop line at $\theta = \theta_2$ the flow consist of three regions. *V* is the fully ventilated zone whose eastern edge is the shadow zone boundary etched by the *solid line* representing the streamline which issued from the eastern boundary at the first outcrop line. It crosses the second outcrop line at longitude $\phi*$. East of this zone subducting fluid in layer 2 delineates a region *R* where subduction in layer 2 takes place over resting fluid in layer 3. The western boundary of *R* is determined by the streamline in *layer 2* that passes through $\phi*$ (the dotted curve). Region *M* is a hybrid region in which fluid in layer 2 is subducted west of $\phi*$, where layer 3 is in motion, but later finds itself in a zone where layer 3 is at rest. *Circled numerals* identify the layer in which the curves represent streamlines. *Dotted lines,* streamlines in layer 2, *solid curves,* are streamlines in layer 3

notation. The interesting new features of the solutions arise when fluid elements in layers 2 and 3 flow southward and subduct under layer 1.

The solution again divides into different regions. First let us consider the region sufficiently far west in which both layers 2 and 3 are in motion and subduct beneath layer 1. This is the region V in Fig. 4.7.2. In this zone potential vorticity is conserved in layers 2 and 3. From (4.3.8), (4.3.13), and (4.3.18) we have for each subducted layer:

$$
\begin{aligned}
q_3 &= \frac{f}{z_3 - z_4} = Q_3(\gamma_3 z_4), \\
q_2 &= \frac{f}{z_2 - z_3} = Q_2(\gamma_3 z_4 + \gamma_2 z_3).
\end{aligned}
\tag{4.7.5a,b}
$$

The function $Q_3$ has already been determined from the subduction of layer 3 under layer 2 at the outcrop latitude $\theta = \theta_3$ where $f = f_3$, i.e., from (4.7.4):

$$Q_3(\gamma_3 z_4) = -\frac{f_3}{z_4} \tag{4.7.6a}$$

so that:

$$z_3 = \left(1 - \frac{f}{f_3}\right) z_4. \tag{4.7.6b}$$

This relation is carried by the fluid in layer 3 throughout the ventilated region of layer 3 no matter how many more outcrop lines of upper layers are passed further south.

For layer 2 the potential vorticity of the subducted fluid is determined at the outcrop line of layer 2 at $\theta = \theta_2$ where $f = f_2$. At this location $z_2$ is zero and thus on $\theta = \theta_2$:

$$Q_2(\gamma_3 z_4 + \gamma_2 z_3) = \frac{f_2}{-z_3}. \tag{4.7.7}$$

From (4.7.6b) $z_3$ can be written in terms of $z_4$, so that on the outcrop line of layer 2 (4.7.7) becomes:

$$Q_2\left[z_4\{\gamma_3 + \gamma_2(1 - f_2/f_3)\}\right] = -\frac{f_2}{z_4[1 - f_2/f_3]} \tag{4.7.8}$$

thus, as a function of its argument, call it $x$ momentarily, the form of $Q_2(x)$ is determined as:

$$Q_2(x) = -\frac{\gamma_3 f_2}{x} \frac{[1 + \gamma_2/\gamma_3(1 - f_2/f_3)]}{[1 - f_2/f_3]} \tag{4.7.9}$$

In the present case $x = \gamma_3 z_4 + \gamma_2 z_3$, and with the use of (4.7.4) and the fact that $z_4 = -h$, where $h$ is the total depth of the moving fluid, $h_1 + h_2 + h_3$, it follows that:

$$\frac{f}{h_2} = \frac{f_2}{h(1 - f_2/f_3)} \frac{[1 + \gamma_2/\gamma_3(1 - f_2/f_3)]}{[1 + \gamma_2/\gamma_3(1 - f/f_3)]}. \tag{4.7.10}$$

Solving for $h_2$ in terms of $h$ and then using (4.7.6) we can find each of the layer depths in terms of the total depth $h$. In particular:

$$\begin{aligned} h_3 &= \frac{f}{f_3} h \\ h_2 &= h \frac{f}{f_2} \frac{(1 - f_2/f_3)[1 + \gamma_2/\gamma_3(1 - f/f_3)]}{[1 + \gamma_2/\gamma_3(1 - f_2/f_3)]} \\ h_1 &= h\left\{1 - \frac{f}{f_3} - \frac{f}{f_2} \frac{(1 - f_2/f_3)[1 + \gamma_2/\gamma_3(1 - f/f_3)]}{[1 + \gamma_2/\gamma_3(1 - f_2/f_3)]}\right\}. \end{aligned} \tag{4.7.11a,b,c}$$

In distinction to the two-layer model, the ratio of the layer thicknesses is now a function of the stratification as well as a function of the position of the outcrop lines.

If (4.7.11) is used in the general Sverdrup relation (4.3.15), using again the relations, $z_4 = -h,\ z_3 = -(h_1 + h_2)$ and $z_2 = -h_1$, we can easily solve for h and thus all the layer depths and velocities in the ventilated region. Luyten et al. show that doing so yields:

$$h = \left[\frac{\left[D_0^2 + H_3^2\right]}{F(f)}\right]^{1/2} \tag{4.7.12}$$

where:

$$\begin{aligned} F(f) = 1 + \Gamma_{23}\left(1 - \frac{f}{f_3}\right)^2 \\ + \Gamma_{13}\left\{1 - \frac{f}{f_3} - \frac{f}{f_2}\left(1 - \frac{f_2}{f_3}\right)\frac{\left[1 + \Gamma_{23}(1 - f/f_3)\right]}{\left[1 + \Gamma_{23}(1 - f_2/f_3)\right]}\right\}^2 \end{aligned} \tag{4.7.13}$$

and:

$$\Gamma_{ij} = \gamma_i/\gamma_j. \tag{4.7.14}$$

In the general case when there are $N$ $M$ moving layers in a region, if we are north of the outcrop of the $n^{\text{th}}$ interface, i.e., where $z_n$ vanishes, we can always relate the interface depths $z_j, j > n$, linearly to the total depth $z_{N+1}$ as we did in the case discussed above where $z_3$ is related to $z_4$ by (4.7.6). If we add additional interfaces, we can iteratively continue the process and find a general representation of the solution in the following form. For simplicity of exposition only, consider the case in which all the density jumps are equal so that all the $\gamma_i$'s are equal. Consider layer $n$. Its potential vorticity is related to the pressure field and the interface depths by:

$$\frac{f}{z_n - z_{n+1}} = Q_n\big(z_{N+1} + z_N + z_{N-1} \ldots + \ldots z_{n+2} + z_{n+1}\big). \tag{4.7.15}$$

At the outcrop line of layer $n$ $z_n$ vanishes by definition. We hypothesize, based on our experience with the two-and three-layer models, that each $z_j, j > n$, is related to $z_{N+1}$, the depth of the base of the lowest moving layer, by the linear relation:

$$z_j = \alpha_j z_{N+1} \tag{4.7.16}$$

where the $\alpha_j$'s are functions of latitude and the stratification parameters. Then, at the outcrop line for layer $n$:

$$-\frac{f_n}{z_{n+1}} = -\frac{f_n}{\alpha_{n+1} z_{N+1}}$$
$$= Q_n \left[ z_{N+1} \{ 1 + \alpha_N(f_n) + \alpha_{N-1}(f_n) + \ldots + \alpha_{n+1} \} \right] \tag{4.7.17}$$

from which we can determine $Q_n$, so that:

$$\frac{f}{z_n - z_{n+1}} = -\frac{f_n}{z_{N+1}\alpha_{n+1}(f_n)} \times \left[ \frac{1 + \alpha_N(f_n) + \alpha_{N-1}(f_n) + \ldots + \alpha_{n+1}(f_n)}{1 + \alpha_N(f) + \alpha_{N-1}(f) + \ldots + \alpha_{n+1}(f)} \right]. \tag{4.7.18}$$

Solving for $z_n$ in terms of $z_{N+1}$ yields:

$$\alpha_n(f) = \frac{z_n}{z_{N+1}}$$
$$= \alpha_{n+1}(f) - \frac{f}{f_n}\alpha_{n+1}(f_n)\frac{[1 + \alpha_N(f) + \alpha_{N-1}(f) + \ldots \alpha_{n+1}(f)]}{[1 + \alpha_N(f_n) + \alpha_{N-1}(f_n) + \ldots \alpha_{n+1}(f_n)]}. \tag{4.7.19}$$

We know that:

$$\alpha_N = \left(1 - \frac{f}{f_N}\right), \quad \alpha_{N+1} = 1 \tag{4.7.20}$$

so that (4.7.19) gives a general recursion relation for the $\alpha_i$'s for any number $M \leq n \leq N + 1$. With the $a_j$'s we can use the general form of the Sverdrup relation (4.3.15) and therefore generate the solution for an arbitrary number of layers, passing, if we so wished, to the continuous limit.

A practical difficulty exists, however, in carrying through the solution with a large number of layers, and this is related to an important *qualitative* feature of the solution which develops as layers are added. The algorithm outlined above holds only in the region in which all the layers are moving and ventilated. The solution structure in the remaining domain of the circulation becomes increasingly more complex. This can be seen clearly in the three-layer model.

With the solution (4.7.12) we can find the shadow zone boundary in layer 3 after it passes southward of the outcrop line where $f = f_2$. To find the shadow zone boundary south of $\theta = \theta_2$ we need only continue to follow the streamline given by $h = H_3$. Using (4.7.12) this yields:

$$D_0^2(\Phi_s, \theta) = (F(f) - 1))H_3^2. \tag{4.7.21}$$

Using the definition of $F(f)$, it is easy to show that the shadow zone boundary swings to the west with respect to what the boundary would be if the formula for the shadow zone boundary in the region $\theta > \theta_2$ (4.7.3) were simply extrapolated southward. The shadow zone boundary strikes the outcrop line for layer 2 at the point $\phi*$ on the outcrop latitude as shown in Fig. 4.7.2 and then proceeds southwestward. The shadow zone boundary is a streamline for fluid in layer 3; it is the eastern bounding streamline for the moving, ventilated fluid in that layer. It is *not* a streamline for fluid in layer 2. In fact, as our discussion of the $\beta$-spiral in Section 4.5 showed, the streamlines of layer 2 are swiveled counterclockwise with respect to those in layer 3. In particular, the streamline in layer 2 which passes the point $(\phi *, \theta_2)$ on the outcrop line follows a curve shown schematically by the dotted curve in Fig. 4.7.2. The solution that we obtained above in which all three layers were assumed to be in motion is valid only in the region, labeled $V$, west of the extended shadow zone boundary.

East of that region there are *two* further regions in which the structure of the solution differs from that which we have already found. These are the regions labeled $R$ and $M$ in the figure. In region $R$ layer 3 is at rest, and fluid in layer 2 subducts under layer 1 at the outcrop line where $f = f_2$. Since there are only two layers in motion in this region, and the base of layer 3 is flat at the outcrop line $f = f_2$, the solution in this region is easily found. It is after all a repeat of the same subduction process, now involving layers 1 and 2, which previously occurred at $f = f_3$ involving layers 2 and 3. The solution is of the same form with one major difference. While the depth of layer 3 on the eastern boundary, $H_3$, differs from zero, the depth of layer 2 on the eastern boundary, $H_2$, must be zero since it is zero further north at the outcrop line $\theta = \theta_3$. This in turn implies that there is no shadow zone in layer 2. Shadow zones can only occur in layers that have nonzero thickness on the eastern wall, and the only layer that has a non-zero thickness is layer 3.

Thus in $R$, repeating the analysis of the earlier subduction process yields:

$$h_2 = \frac{f}{f_2}\tilde{h}, \quad h_1 = \left(1 - \frac{f}{f_2}\right)\tilde{h} \tag{4.7.22}$$

where:

$$\tilde{h} = h_1 + h_2. \tag{4.7.23}$$

The Sverdrup integral (4.3.15) then yields:

$$\tilde{h} = \left(\frac{\gamma_3}{\gamma_2}\right)^{1/2} \frac{D_0}{\left[1 + \Gamma_{12}(1 - f/f_2)^2\right]^{1/2}}. \tag{4.7.24}$$

Note that the upper two layers both have vanishing thickness on the eastern boundary where $D_0$ is zero.

Region $M$ consists of fluid which in layer 2 has subducted under layer 1 in the region west of $\phi *$, i.e. in the region where layer 3 is in motion and then enters the region $M$ where layer 3 is at rest. It is thus a hybrid region. Its potential vorticity relation is determined as in region $V$ because where it crosses the outcrop line layer 3 is in motion and thus $Q_2$ is given by (4.7.9). However, only the upper two layers are in motion when the fluid enters region $M$. Thus in layer 2, writing (4.7.9) in terms of the layer thicknesses:

$$\frac{f}{h_2} = \frac{f_2}{(1 - f_2/f_3)} \left[ \frac{1 + \gamma_2/\gamma_3(1 - f_2/f_3)}{h + \gamma_2/\gamma_3(h_1 + h_2)} \right]. \tag{4.7.25}$$

Now layer 3 in region $M$ (as opposed to the outcrop line) is at rest. This implies that the relation between $z_3$ and $z_4$ given by (4.7.6) is no longer valid since we are outside the ventilated region in layer 3. Fluid in layer 3 which has subducted and satisfies (4.7.6) never reaches region $M$. Thus (4.7.10) no longer follows from (4.7.25). Instead, $h$ in (4.7.25), must be replaced by the constant $H_3$. One can then solve for both $h_1$ and $h_2$ in terms of $(h_1 + h_2)$ while $h$ is equal to the constant $H_3$ everywhere in $M$. Having found all the variable layer depths in terms of $(h_1 + h_2)$ the Sverdrup relation is used to complete the solution in the hybrid region $M$.

The total solution, excluding the constant potential vorticity pool in the west, which in general undergoes a similar process of complexification, now consists of three regions: the shadow zone, a fully ventilated zone, and a hybrid region which is partly ventilated. As more layers are added with more outcrop lines, the foliation of the solution increases each time one of the critical curves (which is a streamline in one of the layers) impinges on an outcrop line. The total domain of the solution becomes increasingly broken into smaller fragments in each one of which an analytical solution is possible. The value of an explicit analytical solution rapidly becomes debased by the need to keep track, in tabular form, of the region to which each analytical representation of the solution applies. After a certain point it becomes easier to discretize the entire domain and attribute to each point its appropriate potential vorticity by following numerically the trajectories of potential vorticity conserving fluid elements. This is, in fact, the basis of the numerical treatment of the continuous model of the ventilated thermocline that we consider in Section 4.11.

Luyten et al. present their three-layer solution as applied to the North Atlantic subtropical gyre. The density layers which they chose correspond to values of $\gamma_n = 0.9, 0.6$, and 0.55 for density layers with mean values of $\sigma_\theta$ of 25.70, 26.60, and 27.20 for layers 1, 2, and 3, respectively. They chose outcrop lines according to a recipe that we discuss more fully in Section 4.9 and settled on idealized, zonally oriented outcrop lines at latitudes $\theta = 51.5°$, 37.5°, and 25°N. Only the second and third outcrop lines lie in the subtropical gyre and correspond to the lines $\theta_3$ and $\theta_2$ in the theory. The Ekman pumping that was used was the zonal average of the observed Ekman pumping which changed sign from the subpolar to subtropical gyre at 45°N.

Figure 4.7.3 shows the regions of the flow. The foliation of the domain in the eastern region is recognizable from our discussion above. The reader notes that a similar foliation occurs in the western pool region where the pool boundary is strongly altered in crossing the second outcrop line. Figure 4.7.4 shows meridional and zonal cross-sections of the density structure at selected longitudes of 50° and 30°W and at latitudes of 16° and 30°N. Again, note the region in which the base of the thermocline is flat, which coincides with the shadow zone. In each panel the upper curve is $-D_0^2$.

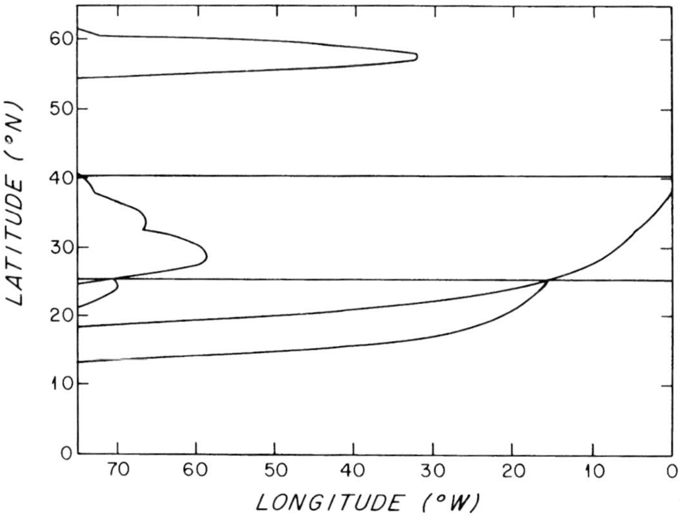


**Fig. 4.7.3.** Different dynamical domains of the flow in the three-layer model for the example discussed by Luyten et al. (1983)

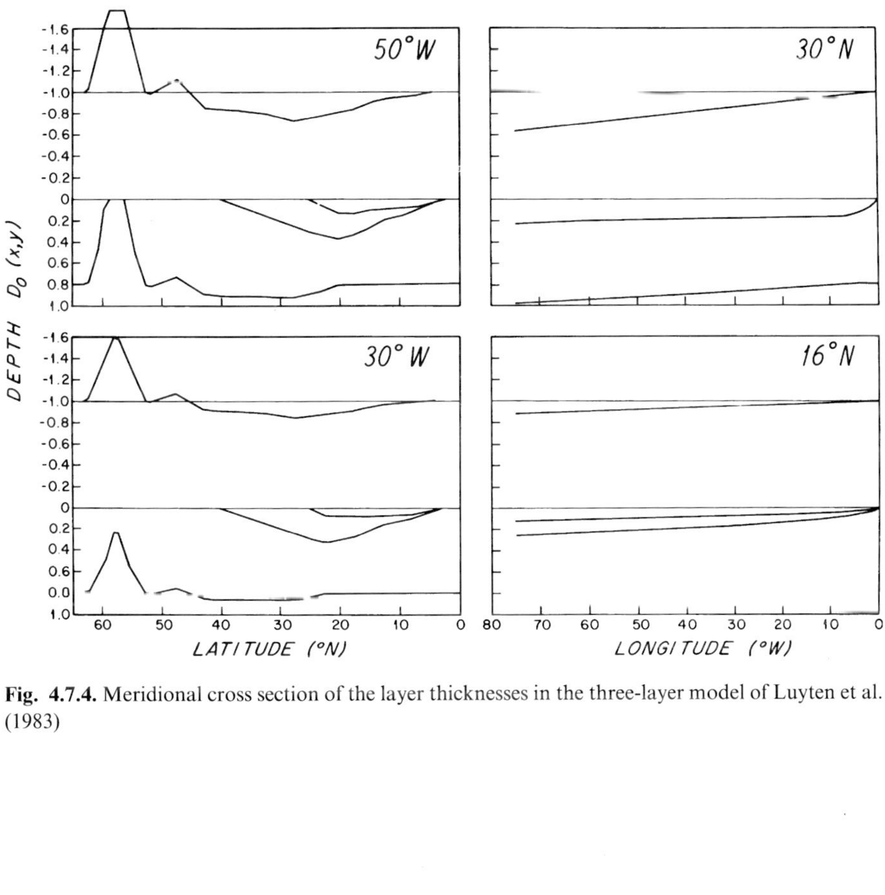


**Fig. 4.7.4.** Meridional cross section of the layer thicknesses in the three-layer model of Luyten et al. (1983)

Aside from the increased complexity of the solution and its higher resolution the basic structure of the solution is qualitatively similar to that already described by the simpler two-layer model.

## 4.8 The Subpolar Gyre

We saw in Chapter 3 that within quasi-geostrophic theory there is no essential difference in the dynamics of the subpolar and subtropical gyres. If $w_E$ changes sign from negative to positive, the quasi-geostrophic equations admit a solution for the subpolar gyre which is identical to the subtropical gyre under a reflection around the latitude of zero Ekman pumping and a change of sign of the streamfunction. With the planetary scale equations, which allow an order one variation in the interface depths, the situation is quite different. The mathematical symmetry no longer exists, and this reflects a fundamental distinction in the physics of the two gyres.

It is not at all clear whether the adiabatic model that we are using is appropriate for the subpolar gyre, or even whether the Sverdrup balance, in which the interaction with the bottom is ignored, is valid there. A diagnostic study by Luyten et al. (1985) suggests the importance of cross-isopycnal fluxes and of bottom topography and therefore casts doubt on both of the assumptions of the present model. In Chapter 5 we discuss the circulation of the subpolar gyre under the influence of nonadiabatic cooling by cross-isopycnal fluxes. However, it suffices for now to demonstrate the distinctive dynamical character of the subpolar gyre by extending the adiabatic ventilated thermocline solution northward across the zero Ekman pumping line.

In the region just north of the latitude of zero Ekman pumping, at $\theta = \theta_0$, where $f = f_0$, only layer 3 in our three-layer model is directly acted upon by the Ekman vertical velocity, which in the subpolar gyre is an upward *suction* into the mixed layer of the fluid beneath. Suppose we assume, as we did for the subtropical gyre, that in the region near $\theta = \theta_0$ only layer 3 is in motion. Then (4.7.1) is valid in this region as well:

$$h = h_3 = \sqrt{D_0^2 + H_3^2} \tag{4.8.1}$$

where now however, $D_0^2$, given by (4.7.2), is *negative* because $w_E$ is positive in the subpolar gyre. As we progress away from the line of zero Ekman pumping $D_0^2$ tends to get larger in magnitude, i.e., more negative. The interface, $z_4$, between layers 3 and 4, is pulled upwards, and layer 3 becomes increasingly thinner. If the Ekman suction is strong enough, the interface surfaces, and layer 3 vanishes along the line $\Phi_3(\theta)$, given by:

$$D_0^2(\Phi_3, \theta) = -H_3^2 \tag{4.8.2}$$

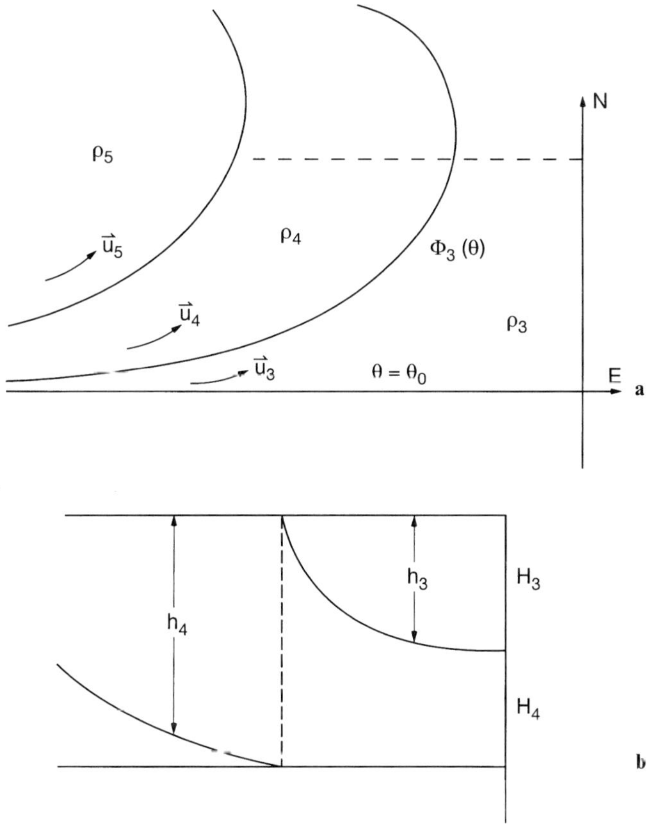


**Fig. 4.8.1.** **a** Outcrop line of layer 4 in the subpolar gyre produced by Ekman suction acting on layer 3. **b** Zonal cross section of the layer thicknesses along the *dotted line in* **a**. Note that the thickness of layer 3 vanishes as the square root of the distance to the outcrop line

For example, when the Ekman suction velocity is independent of longitude, $\Phi_3(\theta)$ is given by:

$$R\cos\theta[\phi_e - \Phi_3] = \frac{\beta H_3^2 \gamma_3}{2f^2 w_E(\theta)}, \tag{4.8.3}$$

where the left side of (4.8.3) represents the linear distance to the eastern boundary. The precise form of the curve on which the interface comes to the surface depends on the structure of the Ekman suction. However, it is clear that for all finite Ekman velocities the curve given by (4.8.3) does not reach the eastern boundary. If $w_E$ increases northward from the intergyre boundary at $\theta = \theta_0$, the curve approaches the eastern boundary and if the Ekman suction decreases further northward the curve swinges back to the northwest, as shown in the schematic in Fig. 4.8.1a. Examination of Fig. 4.7.4 shows the solution found by Luyten et al. exhibiting the surfacing of the interface in the 50°W meridional cross section. Thus in the subpolar gyre the outcrop lines cannot be arbitrarily assigned as in the subtropical gyre. Their positions are determined by the Ekman suction distribution.

It might be thought that the surface density distribution should be a boundary condition that can be consistently applied everywhere to the circulation problem, and this is in fact true. Why then does it appear that the subpolar gyre determines its own surface density? The crucial physical distinction is that in the subtropical gyre the surface density signal is advectively carried downward by the motion and determines the density structure below the surface. In the subpolar gyre the reverse is true. The dynamics of the subsurface flow in the presence of the Ekman suction brings fluid with density already determined at depth towards the surface. If this density is not equal to the surface density of an applied boundary condition (and there is no reason why it should be) the discrepancy between the two densities must be smoothed away in a dissipative boundary layer near the upper surface as described by Pedlosky (1987a). This boundary layer is a balance between downward diffusion of the surface density signal and *upward* advection of subsurface density. This mathematical balance, which is needed only in the subpolar gyre, can also exist mathematically only in the subpolar gyre where the Ekman velocity is upward. Within the context of a purely nondissipative theory, which ignores this thin, dissipative sublayer, the surface boundary condition on the density field must be relaxed in the subpolar gyre and the solution allowed to determine its own surface density.

Along the subpolar outcrop line the thickness of layer 3 vanishes as the square root of the distance to the outcrop line, as shown in Fig. 4.8.1b, i.e., as $\sqrt{\phi - \Phi_3}$. This means that the velocity along the outcropping line is weakly singular with a geostrophic velocity along the outcropping line which goes as $(\phi - \Phi_3)^{-(1/2)}$. The transport, i.e., the velocity times the depth remains bounded. The reader should check that in the subtropical gyre, on the other hand, the layer depth increases linearly from the outcrop line, and that no singularity in the velocity occurs there.

North and west of the outcrop line layer 4 is now exposed directly to Ekman suction and is forced into motion. The solution in this region can easily be found if we assume that layer 5 and deeper layers are at rest. Thus in the region where layer 4 is exposed to Ekman pumping, the application of the Sverdrup balance (4.3.15) yields, after using the relation $Z_5^2 = (H_3 + H_4)^2$;

$$h_4^2 = \frac{\gamma_3}{\gamma_4}(D_0^2 + H_3^2) + (H_3 + H_4)^2. \tag{4.8.4}$$

Note that the first term on the right side of (4.8.4) can be written using (4.8.2) and the definition of $D_0^2$ as:

$$D_0^2 + H_3^2 = -\frac{2f^2}{\beta\gamma_3} \int_{\phi}^{\Phi_3} w_E(\phi', \theta) R \cos\theta \, d\phi'. \tag{4.8.5}$$

This implies that on the outcrop line $h_4$ is a constant. Thus there is no flow across the outcrop line in layer 4 from layer 3 in this solution.

If the Ekman suction is strong enough, the thickness of layer 4 also vanishes on a curve $\Phi_4(\theta)$ which lies everywhere within, i.e., northwestward of, the previous outcrop line for layer 3. The same is subsequently true for deeper layers. One isopycnal interface after another is pulled to the surface leading to a horizontal "jelly roll" pattern, as shown in Fig. 4.8.1a, for the flow in which only the exposed density surface is in motion.

The highly artificial character of this solution is evident. Quite apart from the question of whether adiabatic dynamics is valid in the subpolar gyre, where strong buoyancy loss to the atmosphere is observed, the direction of information flow which is upward and from the west leads us to expect that the solution obtained is nonunique. In distinction to the subtropical gyre, where information is passed downward and southward from the outcrop lines, and where alterations of the solution in the west did not affect the solution already found in the eastern part of the ocean, our circulation solution for the subpolar gyre is very vulnerable to effects of altering the solution in the west. If, for example, we imagine flow in layer 4 to differ from zero near the western boundary and to enter the subpolar gyre from the western boundary, traveling along lines of constant $q_4$, we would have to recalculate the circulation that we have assumed in layer 3. Among other things this would alter the position of the outcrop line, which is fundamental to the solution. If we then ask what the effect of possible motion in layer 5 is, this would again require an entire reiteration of the total solution. The fact that information flows eastward along the intergyre boundary contaminates the calculation of the position of the outcrop line and renders the solution fundamentally nonunique. Only a full solution resolving the western boundary current can resolve the issue. At this time such solutions are lacking, and the state of the theory of the subpolar gyre must be considered unsatisfactory.

## 4.9 Ventilation and Homogenization: A Unified Theory

We have discussed two mechanisms by which the thermocline can be set into motion. The first, discussed in Chapter 3, involves the Rhines and Young recirculation theory for fluid layers that are isolated from sources of potential vorticity such as subduction and ventilation. In such unventilated layers the motion is deduced by appealing to the principal of homogenization of potential vorticity within closed isolines of potential vorticity. The wrapping of the isolines and their closure is produced by the motion itself. We saw that observational and numerical studies seem to validate these ideas, at least for layers which are deep enough in the subtropical gyre to remain unaffected directly by Ekman pumping in the gyre. On the other hand, shallower density layers which outcrop in the subtropical gyre are ventilated and set into motion by the process of subduction even in regions where they are shielded from the direct action of Ekman pumping by overlying layers. These two theoretical ideas are complementary rather than dichotomous since they apply to different levels in the thermocline. At first glance one should be able to unite the two theories by simply stacking the ventilated thermocline on top of the homogenized, unventilated layers in which the fluid recirculates without direct contact with the Ekman layer. A little reflection shows that although this is roughly correct, the situation is actually a good deal more complex. This is because the interaction between the two regions in the vertical is nonlinear. At each geographical position the sum of the transports in the ventilated and unventilated layers must equal the Sverdrup transport. The expression for the Sverdrup transport in terms of the interface depths, for example, is given by (4.3.15) and thus nonlinearly couples all of the layer depths. If we wish to consider a unified model containing both mechanisms we must *simultaneously* consider ventilation and recirculation. This unification was achieved by Pedlosky and Young (1983) and more recently extended by Liu et al. (1993).

We discuss the relevant ideas here in the simplest context that contains both mechanisms. Thus we return to the model of Section 4.4 whose configuration is given in Fig. 4.4.1. Layer 3 in the earlier treatment of Section 4.4 was assumed to be at rest since it was not directly forced anywhere in the subtropical gyre. We saw in Chapter 3 that although such a presupposition can lead to consistent solutions, the resulting flows are not necessarily the most likely to be observed if closed potential vorticity contours appear in the deeper, unventilated layer. In this section we investigate whether layer 3 can be put into motion, and if so what that motion would be. Since layer 3 lies under two ventilated layers, one of which subducts at $\theta = \theta_2$, we are therefore taking up the question of whether the ventilated thermocline solution already found in Section 4.4 is unique, or whether layer 3 can move and so affect and modify the circulation in layers 1 and 2. At the same time, by discussing the possibility of recirculation of the deep layers in the context of planetary scale dynamics we are also relaxing the constraints of Chapter 3 required by quasi-geostrophy and can consistently allow large excursions, over the scale of the gyre, of the interface depths.

As we reconsider the model of Fig. 4.4.1, we again start our analysis in the region north of the first outcrop line at $\theta = \theta_2$ where $f = f_2$. If layer 3 is going to be set into motion by the process of potential vorticity homogenization, this can occur only if the potential vorticity isolines in layer 3 are sufficiently distorted, as described in Chapter 3, so that they elude the eastern boundary and curl around instead to intersect the western boundary or close on themselves in the interior. Near the eastern boundary, as we showed above, the distortions of the interface of deeper layers are small, and we can anticipate that here, *generally speaking*, the potential vorticity isolines strike the eastern boundary so that layer 3 will remains at rest near the eastern boundary. If we start with this hypothesis we can construct the solution sequentially as follows:

North of the outcrop line and sufficiently close to the eastern boundary only layer 2 is in motion. The solution is given by (4.4.6), i.e.:

$$h_2^2 = D_0^2 + H_2^2 \tag{4.9.1}$$

where $D_0^2$ is here defined as in Section 4.4 by (4.4.5). Where layer 3 is at rest the depth of its base is at $z = -(H_2 + H_3)$, where $H_2$ and $H_3$ are the constant thicknesses of layers 2 and 3 on the eastern boundary. Thus, where layer 3 is assumed to be at rest the thickness of layer 3 is given by:

$$h_3 = H_2 + H_3 - h_2. \tag{4.9.2}$$

The potential vorticity in layer 3, where it is at rest is thus:

$$q_3 = \frac{f}{h_3} = \frac{f}{H_2 + H_3 - \left(D_0^2 + H_2^2\right)^{1/2}}. \tag{4.9.3}$$

$D_0^2$ vanishes as $\phi \to \phi_e$ so that near the eastern boundary all the $q_3$ isolines reduce to lines of constant $f/H_3$. They become coincident with latitude circles and thus strike the eastern wall. By the familiar arguments of Chapter 3 this implies that for nondissipative flow in layer 3 the velocity in layer 3 must be zero in regions covered by isolines emanating from the eastern boundary. For sufficiently large forcing, however, the isolines become distorted and further to the west they significantly depart from latitude circles.

Consider an isoline of $q_3$ that emanates from the eastern boundary at a latitude $f'$ where $\theta = \theta'$. Then the isoline of potential vorticity is given by the curve:

$$q_3 = \frac{f'}{H_3} = \frac{f}{H_2 + H_3 - \sqrt{D_0^2 + H_2^2}} \tag{4.9.4}$$

whose parametric equation relating $\phi$ to $\theta$ along the isoline can be written as:

$$D_0^2(\phi, \theta) = 2H_2H_3(1 - f/f') + H_3^2(1 - f/f')^2. \tag{4.9.5}$$

If, for simplicity of discussion only, the Ekman pumping is independent of longitude the equation for each isoline becomes:

$$R[\phi_e - \phi]\cos\theta = \frac{2H_2H_3(1 - f/f') + H_3^2(1 - f/f')^2}{(2f^2/\gamma_2\beta)w_E(\theta)}. \tag{4.9.6}$$

As $f$ approaches $f'$ the numerator of the right side of (4.9.6) vanishes and therefore $\phi$ approaches $\phi_e$, and the isoline intersects the eastern boundary. These are the blocked isolines of potential vorticity on which no flow is allowed. On the other hand, if $f'$ is equal to $f_0$, i.e., if the origin of the isoline is at the northern boundary of the gyre where the Ekman pumping vanishes, both the numerator and the denominator of the fraction in (4.9.6) vanishes as $\theta' = \theta_0$ is approached. The limit must be determined by expanding both numerator and denominator in a Taylor series about $\theta' = \theta_0$, and in that limit we obtain:

$$R[\phi_e - \phi_r] \cos \theta_0 = \frac{-\gamma_2 H_2 H_3 \beta^2}{f_0^3 \{\partial w_E / R \partial \theta\}_{\theta_0}} \tag{4.9.7}$$

where $\phi_r$ is the longitude of the intersection of the isoline with the northern boundary of the gyre at a distance from the eastern boundary given by (4.9.7). When compared to (3.9.8) which is the result from quasi-geostrophic theory, we see that the results for the intersection point are identical. This occurs since the point $\phi_r$ depends only on the undisturbed depth of the layers. It is after all the degree of distortion of the layer thicknesses that distinguishes the quasi-geostrophic theory from planetary geostrophy, and these distortions do not enter the calculation of $\phi_r$.

The isoline emanating from this point on the northern boundary of the gyre carves out a region in the northwest corner of the basin inside of which the potential vorticity isolines curve back and intersect the western boundary twice. The situation is *qualitatively* identical to the quasi-geostrophic case, and the reader should glance at Fig. 3.9.1 to recall the geometry of the curves. The critical dividing isoline between the blocked region of isolines intersecting the eastern boundary and the region of free flow where the isolines close through the western boundary is given by the isoline on which the potential vorticity is that of the emanating point starting at $\theta = \theta_0$ and $\phi = \phi_r$. On this isoline the potential vorticity is $f_0/H_3$ so that the boundary between the region where layer 3 must be at rest, and the pool region where layer 3 may be in motion is given by (4.9.5) with $f'$ replaced by $f_0$.

East of this contour and north of the outcrop line at $f = f_2$ the solution is still given by (4.9.1) in which only layer 2 is in motion and thus carries the full Sverdrup transport. The solution of the ventilated thermocline east of the closed geostrophic contour will remain unaffected by any motion which occurs in layer 3. This, follows physically, from our discussion in Chapter 3 on the propagation of signals by baroclinic Rossby waves. Information about events in the western part of the basin is unable to propagate further eastward than the pool boundary. Information tends to be carried eastward essentially by the Sverdrup advection but east of the boundary of the pool region this information flow is overwhelmed by the stronger westward signal speed of the Rossby waves. The mathematical generalization of this balance beyond quasi-geostrophic dynamics follows from the following:

Consider the total horizontal transport in the Sverdrup interior:

$$\vec{U}_s \equiv \vec{u}_s h = \sum \vec{u}_n h_n \tag{4.9.8}$$

where the sum is over all moving layers. We define the zonal and meridional components of the transport as $U_s$ and $V_s$, respectively, and thus $u_s$ and $v_s$ can be thought of as the vertical average of the interior velocity or equivalently, its barotropic component. The continuity equation summed over all the layers yields:

$$\nabla \cdot \vec{U}_s = -w_E. \tag{4.9.9}$$

On the other hand, the Sverdrup relations allow us to write:

$$V_s = \frac{f}{\beta} w_E = \frac{\gamma_2}{2 f R \cos\theta} \frac{\partial D_0^2}{\partial \phi}. \tag{4.9.10}$$

This allows us to use (4.9.9) to solve for $U_s$:

$$U_s = h u_s = -\frac{\gamma_2}{2 f R} \frac{\partial D_0^2}{\partial \theta}. \tag{4.9.11}$$

In layer 3, where it is motionless, the northward gradient of potential vorticity is, using (4.9.3):

$$\begin{aligned} \frac{1}{R}\frac{\partial q_3}{\partial \theta} &= \frac{1}{h_3}\left\{\beta + \frac{f}{2h_3}\frac{1}{R}\frac{\partial D_0^2}{\partial \theta}\frac{1}{\sqrt{D_0^2 + H_2^2}}\right\} \\ &= \frac{1}{h_3}\left\{\beta + \frac{f}{2h_2h_3}\frac{\partial D_0^2}{R\partial \theta}\right\} \\ &= \frac{1}{h_3}\left\{\beta - \frac{f^2 h u_s}{h_2 h_3 \gamma_2}\right\}. \end{aligned} \tag{4.9.12}$$

Along the northern boundary of the gyre where $w_E$ is zero and where $h_2$ and $h_3$ take on the constant values $H_2$ and $H_3$,

$$\frac{f}{2h_2h_3}\frac{\partial D_0^2}{R\partial\theta} = -\frac{f^3}{\gamma_2 \beta H_2 H_3} R\cos\theta(\phi_e - \phi)\frac{\partial w_E}{R\partial\theta}. \tag{4.9.13}$$

If this result is used in (4.9.12) and compared with (4.9.7) it follows that the existence of the critical point $\phi_r$ where the isolines of $q_3$ emanate from the northern boundary of the gyre, is also a location where the northward gradient of potential vorticity vanishes. Since the eastward component of the gradient of $f/h_3$ also vanishes everywhere along the latitude circle $\theta = \theta_0$, the point $(\phi_r, \theta_0)$ is a point where the total gradient of the potential vorticity vanishes. This point occurs where the bracket in (4.9.12) vanishes so that the critical point also corresponds to the location on the northern boundary of the gyre where:

$$u_S = c_r = \frac{\beta \gamma_2 h_2 h_3}{f^2 h} \tag{4.9.14}$$

where $c_r$ is the baroclinic Rossby wave speed. *Thus the critical point occurs if and where the averaged eastward Sverdrup flow on the northern boundary of the gyre balances the westward propagation of the baroclinic Rossby waves.* The result (4.9.14) is the generalization of the quasi-geostrophic result of Chapter 3 to a case in which $c_r$ depends on the local values of the layer thicknesses and not their constant, rest values. On the northern boundary of the gyre these coincide.

The zonal gradient of $q_3$, again using the solution (4.9.3) for the region of resting fluid is;

$$\frac{1}{R\cos\theta}\frac{\partial q_3}{\partial\phi} = \frac{hf^2}{\gamma_2 h_2 h_3^2} v_s \tag{4.9.15}$$

which vanishes at the northern boundary of the gyre. The slope of the $q_3$ isolines in the horizontal plane is, collecting the above results, given by:

$$\frac{1}{\cos\theta}\left(\frac{\partial\theta}{\partial\phi}\right)_{q_3} = \frac{v_s}{u_s - c_r}. \tag{4.9.16}$$

Thus at the critical point the slope becomes indeterminate, and the critical value of the potential vorticity, $f_0/H_3$, simultaneously corresponds to the value of $q_3$ all along the latitude circle of the northern boundary of the gyre as well as the boundary of the pool region. This boundary enters the gyre and isolates the pool in the northwest corner from the rest of the circulation. The splitting of this isoline at the critical point is a consequence of the indeterminacy of the direction of information propagation where both $v_s$ and $u_s - c_r$ vanish.

Within the pool region layer 3 can be in motion. We argued in Chapter 3 that although it is consistent to have a solution in which layer 3 remains at rest, the existence of closed $q_3$ contours, which allow free geostrophic motion, makes it likely that the solution without motion is unstable and is driven to an $O(1)$ velocity by the slightest frictional or eddy coupling with the overlying layer. Of course, the contours are really only "semiclosed" in the sense that they must actually close through a western boundary layer in which the dynamics may be much different than the interior adiabatic dynamics. However, as in the previous chapter, we assume that the closure through the western boundary current is dynamically equivalent to a closure in the interior.

To determine this motion we once again appeal heuristically to the notion that the potential vorticity becomes homogenized in any region, as argued in the quasi-geostrophic theory, once that region is encircled by a contour of constant potential vorticity and after diffusion of potential vorticity spreads that constant value through the girdled domain. If we take this as a working hypothesis, it would imply that north of the outcrop line and *west* of the boundary of the pool:

$$h_3 = \frac{f}{f_0} H_3. \tag{4.9.17}$$

The Sverdrup relation, (4.3.15), written in terms of layer thicknesses is:

$$h_2^2 + \frac{\gamma_3}{\gamma_2}(h_2 + h_3)^2 = D_0^2 + H_2^2 + \frac{\gamma_3}{\gamma_2}(H_2 + H_3)^2. \tag{4.9.18}$$

When (4.9.17) is used, (4.9.18) yields the solution for $h_2$ and thus the total depth in the pool region. Note that although the *thickness* of layer 3 is a function only of latitude within the pool, the *base* of layer 3 is at a depth $h_2 + h_3$ and is therefore a function of longitude as well. Thus the pressure field in layer 3 in the pool is a function of both latitude and longitude, and the flow has both meridional and zonal components. Figure 4.9.1 shows the solution as determined by Pedlosky and Young (1983) for the case in which the outcrop line is moved all the way to the southern edge of the gyre so that layer 2 covers the whole gyre and thus the solutions (4.9.17) and (4.9.18) are valid in the whole basin (where layer 3 is in motion). The thicknesses of the layers are shown as well as the streamlines. The pool region can be easily discerned in the lower layer thickness field as the region where the thickness is a function only of latitude.

This solution is qualitatively similar to the quasi-geostrophic calculation of Section 3.9.2 with the only difference being the relaxation of the constraint of small isopycnal displacements. Qualitative effects of ventilation only become obvious and interesting when more ventilated layers are added, for example, when the outcrop line at $\theta = \theta_2$ lies within the basin.

Consider the region south of the outcrop line where there are now three layers in motion. Once again the solution domain is split into different zones in which the solution differs. Figure 4.9.2 shows a schematic diagram, in plan view, of the domains of the solution. Starting at the northern boundary of the gyre, where fluid in layer 2 with density $\rho_2$ is exposed to Ekman pumping, there are two regions of flow. In the east there is a region in which only layer 2 is in motion, and layer 3 is at rest. Beyond the pool boundary, $\phi = \Phi_p(\theta)$, is the pool region in layer 3 labeled "pool$_3$" in which the potential vorticity is homogenized to $f_0/H_3$ and where the total solution is given by (4.9.18) while the boundary of the region is given by (4.9.5).

After layer 2 subducts, all three layers are in motion in parts of the gyre. East of the pool of uniform $q_3$ (whose extension south of the outcrop line we must find) layer 3 is at rest, and the solution is similar to that for the two-layer model found in Section 4.4, although we must be more precise about the domain of its validity. There are actually three critical curves, as shown in Fig. 4.9.2, and the solution changes across them. The eastern most curve is the boundary between the shadow zone boundary in layer 2 and the ventilated zone in layer 2. East of this curve both layer 2 and layer 3 are at rest, and the total Sverdrup transport is carried in the upper layer. The solution in the shadow zone is exactly as found in Section 4.4 and is given by (4.4.23) while the boundary itself is given by (4.4.22). Again, *the solution in the eastern part of the gyre is unaffected by alterations in the structure of the flow in the west.*

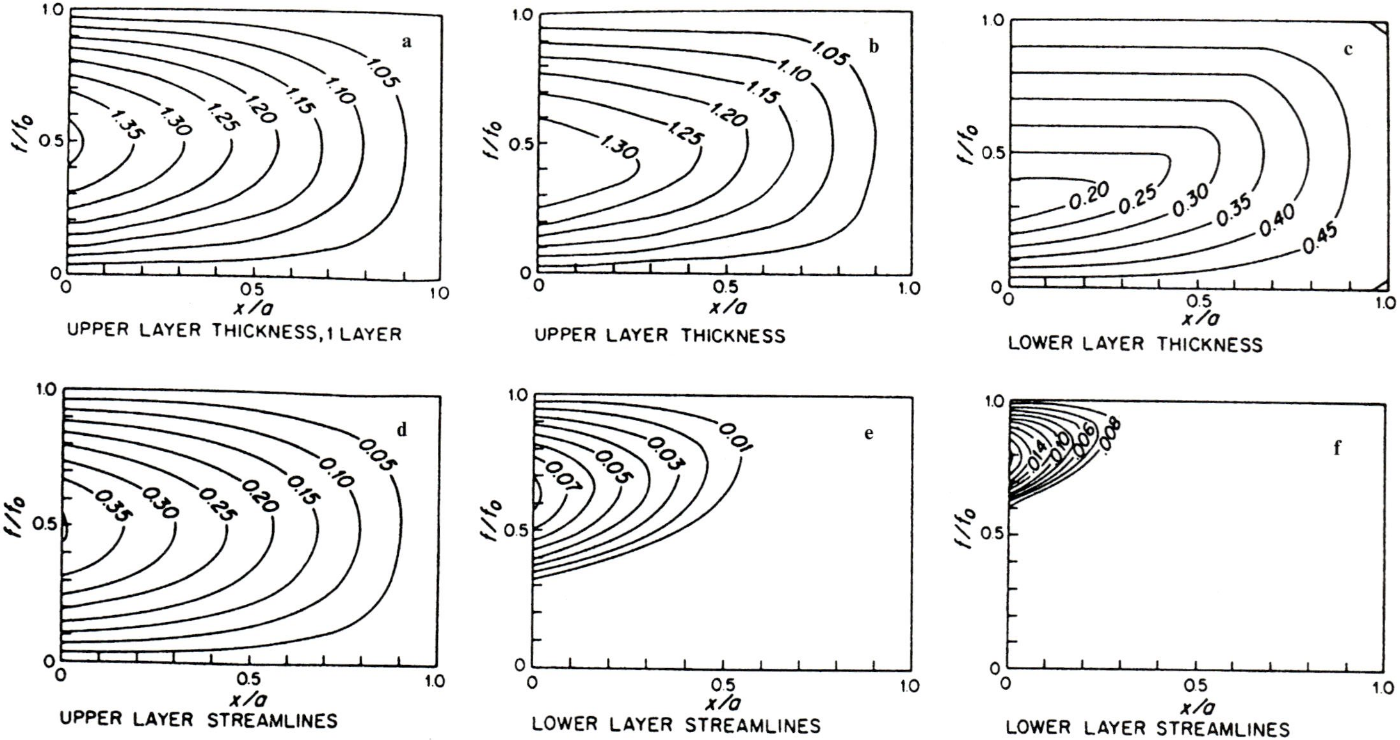


**Fig. 4.9.1a–f.** Solution for the two-layer model when the lower layer possesses close potential vorticity contours. **a** Upper layer thickness field when only the upper layer is in motion, and the lower layer is chosen to be at rest. In this case the contours are also streamlines. **b** Upper layer thickness when the lower layer pool is in motion with constant potential vorticity in the pool. **c** Lower layer thickness. The pool region is visible as the region in which the isolines of lower layer thickness are latitude circles. **d** Upper layer streamlines with the lower layer in motion. Outside the pool the streamlines are as in **a**. **e** Lower layer streamlines. The calculation is for $H_1 = 0.5H_2$. and $\gamma_2 = 1.25\gamma_3$. **f** $H_1 = H_2$, i.e., the thicknesses are equal and the pool shrinks

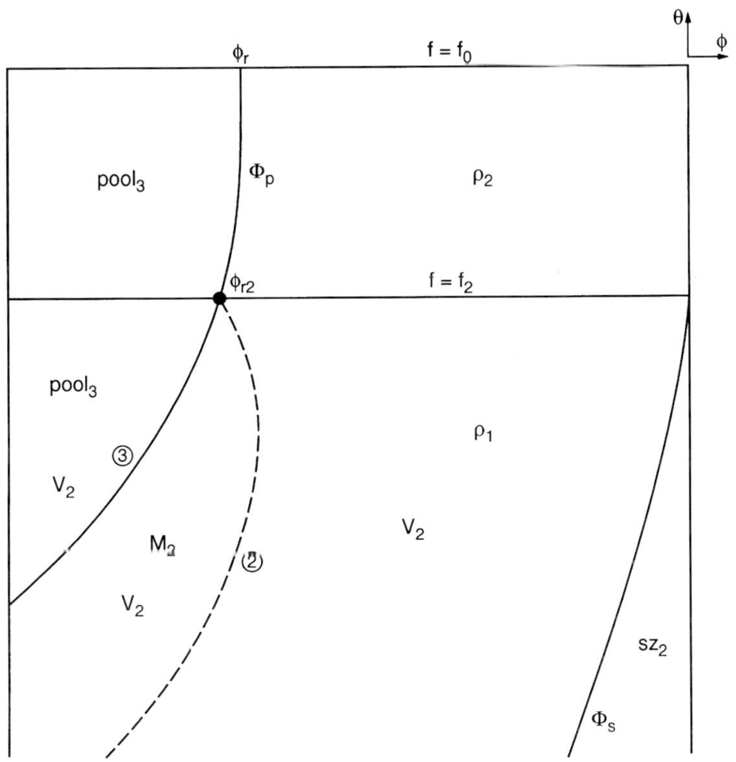


**Fig. 4.9.2.** Domains of solution of the three moving layer model with an outcrop line at $\theta = \theta_2$. The regions of the domain diagram are explained in the text

West of the shadow zone boundary the flow in layer 2 is nonzero. This is the region labeled $V_2$ in the figure. The fluid in layer 2 is ventilated as fluid subducts at the outcrop line $\theta = \theta_2$. There are two critical curves to consider in this region. The first, and qualitatively more important, is the streamline in layer 3 which consists of the continuation of the pool boundary $\Phi_p$ south of the outcrop line. East of this boundary the flow in layer 3 is at rest while layers 1 and 2 are in motion. The pool boundary strikes the outcrop line at the longitude $\phi_{r2}$ before proceeding southward. The second critical curve is a subducting streamline in layer 2 which also originates at $\phi_{r2}$ and proceeds southward. By the $\beta$-spiral result of Section 4.5 we know this streamline lies eastward, i.e., shifted counterclockwise with respect to the streamline in layer 3. The critical streamline in layer 3 that forms the continuation of the pool boundary is shown as a solid curve and is labeled as (3) while the critical streamline in layer 2 is shown as a dashed curve labeled as (2). The situation now is similar to that confronted in discussing the regions $V, M$, and $R$ in the three-layer ventilated model of Section 4.7. East of the dotted curve layer 3 is at rest, and the subducted fluid in layer 2 pours across the outcrop line east of the point $\phi_{r2}$. Thus its potential vorticity is determined along that portion of the outcrop line in which layer 3 is at rest. This is identical to the situation in layer 2 discussed in Section 4.4, and the solution in $V_2$ east of the dotted curve is therefore identical to that given by (4.4.18). Thus east of the dotted curve the entire solution is no different from the solution in which layer 3 is completely at rest and is given by the results of Section 4.4.

However, for fluid in layer 2 that subducts *west* of the intersection of the pool boundary with the outcrop line, i.e., west of $\phi_{r2}$ the potential vorticity of the subducting fluid is determined where layer 3 in motion so that the relationship between $q_2$ and $\pi_2$ is determined differently west of $\phi_{r2}$ than it is to the east of that point. Thus in the pool region where layer 3 is in motion and layer 2 has subducted:

$$q_2 = \frac{f}{h_2} = Q_2\{\gamma_3(h_1 + h_2 + h_3) + \gamma_2(h_1 + h_2)\} \tag{4.9.19}$$

after writing the dynamic pressure fields in terms of the layer thicknesses. On the outcrop line $h_1$ is zero and in the region west of $\phi_{r2}$, within the pool region, $h_3 = f/f_0 H_3$. Thus on this western segment of the outcrop line:

$$\frac{f_2}{h_2} = Q_2\{(\gamma_3 + \gamma_2)h_2 + \gamma_3(f_2/f_0)H_3\}. \tag{4.9.20}$$

The function $Q_2$ can therefore be determined as a function of its argument, $X$, as:

$$Q_2(X) = \frac{f_2(\gamma_2 + \gamma_3)}{X - \dfrac{\gamma_3}{\gamma_2 + \gamma_3}\dfrac{f_2}{f_0}H_3}. \tag{4.9.21}$$

This, in turn yields, using (4.9.19):

$$h_1 = h_2\left(\frac{f_2}{f} - 1\right) + \frac{\gamma_3}{\gamma_2 + \gamma_3}\left[\frac{f_2}{f_0}H_3 - h_3\right] \tag{4.9.22}$$

for all streamlines in layer 2 that subduct west of the point $\phi_{r2}$. In the pool region, labeled pool$_3$ in the figure (4.9.17) applies, and $h_3$ is known. Then (4.9.22) can be used in conjunction with the Sverdrup relation (4.3.15) which, written in terms of layer depths, is:

$$\frac{\gamma_3}{\gamma_2}(h_1 + h_2 + h_3)^2 + (h_1 + h_2)^2 + \frac{\gamma_1}{\gamma_2}h_1^2 = D_0^2 + \frac{\gamma_3}{\gamma_2}(H_2 + H_3)^2 + (H_2)^2 \tag{4.9.23}$$

which with (4.9.22) completes the analytical solution in the pool region. The algebra has grown sufficiently complex that the details of the solution are not given. One important result is obtained rather easily. Consider the boundary of the pool region, given by the curve $\phi = \phi_p(\theta)$. In the region *east* of that boundary layer 3 is at rest and its base is flat. Thus east of the pool boundary, and by continuity of the solution for the layer thicknesses, on the pool boundary itself:

$$-z_4 = h_1 + h_2 + h_3 = H_2 + H_3 \tag{4.9.24}$$

while within the pool, and hence also on its boundary (4.9.17) applies. Using both of these continuity conditions allows us to rewrite (4.9.23) *on the pool boundary* as:

$$D_0^2(\Phi_p, \theta) = 2H_2H_3\left(1 - \frac{f}{f_0}\right) + H_3^2\left(1 - \frac{f}{f_0}\right)^2 + \frac{\gamma_1}{\gamma_2}h_1^2. \tag{4.9.25}$$

When this result is compared with the equation for the pool boundary in the region north of the outcrop line (4.9.5), we see that the two equations are similar except that now there is an additional term on the right side of (4.9.25) that is always positive and proportional to the square of the thickness of the new layer which appears south of the outcrop line. Thus $D_0^2$ is larger that it would be if the new ventilated layer were not added. In order for $D_0^2$ to become larger at the same latitude than it would otherwise be, it is necessary to extend the longitude extent of the region of integration in the definition of $D_0^2$. In turn this means that $\Phi_p$ is further west than it would be if layer 1 were not added. The pool region therefore *shrinks* in consequence of the addition of new ventilated layers. This occurs because as new ventilated layers are added to the flow, here for example layer 1, the layer below it carries less of the Sverdrup transport and hence has a lower velocity, and therefore its base is distorted less. This means that the potential vorticity isolines in the deeper unventilated layers are less distorted by the flow in the layer above it, and the domain of closed potential vorticity contours correspondingly shrinks.

Using (4.9.24) and (4.9.17) we are able to write, on the pool boundary:

$$h_2 = H_2 + H_3(1 - f/f_0) - h_1 \tag{4.9.26}$$

which with the general result (4.9.22) allows $h_1$ on the pool boundary to be written explicitly as:

$$h_1 = \left(1 - \frac{f}{f_2}\right)\left[H_2 + H_3\left\{\frac{\gamma_2}{\gamma_2 + \gamma_3}\frac{f}{f_0} + \left(1 - \frac{f}{f_0}\right)\right\}\right] \tag{4.9.27}$$

which, with (4.9.25) allows the pool boundary to be calculated explicitly.

In region $M_2$ layer 2 is ventilated and in motion and the moving fluid in layer 2 has subducted west of the point $\phi_{r2}$ so that (4.9.21) and (4.9.22) are valid. However the fluid in layer 3 is at rest in this hybrid region so that (4.9.24), instead of (4.9.17), holds throughout this region. With the Sverdrup relation the solution can be completed. The reader is referred to Liu et al. (1983) for the algebraic details.

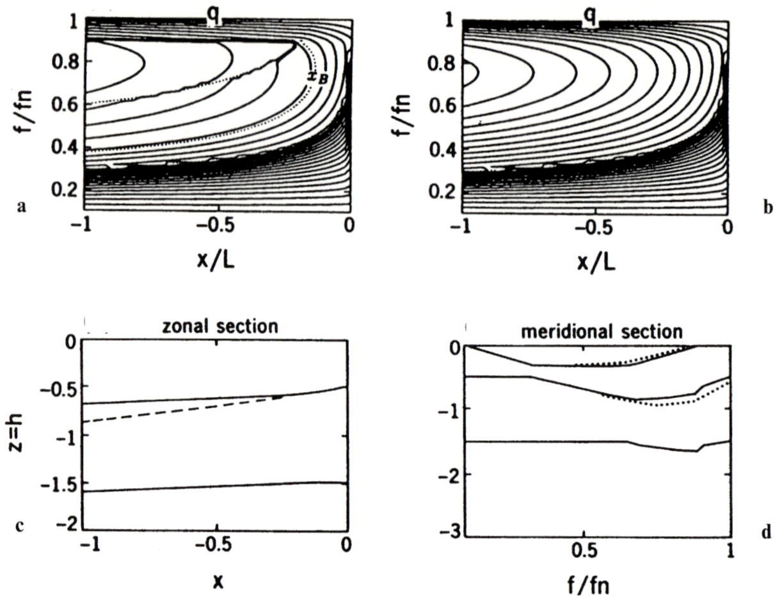


**Fig. 4.9.3a–d.** Solution of the three moving layer model from Liu et al. The Ekman pumping is of the form

$$w_E \propto -\sin\left(\pi(f - f_s/(f_0 - f_s))\right)$$

where $f_0$ is the value of f at the northern boundary of the gyre and $f_s$ is its value at the southern boundary of the gyre. In these calculations, $f_s = 0.1 f_n$. $H_2/H_3 = 0.5$, and the density jumps across each interface have been taken equal. **a** Potential vorticity in layer 2 for the case in which layer 3 is in motion. A Cartesian coordinate frame is used; $L$, width of the basin; $x$, zonal distance. The critical line marked $x_b$ is the dashed curve in Fig. 4.9.2. **b** Potential vorticity isolines in layer 2 when layer 3 is taken to be at rest. **c** Zonal section of the layer depths at the outcrop line. When the motion layer 3 is considered, the layer depths are shown as *solid lines*. When layer 3 is at rest the thickness is shown as *dashed lines*. **d** Meridional section of layer thicknesses

Figure 4.9.3 shows an interesting facet of the full solution as determined by Liu et al. In the upper panels the potential vorticity in layer 2 of the solution is shown (a) with and (b) without the constant potential vorticity motion in layer 3. Thus panel b corresponds to the solution of Luyten et al. while panel a shows the effect of the recirculating lower layers on the ventilated thermocline solution. In both solutions the solution in the east is the same. There is a region of rapid variation of potential vorticity in the region of the shadow zone where the $q_2$ contours are very dense. There is a drop-off of the gradient in the ventilated region as commented upon above in Section 4.4. The interesting thing to note in the figure is that in the western region in which the ventilated layer lies over the moving fluid in layer 3 in the pool region, the potential vorticity variation in layer 2 is even further reduced. The presence of the pool region acts to further *reduce the potential vorticity gradient of the subducted, ventilated fluid.* The reason for this can be understood easily by examining the zonal section of the layer thicknesses at the outcrop line where the potential vorticity in layer 2 is determined. This is shown in panel c of the figure. Only layers 2 and 3 are present at the outcrop line. The dotted curve is the base of layer 2 in the case when layer 3 is *not* in motion while the solid curve is for the case where the presence of motion in the pool has been taken into account. Again, because layer 2 now shares the Sverdrup transport with layer 3 the shear across the interface between the two layers is reduced, and the east-west variation of the thickness of layer 2 is thus reduced at the outcrop line. Since the potential vorticity of the subducting fluid in layer 2 at the outcrop line is $f_2/h_2$, the variation of $q_2$ *along* the outcrop line, and hence *everywhere south of the outcrop line* in the region west of the dotted curve in Fig. 4.9.2 is reduced. Panel d shows a meridional section slicing through the pool region of the layer depths in the two cases with (solid curves) and without (dashed curves) the homogenized motion of layer 3. Liu et al. show that the reduction in the potential vorticity in the ventilated layer above the pool is substantial. For the parameters chosen the reduction is estimated to be of a factor of 2 or 3.

Thus the nonlinear interplay of the ventilated zone and the homogenized region is substantial. The ventilated layers reduce the size of the pool region and also reduce the circulation there below that which would be calculated in the absence of ventilation. At the same time the effect of the pool region substantially reduces the potential vorticity gradients in the ventilated zone. Although the variability of the potential vorticity in the ventilated region and the determination of this variation at the outcrop line form a conceptually important part of the dynamics and of the solution, the flow in the subducted layers of the ventilated region have weak gradients compared to the shadow zone or the layer directly in contact with the Ekman pumping.

This has important interpretive significance because it implies that the observation of a region of very low potential vorticity gradient does not in itself specify the process that produced it. It could be due to the process of homogenization, as is likely the case for deep layers, but for layers that are also ventilated there are additional mechanisms, as we have seen in this section, that can produce zones of weak potential vorticity gradients. This often makes the observational inference of process problematical.

There are also other effects which tend to render the potential vorticity gradients weak in subducting layers. We have, for the sake of analytical simplicity, considered only the cases of outcrop lines that are oriented along latitude circles. This is a reasonable approximation, but observations show that the outcrop lines tend to tilt from northwest to southeast in the gyre. Figure 3.11.3a, for example, shows the outcrop area for the $\sigma_\theta = 26.4$ surface stippled in gray. The relevant winter time outcrop line for the $\sigma_\theta = 26.5$ surface is also shown in Fig. 3.11.4a. In both cases the outcrop line slants northward as one moves to the west. Since the potential vorticity of the $n$th layer along its outcrop line is $f/h_n$, the general increase in $h_n$ to the west along the outcrop line tends to be offset by the increase in f along such a slanting outcrop line and thus reduces the variation in the potential vorticity of the subducted fluid. This leads to the ventilated layer having even weaker gradients in the region on ventilation than hitherto calculated. Indeed, it is possible that a sloping outcrop line can lead to *ventilated,* subducting fluid having completely uniform potential vorticity. We can easily find the shape of such a line.

Suppose we consider the two layer model again and consider the potential vorticity at the first outcrop line. Let the position of the outcrop line be given by:

$$f = f_2(\phi). \tag{4.9.28}$$

Then for the potential vorticity to be constant along the outcrop line we would have

$$\frac{f}{h_2} = \frac{f_2}{h} = \frac{f_e}{H_2} \tag{4.9.29}$$

where $f_e$ is the value of $f_2$ on the eastern boundary. Along the outcrop line $h$ is equal to $h_2$ and is given by (4.9.2) so an implicit equation for $f_2(\phi)$ is given by:

$$\frac{f_2(\phi)}{f_e} = \left(\frac{D_0^2(\phi, f_2)}{H_2^2} + 1\right)^{1/2} \tag{4.9.30}$$

which can be used to solve for $f_2(\phi)$. Figure 4.9.4 shows the outcrop line for the case in which the Ekman pumping is again sinusoidal and $f_e/f_0$ is 0.4. The outcrop line slopes most strongly in the northwest-southeast direction near the eastern boundary as in the observations, rendering the subducted fluid in this idealized case entirely uniform. Of course this does not explain why the outcrop line should slope in that sense. To the degree that the clockwise circulation itself advects the surface density field in a clockwise fashion we might expect such a tilt. Pedlosky et al. (1984), in a simple model in which the surface density field was determined by the flow and its interaction with atmospheric heat fluxes, suggested the way in which the coupled dynamics of an active mixed layer could lead to such slopes of the outcrop line.

When the recirculation theory and the ventilation theory are unified, an elegant structure is predicted for the thermocline. Starting with the deepest unventilated layers that do not outcrop in the subtropical gyre, the circulation occupies a small bowl in the northwest corner of the basin, as we saw in Chapter 3. As we proceed higher in the water column, this bowl expands in size in both latitude and longitude but is always detached from the eastern boundary. Within the bowl potential vorticity is uniform. At higher levels and lower densities, whose surfaces outcrop in the gyre, the circulation is ventilated and laterally more extensive. The potential vorticity is inhomogeneous, although it may be weakly so, in the ventilated regions. In the northwest portions of the ventilated zones regions of homogenized potential vorticity may occur. These are more extensive on the deeper ventilated layers, and the

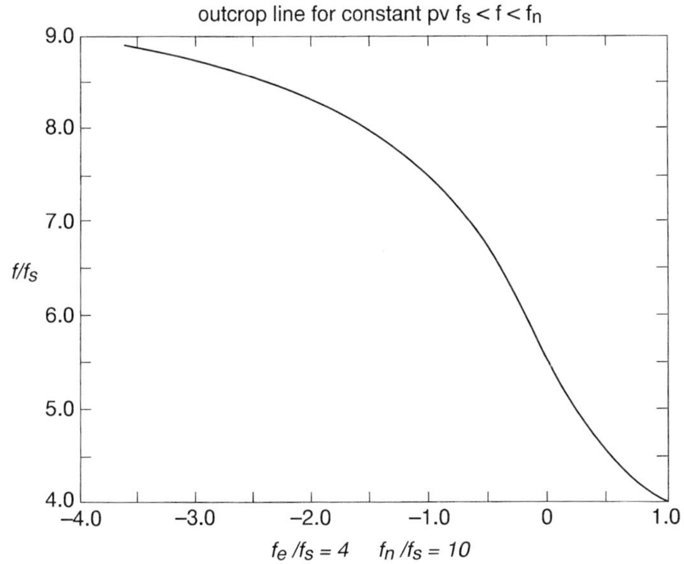


**Fig. 4.9.4.** Position of an outcrop line in a two-layer model such that the subducting fluid has uniform potential vorticity. The parameters are chosen such that the position of the outcrop line on the eastern boundary has a value of $f$ which is 0.4 times the value of $f$ at the northern boundary of the gyre. The layer thickness varies by about 60% along the outcrop line

homogenized regions shrink and finally disappear as surfaces of lighter density are reached that outcrop in the southern portion of the gyre, south of the maximum of the function $D_0^2$. Below the uppermost layer the entire circulation is detached from the eastern boundary by the shadow zone in which no ventilation occurs. Any motion below the most shallow layer in the shadow zone must be driven by nonadiabatic or frictional/eddy effects.

## 4.10 Effect of Finite Mixed Layer Depth

In each of the theories of the thermocline circulation discussed so far the approximation has been made that the mixed layer, in which the Ekman transport occurs, and in which nonadiabatic influences are strong, occupies a negligible fraction of the water column. The mixed layer is the region in which the density field is rendered *vertically* uniform by the action of turbulent stirring by the wind and by convection from the sea surface. The theory for the structure of the mixed layer itself is still tentative, and the observational evidence to distinguish the processes responsible for the structure of the density and velocity fields is still incomplete. See, for example, Price et al. (1986) and Price et al. (1987).

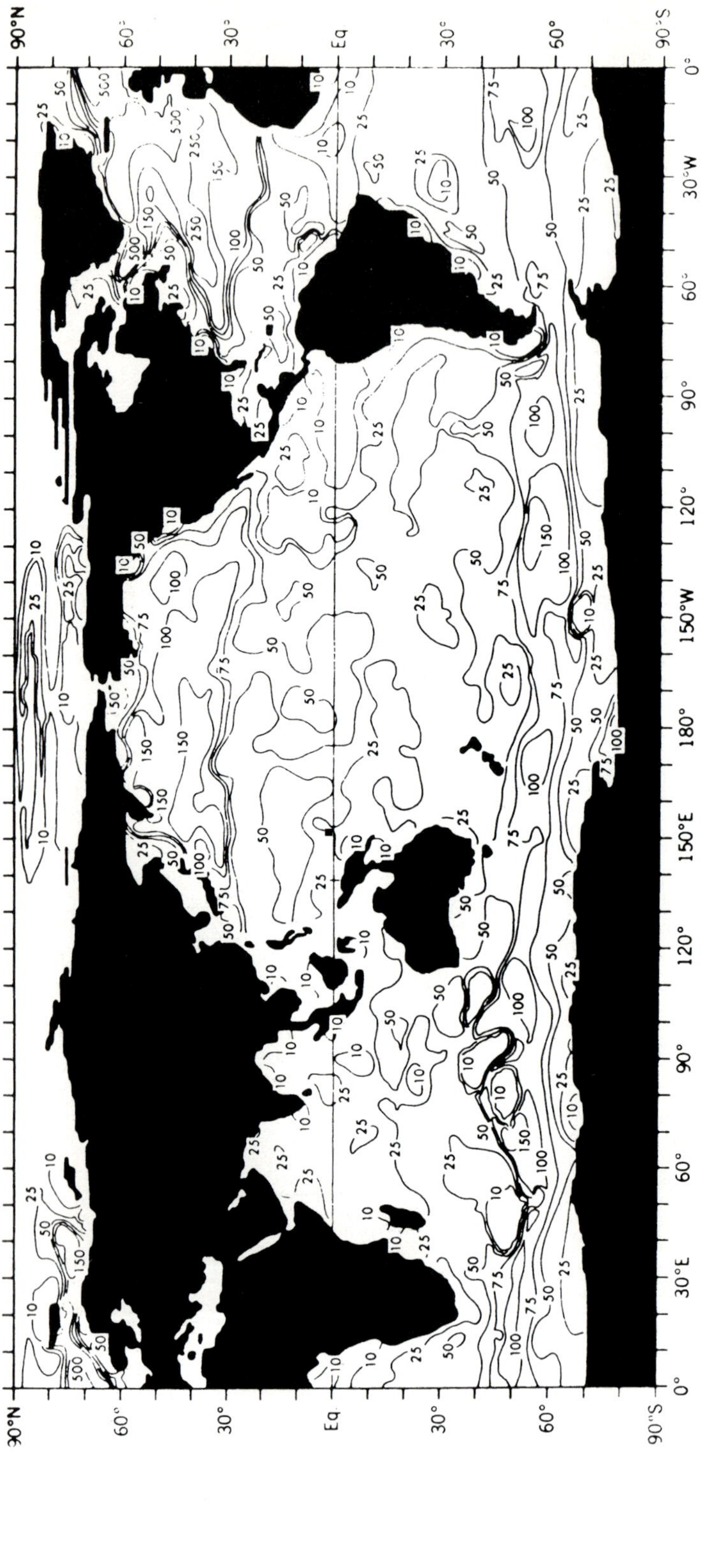


**Fig. 4.10.1.** Mixed layer thickness in the world's oceans in March. (From Levitus 1982)

The thermocline circulation has until now been identified with the adiabatic, nearly frictionless region below the mixed layer. This assumes implicitly that the mixed layer is so thin that the Sverdrup transport captured within the mixed layer is negligible. Yet observations, as presented in Fig. 4.10.1, show that the mixed layer constitutes a nonnegligible fraction of the total water column involved in the circulation of the thermocline. We see, for example, that in the North Atlantic the mixed layer depth in late winter is several hundred meters. Moreover, the depth of the mixed layer varies strongly, from about 50 m in lower, subtropical latitudes to nearly 400 m at the northern boundary of the gyre. The *variation* of the mixed layer depth as well as its substantial value is dynamically important.

Suppose we wish to determine the amount of fluid that ventilates the thermocline from the mixed layer compared to the flux of fluid involved in the horizontal recirculation in the gyre. Were the mixed layer of constant thickness, and relatively thin, the total vertical mass flux from the Ekman layer would be of the order of $w_E L_x L_y$, where $L_x$ and $L_y$ are the characteristic scales in the $x$ and $y$ direction of the gyre. At the same time, the horizontal mass flux in the gyre is of the order $v_s L_x H$ where $H$ is the depth of the thermocline and $v_s$ is the characteristic meridional Sverdrup velocity. Using the Sverdrup balance, we have $v_s H = O(f w_e/\beta)$, and therefore the ratio of the recirculation transport to the flux pumped out of the Ekman layer is given by the Rhines ratio:

$$R_h = \frac{f}{\beta L_y}. \tag{4.10.1}$$

This ratio is normally larger than 1 (the $\beta$ plane approximation is, after all, based on $R_h^{-1}$ being small) and one would conclude that most of the mass transport in the gyre is associated with recirculation rather than ventilation.

However, the effect of the spatial variation of the mixed layer thickness alters the argument radically. Consider the situation shown in Fig. 4.10.2. When the mixed layer has a nonnegligible thickness, $h_m$, which varies with latitude, say, the outcropping of isopycnal layers produce a configuration of the density interfaces as shown schematically in the figure. Within the mixed layer the density surfaces are vertical since there can be, by definition, no vertical variation of density in the layer. An isopycnal interface rises towards the surface in the thermocline, makes contact with the base of the mixed layer, and then rises vertically to the sea surface within the mixed layer. The Ekman layer normally,occupies a small fraction, of the mixed layer and pumps into the top of the mixed layer a vertical velocity $w_E$. Within the mixed layer and below the Ekman layer the velocity is geostrophic. Therefore a horizontal, meridional velocity in the mixed layer, $v_m$, is forced by the stretching of planetary vorticity filaments as in the Sverdrup balance, and we can estimate this velocity as $(f/\beta H)w_E$. That is, it is of the same order as the velocity in the thermocline. Since the base of the mixed layer becomes shallower to the south, some of this

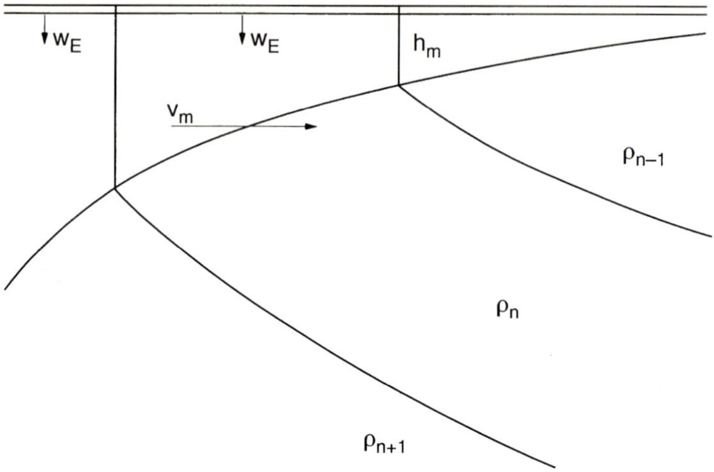


**Fig. 4.10.2.** Schematic presentation of the intersection of the layer interfaces with a mixed layer of variable depth. Fluid enters the mixed layer both horizontally and vertically

horizontal flow escapes the mixed layer and enters the thermocline *horizontally*. We can estimate this amount as:

$$v_m \frac{\partial h_m}{\partial y} L_y L_x = \frac{f}{\beta H} w_E \frac{\partial h_m}{\partial y} L_y L_x \tag{4.10.2}$$

whose ratio with respect to the fluid pumped out of the Ekman layer is:

$$\frac{v_m(\partial h_m/\partial y) L_y L_x}{w_E L_y L_x} = \frac{f}{\beta L_y} \frac{\Delta h_m}{H} \tag{4.10.3}$$

where $\Delta h_m$ is the total variation of the mixed layer depth over the scale $L_y$. Since the total variation of mixed layer depth is substantial (4.10.3) represents a considerable amplification of the fluid flux entering the thermocline from the mixed layer. Since it is during this process that the potential vorticity of the subducted fluid is set, the sloping mixed layer strongly alters the fraction of the fluid whose potential vorticity is affected by ventilation compared to that which recirculates and thus has its potential vorticity determined by, say, homogenization.

The mixed layer of variable depth also has important effects on the *structure* of the ventilated thermocline. In the theories considered so far only the lowest ventilated layer, which we have identified with layer 3 in the model of Section 4.7, has an eastern shadow zone. All the layers above this have zero thickness on the eastern boundary and the flow in these layers can slide along the eastern boundary, without being forced by potential vorticity conservation to separate from the eastern wall. The potential vorticity of each of these layers is infinite at the wall, and hence keeping the potential vorticity constant as well

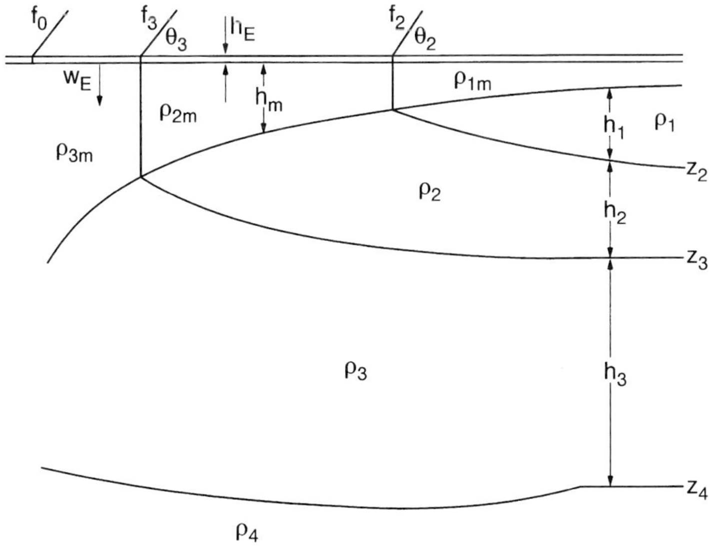


**Fig. 4.10.3.** Three-layer model with variable mixed layer thickness. The density in the mixed layer is a continuous function of latitude

as the layer thickness constant is a condition which is automatically satisfied. When the mixed layer has finite thickness on the eastern boundary, the structural situation changes dramatically, and shadow zones in all the geostophic layers become possible. We follow here the discussion of Pedlosky and Robbins (1991) to investigate this change.

Consider Fig. 4.10.3 in which a three-layer ventilated thermocline is shown under a mixed layer of variable thickness $h_m$. We specify below both the thickness and the density field in the mixed layer. That is, we are not considering a theory for the mixed layer. Rather, here we are interested in the role of the mixed layer in affecting the structure of the thermocline.

In the figure we see the density interfaces, separating layers of uniform density, rise to meet the base of the mixed layer and then rise vertically to the sea surface. We define the density in the mixed layer, $\rho_m$, in the region where layer $j$ is in contact with the mixed layer as $\rho_{mj}$. If this density were always equal to the density of the layer below it, there would be a jump in density across the vertical interface in the mixed layer. Such a density discontinuity would imply a jump in pressure by the hydrostatic balance. Since the velocity in the mixed layer below the Ekman layer (whose depth $h_E$ is assumed to be much less than $h_m$) is in geostrophic balance the jump in pressure would lead to an infinite geostrophic velocity. To avoid this problem Pedlosky and Robbins specified a mixed layer density that is continuous, and such that $\rho_{mj} = \rho_j$ at $\theta = \theta_{j+1}$, and they chose the variation of $\rho_{mj}$ such that $\rho_{mj}$ is less than $\rho_j$ further south. That is, the mixed layer density matches the density of the uppermost layer at its most northern extent and, to maintain the static stability of the system, is always less than the density of the layer beneath it. Thus, at $\theta = \theta_3$ the mixed layer density is $\rho_2$ and decreases smoothly southward to become $\rho_1$ at $\theta = \theta_2$. Both the mixed layer depth and density are taken to be functions only of latitude for simplicity, as are the outcrop lines.

The dynamic pressures in each of the thermocline layers as well as in the mixed layer are:

$$\begin{aligned}
\pi_3 &= \rho_0[\gamma_3 h] \\
\pi_2 &= \rho_0[\gamma_3 h + \gamma_2(h_m + h_1 + h_2)] \\
\pi_1 &= \rho_0[\gamma_3 h + \gamma_2(h_m + h_1 + h_2) + \gamma_1(h_m + h_1)] \\
\pi_m &= \rho_0\left[\gamma_3 h + \gamma_2(h_m + h_1 + h_2) + \gamma_1(h_m + h_1) + \gamma_1 h_m - \frac{\rho_m}{\rho_0} gz\right]
\end{aligned} \tag{4.10.4}$$

where we define:

$$\begin{aligned}
h &= h_1 + h_2 + h_3 + h_m \\
\gamma_j &= \left(\frac{\rho_{j+1} - \rho_j}{\rho_0}\right) g \\
\gamma_{jm} &= \left(\frac{\rho_j - \rho_{mj}}{\rho_0}\right) g
\end{aligned} \tag{4.10.5a,b,c}$$

The total Sverdrup transport can be calculated geostrophically from (4.10.4) and this leads to the constraint:

$$\begin{aligned}
&h^2 + \frac{\gamma_2}{\gamma_3}(h_m + h_1 + h_2)^2 + \frac{\gamma_1}{\gamma_3}(h_m + h_1)^2 + \frac{\gamma_{1m}}{\gamma_3} h_m^2 \\
&\quad = D_0^2 + H^2 + \frac{\gamma_2}{\gamma_3}(H_m + H_1 + H_2)^2 + \frac{\gamma_1}{\gamma_3}(H_m + H_1)^2 + \frac{\gamma_{1m}}{\gamma_3} H_m^2
\end{aligned} \tag{4.10.6}$$

where uppercase letters denote the value of the thickness fields on the eastern boundary.

The last terms on both the right and left sides of (4.10.6) have been added for reasons of symmetry. Since $\mathrm{h_m}$ is taken to be a function only of latitude, $h_m = H_m$, and so these two terms actually cancel.

In the region north of the first outcrop line the only moving layer beneath the mixed layer is layer 3. The Sverdrup balance reduces then to:

$$h^2 + \frac{\gamma_{3m}}{\gamma_3} h_m^2 = D_0^2 + H^2 + \frac{\gamma_{3m}}{\gamma_3} H_m^2. \tag{4.10.7}$$

To obtain (4.10.7) we set $h_1 = h_2 = 0$ in (4.10.6) and use the fact that $\gamma_1 + \gamma_2 + \gamma_{1m} = \gamma_{3m}$ in this region.

This allows us to solve for h in the region north of $\theta = \theta_3$ as in Section 4.4. However, consider the boundary condition on the eastern boundary $\phi = \phi_e$. The zonal velocity in the layer is:

$$u_3 = -\frac{\gamma_3}{fR}\frac{\partial h}{\partial \theta} \tag{4.10.8}$$

while in the mixed layer:

$$u_m = -\frac{\gamma_3}{fR}\frac{\partial h}{\partial \theta} - \frac{1}{fR}\frac{\partial}{\partial \theta}(\gamma_{3m}h_m) + \frac{gz}{\rho_0 fR}\frac{\partial \rho_{3m}}{\partial \theta}. \tag{4.10.9}$$

There is no way to have both $u_3$ and $u_m$ equal to zero on the eastern boundary if the mixed layer density and depth are functions of latitude. If, however, we calculate the total zonal transport of the two layers, we obtain:

$$U_3 + U_m = u_3 h_3 + \int_{-h_m}^{0} u_m dz = -\frac{\gamma_3}{2fR}\frac{\partial}{\partial \theta}\left[h^2 + \frac{\gamma_{3m}}{\gamma_3}h_m^2\right]. \tag{4.10.10}$$

Since we can no longer insist that the normal velocity to the boundary vanishes in each layer on the eastern wall, we appeal to the following expedient. At the wall unspecified nonadiabatic effects, which maintain the mixed layer density and depth gradient with latitude, force a zonal velocity into (or out of) the eastern boundary. The physics here is beyond the scope of the present theory. We therefore suggest that similar dissipative physics furnishes an upwelling or downwelling layer at the eastern boundary such that the zonal transport in the mixed layer at the eastern boundary is balanced by an equal and opposite transport in the layer directly beneath it. Thus in the present case setting the zonal transport given by (4.10.10) equal to zero at the eastern boundary, implies that:

$$H^2 + \frac{\gamma_{3m}}{\gamma_3}H_m^2 = \text{constant} \quad \text{on} \quad \phi = \phi_e. \tag{4.10.11}$$

At the northern boundary of the gyre, where $f = f_0$, the mixed layer density has been set equal to $\rho_3$ so that $\gamma_{3m}$ vanishes there. Thus the constant in (4.10.11) is simply the depth of layer 3, i.e., $H$, evaluated at $f = f_0$. Thus:

$$H(f) = \sqrt{H^2(f_0) - \frac{\gamma_{3m}}{\gamma_3}H_m^2}. \tag{4.10.12}$$

On the eastern boundary the depth of the thermocline, $h = H$, is now no longer constant but varies with latitude according to (4.10.12), $h$ itself is given by (4.10.7), i.e.:

$$h = \sqrt{D_0^2 + H^2}. \tag{4.10.13}$$

South of the outcrop line at $\theta = \theta_3$, but north of the second outcrop line at $\theta = \theta_2$, $h_1 = 0$, and there are two adiabatic layers as well as the mixed layer in motion. The Sverdrup balance becomes:

$$h^2 + \frac{\gamma_2}{\gamma_3}(h_m + h_2)^2 + \frac{\gamma_{2m}}{\gamma_3}h_m^2 = D_0^2 + H^2 + \frac{\gamma_2}{\gamma_3}(H_m + H_2)^2 + \frac{\gamma_{2m}}{\gamma_3}H_m^2 \tag{4.10.14}$$

Again, on the eastern boundary the zonal velocity in the mixed layer differs from zero, and we assume that this flow is returned zonally in the region beneath the mixed layer. The distribution of that return flow between layers 2 and 3 is unknown since we are not actually dealing with the dynamics of the hypothesized eastern upwelling (or downwelling) layer, and we assume again that the flow is returned in the layer just beneath the mixed layer. That is, the zonal velocity in layer 3 is now zero at the eastern wall, and the zonal transport in layer 2 balances that of the mixed layer. The first of these conditions requires that on $\phi = \phi_e$:

$$H = H(f_3) \tag{4.10.15}$$

as determined by (4.10.12), while using arguments similar to that leading to (4.10.10) and (4.10.11) we obtain as a consequence of the second condition:

$$(H_m + H_2)^2 = H_m^2(f_3) - \frac{\gamma_{2m}}{\gamma_2}H_m^2(f). \tag{4.10.16}$$

On the outcrop line $H_2$ is zero. In the theory with a mixed layer of negligible or constant thickness the depth of the second layer remains zero on the eastern wall to satisfy the no normal flow condition there. That condition is relaxed here and replaced by the condition of no net zonal flux at the wall, and this forces $H_2$ to depart from zero. This has important consequences below when we discuss the solution south of the next outcrop line at $\theta = \theta_2$.

The solution in the region north of that outcrop line is determined in the usual way by considering the subduction of fluid at the outcrop line at $\theta = \theta_3$. Potential vorticity is conserved in layer 3, and thus we have again:

$$q_3 = \frac{f}{h_3} = Q_3(h). \tag{4.10.17}$$

For all streamlines in layer 3 which originate at the outcrop line, $h_2$ vanishes at $\theta = \theta_3$. Thus on such streamlines:

$$h = h_3 + h_m(f_3) \tag{4.10.18}$$

at the outcrop line. For all latitudes south of the outcrop line we can therefore determine the function $Q_3(h)$ in the usual way to obtain:

$$Q_3(h) = \frac{f_3}{h - h_m(f_3)} \tag{4.10.19}$$

so that, from (4.10.17):

$$h_3 = \frac{f}{f_3}[h - h_m(f_3)] \tag{4.10.20}$$

and:

$$h_m + h_2 = -z_3 = h\left(1 - \frac{f}{f_3}\right) + \frac{f}{f_3}h_m(f_3). \tag{4.10.21}$$

With the Sverdrup relation (4.10.14), the solution can be completed. Pedlosky and Robbins present first the circulation in this region for the case where the latitude of the second outcrop line is moved to the equator so that the solution presented here fills the gyre, *except for the shadow zone*. There is, of course, a shadow zone near the eastern boundary, just as found in Section 4.4, in which the fluid in layer 3 is at rest. The boundary of the shadow zone is found in just the same manner as in the earlier theory. That is, the streamline emanating from the intersection of the outcrop line with the eastern wall, on which $h = H(f_3)$, is followed using the solution described above. The solution in the shadow zone has only layer 2 in motion (as well as the mixed layer) and so in that region $h = H(f_3)$ and the Sverdrup relation (4.10.14) yields the shadow zone solution directly:

$$h_2 = \sqrt{\frac{\gamma_3}{\gamma_2}D_0^2 + (H_m + H_2)^2 - H_m^2}. \tag{4.10.22}$$

where $H_m + H_2$ is given by (4.10.16). Thus on the eastern wall $h_2 = H_2 \neq 0$.

Figure 4.10.4 shows the solution in the region south of the outcrop line for a sinusoidal Ekman pumping. Panel c shows the variable part of the mixed layer density and the variation of the mixed layer depth used in the calculation. Panel a shows the streamlines in layer 3 in the case in which the mixed layer is negligibly thin. Panel b shows the streamlines in the case in which the mixed layer is variable. In this case the shadow zone broadens linearly, instead of parabolically, with distance from the outcrop line, and in fact all the streamlines are shifted westward. As a consequence of this shift the constant potential vorticity pool region in the west is absent when the mixed layer is variable, at least for this outcrop position. Outcrop positions farther northward would again yield a western pool region, but it would be reduced in size. In panels d and e we see the flow in layer 2. Panel d describes the flow for a negligible mixed layer thickness while panel e shows the flow in the case where the mixed layer depth varies. Note that only a very weak flow issues from the eastern boundary. The variations in density and mixed layer depth tend to offset each other, and therefore there is only a very weak zonal flux of fluid in the mixed layer at the wall requiring compensation. Thus to a large extent layer 2 develops a nonzero thickness at the wall due mostly to the kinematic shrinking of the mixed layer and, to first order, the constancy of the depth of the base of layer 2 as shown in Fig. 4.10.4f.

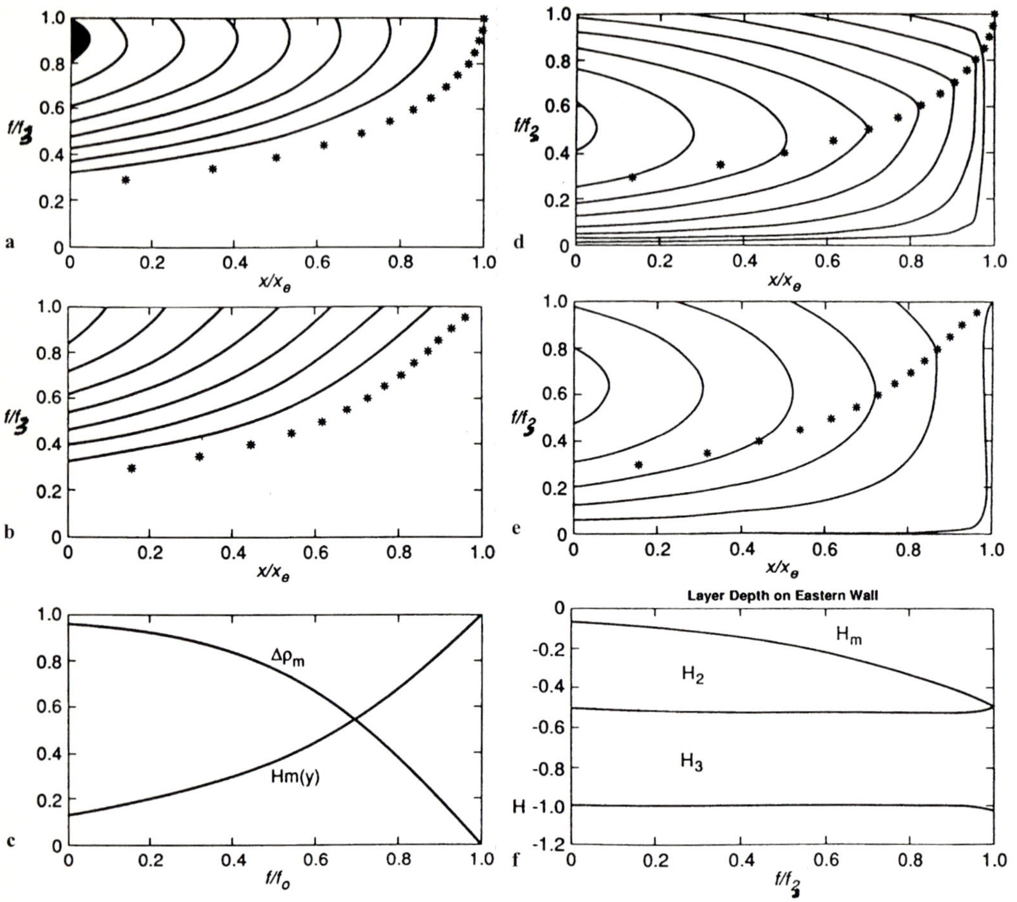


**Fig. 4.10.4a–f.** Solution of the model with two adiabatic layers and a variable mixed layer. **a** Solution in the lower layer south of the outcrop line in the case when the mixed layer is absent. **b** As in **a** but for a variation in mixed layer density and thickness, as shown in **c**. **d** Streamlines in the upper adiabatic layer in the absence of mixed layer variations. **e** Streamlines when the mixed layer varies. **f** Layer thicknesses on the eastern wall. (From Pedlosky and Robins 1991)

Since the solution in the region below the mixed layer is now known, it is possible to find the flow across the sloping mixed layer into the thermocline. Pedlosky and Robbins show this is equal to:

$$w* = w_E - \frac{\beta}{f}\int_{-h_m}^{0} v_m dz + v_2 \frac{1}{R}\frac{\partial h_m}{\partial \theta}. \tag{4.10.23}$$

The first term represents the vertical velocity out of the Ekman layer at the top of the mixed layer. This vertical velocity diminishes in magnitude with depth within the mixed layer since within the mixed layer, $\partial w/\partial z = (\beta/f)v_m$

and $v_m < 0$. The sum of the first two terms yields the *vertical* velocity at the base of the mixed layer. The third term yields the flow which leaves the mixed layer due to the horizontal velocity (which is continuous across the mixed layer base) multiplied by the slope of the mixed layer. With the solution obtained as described above an explicit representation of this ventilation flux can be found. Pedlosky and Robbins show that *near the outcrop line* the augmentation of the ventilation of the thermocline by the effect of the sloping mixed layer is given by:

$$\frac{w*}{w_E} - 1 = -\frac{h_m^2}{h}\frac{\partial}{\partial f}\left(\frac{f}{h_m}\right) \tag{4.10.24a}$$

in the ventilated zone and;

$$\frac{w*}{w_E} - 1 = -h_m \frac{\partial}{\partial f}\left(\frac{f}{h_m}\right) \tag{4.10.24b}$$

in the shadow zone. The augmentation depends on the northward gradient of the potential vorticity, $f/h_m$, in the mixed layer. There is a competition between the reduction of the vertical velocity at the base of the mixed layer due to the second term in (4.10.23) and the increase in ventilation due to the slope of the mixed layer base, and this competition is measured by the potential vorticity gradient of the mixed layer. If the mixed layer deepens rapidly enough northward, the gradient is negative and the augmentation is positive. The enhancement is greatest in the shadow zone since $h > h_m$. Pedlosky and Robbins estimated, from the use of maps such as in Fig. 4.10.1, that this would lead to augmentation of the ventilation over the Ekman pumping by a factor of 1.6 in the ventilated zone and a factor of 4.0 in the shadow zone. Thus, the mixed layer variation strongly increases the amount of fluid passing through the mixed layer before being subducted.

The solution may be continued south of the second outcrop line at $\theta = \theta_2$. The solution is in many respects qualitatively similar to that found in Section 4.7 except for important changes in the solution in the shadow zone of layer 3. In this region layer 2 subducts under layer 1, and layer 3 is at rest. We saw in Section 4.7 that when there is a mixed layer of negligible thickness, this subduction resembles that taking place in layer 3 further north. When layer 2 has zero thickness on the eastern boundary, however, there is no shadow zone in layer 2. In the absence of a mixed layer of variable and substantial thickness only layer 3 has a shadow zone. Now, in the presence of the variable mixed layer depth, layer 2 has a nonzero thickness on the eastern boundary given by (4.10.16) and (4.10.22). With a finite thickness of layer 2 on the eastern wall it is not possible for a fluid column to move directly along the eastern boundary and both preserve potential vorticity and have no zonal velocity, as is required south of $\theta = \theta_2$. *A shadow zone therefore develops in layer 2.* The details of the calculation are given by Pedlosky and Robbins, but the qualitative point is simple. Once the mixed layer has variable depth, it produces thermocline layers

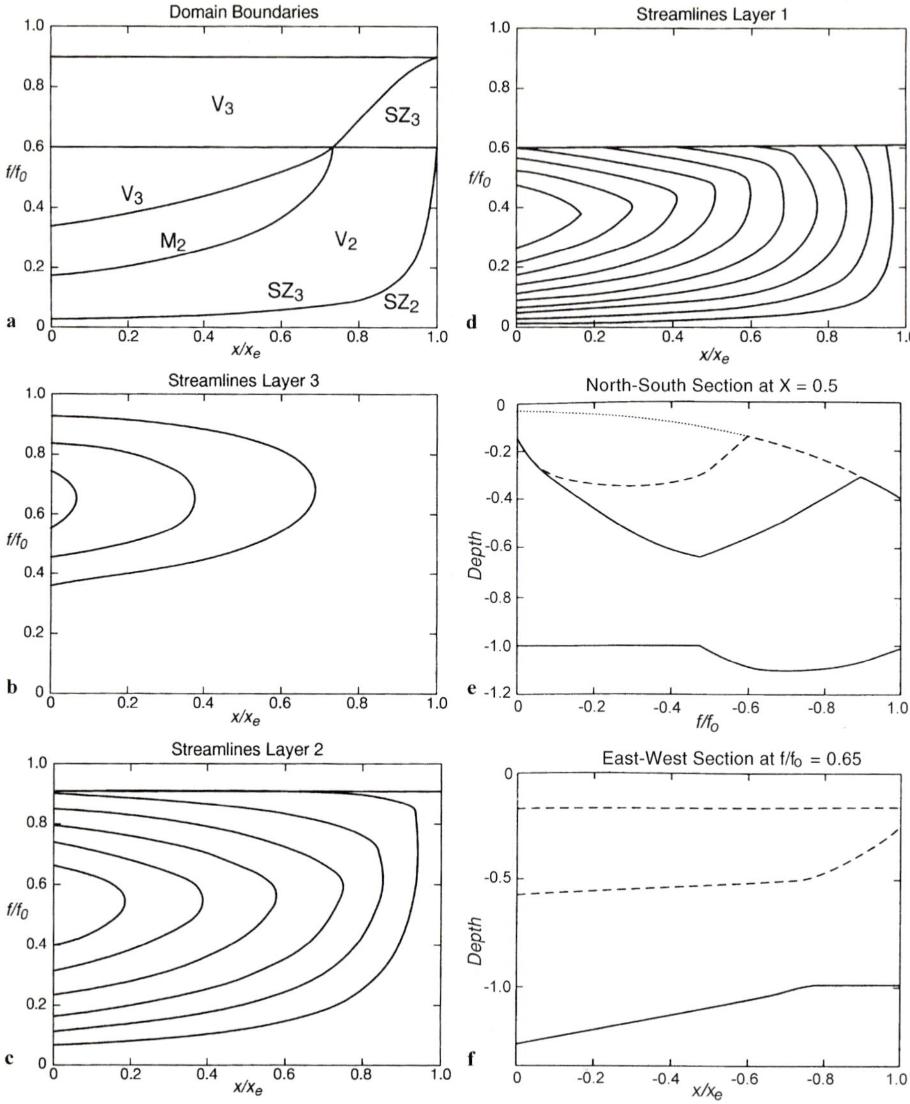


**Fig. 4.10.5a–f.** Solution for the three layer model with a variable thickness mixed layer. **a** Domains of the flow described in the text. Note the appearance of a shadow zone in layer 2. **b** Streamfunction in layer 3. **c** Streamfunction in layer 2. **d** Streamfunction in layer 1. **e** Meridional cross section showing layer thicknesses as a function of latitude half way across the basin. **f** Zonal cross section of layer thicknesses

which have finite depths on the eastern boundary, and each of these layers has a shadow zone. *Shadow zones therefore generally occur in all layers.* The stratification on the eastern boundary of the ocean is finite rather than singular since now it is not necessary that all the isopycnals (except the lowest ventilated one) rise to the surface. The shadow zones spread from the wall into the eastern region of the gyre, altering the structure of the circulation.

Figure 4.10.5 shows the streamlines of the three-layer flow as calculated by Pedlosky and Robbins. Panel a shows the solution domains of the flow. South of the first outcrop line the solution is split in layer 3 into a ventilated zone, $V_3$, and a shadow zone, $SZ_3$. South of the next outcrop line the solution domain fragments further. Most of the subdomains are discussed above in Section 4.7. The details of the solution alter in the presence of the variable mixed layer in the western region, but the solution is similar to that found above. There is the ventilated zone in layer 3, $V_3$, as well as the hybrid zone in layer 2, $M_2$, in which layer 2 is ventilated on the outcrop line where layer 3 is in motion but then passes into the hybrid region where layer 3 is at rest. East of this region is the simple ventilated region in layer 2, $V_2$, where layer 3 is at rest, and where the potential vorticity in layer 2 is determined on the eastern portion of the outcrop line in the shadow zone of layer 3. Then there is the new region, $SZ_2$, which is the shadow zone in layer 2, and in which only layer 1 is in motion beneath the mixed layer. Panels b–c show the streamlines in layers 1, 2 and 3, respectively. In particular the shadow zone in layer 2 is evident. Figure 4.10.5e, and f shows latitude and zone cross sections of the solution in the region in which all three layers exist. The shadow zone in layer 2 is so narrow that the sections do not display it clearly.

If more layers are added to improve the vertical resolution of the model, each new layer has a shadow zone. Each shadow zone boundary in turn produces new subdomains of the flow when it encounters successive outcrop lines further southward. It is not hard to see that the improvement of the model in the direction of greater vertical resolution would soon lead to analytic complexity great enough to vitiate the advantages normally associated with analytic representations of the motion. The principal advantage of an analytical solution is the help its compact representation of the physics gives to develop our physical insight. An analytic solution fragmented into ever smaller zones, requiring a map and a table to reference and locate each domain, soon exhausts its instructive value as well as our patience.

The step to improved resolution requires going finally to a model with continuous distributions of density and velocity. Since the solution can be expected to differ physically and mathematically in the various subdomains, it is unlikely that an analytical solution will emerge, and a numerical treatment is in fact necessary. The formulation and the execution of such a model and its algorithm is taken up in the next section. We see there that it closely follows the essentials of our discussion of the layer models.

## 4.11 Continuous Models of the Ventilated Thermocline

The passage from a layer model of the ventilated thermocline to a continuous model is both important and technically difficult. If for no other reason, it is important to be able to demonstrate that the layer models are physically relevant in the sense that with increased resolution the layer solutions smoothly tend to the solutions for a continuous fluid. As opposed to the continuous quasi-geostrophic model of Chapter 3, the very nonlinear nature of the problem on the planetary scale, in which the isopycnals suffer large vertical excursions over the gyre, makes the mathematical problem a great challenge. This is, after all, why the layer models with their few vertical degrees of freedom are attractive.

Huang (1988, 1989a,b, 1991) was the first to take up this problem and in a series of remarkably fruitful papers described a method of dealing with the continuous problem. Important contributions have also been made by Williams (1989). As our discussion in Section 4.10 suggests, the problem must ultimately be dealt with numerically. However, the analytical formulation of the problem by Huang and Williams is of great conceptual value in itself. With the advantage of hindsight we see that the mathematical character of the problem in its final form is a straightforward extension of the layer models as the number of layers becomes infinite. This is a very important result. It verifies that the layer models smoothly tend to the solution for a continuous fluid. Since the layer models are the exact representation of conceptually realizable, if highly idealized, physical systems, the fact that they are also the finite difference approximations of a continuous model invests even the simplest layer model with a high degree of physical significance. It implies that a low-order layer model is simultaneously an exact representation of a real but crude physical system as well as a crude finite-difference representation, in the vertical, of the exact continuous model. As the number of layers increases both the physical and mathematical representations *continuously* improve.

The relationship between the layer models and the continuous representation of the solution is best seen in a coordinate system in which the vertical coordinate used is the density itself. This is after all what we have already used in the layer models where all fields are represented as function of latitude, longitude, and the *layer index n*. The depths of the layer interfaces, $z_n$, or their thicknesses, $h_n$, become the dependent variables. Using density as a vertical coordinate is simply a way of maintaining this representation in the continuous limit.

As long as there is a 1:1 relation between depth and density, the density may be used as a vertical coordinate. This is certainly the case as long as the ocean is stably stratified, i.e., as long as $\partial\rho/\partial z < 0$. Thus any variable, $u$, say, can then be represented as a function of time, horizontal position, and density, i.e.,

$$u = u(t, \phi, \theta, \rho). \tag{4.11.1}$$

Since the depth of any density layer can be written:

$$z = z(t, \phi, \theta, \rho) \tag{4.11.2}$$

we can equally well write (4.11.1) as:

$$u = u[t, \phi, \theta, z(t, \phi, \theta, \rho)] \tag{4.11.3}$$

which is useful in deriving relations between derivatives in the system of density coordinates and the more standard $z$ coordinate system. Thus, for example:

$$\left.\frac{\partial u}{\partial \phi}\right)_\rho = \left.\frac{\partial u}{\partial \phi}\right)_z + \left.\frac{\partial u}{\partial z}\frac{\partial z}{\partial \phi}\right)_\rho \tag{4.11.4}$$

using the chain rule and (4.11.3). In particular:

$$\frac{\partial u}{\partial \rho} = \frac{\partial u}{\partial z}\frac{\partial z}{\partial \rho} \tag{4.11.5}$$

so that (4.11.4) is equivalent to:

$$\left.\frac{\partial u}{\partial \phi}\right)_z = \left.\frac{\partial u}{\partial \phi}\right)_\rho - \frac{\partial u}{\partial \rho}\left(\frac{\partial z/\partial \phi}{\partial z/\partial \rho}\right). \tag{4.11.6}$$

The continuity equation in $z$ coordinates is:

$$\nabla \cdot \vec{u} + \frac{\partial w}{\partial z} = 0 \tag{4.11.7}$$

where the first term is the two-dimensional horizontal divergence in which each derivative with respect to latitude and longitude is taken at constant $z$. In density coordinates, using (4.11.4), (4.11.5), and (4.11.6) we obtain:

$$\frac{\partial}{\partial t}\left(\frac{\partial z}{\partial \rho}\right) + \nabla \cdot \left(\vec{u}\frac{\partial z}{\partial \rho}\right) + \frac{\partial w*}{\partial \rho} = 0. \tag{4.11.8}$$

In (4.11.8) the horizontal derivatives in the divergence term and the time derivative are each taken in a surface of constant density. The function $w*$ is defined as:

$$w* = w - \frac{dz}{dt} = w - \frac{\partial z}{\partial t} - \frac{u}{R\cos\theta}\frac{\partial z}{\partial \phi} - \frac{v}{R}\frac{\partial z}{\partial \theta} \tag{4.11.9}$$

$w*$ is the cross-isopycnal velocity, and we recall that $\partial z/\partial \phi$, for example, is given by $-(\partial \rho/\partial \phi)/(\partial \rho/\partial z)$ in $z$ coordinates.

The "vertical" velocity in density coordinates is:

$$\omega = \frac{d\rho}{dt} = w* \frac{\partial \rho}{\partial z}. \tag{4.11.10}$$

If the motion is adiabatic $\omega$ or $w*$ vanishes, and the motion becomes two-dimensional in density coordinates. Note the similarity between (4.11.8) and the layer version of the equation for mass conservation (4.2.1). The similarity becomes even stronger if a finite difference version of (4.11.8) in density coordinates is used so that $\partial z/\partial\rho \rightarrow \Delta z/\Delta\rho$ and $\Delta z$ is identified with the layer thickness of a particular density layer that has a density interval $\Delta\rho$.

The pressure gradient in $z$ coordinates can be written using the hydrostatic relation, and our previous results relating derivatives in the two frames yields:

$$\left.\frac{\partial p}{\partial\theta}\right)_z = \left.\frac{\partial p}{\partial\theta}\right)_\rho + \left.\frac{\partial p}{\partial\rho}\frac{\partial\rho}{\partial\theta}\right)_z$$

$$\begin{aligned} &= \left.\frac{\partial p}{\partial\theta}\right)_\rho + \left.\frac{\partial p}{\partial z}\frac{\partial z}{\partial\rho}\frac{\partial\rho}{\partial\theta}\right)_z \\ &= \left.\frac{\partial p}{\partial\theta}\right)_\rho + \rho g\left.\frac{\partial z}{\partial\theta}\right)_\rho \end{aligned} \tag{4.11.11}$$

which can be rewritten in general vector form as:

$$\nabla p)_z = \nabla\pi)_\rho \tag{4.11.12}$$

where $\pi$ is the Montgomery function:

$$\pi = p + \rho g z \tag{4.11.13}$$

which is the same function which naturally appears in the layer models. In fact if $\pi$ is differentiated with respect to $\rho$ and the hydrostatic approximation is used we obtain:

$$\frac{\partial\pi}{\partial\rho} = gz. \tag{4.11.14}$$

This is the hydrostatic relation in density coordinates and a comparison with (4.3.7) shows that the latter could be considered a finite difference approximation to (4.11.14).

The geostrophic relation, in density coordinates becomes, using (4.11.12):

$$f\hat{k}\times\vec{u} = -\frac{1}{\rho_0}\nabla\pi \tag{4.11.15}$$

which is the continuous version of (4.3.11). Just as horizontal derivatives in the layer model are taken within a given layer, the horizontal derivative in (4.11.15), for example, is taken in a fixed density surface.

The planetary vorticity equation follows from taking the curl of (4.11.15). i.e.:

$$\beta v = -f\nabla\cdot\vec{u}. \tag{4.11.16}$$

With the mass conservation equation (4.11.8) this becomes:

$$\frac{dq}{dt} = \left(\frac{q}{\partial z/\partial \rho}\right)\frac{\partial w*}{\partial \rho} \tag{4.11.17}$$

where:

$$q = -\frac{f}{\partial z/\partial \rho}. \tag{4.11.18}$$

The minus sign is introduced into the definition (4.11.18) since $\partial z/\partial \rho$ is negative for a stably stratified ocean. The minus sign then defines the potential vorticity as positive in the northern hemisphere as in the layer model's definition (4.3.1) which again could be interpreted as a finite difference version of (4.11.18). Indeed, (4.11.17) is identical, from this point of view, with the layer model's potential vorticity equation (4.3.26) when midocean friction or eddy momentum mixing is ignored.

For steady, adiabatic motion the function $\vec{u}\,\partial z/\partial \rho$ is horizontally nondivergent and is derivable from a streamfunction, i.e.:

$$\vec{u}\frac{\partial z}{\partial \rho} = \hat{k} \times \nabla \psi. \tag{4.11.19}$$

It follows that both $q$ and $\pi$ are constant on streamlines in each density surface when the motion is adiabatic and so both $q$ and $\pi$ are functions only of $\psi$ and $\rho$. Eliminating $\psi$ between $q$ and $\pi$ yields:

$$q = -\frac{f}{\partial z/\partial \rho} = Q(\pi, \rho). \tag{4.11.20}$$

This is the same statement as (4.3.18) where now the value of the density, $\rho$, substitutes for the density layer index, $n$.

The physical model is shown in Fig. 4.11.1. This is similar to the model discussed in Section 4.10. Now the density is a continuous function of position everywhere. A mixed layer of variable thickness $h_m$ lies over a region in which the thermocline motion is assumed to conserve potential vorticity. The model is steady, and the density layers fall into three classes. There is a surface $z = -D(\phi, \theta)$ below which the fluid is at rest, and where the density is a function only of $z$ and is given by the relation:

$$z = Z_D(\rho) \tag{4.11.21}$$

which expresses the stratification of the resting ocean. Above this surface are density surfaces which do not outcrop but are set into motion, as discussed in Section 4.8, as recirculating fluid. These layers have homogenized potential vorticity. Above these layers are layers which outcrop in the gyre and are

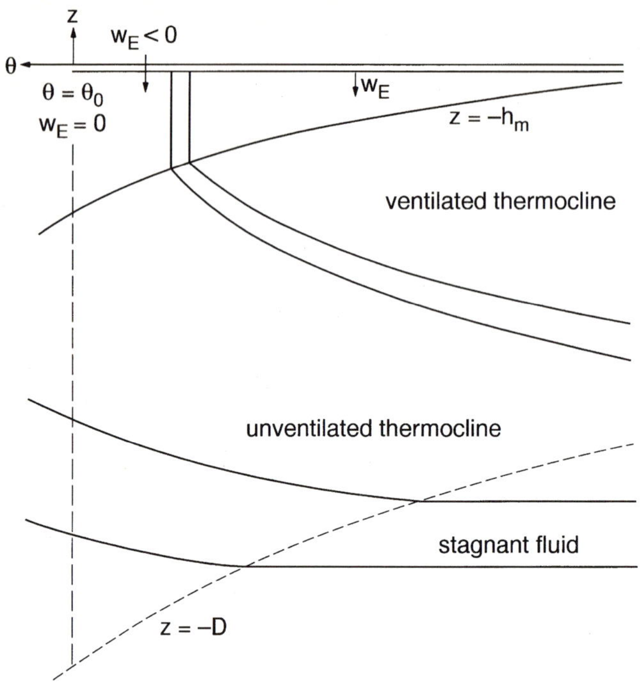


**Fig. 4.11.1.** Continuous model. The adiabatic region of motion lies between $z = -h_m$, the base of the mixed layer, and $z = D$, the base of the bowl containing the motion. The water column is divided between those density surfaces which outcrop in the subtropical gyre (ventilated) and those which do not but are nontheless in motion (unventilated). Below $z = -D(\phi, \theta)$ lies the resting abyss with a given density-depth relation. The density and depth of the mixed layer are specified. A thin Ekman layer at the top of the mixed layer pumps fluid with a velocity $w_E$ downward south of the latitude $\theta_0$

ventilated. Those density surfaces intersect the base of the mixed layer at $z = -h_m$ and then rise vertically to the Ekman layer very near the sea surface where $w = w_E$.

In the region below the mixed layer the potential vorticity is conserved, and in principle, if the function $Q$ is determined, the problem for determining $\pi$ and $z$, can be written as the second-order system of equations:

$$\frac{\partial z}{\partial \rho} = -\frac{f}{q}$$
$$\frac{\partial \pi}{\partial \rho} = gz \tag{4.11.22a,b}$$

or equivalently:

$$\frac{\partial^2 \pi}{\partial \rho^2} = -\frac{gf}{q(\pi, \rho)}. \tag{4.11.23}$$

As a differential equation in $\rho$ the boundary conditions for the system (4.11.22) are rather unusual mathematically but rather straightforward physically. At the base of the thermocline, i.e., at $z = -D$, the density and the pressure must be continuous with that of the resting fluid. Thus for each $\phi$ and $\theta$ in the gyre, at $z = -D$, where the density is $\rho_D$ (and which is a function of horizontal position):

$$\pi = \pi_D(\rho); \quad z = Z_D(\rho) \quad \text{at} \quad \rho = \rho_D. \tag{4.11.24}$$

Of course we do not know ahead of time where the base of the thermocline is, and therefore where the boundary is in either $z$ or $\rho$. That is, we are dealing in density coordinates with a *free* boundary problem, and an important part of the solution is the determination at each latitude and longitude of the density, $\rho_D$, which sets the base of the thermocline. Since the density is continuous, the velocity (but not the vertical shear) must be continuous across the surface $z = -D$, and thus the velocity vanishes there as it does in the abyss. Therefore it follows from (4.11.24) that $\nabla\pi = 0$ at $\rho = \rho_D$.

Within the mixed layer the density surfaces are vertical. That is, the density is equal to its surface value, $\rho_s(\phi, \theta)$, throughout the mixed layer. Thus at the base of the mixed layer the boundary condition on the system (4.11.22) is:

$$z = -h_m(\phi, \theta) \quad \text{at} \quad \rho = \rho_s(\phi, \theta) \tag{4.11.25}$$

Each of these equations has its layer equivalent. If we take note of (4.3.18), the layer equation statements of potential vorticity conservation and the hydrostatic relation between the interface depths and the pressure (4.3.7) are identical to the finite difference version of (4.11.22a,b) although we have not yet indicated how $q$ is to be determined in (4.11.22a).

The Sverdrup relation (4.3.15) also has an exact equivalent in the continuous model. To demonstrate this we must start with the mixed layer and calculate its transport. As opposed to the treatment in the previous section, we now allow both the mixed layer depth and its density to be functions of longitude as well as latitude.

Within the mixed layer the density, being independent of depth, allows us to write the pressure as:

$$p_m = p_s(\phi, \theta) - \rho_s g z \tag{4.11.26}$$

where $p_s$ is the (unknown) surface pressure. This means that the surface pressure is given by:

$$p_s = p_m + \rho_s g z = \pi_m \tag{4.11.27}$$

Therefore $\pi_m$ within the mixed layer is independent of depth and equal to the surface pressure. By continuity of pressure, at the base of the mixed layer:

$$\pi = \pi_m(\rho_s) \quad \text{at} \quad \rho = \rho_s \tag{4.11.28}$$

Since the velocity field in the mixed layer beneath the Ekman layer is geostrophic, the meridional velocity in the mixed layer is given by:

$$v_m = \frac{1}{\rho_0 \, fR\cos\theta}\frac{\partial p_m}{\partial\phi} = \frac{1}{\rho_0 \, fR\cos\theta}\left\{\frac{\partial p_s}{\partial\phi} - gz\frac{\partial\rho_s}{\partial\phi}\right\} \tag{4.11.29}$$

whose integral over the mixed layer depth gives the meridional transport of the mixed layer:

$$V_m = \int_{-h_m}^{0} v_m dz = \frac{1}{\rho_0 \, fR\cos\theta}\left\{h\frac{\partial p_s}{\partial\phi} + \frac{gh^2}{2}\frac{\partial\rho_s}{\partial\phi}\right\}. \tag{4.11.30}$$

The Sverdrup balance, over the region of moving fluid is:

$$\int_{-D}^{0} v\,dz = \int_{-D}^{-h_m} v\,dz + \int_{-h_m}^{0} v_m\,dz = \int_{-D}^{-h_m} v\,dz + V_m = \frac{f}{\beta}w_E. \tag{4.11.31}$$

The integral over the region below the mixed layer can be written as an integral in $\rho$ over the equivalent interval of density, that is:

$$\int_{-D}^{-h_m} v dz = \int_{\rho_D}^{\rho_S} v\frac{\partial z}{\partial\rho}d\rho = -\int_{\rho_S}^{\rho_D} v\frac{\partial z}{\partial\rho}d\rho. \tag{4.11.32}$$

The order of integration (4.11.32) is interchanged here so that the limits of integration in density run from lower to higher values of $\rho$.

When (4.11.30) and (4.11.15) are used in (4.11.31) we obtain:

$$V_m - \int_{\rho_s}^{\rho_D} \frac{1}{\rho_0 \, fR\cos\theta}\frac{\partial\pi}{\partial\phi}\frac{\partial z}{\partial\rho}d\rho = \frac{f}{\beta}w_E. \tag{4.11.33}$$

If the integral (4.11.33) is integrated by parts, and the boundary conditions are used to evaluate the terms arising from the end points of the integration, we obtain for the Sverdrup balance:

$$\left. h\frac{\partial p_s}{\partial\phi} + \frac{1}{2}gh^2\frac{\partial\rho_s}{\partial\phi} + z\frac{\partial\pi}{\partial\phi}\right]_{\rho=\rho_s} + \frac{g}{2}\int_{\rho_s}^{\rho_D}\frac{\partial z^2}{\partial\phi}d\rho = \frac{\rho_0 f^2 R\cos\theta}{\beta}w_E. \tag{4.11.34}$$

In evaluating the third term in (4.11.34) some care must be taken. The derivative with respect to longitude is taken at *constant density*. Thus:

$$\left.\frac{\partial\pi}{\partial\phi}\right)_\rho = \left.\frac{\partial\pi}{\partial\phi}\right)_z - \frac{\partial\pi}{\partial\rho}\frac{\partial\rho}{\partial\phi} = \frac{\partial p_s}{\partial\phi} + gh_m\frac{\partial\rho_s}{\partial\phi} \tag{4.11.35}$$

at $\rho = \rho_s$ where $z = -h_m$.

The derivative with respect to longitude can be brought outside of the integral in (4.11.34) if proper attention is paid to the variation of the limits of the integral. When that is done, and (4.11.35) is also used in (4.11.34) we obtain:

$$\frac{\partial}{\partial \phi} \int_{\rho_s}^{\rho_D} z^2 d\rho - z^2(\rho_D) \frac{\partial \rho_D}{\partial \phi} = \frac{2f^2 \rho_0 R \cos\theta}{g\beta} w_E. \tag{4.11.36}$$

This can be integrated in longitude from the eastern boundary at $\phi = \phi_e$ to the point at $\phi$ to obtain:

$$\begin{aligned} \int_{\rho_s}^{\rho_D} z^2 d\rho - \int_{\rho_s(\phi_e)}^{\rho_D(\phi_e)} z^2(\phi_e) d\rho - \int_{\phi_e}^{\phi} z^2\,(\rho_D) \frac{\partial \rho_D}{\partial \phi'} d\phi' \\ = -\frac{2f^2 \rho_o}{\beta g} \int_{\phi}^{\phi_e} w_E R \cos\theta\, d\phi'. \end{aligned} \tag{4.11.37}$$

The third integral is given in terms only of the resting fluid density and its variation along the slanting base of the thermocline and can be rewritten as an integral over the density itself:

$$\int_{\phi_e}^{\phi} z^2(\rho_D) \frac{\partial \rho_D}{\partial \phi'} d\phi' = \int_{\rho_D(\phi_e)}^{\rho_D(\phi)} z^2(\rho_D) d\rho_D = \int_{\rho_D(\phi_e)}^{\rho_D(\phi)} z_e^2(\rho) d. \tag{4.11.38}$$

The last step in (4.11.38) recognizes that the isopycnals are flat between the eastern wall and their intersection of the base of the thermocline. Hence the integral can be carried out in terms of the isopycnal depths on the eastern wall. The density interval for the integration is given by the range of abyssal density spanned along the thermocline base from its intersection with the eastern wall to the point under consideration at $\phi, \theta$. When (4.11.38) is combined with (4.11.37) we obtain as the statement of the Sverdrup balance:

$$\int_{\rho_s}^{\rho_D} z^2 d\rho = -\frac{2f^2 \rho_0}{\beta g} \int_{\phi}^{\phi_e} w_E R \cos\theta\, d\phi' + \int_{\rho_s(\phi_e)}^{\rho_D(\phi)} z_e^2 d\rho. \tag{4.11.39}$$

Although the derivation is complicated it is apparent that (4.11.39) *is simply the continuous limit of the statement (4.3.15) of the Sverdrup balance for the layer model.*

The statement of the proper boundary conditions on the eastern wall of the basin presents the same difficulty in the continuous case as in the layer model discussed in the previous section. A mixed layer in which both the layer depth and the mixed layer density are functions of latitude on the eastern boundary gives rise to a flow normal to the boundary. In the layer model we balanced this flow with an opposite zonal flux in the layer just beneath the mixed layer without describing the physics of the boundary layer that would effect the required vertical mass flux. In the continuous model we could similarly specify a mass flux in the region below the mixed layer so that its zonal mass flux balances that of the mixed layer. This would require an arbitrary specification of the distribution of this balancing flow with depth (or density). Several different approaches have thus far been taken to this problem, but none of them is completely satisfactory. Huang (1989a) assumed a structure for the compensating flow on the eastern boundary as a function of density. This introduces, however, an arbitrary element in the calculation. Williams (1989) instead took a mixed layer that had vanishing thickness on the eastern boundary to avoid altogether a zonal mass flux impinging on the eastern wall of the basin. This is a rather unrealistic condition and has important consequences in limiting the flow from the mixed layer into the thermocline. Huang (1989b) finally settled on a boundary condition in which the zonal velocity is forced to vanish on each density surface below the mixed layer at the eastern wall. This implies that on the eastern wall the density surfaces below the mixed layer are horizontal. This condition is reminiscent of the result we found in Section 4.10 and shown in Fig. 4.10.4 where we saw that even if we balance the mixed layer zonal transport with a compensating flow in the layer beneath it, the base of that layer is very nearly independent of latitude along the eastern wall. That is, compared to the overall slopes of the thermocline the slope on the eastern wall induced by the mixed layer transport is negligible. Huang (1989b) argues that the continuous model has the same behavior and thus imposes the condition of no zonal veloctiy on the geostrophic flow. This condition implies that the model unfortunately leaks fluid continuously from the mixed layer out of the basin at the eastern boundary. Huang argues that this is a small effect, and that this deficiency in the mass flux of the mixed layer is more than compensated by the improved realism in the structure of the density field on the eastern boundary.

The condition at the eastern boundary implies that each density layer has a nonzero thickness there as each density surface intersects the sloping mixed layer base at different depths and latitudes, as shown in Fig. 4.11.2. As the mixed layer shallows southward it exposes an adiabatic layer of nonzero thickness on the eastern wall. One of the most important consequences of this boundary condition is that it gives the fluid shadow zones on each density surface just as in Section 4.10. Since the density surfaces intersect the eastern

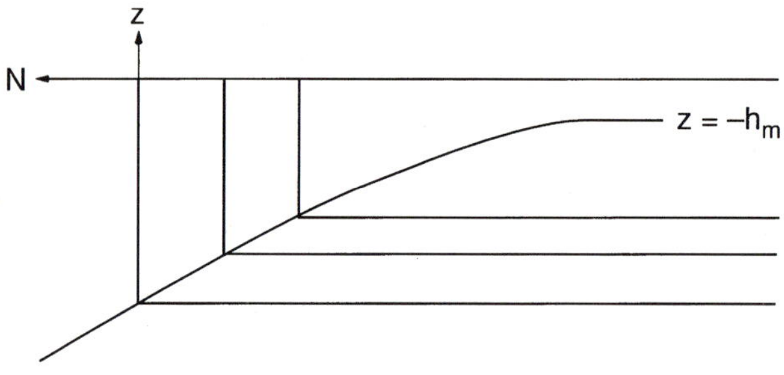


**Fig. 4.11.2.** Structure of the density field along the eastern boundary. The density surfaces are flat beneath the mixed layer. After they intersect, the sloping mixed layer at $z = -h_m$ the isopycnals rise vertically to the sea surface. (After Huang 1989b)

boundary at different depths this boundary condition also avoids the singularity in the density field found in other treatments of the continuous problem where all the density surfaces of the moving thermocline outcrop along the eastern boundary.

The system of equations then is the set (4.11.22a,b) subject to the boundary conditions (4.11.24), (4.11.25), (4.11.27), and the integral condition (4.11.39). This is a very unusual mathematical problem and appears highly overspecified since the differential equation is second order in density, and there are four boundary conditions plus an integral condition to satisfy. The correspondence of the system to the layer model is helpful in understanding that the problem is in fact sensibly posed. Indeed, the conditions on the boundary are not all known and must be actually determined by the solution of the problem. As we have already seen, the base of the thermocline must be determined, i.e., the density $\rho_D$, as a function of latitude and longitude must be found. This is equivalent to the need to find the pool boundaries of each of the unventilated layers in the layer formulation of the problem. Those pool boundaries as a function of layer index are equivalent to the surface $z = -D(\phi, \theta)$ and must be found from the solution. Similarly, the potential vorticity in (4.11.22) is not completely known a priori over the whole water column and must be determined as part of the solution. This is also true of the surface pressure. Hence both of the boundary conditions at the surface, i.e., at $\rho = \rho_s$, are actually used to determine unknown parts of the solution, in particular the potential vorticity of the newly subducted fluid. The Sverdrup integral condition is used to determine the vertical extent of the thermocline at each location, just as we used it in the layer model. It determines the overall depth of the moving fluid, $h$, in the layer model *after* the potential vorticity conservation relations are used to relate all the thicknesses to $h$. The continuous model "works" in exactly the same way as the layer model.

The solution is found numerically by discretizing the governing equations. The solution is started at the density surface that outcrops furthest north in the gyre. For illustration purposes we can consider the outcrop lines to be latitude circles but Huang (1989b) has shown how the argument can be easily generalized to the case with zonal variation of the surface density.

At each horizontal position (4.11.22a,b) are integrated starting from $\rho = \rho_D$, where the value of $\rho_D$ is guessed, on a grid of points as shown in Fig. 4.11.3. The integration proceeds to smaller $\rho$, i.e., higher in the water column. The value of $q$ at each grid point is known from information carried to that point by streamlines on which $q$ is determined upstream. If the layer is unventilated, the value of $q$ is determined by the value of the potential vorticity on the intergyre boundary where the density gradient with depth is known. If the layer is ventilated, the streamline which intersects the grid point carries the potential vorticity which has been determined further northward by the subduction process (we see below how this is done). If the grid point is informed by a streamline which, although in a ventilated layer, emanates from the western boundary, the potential vorticity must be specified on that

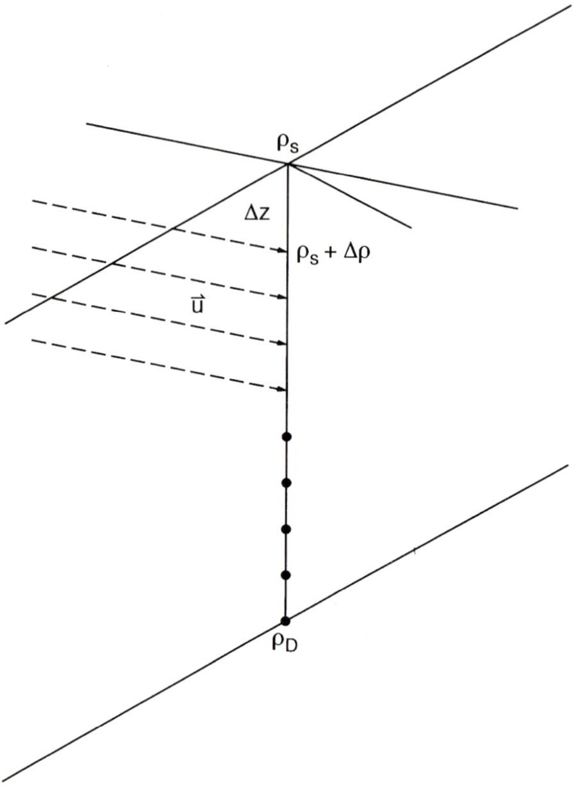


**Fig. 4.11.3.** Grid in density on which the calculation is performed. Starting from the base at $\rho = \rho_D$ (4.11.22) is integrated upwards using information about $q$ brought to the integration points by streamlines (*dashed curves*) arriving from upstream. The final thickness $\Delta z$ for the layer beneath the mixed layer is determined as the last step in the integration and sets the potential vorticity of the subducted fluid as described in the text

streamline as it is not determined by ventilation. Huang (1989b) in his calculation used hydrographic data to specify the potential vorticity for these pool regions in the ventilated and unventilated layers. As the system (4.11.22) is integrated upwards, we arrive at the grid point just beneath the surface. At that point and at all points beneath it $z$ and $\pi$ are determined by the upward integration in terms of the starting value $\rho_D$. The remaining integration to the surface cannot be accomplished with (4.11.22a) since $q$ is not known for the fluid which subducts at the density outcrop where $\rho = \rho_s$. However, the finite difference form of (4.11.22a) for this last grid point allows $q$ to be determined. If $z(\rho_s + \Delta\rho)$ is the known depth of the fluid determined by the integration at the first grid point (in $\rho$) beneath the surface, and since $z = -h_m$ at the grid point $\rho = \rho_s$, it follows that:

$$q(\rho_s) = \frac{f\Delta\rho}{z(\rho_s) - z(\rho_s + \Delta\rho)}. \tag{4.11.40}$$

Note that this determination of $q$ is identical to the calculation in the layer model, for example, the analysis in (4.7.17) in which the potential vorticity of the subducted fluid is determined in terms of the layer depths of the deeper layers. The thickness of the uppermost layer and its potential vorticity are then determined in the layer model in terms of the total depth of the thermocline by the condition of outcropping. In the same way, $q$ is *determined* at the final grid point in the continuous model by (4.11.40), i.e., by the depth of the layer at the grid point at $\rho_s + \Delta\rho$, which is in turn a function of $\rho_D$ from the integration of (4.11.22).

The integration of (4.11.22b) across the final grid interval then also *determines* the surface pressure field from (4.11.27).

The final condition to be met is the Sverdrup integral condition (4.11.39). The integral requires information about the fields, in particular the depth of the thermocline in density space, $\rho_D$, only east of the point under consideration. It is therefore necessary to start the process of calculation at the most eastward point at the given latitude and work westward a step at a time. The solution obtained by the vertical integration of (4.11.22) depends on the assumed starting value of $\rho_D$, the density of the thermocline base. Unless that value has been chosen properly, the Sverdrup condition is not satisfied. The value of $\rho_D$ must be iterated until the Sverdrup balance is satisfied. As we have noted, this is exactly equivalent to the use of the Sverdrup relation in the layer models to determine the base of the moving zone, $h$. Once the solution along a latitude circle (or in the more general case the outcrop line) is determined by marching westward along that line, the geostrophic velocities can be calculated, and the fluid trajectories can be advanced to the position of the next outcrop line. Thus at every depth in the new location the potential vorticity, carried with the fluid, is known (except for the fluid that will be subducted at the new location), and the process of the solution just outlined can be repeated until the whole gyre has been covered. In advancing the potential vorticity from one point to the next along streamlines, $q$ is kept track of numerically in tabular form instead of analytically, but the essential process is the same as in the layer model. Just as in the layer models, the entire process is started at the northern edge of the gyre where (except for the first layer to outcrop just south of the intergyre boundary) the potential vorticity on each density surface is given by the potential vorticity on the intergyre boundary or fixed by the flow out of the western boundary. The direct relation to the layer model calculation is stressed both to clarify the procedure for the continuous model and to emphasize the physical continuity between the layer and continuous models.

Huang (1989b) has applied his model to the subtropical gyre of the North Atlantic. He chose an Ekman pumping velocity that is a function of latitude and longitude so that the zero value of $w_E$ occurs on a line that slopes southwest-northeast in the basin (see Fig. 4.11.4a). The dynamical quantity which enters the theory is the integral of the Ekman pumping (4.11.39) which determines the Sverdrup transport streamfunction. This transport streamfunction is shown in Fig. 4.11.4b. Note that the zero line of the Sverdrup transport slopes less sharply than that of the Ekman pumping and occurs further northward. The boundary of the subtropical gyre could be plausibly associated with either of these zero lines. These two lines coincide when the Ekman

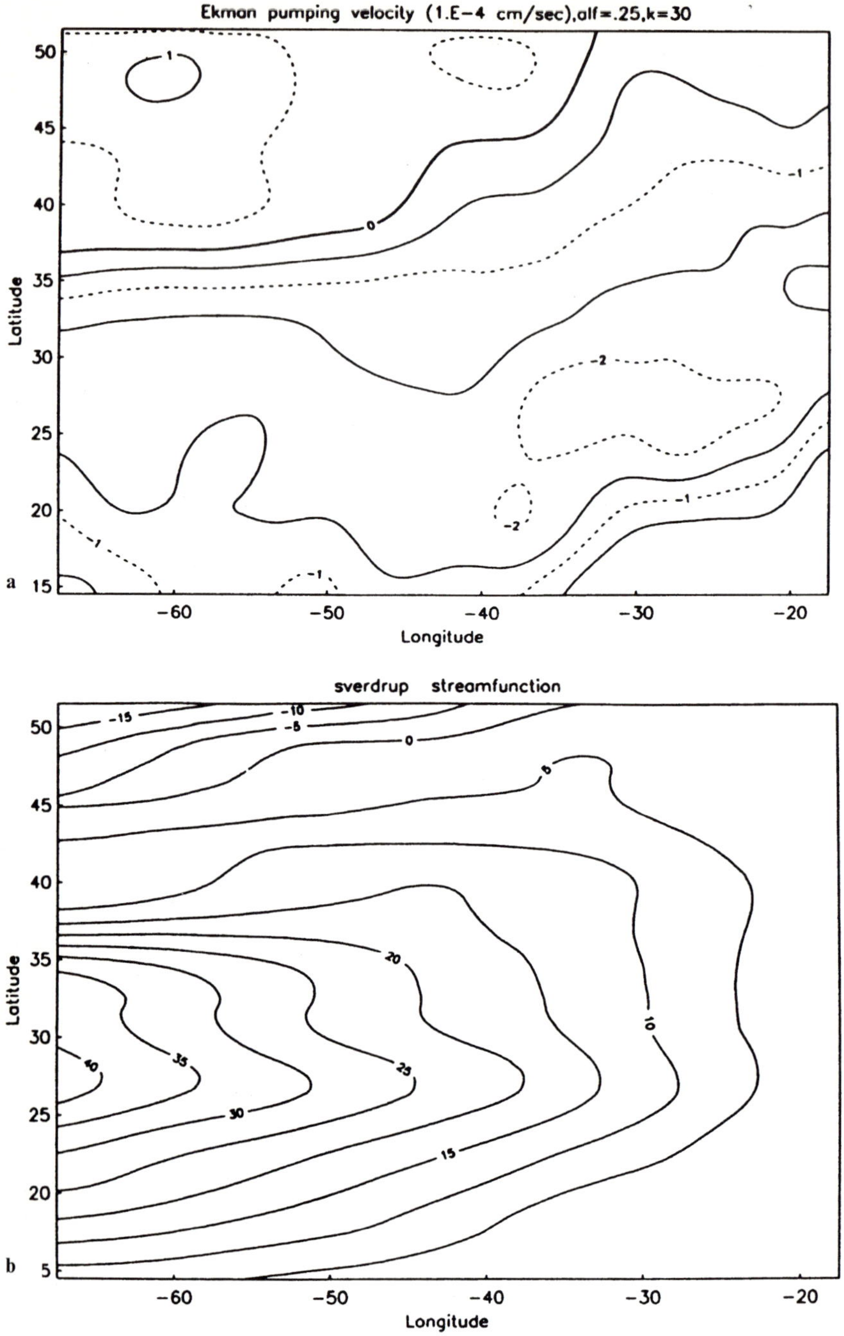


**Fig. 4.11.4a–d.** Fields used by Huang (1989b) to calculate of the thermocline flow in a model of the North Atlantic. **a** Distribution of Ekman pumping. The zero line of $w_E$ slopes southwest to northeast and is taken as the intergyre boundary. *Solid contours* represent downward pumping.

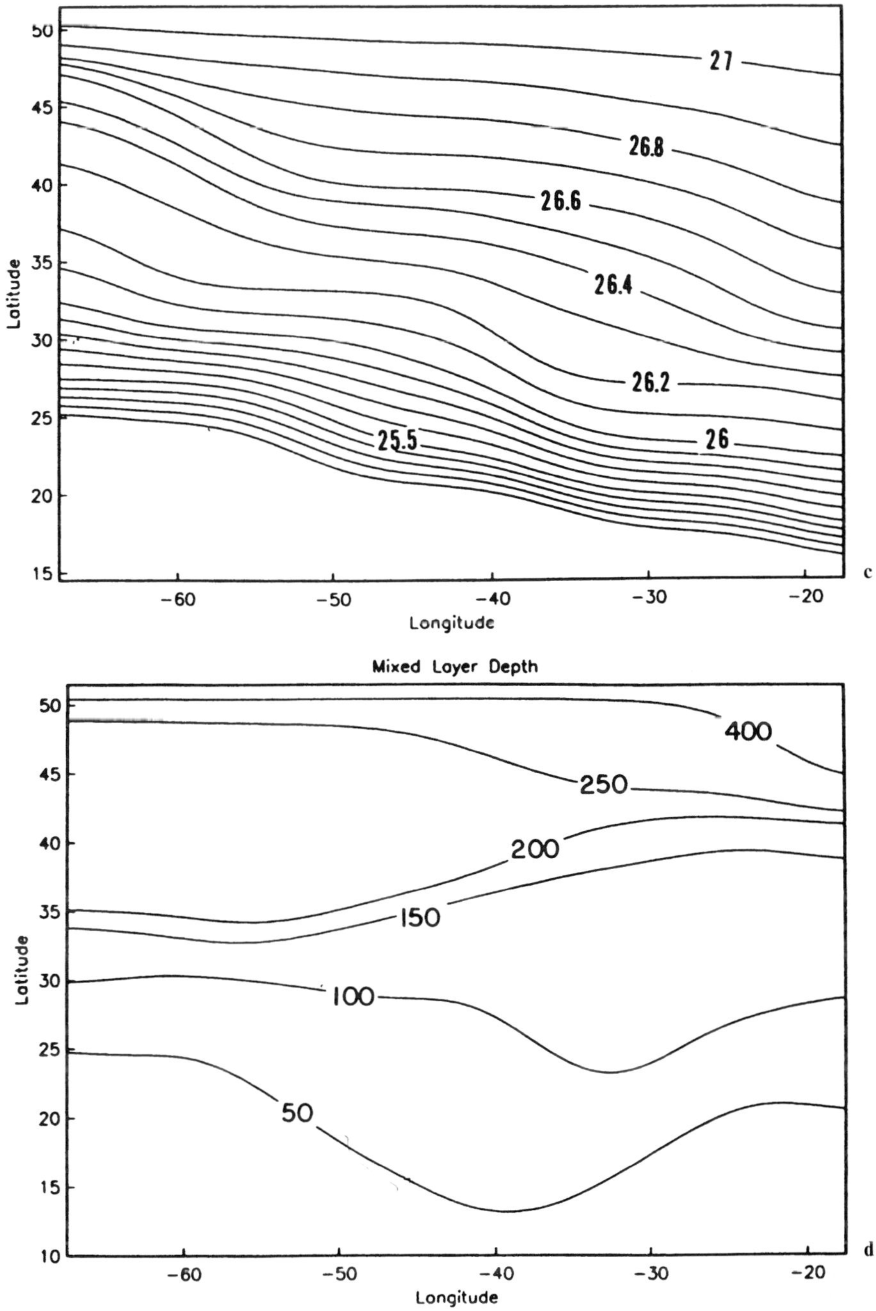


**b** Contours of the barotropic Sverdrup treamfunction for the Ekman pumping of **a.** Units are sverdrups ($10^6$ m$^3$ s$^{-1}$). **c** Surface density field or, equivalently, the outcrop lines of the density surfaces used in the calculation. **d** Depth of the mixed layer (in meters), used in the calculation (From Huang 1989b)

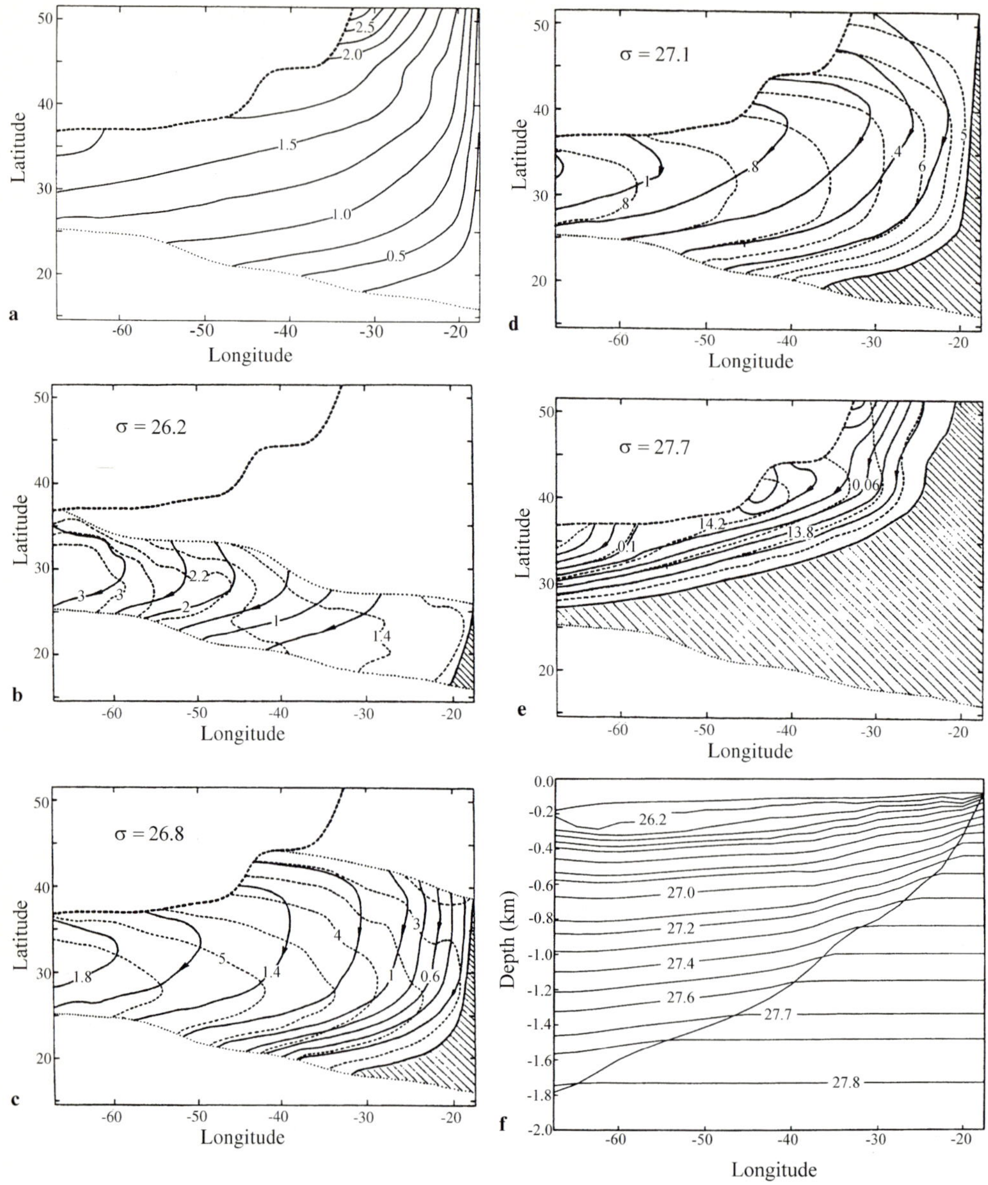


**Fig. 4.11.5a–f.** Results of the North Atlantic calculation of the continuous model by Huang (1989b). **a** Depth of the moving fluid in the thermocline. **b** Flow pattern on the $\sigma_\theta = 26.2$ density surface. *Heavy dotted line,* line of zero Ekman pumping; *dotted line to the north,* winter outcrop line of the layer; *solid lines,* streamlines of the flow on that density layer (in sverdrups); *Dotted lines,* layer depths in 100 of meters; *shaded area,* shadow zone for the layer. **c** as in b but $\sigma_\theta = 26.8$, **d** $\sigma_\theta = 27.1$. and **e** $\sigma_\theta = 27.7$. **f** Density cross section of the solution along the nearly zonal line coincident with the outcrop line of the $\sigma_\theta = 26.1$ surface (see Fig. 4.11.4c)

pumping is a function only of latitude. From a physical point of view we have defined the subtropical gyre as the region in which the Ekman pumping is *downward* so that the subsurface density is determined by the surface values. This would seem to identify the zero line of the Ekman pumping with the boundary of the subtropical gyre. On the other hand, the zero line of the Sverdrup transport divides the horizontal circulation into separate and well-defined gyres. Huang chose the first definition to define the northern gyre boundary. Although the meridional velocity vanishes on the zero line of Ekman pumping, the zonal velocity does not, and therefore there is horizontal flow carrying fluid across the gyre boundary *defined in this way*. Huang specifies the potential vorticity of that flow from hydrographic data.

Figure 4.11.4c shows the position of the outcrop lines for the density surfaces used in the calculation. Note that the outcrop lines slope northwest-southeast. Figure 4.11.4d shows the mixed layer depth. It increases from about 50 m near the southern boundary of the gyre to 400 m in the northeast corner of the basin.

Figure 4.11.5 shows the salient features of the solution. In panel a we see the overall depth of the thermocline as calculated by Huang (1989b). The thermocline increases in depth to the northwest. It is realistically shallow in the southern part of the gyre, with depths less than 1 km. Only north of 30° does the thermocline reach a depth exceeding 1.5 km. On shallow density surfaces, such as the $\sigma_\theta = 26.2$ surface shown in panel b, the shadow zone is very small. Note that on this surface most of the streamlines originate at the outcrop line. That is, on this shallow surface most of the layer is ventilated, and the western pool is very small, as we would anticipate from our discussion of the layer models. The effect of the sloping mixed layer base and the zonal variation of density is evident in panel c. The outcrop line is quite far north but at least half of the transport on this surface occurs in a relatively narrow zone in the eastern part of the basin and enters the thermocline through the sloping mixed layer base. Even on this surface the ventilation process in dominant. Deeper surfaces, $\sigma_\theta = 27.1$ and $\sigma_\theta = 27.7$, are shown in the final panels. In the former most of the flow originates from the gyre boundary . In the latter case the shadow zone occupies most of the density surface, and the flow shrinks to a narrow band of slow moving water and carries very little transport (0.2 sverdrups). Figure 4.11.4f shows a zonal cross section of the density field. The density surfaces are flat in the shadow zone in the east. Note that the density surfaces intersect the eastern wall at nonzero depths.

The continuous model, and its numerical finite-difference form, is a limiting case of the layer models discussed in earlier sections of this chapter. The successful implementation of the continuous model not only allows precise predictions of thermocline structure and velocity, given the surface wind stress and the surface density field, but also verifies the essential validity of the physical processes seen in a more accessible form in the low-order layer models.

A particularly intriguing result of Huang's calculation is his estimate of the total amount of fluid which is ventilated through the mixed layer compared to the circulation captured by deep recirculating fluid in the gyre. Of the roughly 42 sverdrups circulating in Huang's model beneath the mixed layer, nearly 25 sverdrups enters the thermocline through the mixed layer. Another 11 sverdrups enters across the zero Ekman pumping line and about 7 sverdrups recirculates in the deep layers of the gyre without ventilation. If the cross-gyre flow is considered as a kind of ventilation of the gyre, this would leave only 7/42 or 16% of the circulation remaining as an unventilated, recirculating contribution to the flow. It is probably more appropriate physically to count this 11 sverdrups as recirculation *within the gyre* which is really defined by the Sverdrup transport streamfunction. That is, it is probably more realistic to take the zero of the Sverdrup transport function as the defining boundary of the gyre rather than the Ekman pumping velocity itself. If this is done, the partition between the ventilated and unventilated portion of the gyre transport is more evenly balanced, i.e., it is 25 (ventilated) sverdrups vs. 18 (recirculating) sverdrups so that the ventilated portion of the thermocline flow is about 60% of the total. The sloping mixed layer depth has altered the prediction given by the Rhines ratio (4.10.1) and emphasizes the important roles *both* subduction and recirculation play in determining the dynamic structure of the oceanic thermocline.

## 4.12 Observations and Numerical Models

### *Observations*

The unified theory containing the processes of ventilation and recirculation is capable of making detailed predictions of the structure of the steady, stratified, wind-driven circulation in the oceanic interior. The physics of the model is, however, incomplete. The theory as it has been presented ignores any feedback from the western boundary current on the interior. It also neglects the effect of eddy mixing in the interior and the effects of cross-isopycnal motion in the interior. In this sense, as rich as the theory is conceptually, it should be regarded as a type of null hypothesis. That is, the question we can pose is to what extent this adiabatic theory, based on simple, steady, geostrophic, Sverdrup dynamics, explains the observed circulation and density structure of the oceanic interior in spite of the idealizations of the theory. How well does it do?

One of the first difficulties which we face is the fact that the theory also completely neglects the time dependence of the fields of motion. There is a strong seasonal signal in the surface density field as the surface heating and cooling of the ocean by the atmosphere takes place. To a large extent the effect of this variable surface heating is limited to the mixed layer, which in response deepens in the winter and shallows in the summer (see Woods and Barkmann

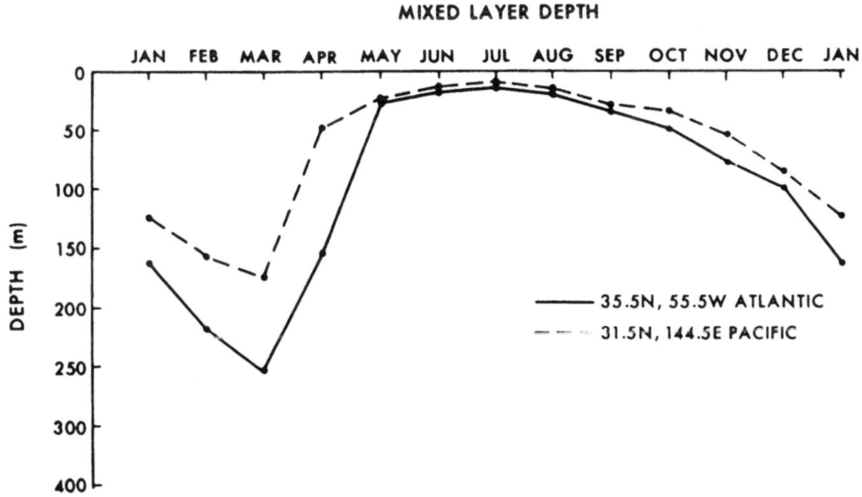


**Fig. 4.12.1.** Depth of the mixed layer as a function of month in the subtropical gyres of the North Atlantic and North Pacific. (From Levitus 1982)

1986; Woods 1988). The annual variation of mixed layer depth is shown in Fig. 4.12.1, from the Levitus *Atlas* (1982). At the latitudes shown in the figure the mixed layer varies between 250 m in late winter, when it is deepest, to a value of less than 50 m in summer, when the mixed layer is shallow and fairly uniform over the gyre. We have discussed in previous sections the spatial variation of the mixed layer and noted how in winter it deepens strongly to the north in the subtropical gyre. Its variation over time is equally significant.

The thermocline below this seasonally variable mixed layer is observed to be, comparatively, steady. Although this is encouraging support for the application of a steady theory to the thermocline, this observation by itself does not explain the relevance of the steady theory. The situation seems at first sight even more paradoxical when we realize that the strong seasonal variation of heating at the surface results in extraordinarily large variations through the seasons in the position of the outcrop lines. As the ocean cools in winter, colder, otherwise deeper water appears on the surface as the boundary of warm, lighter water retreats to the south. Note in Fig. 3.11.4a, for example, how the outcrop line of the $\sigma_\theta = 26.50$ surface moves as much as 15°–20° of latitude or nearly 2000 km from winter to summer. Given the sensitivity of the structure of the ventilated-thermocline steady solutions to the positions of the outcrop lines, it seems remarkable that the whole bowl of density surfaces containing the thermocline flow does not migrate and deform with the seasons as well. Of course, the huge change in potential energy that would be implied by a seasonal alteration for the entire thermocline density field of this magnitude is implausible given the available forcing, but, again, this does not explain the observed steadiness. Even assuming that a steady theory is apt, the large excursions of the outcrop lines through the seasons present us with the problem of deciding which position of the outcrop line should be used for a steady theory. Should the average outcrop position be used or is some particular position of the outcrop line more relevant than others to the steady structure of the thermocline?

The key idea was supplied by Stommel (1979), who gave not only the essential explanation for the steadiness of the structure beneath the mixed layer but also an implied prescription as to which position of the outcrop line should be used for a steady theory. Stommel pointed out that the southward movements of the outcrop lines in the gyre are observed to be much more rapid than the southward motion of the fluid leaving the mixed layer and entering the thermocline. He examined the motion of the outcrop line of the 20–21 °C water in the eastern North Atlantic during the course of the seasons and noted that the southward progression of the outcrop line occurs at an average rate of over 7 cm/s, much larger than the average geostrophic flow. He concluded that any fluid which subducts earlier in the year, during the summer and fall, would be swallowed up by the advancing and deepening mixed layer during the course of the advancing winter. This process is shown schematically in Fig. 4.12.2. In the figure the mixed layer depth is shown at two times, summer with a shallow depth $h_{ms}$, and late winter (March) when its depth is greatest and equal to $h_{mw}$. A sloping line marked $\psi$ shows the idealized path of fluid pumped downward from the Ekman layer after it enters the thermocline. The outcrop line for the particular density surface shown is at $\theta_s$ in the summer and at $\theta_w$, its most southern position, in winter. The position of a fluid element leaving the mixed layer is labeled with an $S$ for its summer position and $W$ for its winter position. As the fluid downwells it is overtaken by the deepening mixed layer and remains encapsulated within the mixed layer until the late winter when fluid at the base of the mixed layer is able to exit the mixed layer and enter the thermocline. Thus the only fluid, *at that latitude*, which enters the thermocline is fluid whose density surface outcrops at $\theta_w$ in late winter. Thus, in spite of the large variation of the surface density at that location over the course of the year only the density associated with the late winter position of the outcrop affects the ventilation of the permanent thermocline. Stommel likened the process, whereby only water of a particular density class is able to enter the thermocline at a given position, to the role of the imaginary "demons" in Maxwell's theory of gases. The deepening mixed layer, allowing only a narrow range of density to subduct at any geographical location, acts as a selective gatekeeper, and it has become common to refer to the selection process of subducted density as "Stommel's demon."

The simple argument of Stommel's suggests that a steady theory of the thermocline can be constructed by choosing the outcrop lines to be those observed in late winter rather than the average position of the outcrop lines. The kinematics of Stommel's demonstration is highly simplified. It is two-dimensional and quasistatic in its construction of flow paths. More sophisticated kinematic analyses by Cushman-Roisin (1987), Marshall et al.

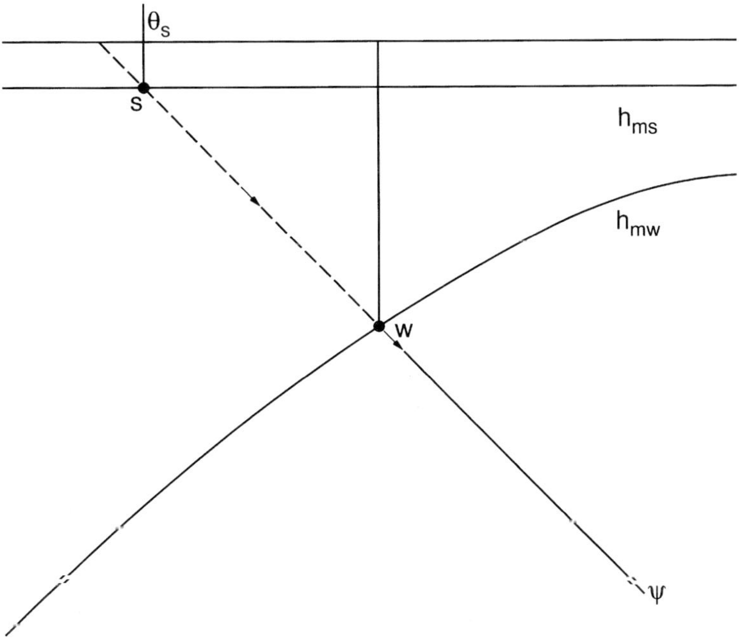


**Fig. 4.12.2.** Schematic illustration of the preference of ventilation of the thermocline by subduction from the late winter time mixed layer. An isopycnal outcrops at latitude $\theta_s$ in the summer into the mixed layer of depth $h_{ms}$. The outcrop position moves southward to $\theta_w$ in the winter faster than a particle sinking on the *line marked* $\psi$. The fluid on this density surface is trapped in the deepening mixed layer until the mixed layer reaches its maximum depth, $h_{mw}$, in late winter when fluid of this density alone enters the thermocline at this latitude

(1993), Williams (1989), and Qiu and Huang (1995) have usefully clarified and sharpened the picture presented by Stommel without, however, altering its essential result. Dynamical models, which allow the mixed layer to evolve with time and are coupled to the thermocline circulation e.g., Liu and Pedlosky 1994; Williams et al. 1994) also demonstrate the same behavior. Attempts to construct steady-thermocline solutions can then simply take as their starting point the surface density distribution of late winter. This is what has been done in the calculations described in the previous sections.

One of the key features of the theory is the adiabatic assumption, i.e., the absence of cross-isopycnal velocities. We mentioned in Chapter 3 the observational evidence by Ledwell et al. (1993) in favor of small values of vertical mixing in the eastern basin of the North Atlantic. A recent study by Jenkins (1994) yields only a slightly larger estimate for the North Atlantic thermocline as a whole, strengthening the support for the adiabatic theory.

Horizontal mixing of potential vorticity along isopycnal surfaces is another possible mechanism that could upset the adiabatic assumption. It is interesting in this regard to note that Armi and Stommel (1983), in their analysis of data from the "$\beta$ triangle" region (a relatively small triangle centered at 27°N, and 32°30′W in the North Atlantic which we would identify with the eastern edge of the ventilalted region) found that advection of potential vorticity alone is sufficient to explain the local potential vorticity balance. Their estimate, using salinity as a tracer, of the lateral mixing coefficient, $\kappa_H$, was $0.5 \times 10^7 \text{ cm}^2/\text{s}$. For scales of the order of 1000 km this would give a ratio of the diffusion of $q$ to its advection of 0.05 for velocities of the order 1 cm/s and hence negligible.

There have been several attempts to directly apply the theory to the actual oceanic thermocline circulation.

We discussed in Section 4.11 the study by Huang (1989b) of the North Atlantic thermocline circulation using the continuous model with a spatially varying mixed layer. The predictions of the theory for the structure of the isopycnals are very encouraging. If Fig. 4.11.5c, for example, is compared with the observations of the same density surface shown in Fig 4.1.3, the agreement is seen to be very good. The variation in depth of the $\sigma_\theta = 26.8$ surface is well captured by the theory. Ventilation is clearly important for this outcropping surface in the theory which predicts a large area covered by ventilated streamlines. Observations of chemical tracers in the ocean reveal that, this density surface is indeed, well ventilated. Sarmiento et al. (1982) presents maps of both salinity and tritium on this surface, and long tongues of both tracers stretching from the outcrop line into the gyre clearly illustrate the ventilation process which is the heart of the theory. A very beautiful representation of the ventilation process, as revealed in the tritium tracer data, has been described by Jenkins (1994). Figure 4.12.3 shows the age of the tracer tritium on several density surfaces in the North Atlantic. Tritium, produced in the atmosphere by the detonation of nuclear weapons in the 1960s enters the ocean and is advected with the flow. Its mutation to $^3He$ gives a measure of its age at any location. In the figure isolines of age of the tracer on selected surfaces are shown. Examining the $\sigma_\theta = 26.8$ surface again, we see the gradual increase in age as we follow the fluid away from the outcrop line. Jenkins argues that the distribution of the tritium tracer is consistent with flow trajectories on the subducted interfaces.

Armi and Stommel (1983) examined in detail the structure of a small portion of the subtropical gyre in the eastern North Atlantic. We commented above on their conclusion that conservation of potential vorticity is apparently saitsfied in the location of the $\beta$ triangle region. In addition, they point out that on the $\sigma_\theta = 26.5$–$27.5$ density layer which outcrops in the subtropical gyre where the Ekman velocity is downward, the potential vorticity gradient is weak even though this layer is clearly ventilated. This layer outcrops in a region where the outcrop line slopes from northwest to southeast, and the relative uniformity of the potential vorticity on this surface is quite possibly due to the subduction of uniform $q$ water as described at the end of Section 4.9.

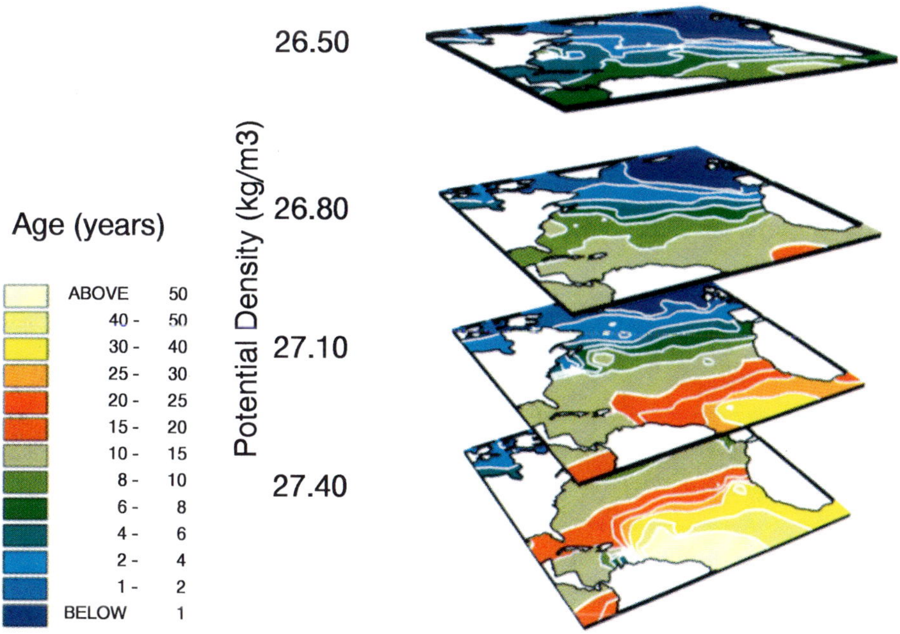


**Fig. 4.12.3.** Age distribution of the tracer tritium and its daughter product $^3He$ on selected surfaces in the North Atlantic. Note the aging of the tracer on the $\sigma_\theta = 26.8$ surface as would be expected from ventilation and subduction. (Jenkins, pers. comm., 1994)

One of the most ingenious applications of the theory is the study by Talley (1985). Talley uses a three-layer ventilated-thermocline model for the North Pacific in which she uses a realistic pattern of Ekman pumping and zonal outcrop lines. Talley carefully includes the effects of the western pools of constant potential vorticity in each of the layers and describes the complex mosaic of flow subdomains (there are eight of them), that occur when both the boundary of the western pools and the shadow zone boundaries strike the southern outcrop line. By using the observed surface distribution of salinity and assuming that it is carried with the flow as a tracer, Talley was able successfully to describe the observed shallow salinity minimum in the North Pacific. The occurrence of the minimum is the result of a complex pattern of stream paths carrying the surface salinity signal downward, and the success of the theory in reproducing the subsurface salinity structure is a remarkable achievement for the idealized model. Even more remarkable is the ability of the simple theory to determine the overall domains of flow in the gyre. Figure 4.12.4 shows the calculation by Talley of the boundary of the western, unventilated pool region on two ventilated layers of the North Pacific,

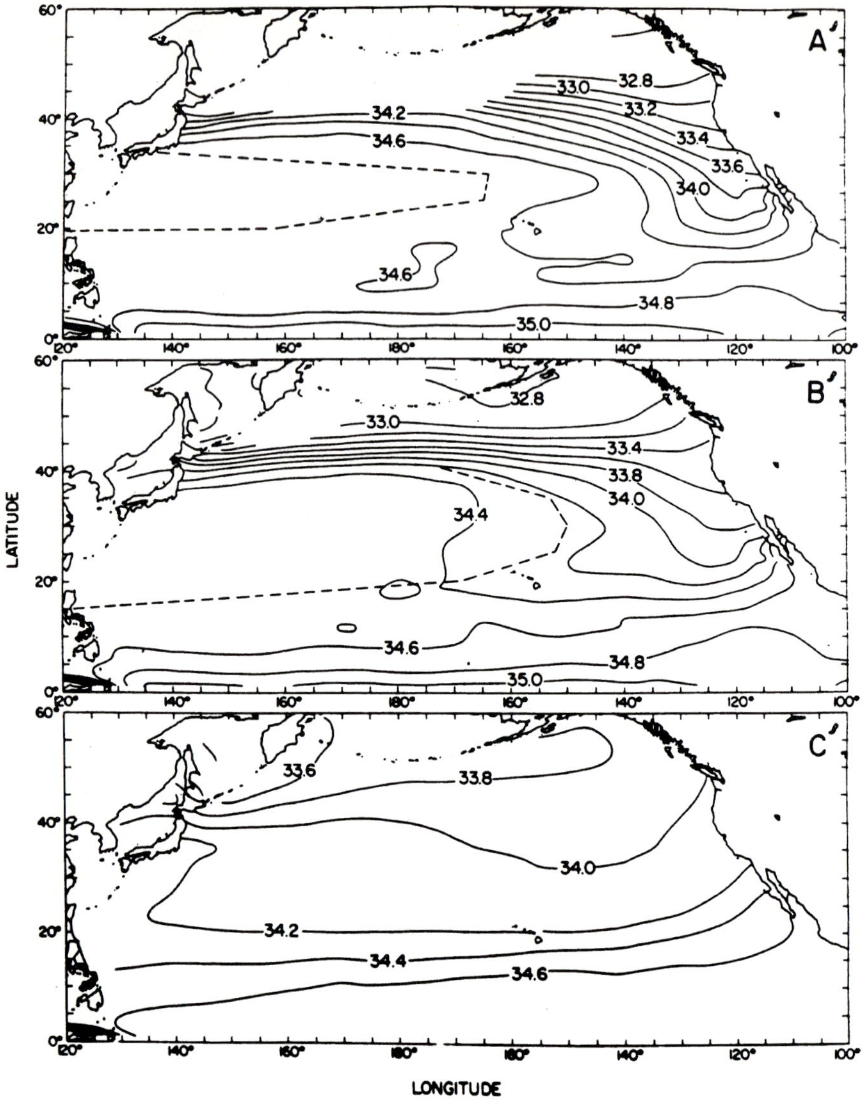


**Fig. 4.12.4.** Salinity on density surfaces in the ventilated thermocline in the North Pacific. *Dashed line*, is the analytical prediction of the western pool of unventilated fluid. Note the way in which the observed salinity contours in the subducting fluid wrap around the region on the ventilated surfaces, $\sigma_\theta = 25.5$ and 26.0 in the subtropical gyre. (From Talley 1985)

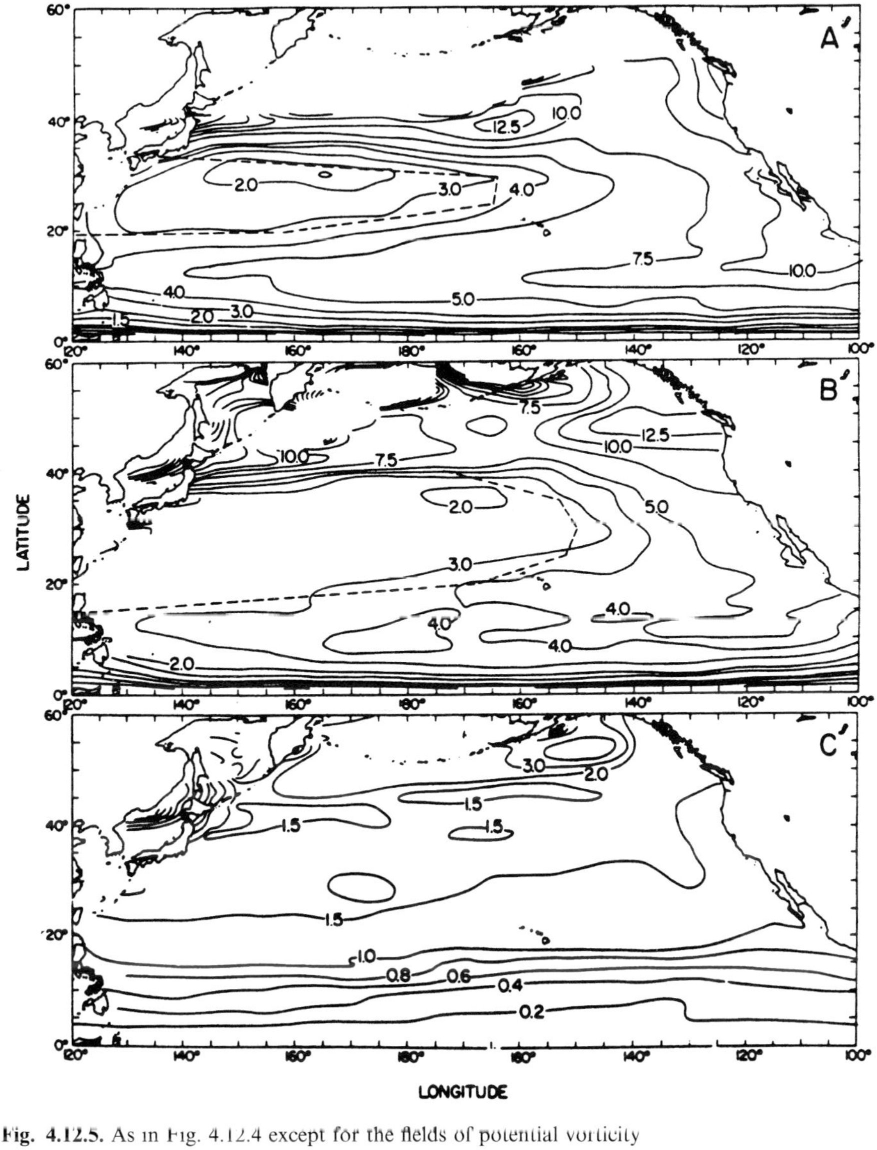


Fig. 4.12.5. As in Fig. 4.12.4 except for the fields of potential vorticity

$\sigma_\theta = 25.5$ and 26.0, along with the observed salinity field. One sees quite clearly the way in which the subducted fluid wraps the ventilated water around the unventilated zone, in excellent agreement with the theoretical predictions. A similar pattern is shown by Talley in the potential vorticity fields on these surfaces and is shown in Fig. 4.12.5. The deeper unventilated layer $\sigma_\theta = 26.8$ shows no such structure, as expected.

De Szoeke (1987) applied the three-layer model to the thermocline of the South Pacific and also found good agreement between the predictions of the theory and the observed isopycnal depths and dynamic topography. In his calculation it appears that most of the basin at the density levels which he considered is covered by the unventilated pool in the western region in an interesting contrast to the North Atlantic circulation where Huang (1989b) showed the dominant importance of ventilation. The success of the theory in both of these limiting cases is strong evidence in favor of the unified theory of recirculation and ventilation.

### *Numerical Models*

Although numerical models are not adequate surrogates for the natural ocean, they do contain processes, in particular the interaction of the interior with the western boundary current, that are absent in the analytical models described in this chapter. It is therefore of great interest to see to what extent the dynamical picture which we have developed can be discerned in numerical experiments. In order to deal with order 1 variations of the depth of the density surfaces numerical models must abandon the quasi-geostrophic approximation. This has historically meant that such models utilize what is called the *primitive equations*. The equations are "primitive" in the sense that no additional approximations are used to refine the equations beyond the hydrostatic approximation. The complexity of these equations means that numerical experiments are costly and time consuming.

After the construction of the analytical, ventilated-thermocline theory of Luyten et al. (1983), Cox and Bryan (1984) and Cox (1985) reexamined their numerical model in the context of simplified basin shapes and boundary conditions that allowed a useful comparison with the theory (see also Bryan 1987, for a discussion of the experiments).

Cox and Bryan (1984) and Cox (1985) describe the results of two central experiments. In the first experiment a relatively course grid is employed (1° meridionally, 1.2° zonally) while in the fine-grid experiment the resolution is improved (to 1/3° meridionally, 0.4° zonally). The coarse-grid experiment does not resolve time-dependent synoptic scale eddies, and in their absence mixing is accomplished by large explicit lateral mixing of momentum and density (although the momentum balance remains geostrophic and the interior is in Sverdrup balance.) The mixing parameterization for both momentum and density mixing is, in the coarse-resolution case, the traditional diffusion representation, with turbulent mixing coefficients of $10^8\ cm^2\ s^{-1}$ and $10^7\ cm^2\ s^{-1}$, respectively. In the fine grid experiment the diffusion is accomplished largely by the resolved eddies, and the explicit mixing is small and is parameterized in terms of a biharmonic representation of the mixing terms that acts selectively on the very small scales. This explicit mixing is not significant in the interior on the energy-containing scales of motion. The calculations are double-gyre

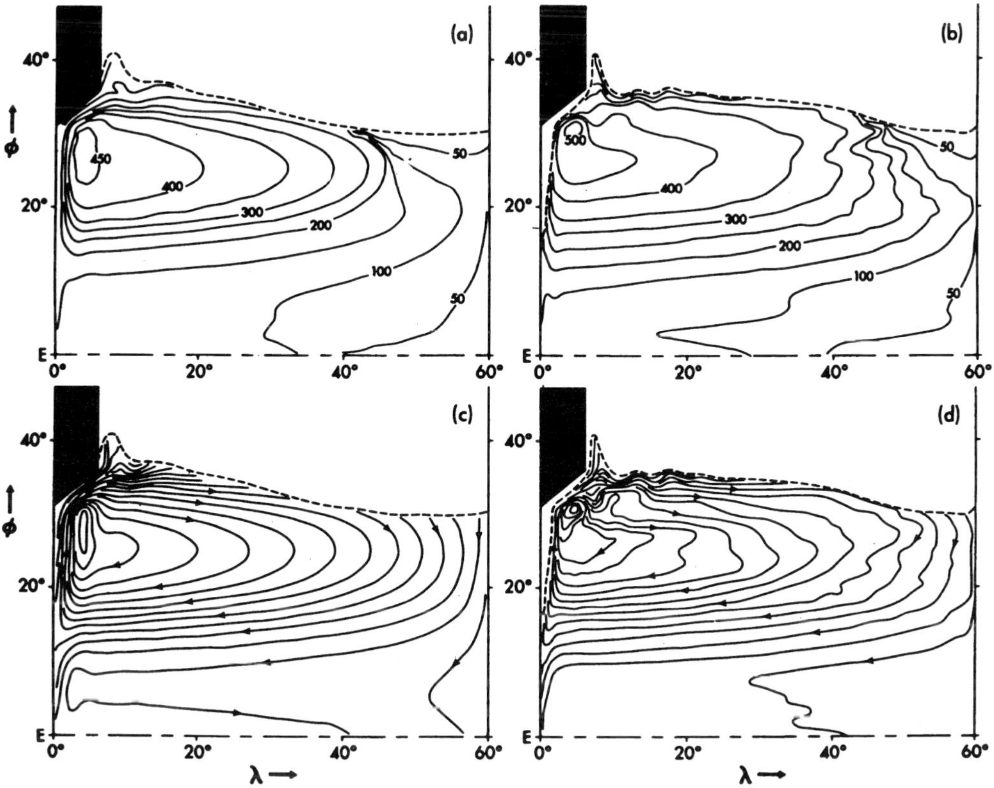


**Fig. 4.12.6.** Depth contours and streamlines on the $\sigma_\theta = 26.0$ surface from the calculations of Cox (1985). **a** Coarse-grid calculation of the depth of the surface. **b** As in **a** except for the fine-resolution case. **c** Montgomery function for the coarse-grid case. **d** Montgomery function for the fine-grid case

models in which the net vorticity input is very small, and therefore the problems of basin scale vorticity balances as described in Chapter 2 are avoided by fluxing potential vorticity from the subtropical gyre to the subpolar gyre rather than through the lateral boundaries. The calculations are carried out with the circulation of the North Atlantic in mind.

Figure 4.12.6 shows some of the principal results of the two calculations. Panels a and c show the depth and Montgomery function on the $\sigma_\theta = 26.0$ density surface for the coarse-resolution case (with no eddies) while panels b and d show the same quantities for the case in which the eddies appear and are resolved by the model. In both cases the results of the numerical model are qualitatively similar to those of the analytical theory. The depth of the density surface is the familiar sloping bowl at whose northern boundary is the outcrop line. The streamlines of the flow, given by the Montgomery function isolines, clearly display the three regions described by the theory. In both the eddy-free and eddy-resolving models there is a band of streamlines in the eastern part of

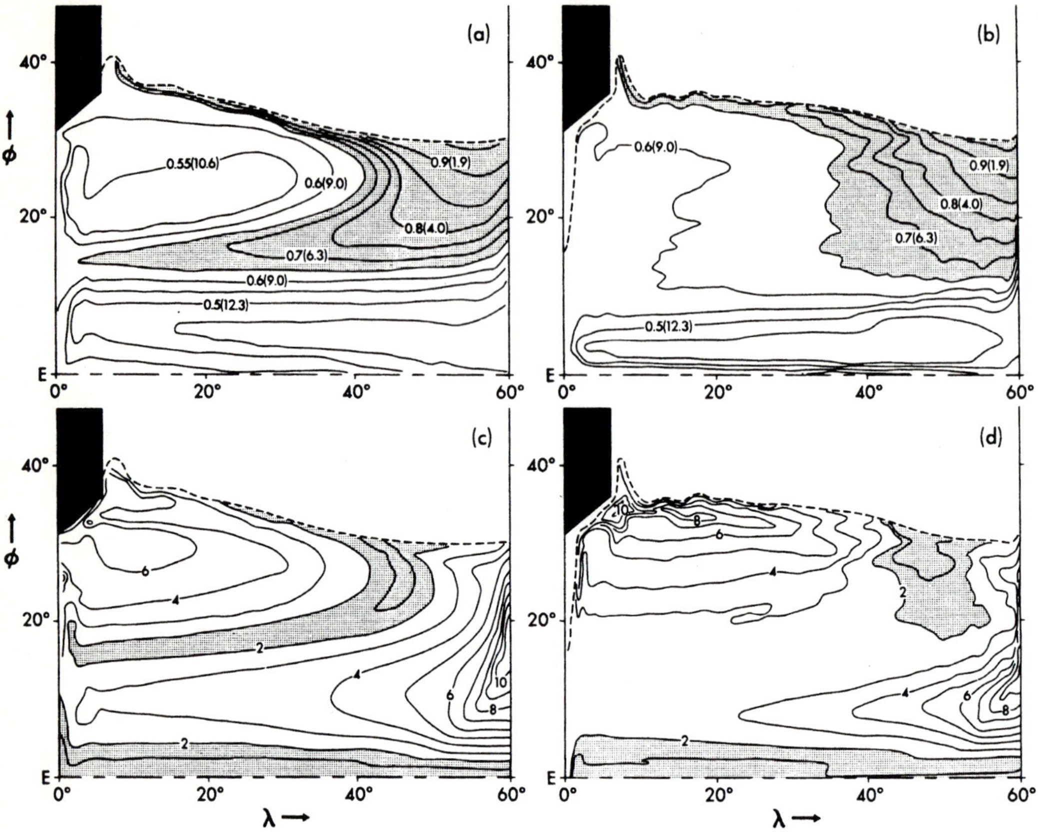


**Fig. 4.12.7a–d.** As in Fig. 4.12.6 except for the fields of a passive tracer and potential vorticity. **a** and **b** are the isolines of tracer concentration and age for the coarse and fine grid resolution calculations, respectively; **c** and **d** are for the potential vorticity

the gyre that originate at the outcrop line. These streamlines flow southward and pull away from the eastern boundary, leaving exposed there a zone of weak flow that we recognize as the region of the shadow zone. Further west in the layer there is a region of streamlines which close through the western boundary layer, and which we identify with the western pool region of unventilated fluid.

A passive tracer with a half-life decay equal to that of tritium (12.3 years) is injected into the surface layer of the model and carried along during the experiment. Panels a and b of Fig. 4.12.7 show the isolines of tracer age (measured from time of injection) on the same density surface as in Fig. 4.12.6. In both the coarse- and fine-resolution experiments it is easy to distinguish the ventilated region, by the tongue of fluid in which the tracer age is less than 6.5 years, from the shadow zone and the unventilated western pool where the tracer age is considerably greater. Panels c and d show the potential vorticity in the two cases, and the three regions of the flowing the subtropical gyre are again easily observed.

Although the overall results of the numerical experiment lend support to the theory, there are significant differences between the numerical results and the analytical theory. In the analytical theory, in which diffusion of density or potential vorticity is absent in the interior, isolines of $q$ on each density surface coincide with streamlines and thread their way from the outcrop line to the western boundary or wrap up in the pool regions and become homogenized. The numerical experiments are diffusive, and the turbulent flux of density, particularly, no longer permits exact conservation of potential vorticity. As a consequence, the potential vorticity isolines are shaped as drawn-out tongues rather than completed pathways, and the streamlines of the flow cross the tongues as a manifestation of the lack of potential vorticity conservation. In the numerical experiments this is particularly striking in the fine-resolution case with its active eddy field. In the coarse-resolution calculation the potential vorticity in the western pool is variable rather than uniform as a consequence of nonadiabatic effects in the model. The eddy-resolving model shows large regions of nearly uniform potential vorticity near the boundary between the ventilated and pool regions. Cox suggested that strong and rapid eddy mixing of potential vorticity *across* the pool boundary with the ventilated region, rather than slow homogenization within the pool region is the mechanism *in the numerical experiment* for the production of large regions of nearly uniform $q$. The model produces regions of large potential vorticity near both the eastern and western boundaries separated by low potential vorticity water, produced by convection, that feeds the ventilated region from the outcrop line. The strong eddy mixing rubs out the tongue of low potential vorticity over a large part of the basin and yield an enormous region of uniform $q$ connecting the eastern and western zones.

Thus the numerical, primitive equation models present yet an additional mechanism for the production of regions of uniform potential vorticity beyond those described by the homogenization theories of Chapter 3 or the quasi-geostrophic numerical models. To what extent the numerical models accurately represent the effect of the eddy field on the general circulation is unclear.

The numerical models emphasize (perhaps even overemphasize) what is also evident in the data, namely that potential vorticity is not exactly conserved. Nonadiabatic effects are clearly important in the flow. However the data, for example, Talley's maps of salinity and potential vorticity, show the tracer tongues more drawn-out and less fragile to eddy mixing than the numerical experiments would seem to indicate.

Overall the theories prove remarkable in capturing the flow structure and the density field of the interior circulation, at least in the subtropical gyre. The western boundary current, which is not included in the analytical theories, does not seem fundamentally to upset the interior dynamics either in the observations or in the numerical models of the general circulation.

On the other hand, the dependence of the theory on potential vorticity conservation and geostrophy provide limitations that must be relaxed if the theory is to be applied to the subpolar gyre, where observational evidence (Luyten et al. 1985) indicates that cross-isopycnal fluxes are important or to the equatorial regime where $f$ is small enough to invalidate the geostrophic approximation. The discussion of the next two chapters centers on these issues.

## References

Armi, L. and Stommel, H. 1983: Four views of a portion of the North Atlantic subtropical gyre. *J. Phys. Ocean.*, **13**, 828–857.

Bryan, K., 1987: Potential vorticity in models of the ocean circulation. *Q. J. R. Meteorol. Soc.*, **113**, 713–734.

Cox, M.D., 1985: An eddy resolving model of the ventilated thermocline. *J. Phys. Ocean.*, **15**, 1312–1324.

Cox, M. D. and Bryan, K. 1984: A numerical model of the ventilated thermocline. *J. Phys. Ocean.*, **14**, 674–687.

Cushman-Roisin, B., 1987: Subduction In: *Dynamics of the oceanic mixed layer*. Eds. Muller P. and Henderson, D. Hawaii Inst. of Geophys. special publication. Univ. of Hawaii, Honolulo.

De Szoeke, R.A. 1987: On the wind-driven circulation of the South Pacific ocean. *J. Phys. Ocean*, **17**, 613–630.

Haynes, P.H. and McIntyre, M.E. 1987: On the evolution of vorticity and potential vorticity in the presence of diabatic heating and frictional or other forces. *J. Atmos. Sci.*, **44**, 828–841.

Huang, R.X., 1988: On boundary value problems of the ideal-fluid thermocline. *J. Phys. Ocean.*, **18**, 619–641.

Huang, R.X., 1989a: The generalized eastern boundary conditions and the three-dimensional structure of the ideal fluid thermocline. *J. Geophys. Res*, **94**, 4855–4865.

Huang, R.X., 1989b: On the three- dimensional structure of the wind-driven circulation in the North Atlantic. *Dyn. Atmos. and Oceans.*, **15**, 117–159.

Huang, R.X., 1991: The three dimensional structure of the wind-driven gyres: ventilation and subduction. *Rev. of Geophys.*, (Supplement) U. S. National report to the International Union of Geodesy and Geophysics., 590–609.

Iselin, C.O'D, 1939: The influence of vertical and lateral turbulence on the characteristics of the waters at mid-depths. *Trans. Amer. Geophys. Un.*, **20**, 414–417.

Jenkins, W.J., 1994: A time series of tritium and $^3$He in the Sargasso Sea. *J. Geophys Res.*, **(submitted).**

Ledwell, J., Watson, A.J. and Law, C. S. 1993: Evidence for slow mixing across the pycnocline from an open-ocean tracer release experiment. *Nature*, **364**, 701–703.

Levitus, S., 1982: *Climatological Atlas of the World Ocean*, NOAA Prof. paper **13**, U.S. Govt. Printing Office, Washington, D.C. 173 pp

Liu, Z.L. and Pedlosky, J. 1994: Thermocline forced by annual and decadal surface temperature variation. *J. Phys. Ocean.*, **24**, 587–608.

Liu, Z.L., Pedlosky, J., Marshall, D. and Warncke, T., 1993: On the feedback of the Rhines-Young pool on the ventilated thermocline. *J. Phys. Ocean.*, **23**, 1592–1596.

Lozier, M.S., Owens, W.B., Curry, R.G. (1996) The climatology of the North Atlantic. Prog Oceanogr 36: 1–44

Luyten, J.R., Pedlosky, J. and Stommel, H., 1983: The ventilated thermocline. *J. Phys. Ocean.*, **13**, 292–309.

Luyten, J.R., Stommel, H., and Wunsch, C., 1985: A diagnostic study of the Northern Atlantic subpolar gyre. *J. Phys. Ocean.*, **15**, 1344–1348.

Marshall, J.C. and Nurser, J.G., 1992: Fluid dynamics of oceanic thermocline ventilation. *J. Phys. Ocean.*, **22**, 583–595.

Marshall, J.C., Nurser, A.J., and Williams, R.G., 1993: Inferring the subduction rate and period over the North Atlantic. *J. Phys. Ocean.*, **23**, 1315–1329.

Montgomery, R.B. 1938: Circulation in upper layers of southern North Atlantic deduced with the use of isentropic analysis. *Papers Phys. Ocean, and Meteorology*, MIT, **6**(2), 55 pp.

Pedlosky, J, 1987a: Thermocline theories. In, *General circulation of the ocean. Springer,* Berlin Heidelberg New York, Eds Abarbanel, H.D.I. and Young, W.R. pp 291

Pedlosky, J, 1987b: Geophysical fluid dynamics 2nd edn. *Springer,* Berlin Heidelberg New York, 710 pp.

Pedlosky, J. and Robbins, P., 1991: The role of finite mixed-layer thickness in the structure of the ventilated thermocline. *J. Phys. Ocean.*, **21**, 1018–1031.

Pedlosky, J. and Young, W.R., 1983: Ventilation, potential-vorticity homogenization and the structure of the ocean circulation. *J. Phys. Ocean.*, **13**, 2020–2037.

Pedlosky, J., Smith W. and Luyten J.R., 1984: On the dynamics of the coupled mixed-layer–thermocline system and the determination of the oceanic surface density. *J. Phys. Ocean.*, **14**, 1159–1171.

Price, J.F., Weller, R.A., and Pinkel, R., 1986: Diurnal cycling: observations and models of the upper ocean response to diurnal heating, cooling and wind mixing. *J. Geophys. Res.*, **91**, 8411–8427.

Price, J.F., Weller, R.A., and Schudlich, R.R., 1987: Wind-driven currents and Ekman transport. *Science*, **288**, 1534 1538.

Qiu, B. and Huang, R.X., 1995: Ventilation of the North Atlantic and North Pacific: subduction vs. obduction. *J.Phys. Ocean.*, **25**, 2374–2390.

Robinson, A.R.R. and Stommel, H., 1959: The ocean thermocline and associated thermocline circulation. *Tellus*, **11**, 295–308.

Sarmiento, J.L., Rooth, C.G. H., and Roether, W., 1982: The North Atlantic tritium distribution in 1972. *J. Geophys. Res.*, **87**, 8047–8056.

Stommel, H., 1979: Determination of water mass properties of water pumped down from the Ekman layer to the geostrophic flow below. *Proc. Natl. Acad. Sci. USA*, **76**, 3051–3055.

Stommel, H., 1984: The delicate interplay between wind-stress and buoyancy input in ocean circulation: the Goldsbrough variations. *Crawfoord Prize Lecture*, published in, *Tellus.*, **36A**, 111–119.

Stommel, H., and Schott, F., 1977: The beta spiral and the determination of the absolute velocity field from hydrographic data. *Deep - Sea Res.*, **24**, 325–329.

Talley, L.D., 1985: Ventilation of the subtropical North Pacific: the shallow salinity minimum. *J. Phys. Ocean.*, **15**, 633–649.

Veronis, G., 1969: On theoretical model of the thermocline circulation. *Deep-Sea Res.*, **16**, 301–323.

Welander, P. 1959: An advective model of the ocean thermocline. *Tellus.*, **11**, 309–318.

Welander, P., 1971: Some exact solutions to the equations describing an ideal-fluid thermocline. *J.Mar.Res.*, **29**, 60–68.

Williams, R.G., 1989: The influence of air-sea interaction on the ventilated thermocline. *J. Phys. Ocean.*, **19**, 1255–1267.

Williams, R.G., Spall, M.A., and Marshall, J.C., 1994: Does Stommel's mixed layer demon' work? *J. Phys. Ocean.*, **25,** 3080–3102.

Woods, J.D., 1988: A Lagrangian mixed-layer model of Atlantic 18° C water formation. *Nature*, **319**, 574–576.

Woods, J.D., and Barkmann, W., 1986: The response of the upper ocean to solar heating. I. *Q. J. R. Meteor. Soc.*, **112**, 1–27.

Young, W.R., and Rhines, P.B., 1982. A theory of the wind-driven ocean circulation. II. Gyres with western boundary layers. *J. Mar. Res.*, **40**, 849–872.

# 5 Buoyancy Forced Circulation and Cross-Gyre Flow

## 5.1 Introduction

This chapter discusses two subjects which appear at first sight unconnected, but which in fact are linked by the underlying question of the way in which information is propagated in the dynamics of the ocean circulation.

The discussion in the previous chapters centers on fundamentally non-dissipative dynamics in which, to the first approximation, the potential vorticity is carried unchanged along streamlines. It is natural then to think of the determination of the potential vorticity, which is the key dynamical quantity in the physics of the circulation, as being effected by the advection with the fluid velocity. As the fluid moves, it does so along the path on which constant values of the potential vorticity are inscribed. However, we have seen that this is not the whole story when it comes to the flow of information in shaping the structure of the circulation. The Sverdrup balance, with its integral along latitude circles, is a reflection of the purely westward propagation of barotropic Rossby waves. Similarly, as we saw in Chapter 3, the shaping of the zones of unventilated motion involves the interplay between advection by the fluid and the westward propagation of baroclinic Rossby waves. Although the potential vorticity equation is a hyperbolic system whose characteristics appear to be given by the fluid velocity, the velocity itself is generally a complicated function of the potential vorticity and its gradients. Therefore the direction of information flow in this mathematical system is not as simple as would obtain were the potential vorticity equation of the more familiar, quasilinear type, i.e., with coefficients in which gradients of the potential vorticity do not appear. The question of information flow is considerably more subtle. Information is propagated simultaneously along streamlines as well as along latitude circles by both barotropic and baroclinic Rossby waves.

If potential vorticity is not conserved, the flow paths and isolines of potential vorticity no longer coincide, and we can no longer associate each streamline with a value of potential vorticity. We can then no longer find a first integral of the potential vorticity equation, as in Chapter 4, where $q = Q(\psi, \rho)$. Alternative formulations are necessary, and the issue of information flow becomes even more central to the physical discussion.

In fact, and as we have noted above, diagnostic studies of the subpolar gyre, for example Luyten et al. (1985), imply that there the cross-isopycnal flux is locally large compared to the Ekman pumping with the consequence that potential vorticity can no longer be considered conserved. To discuss the dynamics of the subpolar gyre, and go beyond the simple model of Section 4.8 we must construct an alternative dynamical framework.

At the same time, one of the implicit assumptions we have used throughout our discussion has been the identification of the *barotropic* boundary of the gyre, defined on the basis of the Sverdrup transport, with the gyre boundary *at every level* (or in every layer). When the Ekman pumping is a function only of latitude, this boundary is given unambiguously as the latitude circle where $w_E$ vanishes. In this case the Sverdrup function $D_0^2$, given by (4.4.5) also vanishes there. The situation is more complex when $w_E$ is a function of longitude as well. This is perhaps the more realistic situation. In the North Atlantic, for example, the line on which the Ekman pumping vanishes stretches from about 40°N on the western side of the ocean to about 55°N on the eastern side (see Fig. 4.11.4a). The zero lines of Ekman pumping and $D_0^2$ no longer coincide. Because of the integration from the east in its definition, the zero line of $D_0^2$ lies to the west and north of the zero line of $w_E$.

In the solutions that we have so far found, the isopycnal depths are all constant on the line where $D_0^2$ is zero and therefore this line is the most obvious defining boundary of the gyre since no geostrophic flow crosses this line in those solutions. In the case in which the line $D_0^2 = 0$ lies west and north of the zero line of Ekman pumping, a zone between the two lines opens up in which the meridional velocity in the subtropical gyre is *northward.* Flow can cross the zero Ekman pumping line when it is tilted with respect to a latitude circle since only the meridional flow is zero there but the Sverdrup zonal flow is nonzero and thus crosses the tilted line of $w_E = 0$. This can lead to very interesting changes in the gyre circulation structure (see, for example, Rhines and Schopp 1991) although the basic theory is not fundamentally altered.

A more fundamental question is whether the gyre flow can actually cross the line $D_0^2 = 0$. This, as we noted, is the natural definition of the gyre boundary when we consider the geostrophic transport or the barotropic velocity. The question arises as to whether a solution also exists in which there is baroclinic flow across the line $D_0^2 = 0$ which at the same time has zero transport across that line. That is, can there be a purely *internal* mode of fluid interchange between the subtropical and subpolar gyres in which, at each latitude and longitude along the boundary, there is no net transport of mass? Such a net interchange in which, say, warm water is going northward while cold water is going southward across the gyre boundary would yield a net heat flux even in the absence of a net mass flux.

We see below that both the problem of nonadiabatic motion, in which potential vorticity is not conserved, and the problem of cross-gyre flow involve the question of the flow of information in the gyre. We begin by discussing the question of cross-gyre flow.

## 5.2 Cross-Gyre Flow

Consider the model of a three moving layer model in Fig. 5.2.1. The situation is similar to that in Fig. 4.4.1 with the important exception that in Section 4.4 we initially considered layer 3 to be at rest. We found eventually that layer 3 could be in motion if potential vorticity isolines in layer 3 are sufficiently distorted to meet the intergyre boundary rather than the eastern boundary of the gyre. In this section we examine a related question, namely, whether the flow normal to the intergyre boundary must be zero at that boundary, or whether crossgyre flow can occur.

To keep the discussion as simple as possible we choose the Ekman pumping velocity to vanish on the latitude circle $\theta = \theta_0$ so that the line where $w_E = 0$ coincides with the line where $D_0^2 = 0$. There may be an outcrop line further south at $\theta = \theta_2$, but this does not enter the present discussion.

We again define the function:

$$D_0^2 = \frac{-2f^2}{\gamma_2 \beta} \int_{\phi}^{\phi_e} w_E R \cos\theta d\phi' \tag{5.2.1}$$

where $\gamma_2 = g(\rho_3 - \rho_2)/\rho_0$ and where $\rho_0$ is the average density of the gyre.

The Sverdrup balance in the region south of zero Ekman pumping line but north of the outcrop line for layer 2 is:

$$v_2 h_2 + v_3 h_3 = \frac{f}{\beta} w_E \tag{5.2.2}$$

The geostrophic relations:

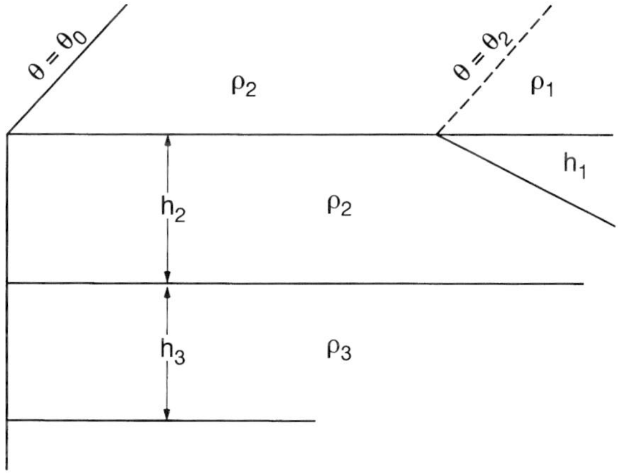


**Fig. 5.2.1.** A schematic of the layer model near the intergyre boundary at $\theta = \theta_0$. Only layer 2 is driven directly by Ekman pumping near the gyre boundary. The thicknesses of these layers are $H_2$ and $H_3$ when the fluid is at rest and these remain the layer thicknesses on the eastern boundary

$$\vec{u}_n = f\hat{k} \times \nabla \pi_n \tag{5.2.3}$$

where $\pi_n$ as defined in (3.2.12) allows us, with the use of the hydrostatic relation between $\pi_n$ and the layer thicknesses, $h_n$, to write the Sverdrup relation as:

$$h^2 + \frac{\gamma_2}{\gamma_3} {h_2}^2 = \frac{\gamma_2}{\gamma_3} D_0^2(\phi, \theta) + H^2 + \frac{\gamma_2}{\gamma_3} H_2^2. \tag{5.2.4}$$

As before, $h$ refers to the total depth of the moving fluid, i.e., $h = h_2 + h_3$ while H and $H_3$ are the constant values of h and $h_3$ on the eastern boundary at $\phi = \phi_e$ (where also $h_2 = H_2$).

On the latitude line $\theta = \theta_0$, $D_0^2$ vanishes, and therefore (5.2.4) reduces to:

$$h^2 + \frac{\gamma_2}{\gamma_3} h_2^2 = H^2 + \frac{\gamma_2}{\gamma_3} H_2^2 \tag{5.2.5}$$

One solution of (5.2.5) is clearly $h = H$ and $h_2 = H_2$. In this case (5.2.2) is satisfied by having both $v_2$ and $v_3$ zero on the line of zero Ekman pumping. This boundary then completely isolates the subpolar gyre from the subtropical gyre. This is the solution on the intergyre boundary that we have used in all the theories above. The question that we investigate now is whether there are *additional* solutions in which (5.2.5), or, equivalently (5.2.2), is satisfied by having $v_2 h_2 = -v_3 h_3 \neq 0$. Although the total transport would vanish, the baroclinic velocity field across the gyre boundary would differ from zero, and the two gyres would interact.

If layer 3 is in fact in motion south of the intergyre boundary as well as on that boundary, potential vorticity in this layer is conserved, and the potential vorticity isolines in layer 3 must coincide with the geostrophic streamlines of the flow in layer 3 which by geostrophy and hydrostatic balance are lines of constant $h$ or, equivalently:

$$q_3 = \frac{f}{h_3} = Q_3(h) \tag{5.2.6}$$

How is $Q_3$ to be determined? We see below that it is directly related to the issue of whether there is flow across the intergyre boundary.

Suppose that on $\theta = \theta_0$, and over some extent of longitude to be determined, $v_2$ and $v_3$ are not zero. Then using (5.2.5), we have on the latitude of zero Ekman pumping,

$$\begin{aligned} h_2 &= \left\{ H_2^2 + \frac{\gamma_3}{\gamma_2}(H^2 - h^2) \right\}^{1/2} \\ &\equiv \tilde{H}_2(h) \end{aligned} \tag{5.2.7}$$

One obvious possibility is that $h = H$ everywhere on $\theta = \theta_0$ so that $\tilde{H}_2 = H_2$ there. However, this is not the only possiblity. Solutions can exist in which $h$ departs from $H$ on the intergyre boundary.

Note that in order to keep $h_2$ real:

$$\frac{h}{H} < \left\{1 + \frac{\gamma_2}{\gamma_3}\frac{H_2^2}{H^2}\right\}^{1/2} < \left\{1 + \frac{\gamma_2}{\gamma_3}\right\}^{1/2} \tag{5.2.8}$$

The variation of $h_2$ might be substantial with respect to $H_2$, but the variation of the total depth of the moving fluid on the intergyre boundary is limited by (5.2.8).

For flow in layer 3 which crosses the intergyre boundary both (5.2.6) and (5.2.7) must apply so that on $\theta = \theta_0$ where $f = f_0$ and where:

$$h_3 = h - h_2 = h - \tilde{H}_2(h) \tag{5.2.9}$$

we have:

$$Q_3(h) = \frac{f_0}{h - \tilde{H}_2(h)}. \tag{5.2.10}$$

Thus if the flow in layer 3 crosses the intergyre boundary, $h_3$ must be related to $h$ by the unforced Sverdrup relation (5.2.7), which in turn determines the potential vorticity on the boundary. If potential vorticity is conserved, this relation is carried into the adjacent gyre. Thus on all streamlines emanating from the intergyre boundary and on which potential vorticity is conserved in layer 3:

$$\frac{f}{h_3} = \frac{f_0}{h - \tilde{H}_2(h)} \tag{5.2.11}$$

or:

$$\begin{aligned} h_3 &= \frac{f}{f_0}\{h - \tilde{H}_2(h)\} \qquad h_2 = h - h_3 \\ &= \left(1 - \frac{f}{f_0}\right)h + \frac{f}{f_0}\tilde{H}_2(h) \end{aligned} \tag{5.2.12}$$

where $\tilde{H}_2(h)$ is given by (5.2.7).

The Sverdrup balance (5.2.4) then becomes:

$$h^2 + \frac{\gamma_2}{\gamma_3}\left\{h\left(1 - \frac{f}{f_0}\right) + \frac{f}{f_0}\tilde{H}_2(h)\right\}^2 = \frac{\gamma_2}{\gamma_3}D_0^2 + H^2 + \frac{\gamma_2}{\gamma_3}H_2^2 \tag{5.2.13}$$

which yields an equation for $h(\phi, \theta)$.

One of the most interesting questions that arises is the extent of the longitudinal interval on the intergyre boundary on which cross-gyre flow can arise. That is, how big is the "window" through which flow from one gyre can enter and influence the other? We anticipate that this window cannot stretch all the way across the gyre boundary. It certainly cannot reach the eastern boundary apart from exceptional cases. This follows from the fact that if $H_2$

and $H_3$ differ from zero, potential vorticity conserving flow along the eastern boundary is inconsistent with the condition of no zonal flow there. This is the same basic reasoning that leads to the existence of the shadow zone described in Section 4.4. Thus we anticipate that there is an eastern edge, or limit, to the cross-gyre window (if the window exists). East of this edge layer 3 is at rest.

To find the eastern edge of the window we can use the fact just noted that east of the window layer 3 is at rest, and therefore in this region the base of layer 3 is flat and is at $z = -(H_2 + H_3) = -H$, and this must be true, by continuity on any hypothesized streamline that forms the eastern edge of the window (see Fig. 5.2.2). If $h = H$ is used in (5.2.7), it follows that $\tilde{H}_2(h) = H_2$ and so from the Sverdrup relation (5.2.4):

$$\frac{\gamma_2}{\gamma_3}\left\{H\left(1-\frac{f}{f_0}\right)+f/f_0 H_2\right\}^2 = \frac{\gamma_2}{\gamma_3}D_0^2 + \frac{\gamma_2}{\gamma_3}H_2^2. \qquad (5.2.14)$$

It therefore follows that the parametric equation for the eastern bounding streamline of the fluid stream pouring across the intergyre boundary is given by:

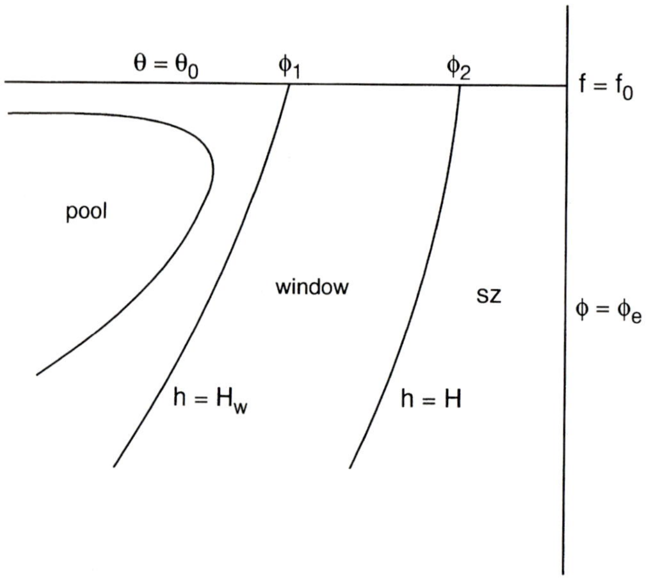


**Fig. 5.2.2.** Streamlines in layer 3 in the subtropical gyre which outline the domains of flow. The window on the intergyre boundary is the interval $\phi_2 \leq \phi \leq \phi_1$. The eastern edge of the window in the subtropical gyre is given by the streamline on which $h = H$ where $H$ is the known thickness of the upper two layers on the eastern boundary of the basin. The western edge of the window is given by the streamline on which $h = H_w$ and this is determined by the analysis of the cross-gyre dynamics

$$D_0^2(\phi, \theta) = 2H_2H_3\left(1 - \frac{f}{f_0}\right) + H_3^2\left(1 - \frac{f}{f_0}\right)^2. \tag{5.2.15}$$

When (5.2.15) is compared with (4.9.5) we see that the eastern boundary of the domain of possible cross-gyre flow is *identical* with the equation for the eastern boundary of the unventilated pool region of uniform potential vorticity. In order to obtain either closed contours within the subtropical gyre or streamlines which lead from one gyre to the next, the conservation of potential vorticity imposes the condition that the flow in the upper layer must so distort the potential vorticity isolines in layer 3 that they are drawn away from the eastern boundary and instead strike the intergyre boundary. From there the protopathways of the flow may be considered either to head westward to close in the subtropical gyre or, as here, to establish a pathway for cross-gyre flow. In either case the most eastern streamline satisfies (5.2.15). The point of intersection of the pathway with the intergyre boundary is at $\phi = \phi_1$ which, in the notation of Section 4.9, *is the same as the position of the Rossby repeller at* $\phi = \phi_r$. Hence for an internal mode to exist the wind forcing must be strong enough to place the Rossby repeller within the basin. We saw in Section 4.9 that this implies a balance between the eastward advection by the Sverdrup transport and the westward propagation of a baroclinic Rossby wave. The position of the repellor (4.9.7) is given by the longitude at which the Rossby wave is arrested, i.e., where (4.9.14) is satisfied.

From this point of view, cross-gyre flow does not represent a *free* internal mode of the circulation. Wind forcing is required first to twist the potential vorticity contours in the lower of the two moving layers so that the $q$ contours lead to the window instead of to the eastern boundary where the flow would otherwise be blocked. Just as a hero of the ancient myths, the wind forcing must be strong enough to sufficiently bend the q-pathways of the oceanic flow to open the gates across the gyres.

What is not so obvious is that there is also a western boundary to the cross-gyre window as well. This was demonstrated first by Pedlosky (1984) and more clearly by Schopp and Arhan (1986). If a derivative with respect to latitude is taken of the Sverdrup relation (5.2.4), we obtain:

$$2h\frac{\partial h}{\partial \theta} + \frac{\gamma_2}{\gamma_3}2h_2\frac{\partial h_2}{\partial \theta} = \frac{\gamma_2}{\gamma_3}\frac{\partial D_0^2}{\partial \theta} \tag{5.2.16}$$

When (5.2.12) is used to relate $h_2$ and its derivative to $h$, we obtain:

$$\frac{\partial h_2}{\partial \theta} = \left(1 - \frac{f}{f_0}\right)\frac{\partial h}{\partial \theta} - \frac{\beta R}{f_0}\left[h - \tilde{H}_2(h)\right] + \frac{f}{f_0}\frac{\partial \tilde{H}_2}{\partial \theta} \tag{5.2.17}$$

Using the definition of $\tilde{H}_2(h)$ given by (5.2.7) we see that on the intergyre boundary where $f = f_0$, $\tilde{H}_2(h) = h_2$. When this is used along with the above two equations, we obtain after some algebra the condition:

$$\frac{2\beta_0 R h_2 h_3}{f_0} = -\frac{\partial D_0^2}{\partial \theta} \tag{5.2.18}$$

where $\beta_0$ is the value of $\beta$ on $\theta = \theta_0$. If both sides of (5.2.18) are multiplied by the factor $(\gamma_2/2f_0 R)$, and comparison is made with (4.9.14), we see that the condition for cross-gyre flow can be written in the revealing form:

$$hc_r \equiv \frac{\gamma_2 \beta_0 h_2 h_3}{f_0^2} = U_s \equiv \frac{\gamma_2}{2f_0 R}\frac{\partial D_0^2}{\partial \theta}. \tag{5.2.19}$$

Thus the condition for cross-gyre flow can be interpreted as requiring that the Rossby wave speed balance the zonal Sverdrup advection velocity over a *range* of longitudes instead of just at one point. That is, the window between the two gyres is only as wide as the zone along the intergyre boundary in which (5.2.19) can be satisfied. At each point in the window the Rossby wave speed balances the eastward Sverdrup velocity. Generally the eastward Sverdrup velocity vanishes on the eastern boundary and then increases in magnitude westward. If the Ekman pumping is independent of longitude, this increase westward is linear with distance from the eastern wall. At the eastern wall $h_2 = H_2$ and $h_3 = H_3$. The zonal Sverdrup flow is zero, but the Rossby wave speed is not zero, and the wave progresses westward until it reaches the point where:

$$U_s(\phi, \theta_0) = \frac{\gamma_2 \beta_0}{f_0^2} H_2 H_3 \tag{5.2.20}$$

Further west than this point (which is $\phi = \phi_r$), $U_s$ is larger in magnitude and for the window to have a nonzero width the Rossby wave speed would also have to increase due to the variations of $h_2$ and $h_3$. In quasi-geostrophic theory this would not be possible because the layer thicknesses in quasi-geostrophic theory are permitted to have only very small *total* variations in depth since formally the lateral scale of the motion is limited as described in Chapter 3. In the planetary scale dynamics of the present treatment there is no such limit and $c_r$ can vary sufficiently to allow a window of non zero width to develop.

In order to examine the extent of the window we use the fact that on the intergyre boundary $\tilde{H}_2(h) = h_2$ and (5.2.19) can thus be written as:

$$U_s(\phi, \theta_0) = \frac{\gamma_2 \beta_0 H^2}{f_0{}^2}\left[\frac{\tilde{H}_2(h)(h - \tilde{H}_2(h))}{H^2}\right] \tag{5.2.21}$$

or, using the definition of $\tilde{H}_2(h)$ in (5.2.7):

$$\frac{U_s(\phi, \theta_0)}{\gamma_2 \beta_0 H^2/f_0{}^2} = \left[\left(\frac{H_2}{H}\right)^2 + \frac{\gamma_3}{\gamma_2}(1 - \xi^2)\right]^{1/2}\left[\xi - \left\{\left(\frac{H_2}{H}\right)^2 + \frac{\gamma_3}{\gamma_2}(1 - \xi^2)\right\}^{1/2}\right] \tag{5.2.22}$$

where:

$$\xi \equiv \frac{h}{H} \tag{5.2.23}$$

The situation described by (5.2.22) is shown schematically in Fig. 5.2.3. On the left is the nondimensional Sverdrup transport, as scaled in (5.2.22), as a function of longitude. The right side of the figure shows the right side of (5.2.22). This represents the Rossby wave speed multiplied by the total depth $h$ and scaled by the same factor as the Sverdrup transport. The figure is shown for the case in which $\gamma_2/\gamma_3 = 4$ and $H_2/H = 1/2$.

Imagine that the flow in the region of the window, west of the Rossby repellor at $\phi = \phi_r$, to be southward in layer 3. Then $h$ must increase westward, and we are therefore interested in the range $\xi \geq 1$. The longitudinal position of the eastern edge of the window is determined by the intersection of the horizontal line where $\xi = 1$ (the scaled Rossby speed in this case is 0.25 for $\xi = 1$) with the sloping line representing the Sverdrup transport. As we move westward in the window, increasing values of $h$, i.e., $\xi$, lead to increasing values of $hc_r$ leading to intersection points with the Sverdrup transport which lie further westward. Finally a value of $\xi$ is reached where the right side of (5.2.22) reaches a maximum. In the case shown, this occurs at $\xi \approx 1.2$, where the function plotted has a maximum of $\approx 0.309$. No solution of (5.2.22) can be found then west of this point, and the intersection of the horizontal line from the max-

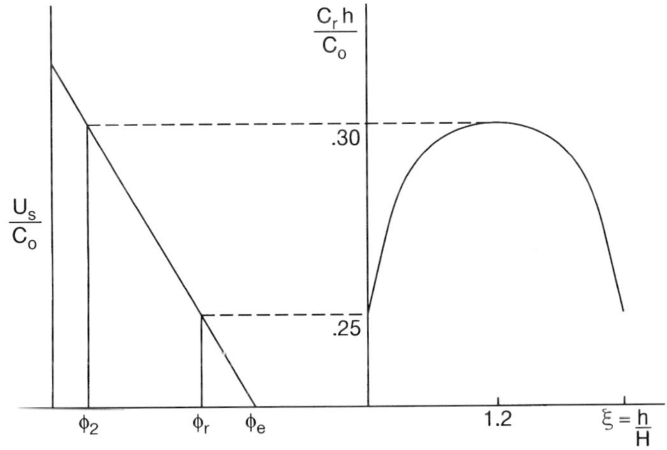


**Fig. 5.2.3.** *left*, Sverdrup zonal transport, scaled by the factor $\gamma_2\beta_0 H^2/f_0^2$ as a function of longitude; *right*, Rossby wave speed (multiplied by $h$ and scaled by the same factor as the Sverdrup transport) as a function of $\xi = h/H$. The position of the cross-gyre window is determined by equating the two functions shown. For southward flow in the lower layer $\xi$ must increase westward. The solution of (5.2.22) yields the range of longitude between $\phi = \phi_1 = \phi_r$ and the western limit of the solution which occurs when the Rossby speed curve reaches its maximum as a function of $h/H$. This point yields the western boundary at $\phi = \phi_2$. In this calculation $\gamma_2 = 4\gamma_3$ and $H_2 = 1/2H$

imum of the Rossby speed curve with the Sverdrup transport line determines the western boundary of the window at $\phi = \phi_2$. A little algebra shows that:

$$\left(\frac{h_2 h_3}{H^2}\right)_{\max} = \left[\frac{1+\Gamma_{23}(H_2/H)^2}{\Gamma_{23}+\left\{1+\sqrt{1+\Gamma_{23}}\right\}^2}\right]\sqrt{1+\Gamma_{23}} \tag{5.2.24}$$

where $\Gamma_{23} = \gamma_2/\gamma_3$.

For an Ekman pumping independent of longitude this would lead to a window of width:

$$R\cos\theta_0(\phi_1-\phi_2) = \left[\frac{(h_2h_3)_{\max}}{H^2} - \frac{H_2H_3}{H^2}\right]\frac{\beta_0^2\gamma_2 H^2}{f_0{}^3(\partial w_E/R\partial\theta)_{\theta=\theta_0}} \tag{5.2.25}$$

There is no a priori reason necessarily to expect the flow in the lower layer through the window to be southward. Northward flow seems equally likely. However, northward flow in the lower layer would require that the base of layer 3 at $z = -h$ *deepens* westward. The right side of (5.2.22) must then increase as $h$ decreases so that the window opens west of its eastern limit when the two sides of (5.2.22) are equated. We can see from the figure that *for the parameters chosen* the Rossby speed would decrease as $h$ decreases from $H$, and that no solution would be possible. This, however, is a function of the parameters which represent the stratification at the inter-gyre boundary, i.e., $\Gamma_{23}$ and the ratio of the layer depths.

Consider the function $I = h_2h_3$ which is the right side of (5.2.22) aside from a constant factor. Its variation as $h$ is varied is given by:

$$\delta I = h_3\delta h_2 + h_2\delta h_3 \tag{5.2.26}$$

On the intergyre boundary, where the Ekman pumping is zero, the Sverdrup relation (5.2.5) allows (5.2.26) to be written:

$$\delta I = \frac{1}{\Gamma_{23}h_2}[\Gamma_{23}h_2^2 + 2h_2h - h^2]\delta h \tag{5.2.27}$$

where we use $h_3 = h - h_2$ to eliminate $h_3$ in arriving at (5.2.27). At the eastern edge of the window each of the layer thicknesses take on their rest values, for example, $h = H$ and $h_2 = H_2$. In order for the window to extend westward $\delta I$ must be positive regardless of the sign of $\delta h$ since $U_s$ increases westward. Thus, if $\delta h > 0$ (southward lower layer flow in the window) the square bracket in (5.2.27) must be positive, which requires that:

$$\frac{H_2}{H} \geq \frac{\sqrt{1+\Gamma_{23}}-1}{\Gamma_{23}} \tag{5.2.28}$$

If $H_2/H$ is less than this critical value, southward flow in the lower layer across the intergyre boundary is not possible. On the other hand, when $H_2/H$ is below this threshold value, so that the square bracket in (5.2.27) is negative, a window is possible if $\delta h < 0$, i.e., if $h < H$ west of $\phi = \phi_r$. This implies that the flow in the lower layer is northward in layer 3, and deep fluid moves from the subtropical to the subpolar gyre. Figure 5.2.4 shows the right side of (5.2.22) in the case when $H_2/H$ is less than the critical value given by (5.2.28). If $h/H$ increases above 1, the value of the right side would fall, and no solution of (5.2.22) would occur. However, if $h$ decreases below $H$, so that there is northward flow in the lower layer a window solution is possible between $h/H = 1$, and the position of the maximum of the curve at $h/H \approx 0.94$ [in this case $\Gamma_{23} = 1$, and thus the critical value given by (5.2.28) is 0.414 while in the figure $H_2/H = 0.2$.

Thus the existence of the window depends only on the same combination of factors required to produce an unventilated deep pool in the gyre as discussed in Chapters 3 and 4. However, the direction of the flow entering across the window instead depends only on parameters reflecting the stratification of the ocean at the intergyre boundary. The longitudinal extent of the window depends on the stratification and the wind forcing as described by (5.2.25).

West of the point $\phi = \phi_2$, i.e., the western limit of the window between the gyres, the potential vorticity contours in layer 3, are still drawn around to reach the intergyre boundary at $\theta = \theta_0$. However, west of this point the Sverdrup advection is too strong to hold the window open, and information from the western boundary arrives to turn off the cross-gyre flow. Flow in the subtropical gyre in the layer 3 is thus still possible in the region west of streamline forming the western boundary of the current crossing the gyre boundary through the window. However, it must remain a recirculation within the subtropical gyre, as indicated in Fig. 5.2.2. One can think of the window as now occupying an eastern band of what would otherwise be the Rhine/Young pool, reducing that pool in size and replacing the outer belt of the pool with a stream of water either entering the subtropical gyre from the north or leaving it on geostrophic contours that have been reshaped by the wind to connect the two gyres.

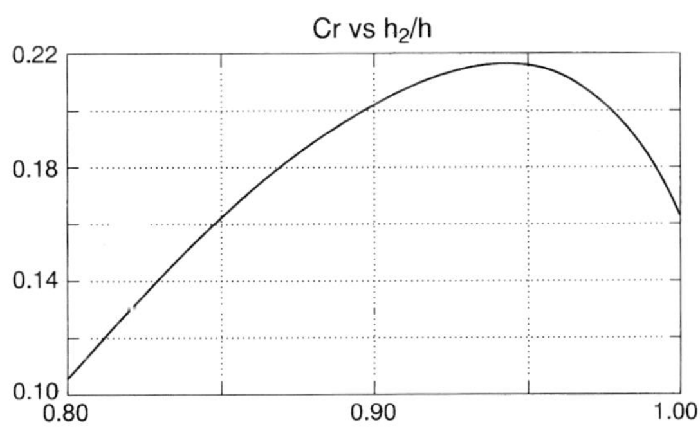


**Fig. 5.2.4.** *Right* side of equation (5.2.22) for the case in which $H_2/H$ is less than the critical value given by (5.2.28). Now at $h = H$ the Rossby wave curve decreases with increasing $h/H$ and only northward flow in the window is allowed in the lower layer

Figure 5.2.5 shows a cross-gyre solution calculated by Schopp and Arhan (1986) for the case of deep northward flow. The pool to the west of this region is kept at rest for simplicity. Schopp and Arhan continued the solution into the subpolar gyre, assuming that potential vorticity is conserved even there, although it is problematic whether this is true. One of the consequences of this continuation is a prediction of the outcrop line in which layer 3 becomes exposed to the Ekman suction of the subpolar gyre. This is shown in panel a of the figure in which the "inverted V" in the subpolar gyre is the outcrop line. Panel b shows the flow of fluid across the gyre from south to north in the narrow band defined by the window. In their calculation Schopp and Arhan take $H_2 = 0$ so that the window extends all the way to the eastern boundary. Panels c–e show zonal sections of the layer thicknesses in the subpolar gyre, on the intergyre boundary, and in the subtropical gyre, respectively. Schopp and Arhan applied their solution to the eastern North Atlantic and estimated a flux of about $4.25 \times 10^6\,\mathrm{m}^3\,\mathrm{s}^{-1}$ (4.25 sverdrups) to cross the intergyre boundary.

Chen and Dewar (1993) have extended the theory of cross-gyre communication to a model with three moving layers. As might be expected, the presence of three layers increases the richness of the structure of the possible cross-gyre communication. Just as in the case of pools of unventilated flow which are nested, one beneath the other, and shrink in size to the northwest, the presence of additional layers allows deeper windows which can appear to the west of the window that we have discussed. Chen and Dewar call attention to the observed structure of potential in layer between the $\sigma_\theta = 27.3$ and 27.6 surfaces in the North Atlantic and shown in Fig. 5.2.6. The contours stretch from the subpolar to the subtropical gyre, and the region shown has been described by McDowell et al. (1982) as a flow linking subpolar water in Irminger Sea with the subtropical gyre. This would be an example of southward flow through the window.

We therefore have, from a purely theoretical point of view, an additional nonuniqueness in the solution for the circulation structure. Solutions with cross-gyre communication are possible whenever the wind forcing is strong enough to produce unventilated pools with geostrophic contours (potential vorticity isolines) which meet the intergyre boundary. However, a solution in which the window, while possibly open, remains closed, is also always a possible steady solution. That is, a solution with $h_2 = H_2$ and $h_3 = H_3$ on the line of zero $D_0^2$ is consistent. The nonuniqueness is a consequence of our use of nondissipative, steady dynamics and could be resolved by relaxing either of these idealizations.

Schopp (1988) described a heuristic model of the time-dependent dynamics of the nonlinear Rossby waves on the intergyre boundary (see also Dewar 1987) and obtained an equation similar to the classical KdV equation (Korteweg and de Vreis 1895) . He chose, as initial conditions on the boundary, to have the interface between the layers initially flat and then to lower the interface at the western boundary (as if there were heating in the west) and raise it at the eastern boundary (cooling there) and held at that position during the

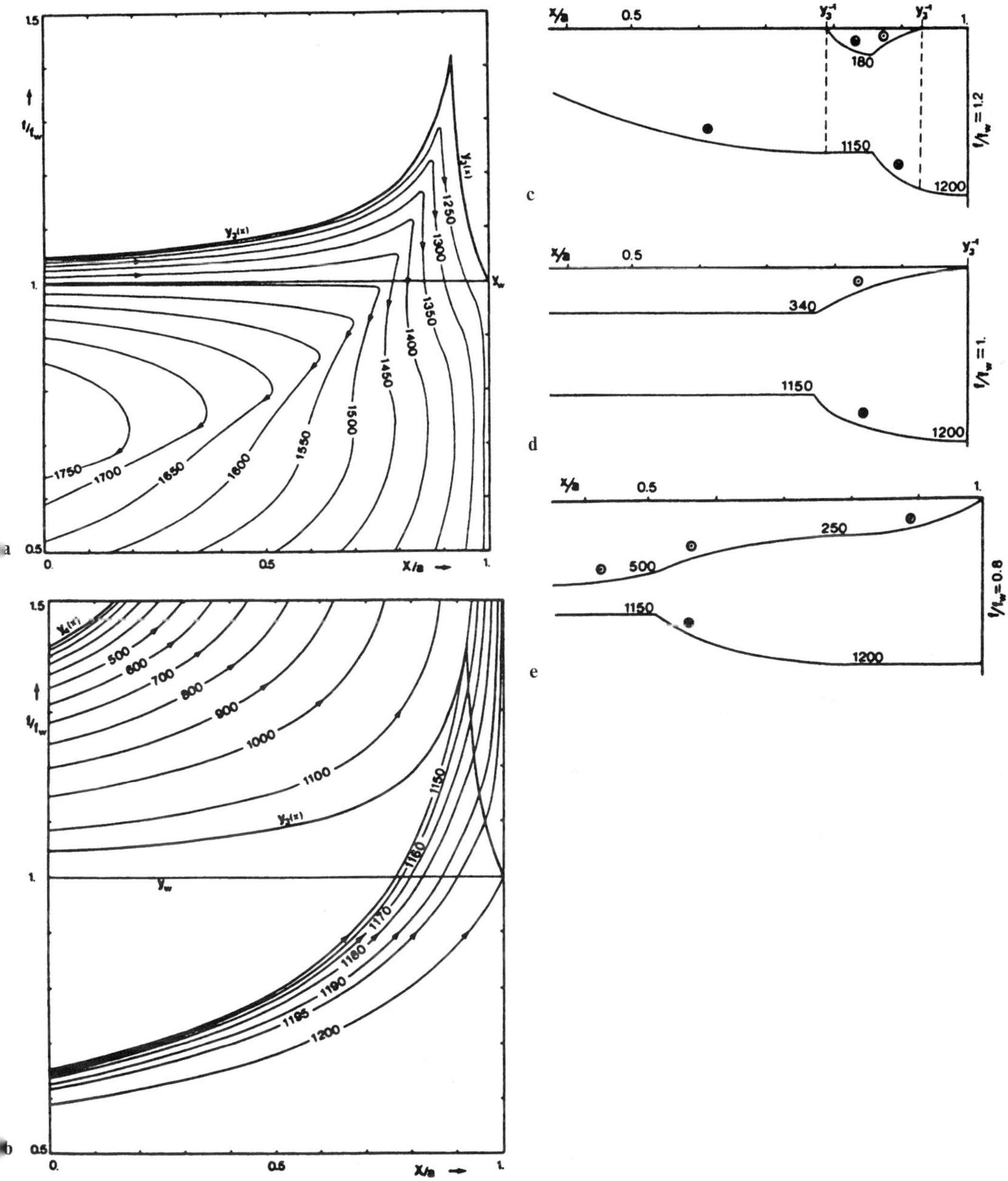


**Fig. 5.2.5a–e.** Crossgyre solution of Schopp and Arhan (1986) in which flow in layer 3 crosses the intergyre boundary moving northward. **a** Streamlines of the flow in the upper layer. The *"inverted V"* in the subpolar gyre shows the outcrop line of layer 3 as determined by the solution. **b** Lower layer streamlines and the flow leaving the subtropical gyre across the window. **c** Zonal section of the two layers in the subpolar gyre showing the zonal position of the outcrop line. **d** Same zonal section on the intergyre boundary. **e** Zonal section in the subtropical gyre. *Numbers* on the contours

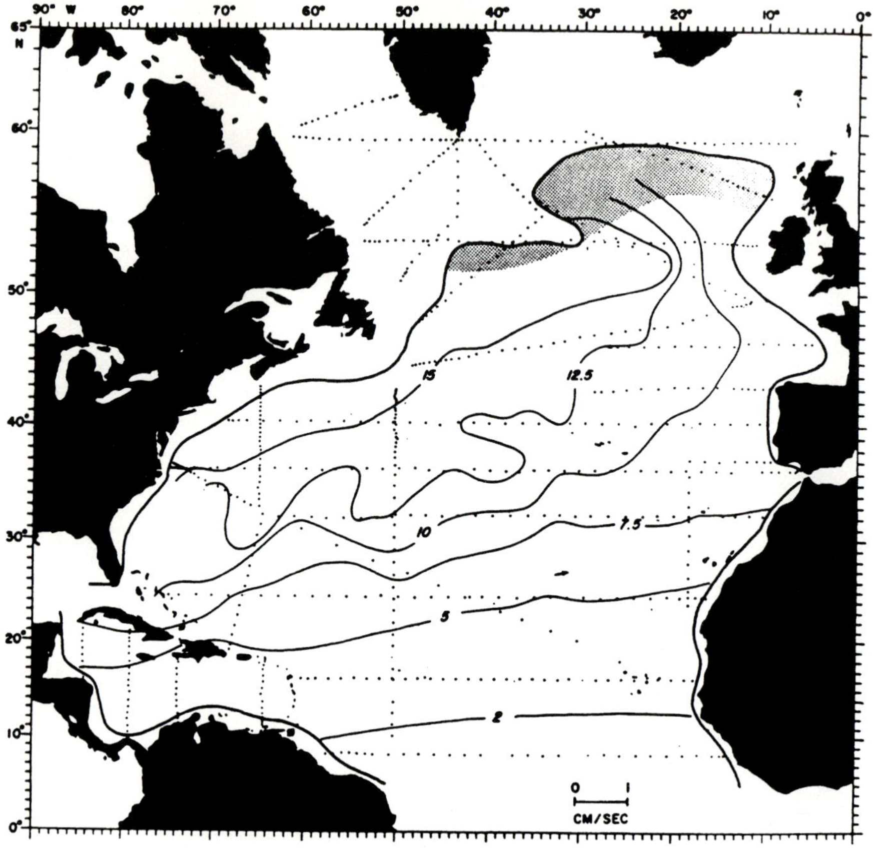


**Fig. 5.2.6.** Potential vorticity between the $\sigma_\theta = 27.3$–$27.6$ density surfaces. The contours stretch from the northeast to southwest, and water mass analysis suggests a southward cross-gyre flow through a window between the subpolar and subtropical gyres. (From McDowell et al. 1982)

calculation. Figure 5.2.7a shows the interface at $t = 0$. The sloping line in the figure has the slope of steady window solution. As time goes on, baroclinic Rossby waves bring signals from the boundary (panel b). From the east the Rossby wave arrives whose westward travel is faster than the Sverdrup advection. From the west the wave signal arrives as the eastward advection of the Rossby wave by the Sverdrup flow overpowers its natural tendency to pro-

**Fig. 5.2.7a-d.** Time-dependent calculation by Schopp (1988) describing the spin-up of the intergyre window from an initial state in which the interface is flat, and there is no cross-gyre flow. **a** At $t = 0$ the interface is flat except for displacements at the eastern and western edges which are held steady throughout the calculation. The linearly sloping curve has the same slope as the steady "window" solution. **b** After 1.6 years. Note the signals from the boundary propagate inward introducing two flat regions for the interface with different depths. **c** After 8 years. **d** After 24 years, showing the emergence of the solution with the steady-state window in which the eastern and western regions with different interface depths are joined by the window solution

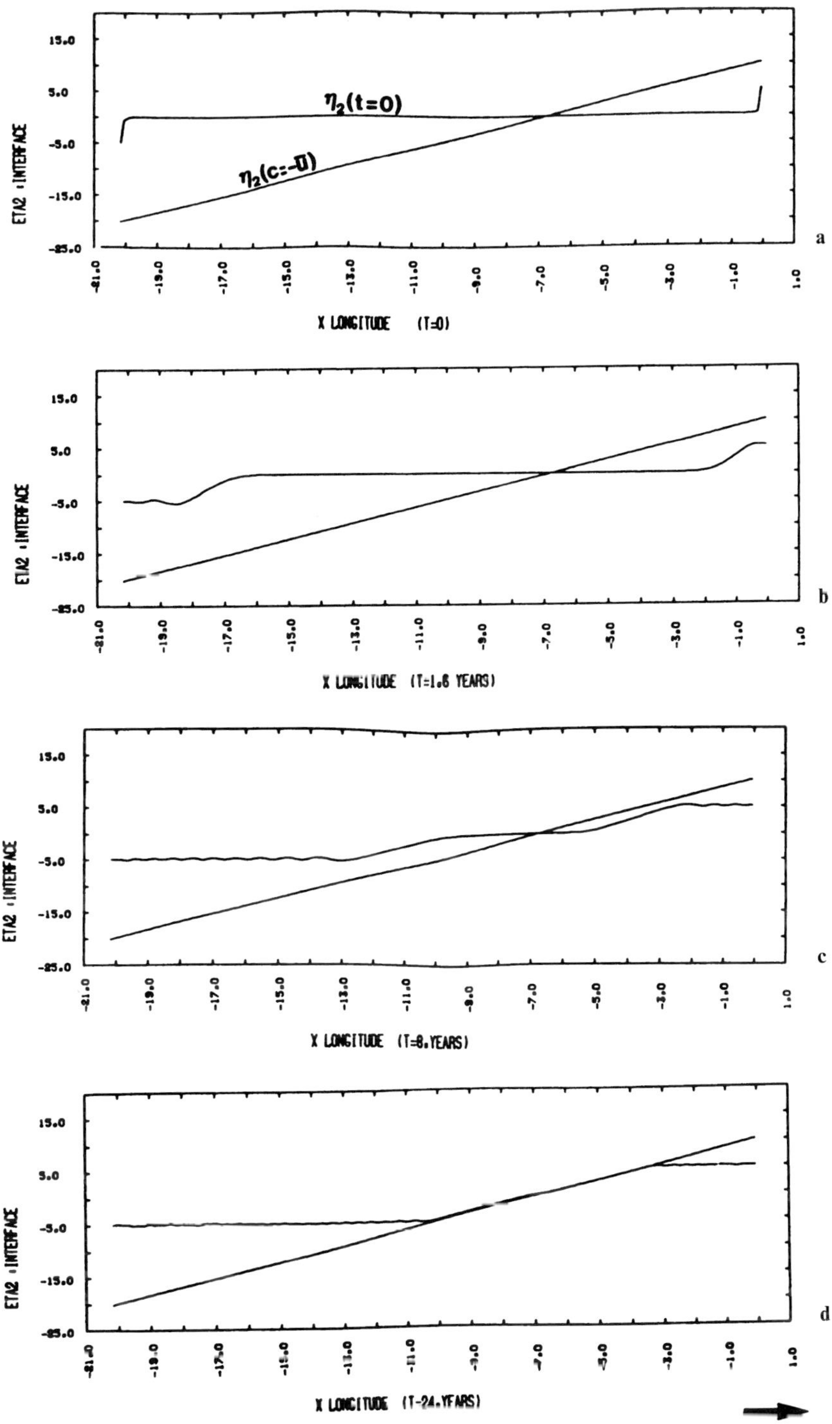
15.0
5.0
-5.0
-15.0
-25.0
ETA2 : INTERFACE
$\eta_2(t=0)$
$\eta_2(c=-\bar{U})$
a
-21.0
-19.0
-17.0
-15.0
-13.0
-11.0
-9.0
-7.0
-5.0
-3.0
-1.0
1.0
X LONGITUDE (T=0)
b
X LONGITUDE (T=1.6 YEARS)
c
X LONGITUDE (T=8.YEARS)
d
X LONGITUDE (T=24.YEARS)

pagate westward. After about 8 years the interface is well on its way, as seen in panel c, to a solution in which the interface is flat, with no cross-gyre flow, except in the region where the steady window is expected. The final steady solution, (panel d) shows the interface whose slope in the window exactly matches the slope anticipated by the steady theory. The implication then is that nonadiabatic dynamics on the gyre boundary, represented here by specified alterations in the stratification, determines the degree to which the two gyres can communicate.

## 5.3 Nonadiabatic Equations in Characteristic Form

### *Formulation*

The presence of cross-isopycnal fluxes of fluid represents the nonadiabatic processes in the body of the fluid which, among other things, destroy the conservative property of the potential vorticity. We have relied heavily on conservation of potential vorticity to deal with the nonlinear dynamics of the wind driven circulation. Therefore the presence of cross-isopycnal motion, as well as yielding an additional mechanism for motion, removes an important and useful constraint on the motion. This section describes a two-layer model including cross-isopycnal velocity following a formulation by Luyten and Stommel (1986a,b). The formulation is analogous to the transformation used by Rhines and Young (1982) for quasi-geostrophic motion and described in Section 3.5. In the quasi-geostrophic limit the transfomation yields a linear differential equation (3.5.12) for the baroclinic field once the barotropic circulation is determined independently from the Sverdrup relation. In the present case, as we see below, a similar approach in the dynamics on the planetary scale, where the layer thicknesses can vary by the order of 1, leads to a quasi-linear differential equation which is a great simplification for analysis.

In the work described below the cross-isopycnal velocity, $w_*$, is specified as a function of geographical location. This is clearly an inadequate representation of the physics since the cross-isopycnal velocity is certainly a result of turbulent processes that depend on the state of the ocean's local stratification and velocity. However, there is as yet no truly convincing parameterization of deep nonadiabatic processes that can be used to link $w_*$ to the large scale density field. Extant theories take the cross-isopycnal velocity as a prescribed function of position *analogous* to the Ekman pumping although of course it is physically not an external quantity as is the wind-driven Ekman pumping.

We recall from (3.2.18) that $w_*$ is related to the motion and the depth of the $n$th interface by:

$$w_*(z_n) = w_n(z_n) - \frac{dz_n}{dt}. \qquad (5.3.1)$$

The position of the interface at the top of the $n^{\text{th}}$ layer is at $z = z_n(\phi, \theta, t)$. From (4.3.16) we have the potential vorticity equation for layer $n$:

$$\frac{dq_n}{dt} = \frac{q_n}{h_n}\{w_*(z_n) - w_*(z_{n+1})\} \tag{5.3.2}$$

where, consistent with the assumption of the validity of the Sverdrup balance in the interior for the barotropic mode, the effect of mechanical dissipation in the interior is ignored. For the interior $q_n = f/h_n$.

Consider the two-layer model described in Section 4.1. We consider here the possibility that layer 2 does not outcrop so that layer 1 has a nonzero thickness everywhere, for example, including the eastern boundary. The Sverdrup balance for the two layers, whether a cross-isopycnal velocity between layer 1 and layer 2 exists or not, can be written:

$$h^2 + \frac{\gamma_1}{\gamma_2} h_1^2 = D_0^2 + H^2 + \frac{\gamma_1}{\gamma_2} H_1^2 \tag{5.3.3}$$

As before $h = h_1 + h_2$ and:

$$D_0^2 = -\frac{2f^2}{\gamma_2 \beta} \int_\phi^{\phi_e} w_E(\phi', \theta) R \cos\theta \, d\phi' \tag{5.3.4}$$

We suppose that between layers 1 and 2 there is a cross isopycnal velocity:

$$w_* \equiv w_*(z_2) \tag{5.3.5}$$

as shown in Fig. 5.3.1.

The steady form of the potential vorticity equation (5.3.2) for layer 2 is:

$$\frac{u_2}{R\cos\theta}\frac{\partial}{\partial\phi}\left(\frac{f}{h_2}\right) + \frac{v_2}{R}\frac{\partial}{\partial\theta}\left(\frac{f}{h_2}\right) = \frac{f}{h_2^2} w_* \tag{5.3.6}$$

which with the geostrophic equations for layer 2:

$$\begin{aligned} u_2 &= -\frac{\gamma_2}{fR}\frac{\partial h}{\partial\theta} \\ v_2 &= \frac{\gamma_2}{fR\cos\theta}\frac{\partial h}{\partial\phi} \end{aligned} \tag{5.3.7a, b}$$

yields the nonlinear partial differential equation:

$$\frac{\partial h}{\partial\theta}\frac{\partial h_2}{\partial\phi} - \frac{\partial h_2}{\partial\theta}\frac{\partial h}{\partial\phi} + \frac{\beta R}{f} h_2 \frac{\partial h}{\partial\phi} = \frac{fR^2 w_*}{\gamma_2}\cos\theta \tag{5.3.8}$$

The complete system for the determination of $h$ and $h_2$ is (5.3.8) and the Sverdrup balance (5.3.3). Note that (5.3.8) is nonlinear in the derivatives of the layer thicknesses.

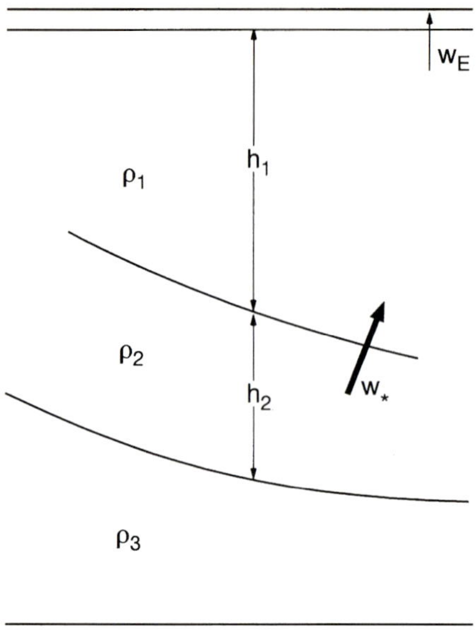


**Fig. 5.3.1.** Schematic presentation of the two moving layer model in which a cross-isopycnal flux is considered positive if fluid from layer $\rho_2$ is transformed into fluid of density $\rho_1$ in layer 1

As a first step in the analysis of (5.3.8) note that, since $h_2 = h - h_1$, the first two terms in (5.3.8) can be written in terms of $h_1$ instead of $h_2$ as:

$$-\frac{\partial h}{\partial \theta}\frac{\partial h_2}{\partial \phi} + \frac{\partial h_1}{\partial \theta}\frac{\partial h}{\partial \phi} + \frac{\beta R}{f} h_2 \frac{\partial h}{\partial \phi} = \frac{f R^2 w_*}{\gamma_2} \cos\theta \tag{5.3.9}$$

Since:

$$\frac{\partial h}{\partial \theta}\frac{\partial h_1}{\partial \phi} - \frac{\partial h_1}{\partial \theta}\frac{\partial h}{\partial \phi} = \frac{1}{2h_1}\left[\frac{\partial h}{\partial \theta}\frac{\partial h_1^2}{\partial \phi} - \frac{\partial h_1^2}{\partial \theta}\frac{\partial h}{\partial \phi}\right] \tag{5.3.10}$$

we can now use the Sverdrup relation to rewrite the nonlinear term in (5.3.8) so that the equation becomes:

$$\left(u_s - c_r\right)\frac{1}{R\cos\theta}\frac{\partial h}{\partial \phi} + v_s \frac{1}{R}\frac{\partial h}{\partial \theta} = -w_* \frac{h_1}{h}\frac{\gamma_1}{\gamma_2} \tag{5.3.11}$$

where:

$$u_s = \frac{\gamma_2}{2fhR}\frac{\partial D_0^2}{\partial \theta}$$
$$v_s = \frac{\gamma_2}{2fhR\cos\theta}\frac{\partial D_0^2}{\partial \phi} \tag{5.3.12a,b}$$

are the average Sverdrup velocity components for the barotropic mode of motion. That is:

$$\vec{U}_s = h\vec{u}_s = \frac{\gamma_2}{2f}\hat{k} \times \nabla D_0^2 \tag{5.3.13}$$

is the known Sverdrup transport. The baroclinic Rossby wave speed $c_r$ is given by:

$$c_r = \frac{\beta\gamma_1 h_1 h_2}{hf^2} = \frac{\beta\gamma_1 h_1(h - h_1)}{hf^2} \tag{5.3.14}$$

The governing equation for $h$ is now a quasilinear equation. The coefficients of the derivatives of $h$ are functions of $h$ (the Sverdrup velocity $\vec{u}_s$ depends on the unknown value of $h$, as does $c_r$), but they do not depend on the derivatives of $h$, and this is a considerable simplification. If the quasi-geostrophic approximation were employed instead, and the layer thicknesses were thus only small departures from the constant values that they have at the eastern boundary, both the Rossby wave speed and the Sverdrup velocities would be completely known, and (5.3.11) would reduce to a linear partial differential equation of the form (3.5.10). In quasi-geostrophic theory, as we saw in Chapter 3, the geostrophic contours are determined completely once the barotropic flow is known from the Sverdrup theory. In the present case the equivalent curves are the characteristics curves, to be defined momentarily, and they are not known (except qualitatively) since they depend on the unknown depth fields $h$ and $h_1$. The governing equations are (5.3.3) and (5.3.11) for $h$ and $h_1$.

We can define the *characteristic curves* of (5.3.11) in the horizontal plane in terms of the parameter $s$, which increases along each curve, by the following ordinary differential equations:

$$\begin{aligned} R\cos\theta\frac{d\phi}{ds} &= u_s - c_r \\ R\frac{d\theta}{ds} &= v_s. \end{aligned} \tag{5.3.15a,b}$$

The parameter $s$ is a measure of distance along the characteristic curves. The reader may find it helpful to think of $s$ as a timelike variable. After a certain time $s$ has elapsed an observer would have moved along a particular characteristic curve in the horizontal plane at a rate which depends on the Sverdrup velocity and the Rossby wave speed. In the meridional direction this characteristic speed is given solely by the Sverdrup velocity while in the zonal direction it is the Sverdrup velocity reduced by the propagation speed of the long baroclinic Rossby wave. When (5.3.15a,b) is used, (5.3.11) can be rewritten:

$$\frac{dh}{ds} = -w_* \frac{h_1}{h}\frac{\gamma_1}{\gamma_2} \tag{5.3.15c}$$

where the total derivative of $h$ is:

$$\frac{dh}{ds} \equiv \frac{\partial h}{\partial \phi}\frac{d\phi}{ds} + \frac{\partial h}{\partial \theta}\frac{d\theta}{ds} \tag{5.3.16}$$

The dynamics has now been reduced to the integration of the set of ordinary differential equations (5.3.15 a,b,c) coupled to the algebraic Sverdrup balance (5.3.3). The integration simultaneously determines the characteristic curves and the layer thicknesses. Normally the integration must be accomplished numerically. Starting with a particular, known value of the layer thicknesses, say, on the eastern boundary, the change in both $\phi$ and $\theta$ can be determined along a characteristic curve in one forward integration step in $s$ of the set (5.3.15a,b). Thus the development of each characteristic curve as it emerges from the boundary follows from the integration. Similarly, new values of $h$ and $h_1$ can be determined through the use of (5.3.15c) and (5.3.3) after the first step in $s$. With these new values the characteristic curves can be developed a step further in $s$ and the process iterated until the whole domain is covered with characteristics on which the layer thicknesses and thus the velocities are known. Information needed for the calculation requires the layer thicknesses at the starting point of the characteristic curve, and the development of the layer thickness fields is determined by integrating along the characteristics. In this sense *information flows along the characteristics determined by the interplay of Sverdrup advection and baroclinic Rossby wave propagation.* This is the formal restatement of the heuristic discussion of Section 3.5. The algebraic Sverdrup balance remains as the ghostly image of the purely westward propagation of the swift barotropic wave whose relatively high speed allows information in the barotropic mode to sweep across the basin and establish the barotropic mode independently along each latitude circle.

If the cross-isopycnal flux is zero, the total depth $h$, from (5.3.15), is constant along each characteristic curve. Since lines of constant $h$ are, geostrophically, streamlines of the flow in the lower layer, it follows that only if the cross-isopycnal flux is zero will the characteristic curves be streamlines. When the cross-isopycnal flux is nonzero the two sets of curves depart from one another, and information flows along the characteristic curves rather than along streamlines.

Although the solution for $h$ from (5.3.15) allows the calculation of $h_1$ from the Sverdrup balance and the subsequent calculation of $h_2$, it is useful to develop explicit equations for the variation of each of the layer thicknesses along the characteristic curves. From the Sverdrup relation:

$$\frac{dh}{ds} = \frac{1}{2h}\frac{dh^2}{ds} = \frac{1}{2h}\left[\frac{dD_0^2}{ds} - \frac{\gamma_1}{\gamma_2}\frac{dh_1^2}{ds}\right] \tag{5.3.17}$$

Since:

$$\frac{dD_0^2}{ds} = \frac{\partial D_0^2}{\partial \phi}\frac{d\phi}{ds} + \frac{\partial D_0^2}{\partial \theta}\frac{d\theta}{ds} \tag{5.3.18}$$

if (5.3.12a,b) and (5.3.15a,b) are used, it follows that:

$$\frac{dD_0^2}{ds} = -2c_r f \frac{hv_s}{\gamma_2} \tag{5.3.19}$$

If these results are used to evaluate the right side of (5.3.17):

$$\begin{aligned}\frac{dh}{ds} &= -c_r \frac{f}{\gamma_2} v_s - \frac{\gamma_1}{\gamma_2} \frac{h_1}{h} \frac{dh_1}{ds} \\ &= -w_* \frac{h_1}{h} \frac{\gamma_1}{\gamma_2}\end{aligned} \tag{5.3.20}$$

where the last step follows from (5.3.15c). If the definition (5.3.14) of $c_r$ is used as well as the Sverdrup balance, $fw_E = \beta h v_s$, we obtain:

$$\frac{dh_1}{ds} = w_* - \left(\frac{h - h_1}{h}\right) w_E \tag{5.3.21}$$

so that the increase in the upper layer thickness along a characteristic curve is a balance between the cross-isopycnal flux into the layer across its base and the Ekman flux out the top of the layer. In the subpolar gyre where the cross-isopycnal flux is negative (warm water is being cooled and transformed to colder water), and where the Ekman velocity is positive, the thickness of the upper layer is diminished along a characteristic curve.

The streamlines in the upper layer follow curves of constant pressure which are curves of constant $h + \Gamma_{12} h_1$ where $\Gamma_{12} = \gamma_1/\gamma_2$. Combining (5.3.21) and (5.3.15 c) we obtain:

$$\frac{d}{ds}\{h + \Gamma_{12} h_1\} = \Gamma_{12}\left(1 - \frac{h_1}{h}\right)\{w_* - w_E\}. \tag{5.3.22}$$

The variation of the potential vorticity of the lower layer along the characteristic curves can be calculated be similar methods to yield:

$$\frac{d}{ds}\left(\frac{f}{h_2}\right) = w_* \frac{f}{h_2^2}\left(1 + \Gamma_{12} \frac{h_1}{h}\right). \tag{5.3.23}$$

When the cross-isopycnal flux is nonzero the potential vorticity varies along the characteristic curves as well as along streamlines. Thus in the presence of non-adiabatic effects the isolines of potential vorticity, the characteristic curves, and the streamlines all differ from each other. They coincide only for adiabatic flow.

If we suppose, as in Section 5.2, that the Ekman pumping is a function only of latitude, then on the latitude where the Ekman pumping vanishes (5.3.15c) reduces to:

$$(u_s - c_r) \frac{1}{R\cos\theta} \frac{\partial h}{\partial \phi} = -w_* \frac{h_1}{h} \Gamma_{12}. \tag{5.3.24}$$

We can divide the latitude band into (at least) two regions. We may expect that to the west, where $u_s > c_r$, that heating ($w_* > 0$) drives southward flow across the gyre boundary in the lower layer (with equal northward return flux above in the upper layer) while cooling does the opposite. In the eastern regions where we may expect $u_s < c_r$, heating leads to deep northward flow across the gyre boundary. If there is no heating or cooling on the intergyre boundary there is no flow across the boundary (i.e., $\partial h/\partial\phi = 0$) unless a third intermediate zone develops in which $u_s = c_r$ which is exactly the "window" described in Section 5.2.

### *Example 1: Subpolar Gyre*

Consider now the example discussed by Luyten and Stommel (1986b) in which a model of the subpolar gyre is driven by both Ekman pumping, $w_E > 0$, and cooling, $w_* < 0$. For simplicity Luyten and Stommel considered the upper layer to be thick enough on the eastern boundary that in spite of loss of fluid to the upper Ekman layer and to the layer beneath it, the upper layer covers the entire area of the subpolar gyre, i.e. no outcropping of layer 2 occurs. The Ekman pumping and the cross-isopycnal velocities are both specified as functions only of latitude. The former is positive, and the latter is negative in the subpolar gyre, and they are both assumed to increase linearly with distance from the intergyre boundary where they are both taken to be zero. There is no geostrophic flow across the boundary between the subpolar and subtropical gyre in this model, and we concentrate on the subpolar gyre for now. Figure 5.3.2 schematically presents the region and the forcing. Note that for the chosen forcing the ratio of $w_*$ to $w_E$ is a constant. The linear increase in Ekman pumping is realistic only for the southern portion of the subpolar gyre, and we therefore consider the domain of integration to represent the southern portion of the gyre and allow its northern boundary to be porous. As we see below, the characteristics at this boundary arrive from the south, and it is thus not necessary to prescribe boundary conditions on the northern edge of the domain of the calculation.

On the eastern boundary the layer thicknesses $H_1$ and $H_2$ are given constants, each different from zero. Therefore, near the eastern boundary $c_r$ is nonzero and since $u_s$ is 0 there, characteristics emanating from the eastern boundary enter the interior of the basin and trend westward. The further development of the characteristic field depends on the solution of the problem, and they cannot be described precisely until the calculation is completed in the manner described above.

On the western boundary the variation of the layer thicknesses with latitude is not constrained except by the Sverdrup relation (3.5.3), which gives only one constraint between the two variables $h$ and $h_1$. For their calculation Stommel and Luyten chose to set the layer thicknesses on the western boundary by assuming that all the flow out of the western boundary current

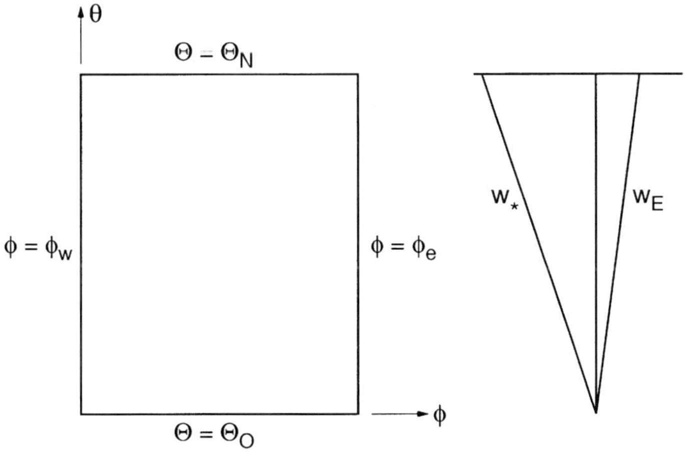


**Fig. 5.3.2.** Schematic presentation of the subpolar basin with the calculation of Luyten and Stommel (1986b). To the right of the basin are shown the distribution with latitude of the cross-isopycnal velocity and the Ekman pumping velocity. The former is negative (cooling) while the latter is positive

occurs in the upper layer. Thus $h$ is a constant along the western boundary in their calculation. This is completely arbitrary and reflects the nonuniqueness already seen in the adiabatic pool region of the ventilated thermocline in the previous chapter. The Sverdrup relation then determines $h_1$ along that boundary. Then, with the layer depths specified on the western boundary, the characteristics leaving the western boundary can be calculated. Since generally the Sverdrup eastward flow is greater there than the westward Rossby wave speed, the characteristics from the western boundary enter the basin from the west and trend eastward.

The results of the Luyten and Stommel's calculation are shown in Fig. 5.3.3, where the characteristics are shown in panel a. Although the solution requires detailed calculation to quantitatively determine the field of characteristic curves and the resulting fields of layer depths, it is possible to reason directly from the equations in characteristic form to qualitatively describe the results of the calculation.

Since $v_s > 0$ everywhere in the subpolar gyre, all the characteristics, after they leave either the eastern or western boundaries, must trend northward. Those from the west move to the northeast while those leaving the eastern boundary trend northwest. Along the gyre boundary to the south where $v_s = 0$ the characteristics entering from the east and west will experience a critical point at that longitude where $u_s - c_r = 0$. This point we have called the Rossby repellor. It is marked with an $R_o$ in Fig. 5.3.3b. From this critical point (in this solution the "window" has collapsed onto this point) a critical characteristic emerges, labeled B in the figure, which divides the basin into western and

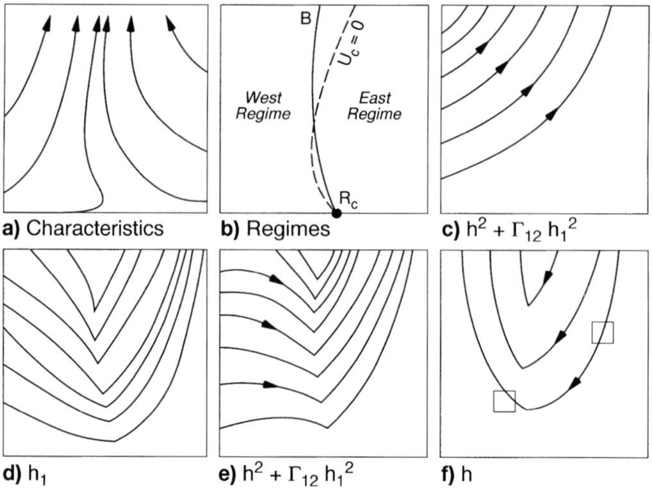


**Fig. 5.3.3.** Subpolar solution of Luyten and Stommel (1986b). **a** The characteristics. **b** The two regimes into which the characterisic curves divide the basin. **c** The Sverdrup transport streamfunction $h^2 + \Gamma_{12}h_1^2$. **d** Contours of upper layer depth. **e** Upper layer streamfunction. **f** Lower layer streamfunction

eastern portions. The western regime is covered by characteristics carrying information from the western boundary while the eastern basin is bathed in characteristics which have emanated from the eastern wall.

On each of the characteristics $h$ increases monotonically along the characteristic curve as required by (5.3.15c) since $w_* < 0$. Since $w_*$ is independent of longitude, the increase in $h$ is greatest on the longest paths, i.e., those along the separating characteristic B. Thus we expect a maximum of $h$ in the center of the northern edge of the region at the termination of the longest paths, and this in fact is what can be seen in panel f of the figure. This leads to a high pressure center placed roughly midway between the eastern and western walls at the northern edge of the calculation region around which a large anticyclonic circulation takes place.

Similarly, (5.3.21) and (5.3.22) imply that both $h_1$ and the upper layer geostrophic pressure field, $h + \Gamma_{12}h_1$, reach a minimum at the intersection of B and the northern boundary so that a large cyclonic circulation occurs in the upper layer, as shown in panel e of the figure. Panel d shows the thickness field of the upper layer. It is important to note how qualitatively different the circulation is in each of the layers when compared to the barotropic, Sverdrup streamfunction in panel. This is related to the great strength of the cooling-induced, cross-isopycnal velocity, which is chosen in the calculation to be as large as, and opposite to, the Ekman suction of 3 cm/day. The barotropic circulation is here a poor indicator of the nature of the circulation in the individual layers, which are strongly driven by baroclinic cooling.

A particularly interesting aspect of the solution is the relationship between the vertical velocity and the cross-isopycnal velocity. Consider the interface between layers 1 and 2, where:

$$w_*(z_2) = w_2(z_2) - \vec{u}_2 \cdot \nabla z_2. \tag{5.3.25}$$

Since $w_2$ vanishes at the base of layer 2 where it must be continuous with the vertical velocity in layer 3 which is at rest, the Sverdrup vorticity balance yields, when integrated over layer 2:

$$w_2(z_2) = \frac{\beta h_2 v_2}{f}. \tag{5.3.26}$$

Thus in the subpolar gyre where $w_*$ is negative the vertical velocity and the isopycnal velocity have the same sign when the fluid in *layer 2 is moving southward.* Where this happens, the horizontal advection of density, represented by the second term on the right side of (5.3.25), is relatively small. As (4.5.5) shows, this term is proportional to the cross-product between the velocity vectors between the two layers, i.e.:

$$\vec{u}_2 \cdot \nabla z_2 = \frac{f}{\gamma_1} \hat{k} \cdot [\vec{u}_1 \times u_2]. \tag{5.3.27}$$

Thus the horizontal advection of density is small in regions where the streamlines in the two layers are nearly parallel, and from (5.3.25) and (5.3.26) this tends to occur in the subpolar gyre in regions of southward flow in layer 2. This is exactly the situation in the region marked by the right-hand box in Fig. 5.3.3f. The flow is southward in layer 2 and northward in layer 1, and the streamlines in the two layers tend to be aligned. In this region warmer, upper layer fluid is moving northward where the cooling is greatest and then sinks and enters nearly vertically [$w_* \approx w_2(z_2)$] into the lower layer and then flows southward. This circulation, in which the motion is overturning in an intuitively obvious sense (the vertical velocity has the same direction as the cross-interface flux) in response to the cooling is called a direct circulation by Luyten and Stommel (1986a), and it is seen to occur here in the eastern regime of the basin.

In the western regime, for example, in the region of the left-hand small box in Fig. 5.3.3f, $v_2$ is positive, and therefore the vertical velocity at the interface must be positive even though the cross-interface flux is negative. This can occur only because there is strong horizontal advection of density, and fluid crosses the interface due to the horizontal velocity across the strongly sloping interface. This, by (5.3.27), is signaled by the near orthogonality of the streamlines of the two layers as the velocity vector spirals with strongly with depth in the western portion of the basin, i.e., to the west of the characteristic B. Luyten and Stommel term this a region of an indirect circulation ($w_*$ and the vertical velocity have opposite signs). Warm water flows southward away from the region of cooling and is converted to cold water by flowing nearly horizontally across the steeply sloping interface between the two layers.

The difference in the behavior in the eastern and western regimes can be traced to the characteristic equations for the pressure fields in the two layers, i.e., (5.3.15c) and (5.3.22). Since the ratio of the Ekman velocity to the cross-isopycnal velocity is a constant, as we step along a characteristic curve the ratio of the change in upper layer pressure compared to the change in the lower layer pressure is proportional to:

$$\frac{\delta(h+\Gamma_{12}h_1)}{\delta h} \propto \frac{h_1}{h_2} \tag{5.3.28}$$

On or near the eastern boundary the ratio of the layer depths is a constant due to the boundary conditions. Thus the pressure changes in the two layers tend to advance in synchrony in the eastern region, resulting in aligned isobars and a thermally direct circulation. It is quite different in the western regime where each of the layer thicknesses on the western wall varies quite differently with latitude as fluid exits in the upper layer in the western boundary current. This results in a spiral of the velocity vector with depth and strong horizontal flow across the interface between the layers.

### *Example 2: Subtropical Gyre with Heating*

The effect of heating and cooling on the circulation of the subtropical gyre can be examined in fundamentally the same way as has been done for the subpolar gyre. There are, however, several interesting dynamical issues which distinguish the two domains. As we saw in Chapter 4 for the adiabatic model, the circulation in the subtropical gyre contains different subregions of the flow such as the shadow zone and the ventilated region. The delineation of these zones in the adiabatic theory depends on the position of critical streamlines which carve out the various subdomains. When the motion is no longer conservative, the streamlines lose this privileged function, and we expect it to be the characteristic curves that carry the information to determine the subdomains.

The presence of outcropping presents an apparent difficulty in the characteristic formulation. If there is an outcrop line, say, along the latitude circle $\theta = \theta_2$, the upper layer thickness vanishes there. It also vanishes at this latitude at the eastern boundary, and if the zonal flow vanishes at the eastern boundary, the layer thicknesses must be constant there so that $h_1$ is thus zero for all latitudes along $\phi = \phi_e$. This implies that the Rossby speed $c_r$ is zero on the eastern boundary. Certainly $u_s$ is also zero there. It would appear from a superficial examination of (5.3.15a,b) that the characteristics from the eastern boundary can not enter the basin in the region south of the outcrop line.

This is, however, not the case. Consider the neighborhood of the eastern boundary where $h_1$ goes to zero. Although both the Sverdrup zonal velocity and the Rossby wave speed vanish there, the former vanishes linearly with distance from the eastern wall while, as we see below, $h_1$ and thus $c_r$, vanish as the square root of the distance from the wall and thus dominate $u_s$ near the wall. Thus, for a characteristic which emanates from the eastern boundary at the latitude $\theta = \theta_*$, in the neighborhood of the eastern boundary (5.3.15a) can be approximated as:

$$R\cos\theta \frac{d\phi}{ds} = -\frac{\gamma_1 \beta h_1}{f^2} \tag{5.3.29}$$

where we use the fact that $h_2 \approx h$ near $\phi = \phi_e$. The variation in $h_1$ along the characteristic can be calculated from (5.3.21). Again, using $h_2 \approx h$, we obtain:

$$\frac{dh_1}{ds} = w_* - w_E \tag{5.3.30}$$

and for simplicity we assume that both the Ekman and cross-isopycnal velocities are independent of longitude. It follows from (5.3.29) and (5.3.30) that:

$$h_1 \frac{dh_1}{d\phi} = R\cos\theta[w_E - w_*]\frac{f^2}{\gamma_1\beta} \tag{5.3.31}$$

or, along the characteristic:

$$h_1 = \left[-\frac{2f^2}{\gamma_1\beta}(w_E - w_*)R\cos\theta(\phi_e - \phi)\right]^{1/2}. \tag{5.3.32}$$

In performing the integral along the characteristic we should really have considered $\theta$ as a function of $\phi$, but if we remain close enough to the starting point of the integration, $\theta$ can be replaced by its initial value, $\theta_*$, and may therefore be considered a constant in the integration. Thus $h_1$ vanishes as the square root of the distance to the boundary. This verifies our starting hypothesis that near the wall $c_r$ dominates $u_s$ which goes to zero linearly with distance to the boundary (it vanishes as $D_0^2$). If (5.3.32) is used in (5.3.29), and if (5.3.15b) is also used, we obtain for the characteristic emanating from the wall at $(\phi_e, \theta_*)$:

$$R\frac{d\theta}{d\phi} = \frac{-v_s}{\left\{-\frac{2\gamma_1\beta}{f^2}(w_E - w_*)\right\}^{1/2}} \frac{(R\cos\theta)^{1/2}}{[\phi_e - \phi]^{1/2}} \tag{5.3.33}$$

If we again set $\theta = \theta_*$ in the coefficients of the right side of (5.3.33), we can integrate the equation easily to obtain, as the equation for the characteristic curve as it leaves the eastern boundary:

$$R(\theta - \theta_*) = \frac{v_s}{\left[-\frac{\gamma_1 \beta}{f^2}(w_E - w_*)\right]^{1/2}} \left[R \cos \theta_* (\phi_e - \phi)\right]^{1/2}. \tag{5.3.34}$$

Hence each characteristic initially enters the region from the eastern boundary as a parabola tangent to the boundary, i.e. with an infinite slope in the horizontal plane as shown in Fig. 5.3.4.

North of the outcrop line, where $h_1$ is identically zero, the characteristics of the system coincide with the streamlines of the Sverdrup flow which are lines of constant $D_0^2$. There is no cross-isopycnal motion in this region by hypothesis, and the solution in this region is identical to the single layer solution (4.4.6). Hence $h$ remains constant along the characteristics in this region which coincide then with streamlines. The streamline on the eastern wall is labeled C in Fig. 5.3.4, and it hugs the eastern wall until the outcrop line is reached. After crossing the outcrop line the characteristic, as we have seen, pulls away from the eastern boundary. The curve it traces is labeled C in the figure. East of that curve is the region of the shadow zone, now defined as the region in which information from the western boundary and outcrop line cannot penetrate. Instead, information in this region is propagated from the eastern boundary

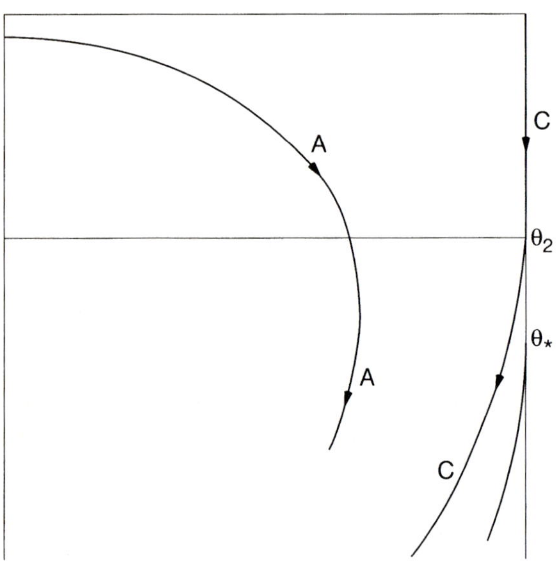


**Fig. 5.3.4.** Schematic presentation of the characteristic curves entering the basin in the subtropical gyre. From the eastern wall, south of the outcrop line at $\theta = \theta_2$, the characteristics from the eastern boundary enter the basin on curves which are initially tangent to the boundary, as can be seen from the characteristic entering from the latitude $\theta = \theta_*$. *A*, North of the outcrop line the characteristics are given by the Sverdrup flow streamlines North of $\theta = \theta_2$ the characteristic on the eastern wall thus remains parallel to the wall. This characteristic pulls away from the wall south of $\theta = \theta_2$, and its position determines the position of the shadow zone

which is the starting point of the integration of the characteristic system (5.3.15). West of the curve C all the characteristics emanate from the western boundary, as shown with the curve A which is also coincident with a streamline in layer 2 in the region north of the outcrop line.

When a characteristic crosses the outcrop line, the continuation of the integration requires that the upper layer depth be calculated as follows. From (5.3.15b) and (5.3.21) it follows, again using $h_1 \approx 0$ near the outcrop line:

$$\begin{aligned}\frac{1}{R}\frac{dh}{d\theta} &= \left(\frac{w_* - w_E}{v_s}\right)\\ &= -\frac{\beta h}{f}\left\{\frac{w_E - w_*}{w_E}\right\}.\end{aligned} \tag{5.3.35}$$

As long as $(w_E - w_*)/w_E > 0$, the thickness of the upper layer increases and is nonzero south of the outcrop line as the upper layer is filled with fluid pumped down from the Ekman layer and entrained from the layer below. If this condition is not met, it is inconsistent to specify the outcrop line at that position. If $w_*$ is negative and large enough to render the ratio negative, the strong cooling that this implies forces the outcrop line to be further south.

Figure 5.3.5 shows the calculation of Luyten and Stommel for the sub tropical gyre. The outcrop line is the curve labeled $Y_s$ in panel f. South of this point $(w_E - w_*)/w_E > 0$ , and the upper layer grows in thickness south of this latitude. The distribution of Ekman pumping and cross-isopycnal velocity is shown in panel g. Note that the cross-isopycnal flux is positive representing heating of the subtropical gyre. The maximum value of the Ekman pumping is taken to be –3 cm/day while the maximum of the cross-isopycnal flux is 3 cm/day. Panel a shows the characteristics calculated in the solution by Luyten and Stommel. The characteristics emanating from the eastern boundary are not drawn accurately in the figure since they should really enter tangentially from the boundary. However, the difference is slight. Panel b shows how the basin is divided in two by the critical characteristic which leaves the outcrop line at the eastern boundary and sets apart the shadow zone from the rest of the basin. Panel c shows the Sverdrup transport streamlines which form a single anticyclonic cell. Panel d shows the contours of upper layer thickness. Panel e shows the circulation in the upper layer. This is qualitatively similar to the circulation in the adiabatic case. The kink in the upper layer streamlines when crossing the characteristic which delineates the shadow zone boundary is evident. The biggest difference in the circulation is apparent in panel f, showing contours of $h$, which are also the lower layer streamlines. The region of ventilation is similar to that in the adiabatic model, with southward motion taking part in the anticyclonic cell in the lower layer. Note that the outcrop line is taken sufficiently far southward in the calculation so that a pool of unventilated fluid in layer 2 is absent.

In the region of the old shadow zone, fluid that was once at rest in the adiabatic theory is now in motion, taking part in a large recirculating cyclonic

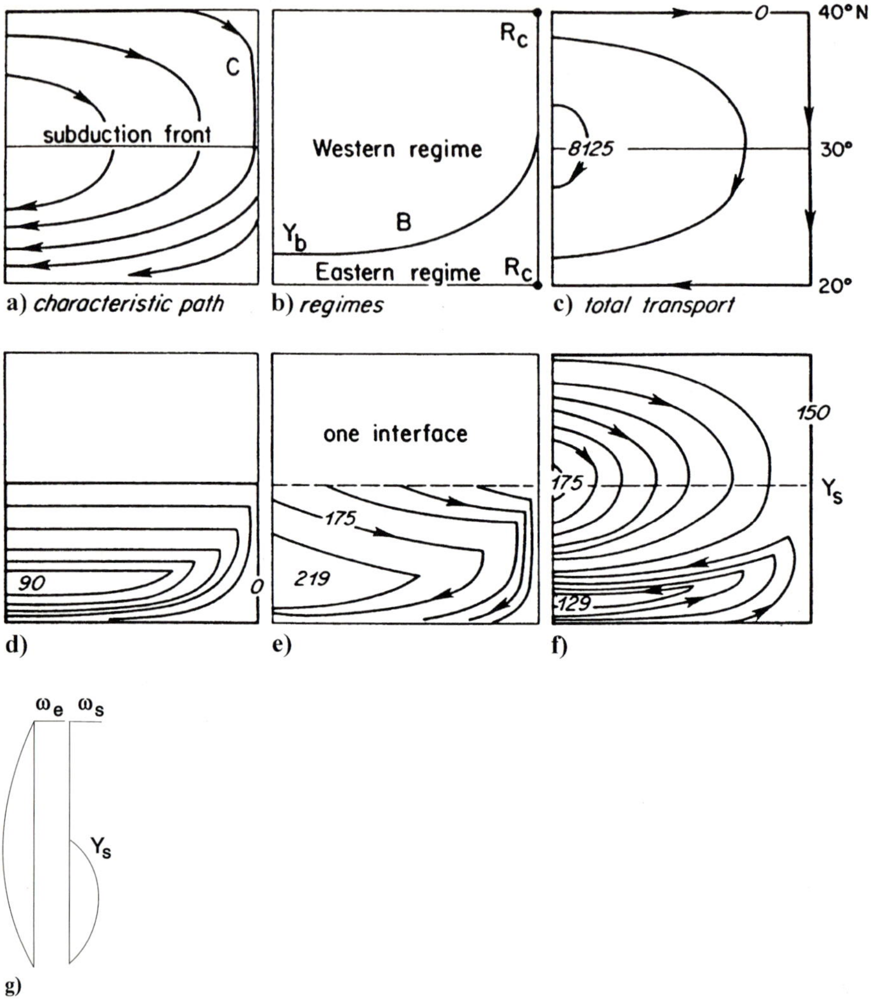


**Fig. 5.3.5a–g.** Results of the calculation of Luyten and Stommel (1986b) for the subtropical gyre. The gyre is driven by a distribution of Ekman pumping and cross-isopycnal fluxes, each a function only of latitude **g**. The depth of the lower layer on the eastern boundary is 150 m, and $\gamma_1 = 1 \mathrm{cm/s}^2$. The width of the basin is 2000 km, and the midlatitude of the basin is at 30°N, where the outcrop line is placed. The maximum amplitude of each forcing velocity is 3 cm/s and of course the Ekman pumping velocity is negative. **a** Characteristic curves of the solution. **b** Western and eastern domains of the solution as determined by the critical characteristic emanating from the intersection of the outcrop line and the eastern boundary. **c** Sverdrup transport streamlines. **d** Depth contours (in meters) of the upper layer. **e** Contours of $h + \Gamma_{12} h_1$ which are the streamlines of the upper layer flow. **f** The contours of $h$ which yield the streamlines of the lower layer flow

cell which is inserted between the ventilated fluid and the eastern wall and is driven by the cross-isopycnal flux, i.e. by the heating. Careful comparison of panels b and f shows that the eastern regime is characterized by northward flow east of the critical characteristic. This northward flow returns as a recirculation *west* of the critical characteristic. The recirculation is detached from the flow of the subducting fluid and spans the shadow zone boundary. That boundary is marked by kinks in the recirculation streamlines. It is important to re-emphasize the fact that the characteristics are not streamlines, and in particular the characteristic separating the eastern and western regimes is one along which $h$ is continuously decreasing along the characteristic (5.3.15c). This is consistent with the northwest flow across the regime boundary observed in panel f. The streamline representing the eastern boundary of the ventilated fluid *lies to the west of the critical characteristic.* The critical characteristic forms the boundary between the northward and southward branches of the *recirculation* in the old shadow zone produced by the cross-isopycnal flux.

Comparing panels e and f we see that the streamlines in this eastern regime are nearly antiparallel in the two layers. Our preceding discussion of the subpolar gyre and equations (5.3.25), (5.3.26), and (5.3.27) suggests that in this region the circulation is *direct,* i.e., that $w_2(z_2)$ and $w_*$ are both positive in this region. Cold water in layer 2 moves northward as it is heated, rises, and enters the warm water layer and then flows southward. With $w_2(z_2)$ positive this implies that $v_2 > 0$, and this is what is observed on the eastern branch of the cyclonic gyre in the shadow zone. In Section 5.4 we shall reexamine the motion in this region analytically to explain the occurrence of southward motion in the western branch of the cyclonic gyre.

In the western regime where the fluid is ventilated the motion is southward, and therefore the vertical velocity and the cross-isopycnal velocity have opposite signs. In this case the motion is *indirect.* Cold water flows southward and enters the upper layer by laterally crossing the strongly sloping density interface.

Since the specification of the heating function, or cross-isopycnal velocity $w_*$, is made a priori, it is difficult, as has been mentioned, to insist on the physical validity of the details of the solution structure. The solution, however, is very suggestive as to how the shadow zone, considered stagnant in the adiabatic theory, can easily be set into motion by heating and cooling. It also clearly illustrates the existence of distinct regions of flow where heating or cooling produces vertical motions that may be counterintuitive, i.e., the regions of indirect circulation.

The characteristic formalism of Luyten and Stommel is well adapted to numerical calculation but is limited to the dynamics of the two-layer model. If more than two layers are present, the characteristics of the complete system become so complex that no real progress can be made. For this reason an alternative representation of the effect of heating and cooling is useful, and this is discussed in the following section.

## 5.4 The Buoyancy- and Wind-Driven Subtropical Gyre: Analytical Solutions

### *Solution in the Ventilated Region*

Stimulated by the numerical calculations of Luyten and Stommel (1986b), Pedlosky (1986) described an analytical approach to the combined wind- and buoyancy-driven circulation. The approach has the added advantage of being generalizable to a system of more than two layers, but for simplicity the present discussion is limited to the same two-layer model described in the last section.

The governing system of equations is the Sverdrup balance (5.3.3) and the potential vorticity equation written in terms of the layer thicknesses, for example, (5.3.8).

When the cross-isopycnal velocity is zero, the solution in the region south of the outcrop line in the ventilated region is given by (4.4.14a,b) in which each individual layer thickness is given in terms of the total depth, $h$, in the form;

$$\begin{aligned} h_1 &= \Phi(f)h \\ h_2 &= (1-\Phi(f))h. \end{aligned} \tag{5.4.1a,b}$$

In the case of adiabatic motion for which the cross-isopycnal flux is zero the function $\Phi(f)$ is determined at the outcrop line and is given by (4.4.14a,b);

$$\Phi(f) = \left(1 - \frac{f}{f_2}\right) \quad \text{for} \quad w_* = 0. \tag{5.4.2}$$

We can generalize this approach to the case when $w_* \neq 0$ in the following way. Suppose we adopt (5.4.1) as a trial solution and insert it into the Sverdrup balance (5.3.3) and the governing partial differential equation for $h_2$, (5.3.8). Note that $H_1 = 0$ and $H = H_2$. Thus (5.3.3) yields:

$$h^2\left[1 + \frac{\gamma_1}{\gamma_2}\Phi(f)^2\right] = D_0^2 + H_2^2 \tag{5.4.3}$$

or:

$$h^2 = \frac{D_0^2 + H_2^2}{1 + \Gamma_{12}\Phi^2} \tag{5.4.4}$$

which should be compared with (4.4.18). In (5.4.4) the notation $\Gamma_{12} = \gamma_1/\gamma_2$ is again been used.

To determine $\Phi$ the proposed solution (5.4.1) should be inserted into (5.3.8). If we use the relation:

$$\frac{\partial h}{\partial \theta}\frac{\partial h_2}{\partial \phi} - \frac{\partial h}{\partial \phi}\frac{\partial h_2}{\partial \theta} = h\frac{\partial h}{\partial \phi}\frac{d\Phi}{\partial \theta} \tag{5.4.5}$$

(5.3.8) becomes:

$$\left[\frac{d\Phi}{d\theta}+\frac{\beta R}{f}(1-\Phi)\right]\frac{\partial h^2/2}{\partial\phi}=\frac{fR^2 w_*}{\gamma_2}\cos\theta \tag{5.4.6}$$

From the Sverdrup balance (5.3.3) and the definition of $D_0^2$ it follows that:

$$\frac{\partial h^2/2}{\partial\phi}=\frac{f^2 R\cos\theta}{\gamma_2\beta}\frac{w_E}{1+\Gamma_{12}\Phi^2} \tag{5.4.7}$$

so that (5.4.6) becomes:

$$\frac{d\Phi}{d\theta}+\frac{\beta R}{f}(1-\Phi)=\frac{\beta R}{f}\left[\frac{w_*}{w_E}\right]\left(1+\Gamma_{12}\Phi^2\right). \tag{5.4.8}$$

Our trial solution (5.4.1) assumes that $\Phi$ is a function of $f$ (or of latitude) alone. Thus (5.4.8) is consistent with this form of solution if and only if the ratio of $w_*$ to $w_E$ is a function only of latitude. This restriction is a rather weak one, and it allows us to describe the forcing functions of the previous section, each one of which was taken to be a function only of latitude.

Since $\Phi$ is explicitly a function of the Coriolis parameter, $f$, it is useful to choose $f$ instead of $\theta$ as the independent variable. Writing:

$$\zeta=\frac{f}{f_2} \tag{5.4.9}$$

(5.4.8) becomes:

$$\zeta\frac{d\Phi}{d\zeta}+(1-\Phi)=b\left(1+\Gamma_{12}\Phi^2\right) \tag{5.4.10}$$

where:

$$b(\zeta)=\frac{w_*}{w_E}. \tag{5.4.11}$$

At the outcrop line, which corresponds to $\zeta=1$, the upper layer thickness must vanish, hence the initial condition for (5.4.10) is:

$$\Phi=0 \quad \text{at} \quad \zeta=1 \tag{5.4.12}$$

Thus the entire problem for the structure of the combined buoyancy- and wind-driven thermocline circulation, in the ventilated region in which (5.4.1) applies, reduces to the solution of a single ordinary differential equation for the structure function $\Phi$, which is a function of latitude alone. In particular, the longitudinal structure of the solution is determined completely by (5.4.4) and depends only on the integral of the Ekman pumping, assuming, as stated above that the *ratio* of cross-isopycnal velocity to Ekman velocity is a function only of latitude. As noted, they may both individually be functions of longitude as well.

In particular, we may calculate $v_2$ directly:

$$v_2 = \frac{\gamma_2}{fR\cos\theta}\frac{\partial h}{\partial\phi} = \frac{f}{\beta h}\frac{w_E}{(1+\Gamma_{12}\Phi^2)} < 0 \tag{5.4.13}$$

since the Ekman pumping is negative in the subtropical gyre. Thus the meridional velocity in layer 2 is always southward in the ventilated zone regardless of the strength of the heating or cooling. The magnitude of the heating and cooling can affect the magnitude of the velocity and the extent of the domain of the ventilated solution, but within that domain the meridional velocity must be negative, and therefore the vertical velocity in layer 2 at the interface between layers 1 and 2 must also be negative. Therefore, if the layer is heated, so that $w_* > 0$, the circulation in the ventilated region must be indirect, as we found in the numerical solution of Luyten and Stommel (1986b) in the previous section. It follows here directly from the form of the ventilated solution (5.4.1).

The streamline which separates the subducted fluid from the recirculating fluid driven by the heating can be found by the same process as in Section 4.4. We examine the path of the streamline which leaves the outcrop line in its most eastern position along the eastern boundary where, at that position, $D_0^2 = 0$ and $h = H_2$. Thus, as before, we obtain as the parametric equation for the eastern boundary of the ventilated region:

$$D_0^2(\phi,\theta) = H_2^2\Gamma_{12}\,\Phi\,(f)^2 \tag{5.4.14}$$

It is important to recall that the boundary determined by (5.4.14) is not the shadow zone boundary determined by the *characteristic* emanating from the intersection of the outcrop line and the eastern boundary. This boundary lies to the east of the eastern edge of the ventilated zone, as was shown in Fig. 5.3.5. The zone in between the two curves is the domain of the southward return flow of the buoyancy-driven cyclonic circulation which is wedged between the ventilated zone and the eastern boundary.

Generally, for a specified ratio of the cross-isopycnal velocity to Ekman velocity, $b(f)$, (5.4.10) must be integrated numerically, starting at $\zeta = 1$ and proceeding southward towards $\zeta = 0$. It is somewhat more natural to consider an interval in which the dependent variable increases, and the change of variable:

$$\mu = 1 - \zeta \tag{5.4.15}$$

allows (5.4.10) to be written:

$$(1-\mu)\frac{d\Phi}{d\mu} = (1-\Phi) - b(1+\Gamma_{12}\Phi^2) \tag{5.4.16}$$

over the interval $0 \le \mu \le 1$, with the initial condition $\Phi(\mu = 0) = 0$.

When there is no cross-isopycnal flux, and $b = 0$, the solution of (5.4.16) which satisfies the initial condition is:

$$\Phi = \mu = 1 - \frac{f}{f_2} \tag{5.4.17}$$

which is the solution (5.4.2) found in Section 4.4.

In the presence of heating ($w_* > 0$) and with the negative Ekman pumping in the subtropical gyre ($w_E < 0$), it follows that b $<$ 0, and thus (5.4.16) implies that $\Phi$ increases with $\mu$ more rapidly than in the adiabatic case. Figure 5.4.1 shows the solution for $\Phi$ for the case in which:

$$b = -0.5\frac{f}{f_2}\left[1 - \frac{f}{f_2}\right] \tag{5.4.18}$$

so that the cross-isopycnal velocity vanishes at the outcrop line and goes to zero at the equator. The line in the figure with a slope of $-1$ is the solution for $\Phi$ in the absence of heating. The line above it is the solution for $b$ given as in (5.4.18). We see that $\Phi$ is, as expected, everywhere greater than the value which it would have in the absence of heating. This has several important consequences.

The trajectory for any streamline in the ventilated region is given by the condition that $h$ be constant, or:

$$D_0^2(\phi, \theta) = H_2^2\Gamma_{12}\Phi^2 + D_0^2(\phi', \theta_2)(1 + \Gamma_{12}\Phi^2) \tag{5.4.19}$$

where $\phi'$ is the longitudinal position of the trajectory at its starting location on the outcrop line. The equation for the eastern edge of the ventilated region (5.4.14) is a special case of (5.4.19) as $\phi' \to \phi_e$ where $D_0^2$ vanishes. Since heating increases the magnitude of $\Phi$ at each latitude, the right side of (5.4.19) becomes larger than would be the case for adiabatic motion. Since $D_0^2$ is an increasing function of distance from the eastern boundary, the only way in which the left side of (5.4.19) can increase at each latitude is if the trajectory of the streamline is shifted further westward. Heating, then, moves all the ventilated streamlines westward of their adiabatic positions. In particular, the eastern boundary of the ventilated regime is also moved westward. The eastern unventilated region then increases in size while at the same time the size of any unventilated pool in the western part of the circulation decreases in longitudinal extent.

When $\Phi(f) = 1$, the thickness of the lower layer goes to zero. For adiabatic motion this occurs only at the equator although we recognize that the solution based on the geostrophic approximation may not be valid there. Figure 5.4.1 shows that $\Phi$ equals 1 at a value of $f > 0$ when there is heating. If this occurs, it means that the flux of fluid from layer 2 into the upper layer and the general thinning of the layer as fluid moves westward into the western boundary layer, combine to completely exhaust the volume of layer 2 at the latitude in which $\Phi = 1$. In Fig. 5.4.1 this occurs at $f/f_2 = 0.14$. The lower layer then *pinches off* at that latitude, and layer 1 comes into direct contact with layer 3. Our solution is not valid of course south of this point, but one can easily imagine that if the heating persists, it would then lead to a cross-isopycnal flux

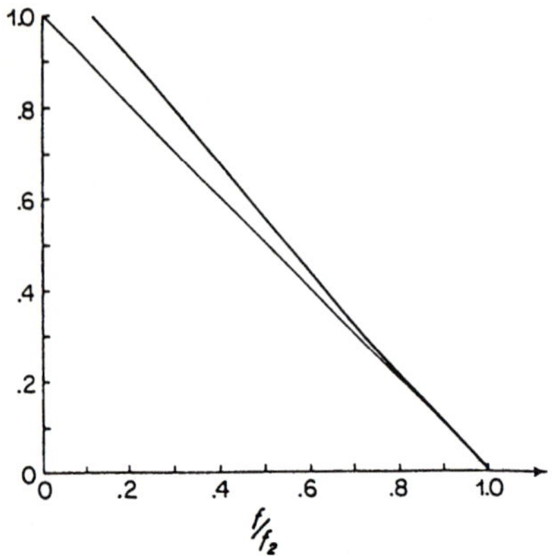


**Fig. 5.4.1.** Solution for the structure function $\Phi(f)$ for the heating distribution in the subtropical gyre (5.4.18). (From Pedlosky 1986)

across the interface between layers 1 and 3. Layer 3 would thus be set into motion south of the pinch off latitude. This process is reminiscent of a similar phenomenon that we found in the adiabatic solution for the subpolar gyre of the previous chapter in which it was the Ekman suction which exhausted the volume of the upper layer, leaving the next lower layer exposed to Ekman pumping which then set it into motion.

In order for the pinch-off to occur it must occur within the boundary of the ventilated region. This boundary is given by (5.4.14). It generally sweeps southwestward across the basin starting at the eastern boundary on the outcrop line. The boundary intersects the latitude circle where pinch-off occurs at the position where $\Phi = 1$. Thus, if the intersection point for pinch-off occurs at the latitude $\theta = \theta_{po}$, this point is determined by (5.4.14) with $\Phi = 1$, i.e:

$$D_0^2(\phi_{po}, \theta_{po}) = H_2^2 \Gamma_{12} \tag{5.4.20}$$

In order for pinch-off to occur in the basin the longitude, $\phi_{po}$, at which this intersection takes place must lie within the basin, i.e., $\phi_{po} \geq \phi w$. Since $D_0^2$ is a decreasing function of longitude, it follows that to have the pinch-off point within the basin:

$$D_0^2(\phi_w, \theta_{po}) \geq D_0^2(\phi_{po}, \theta_{po}) = H_2^2 \Gamma_{12}. \tag{5.4.21}$$

Thus, for example, if the Ekman pumping is independent of longitude, (5.4.21) implies that for pinch-off to occur,

$$R \cos \theta_{po}(\phi_e - \phi_w) \geq \frac{H_2^2 \gamma_1 \beta_{po}}{2 f_{po}^2(-w_E(\theta_{po}))} \tag{5.4.22}$$

where the subscript *po* refers to the latitude of pinch-off. Thus wider basins are more likely to experience pinch-off, and it is also favored by weak stratification and strong wind forcing.

The solution (5.4.4) is valid only west of the boundary given by (5.4.14), the region which in the adiabatic solution is the ventilated zone. To complete the solution we must find the solution in the eastern unventilated region.

### *The Solution in the Unventilated Region*

There is no analytical solution currently available for the region east of the critical streamline given by (5.4.14), but analytical progress of a revealing sort can be made by considering perturbation solutions for small $b$, i.e., for cases in which the buoyancy forcing is small compared to the wind driving. Normally such perturbation solutions which give rise to small changes in the solution, contain little of qualitative interest. However, in the region of the old shadow zone the small velocities that are driven in layer 2 by the heating comprise the only circulation that exists there since the region is at rest for adiabatic motion. Therefore a perturbation solution around the state of rest in layer 2 is of considerable qualitative interest.

In the ventilated zone the solution for $\Phi$ is of the form,

$$\Phi = 1 - \frac{f}{f_2} + O(b) \tag{5.4.23}$$

and as we noted this leads to a westward shift of $O(b)$ in the eastern boundary of the ventilated zone. For $b = 0$ the velocity in layer 2 is zero in the unventilated region and the solution in the eastern region is given by the adiabatic shadow zone solution:

$$h_1^{(0)} = \left[\frac{\gamma_2}{\gamma_1} D_0^2\right]^{1/2}, \qquad h^{(0)} = H_2 \tag{5.4.24}$$

where the superscript (0) refers to the first term in an implicit expansion of the solution in a parameter which measures the magnitude of $b$. The potential vorticity in layer 2 associated with this $O(1)$ solution is:

$$q_2^{(0)} = \frac{f}{H_2 - h_1^{(0)}} \tag{5.4.25}$$

The contours of the lowest order potential vorticity are shown schematically in Fig. 5.4.2.

The contours are shown by the dashed lines in the figure. Also shown is the boundary of the shadow zone when $b = 0$. This is the curve for which $h^{(0)} = H_2$.The isolines of $q_2^{(0)}$ cover the region between this curve and the eastern boundary. The effect of heating shifts the eastern boundary of the ventilated zone westward to the curve on which $h = H_2$. The narrow space between them has a width of $O(b)$ and is not reached by $q_2^{(0)}$ isolines of the family emanating from the eastern wall. Each of these curves is identified by its

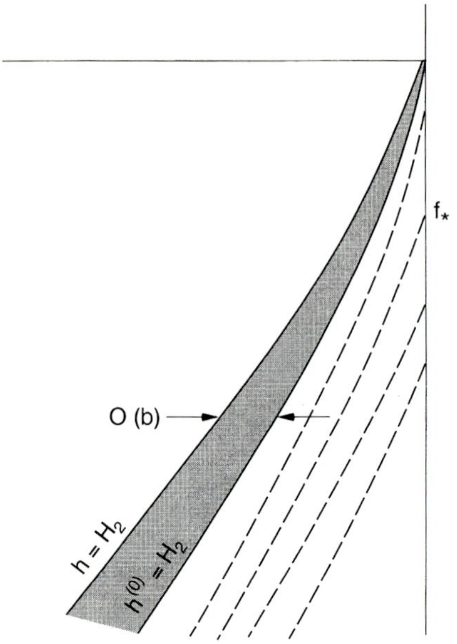


**Fig. 5.4.2.** Region of the perturbation solution in the eastern unventilated region. The western boundary of the adiabatic shadow zone is given by the curve $h^{(0)} = H_2$. Between this curve and the eastern boundary the domain is covered by the isolines of potential vorticity, $q_2^{(0)}$ of the adiabatic solution. The effect of small heating moves the boundary westward a distance of $O(b)$ to the line $h = H_2$, opening up a region (*gray*), which is *not* covered by the isolines of potential vorticity in the adiabatic shadow zone

starting point on the eastern boundary at the latitude $\theta_*$ for which $f = f_*$. The most northern of these starting points has $f = f_2$, and the $q_2^{(0)}$ isoline emanating from there yields the $h_2^{(0)} = H_2$ isoline. Thus the shaded area is a sliver of unventilated fluid in the basin which is not covered by the adiabatic potential vorticity isolines coming from the eastern wall. This becomes important because information about the $O(b)$ solution is propagated along these isolines, and the region of the sliver is therefore bypassed by these characteristics. The solution in the sliver consequently differs from that in the region covered by the adiabatic characteristics. In the numerical solution of Luyten and Stommel discussed in the previous section the sliver region is covered with characteristics which come from the western region.

The parameter $f_*$ is related to $q_2^{(0)}$ by the relation, determined on the eastern boundary:

$$q_2^{(0)} = \frac{f}{H_2 - h_1^{(0)}} = \frac{f_*}{H_2} \tag{5.4.26}$$

so that:

$$f_* = f\frac{H_2}{h_2^{(0)}} = f_*(\phi, \theta) \tag{5.4.27}$$

The last equality in (5.4.27) states that every point within the region covered by the $q_2^{(0)}$ isolines is uniquely related to a starting value of $f = f_*$ by tracking the $q_2^{(0)}$ isoline associated with that point back to the eastern boundary. Thus $f_*$ is constant on each $q_2^{(0)}$ isoline.

The solutions for $h$ and $h_2$ are written in the form:

$$\begin{aligned} h &= h^{(0)} + h^{(1)} + \cdots \\ h_2 &= h_2^{(0)} + h_2^{(1)} + \cdots \end{aligned} \tag{5.4.28}$$

where the superscript (1) variables are $O(b)$ smaller than the superscript (0) variables which are given by (5.4.24). If this series is put into (5.3.8), and the fact that $h^{(0)}$ is a constant in the shadow zone is used, we obtain at order $b$:

$$\frac{1}{R^2\cos\theta}\left\{\frac{\partial h^{(1)}}{\partial\phi}\frac{\partial q_2^{(0)}}{\partial\theta} - \frac{\partial h^{(1)}}{\partial\theta}\frac{\partial q_2^{(0)}}{\partial\theta}\right\} = \frac{f^2 w_*}{\gamma_2\left[h_2^{(0)}\right]^2}. \tag{5.4.29}$$

The right side is $O(b)$ by virtue of the smallness of the cross-isopycnal velocity.

Equation (5.4.29) is a linear equation for $h^{(1)}$, and its solution can be easily found by integrating along the characteristics of the equation which coincide with lines of constant $q_2^{(0)}$. Thus we introduce the characteristic curves as solutions of:

$$\begin{aligned} R\cos\theta\frac{d\phi}{d\tau} &= \frac{1}{R}\frac{\partial q_2^{(0)}}{\partial\theta} \\ R\frac{d\theta}{d\tau} &= -\frac{1}{R\cos\theta}\frac{\partial q_2^{(0)}}{\partial\phi} \end{aligned} \tag{5.4.30a,b}$$

where $\tau$ is a parameter measured along the characteristic curves from the eastern wall where $\tau = 0$. In terms of these characteristic curves:

$$\frac{dh^{(1)}}{d\tau} = \frac{f^2}{\left(h_2^{(0)}\right)^2}\frac{w_*}{\gamma_2} = \left(q_2^{(0)}\right)^2\frac{w_*}{\gamma_2} \tag{5.4.31}$$

Integrating (5.4.31) along each characteristic curve from its starting point on the eastern wall where $h^{(1)}$ vanishes yields:

$$h^{(1)} = \int_0^\tau \left(q_2^{(0)}\right)^2\frac{w_*}{\gamma_2}\,d\tau' \tag{5.4.32}$$

The integral in (5.4.32) can be rewritten in terms of an integration with respect to latitude $\theta$, noting that along each characteristic $\phi$ can be written as a function of $\theta$ since the equation for the characteristic is known, i.e., $q_2^{(0)}(\phi,\theta) = f_*/H_2$. Thus with the aid of (5.4.30b):

$$\begin{aligned} h^{(1)} &= \int_{\theta_*}^{\theta} \left(q_2^{(0)}\right)\frac{w_*}{\gamma_2}\frac{d\tau}{d\theta'}\,d\theta' \\ &= \int_{\theta_*}^{\theta} \left(q_2^{(0)}\right)^2\frac{w_*}{\gamma_2}\frac{R\,d\theta'}{\left\{-\dfrac{1}{R\cos\theta}\dfrac{\partial q_2^{(0)}}{\partial\theta}\right\}} \end{aligned} \tag{5.4.33}$$

If $q_2^{(0)}$ is written in terms of $h_1^{(0)}$ using (5.4.26) we obtain:

$$h^{(1)} = \int_{\theta}^{\theta_*} f \frac{w_*}{\gamma_2} h_1^{(0)} \frac{R^2 \cos\theta'}{\frac{1}{2}\frac{\partial}{\partial\phi}\left(h_1^{(0)}\right)^2} d\theta' \tag{5.4.34}$$

Finally, if (5.4.24) and the definition of $D_0^2$ is used, we obtain:

$$h^{(1)} = \int_f^{f_*} \Gamma_{12} \frac{w_*}{w_E} h_1^{(0)} \frac{df'}{f'} \tag{5.4.35}$$

where the relation $Rd\theta = df/\beta$ is used to change the variable of integration from latitude to Coriolis parameter. Since by (5.4.26):

$$h_1^{(0)} = H_2 - \frac{f}{q_2^{(0)}} \tag{5.4.36}$$

the integral, which is along a path of constant $q_2^{(0)}$ can be conveniently written as:

$$h^{(1)} = \int_f^{f_*} b\Gamma_{12} \left\{ H_2 - \frac{f'}{q_2^{(0)}} \right\} \frac{df'}{f'} = \int_f^{f_*} b\Gamma_{12} H_2 \left\{ 1 - \frac{f'}{f_*} \right\} \frac{df'}{f'} \tag{5.4.37}$$

The integration in (5.4.37) is along a path of constant $f_*$. If $b$ were a function of longitude, $\phi$ would have to be written as a function of $f$, i.e. $\theta$, along the path. If, however, $b$ is a function of $\theta$ alone, as in the cases discussed until now, the integral in (5.4.37) is independent of path and can be considered an integral in $f$ alone with $f_*$ held fixed during the integration. Thus, for example, in the simple case where $b$ is a constant:

$$h^{(1)} = b\Gamma_{12} H_2 \left[ ln\left(\frac{f_*}{f}\right) - \left(1 - \frac{f}{f_*}\right) \right] \tag{5.4.38}$$

Since $f_*$ is a function of longitude, $h^{(1)}$ is also a function of longitude. It follows from the general solution (5.4.37) that:

$$\frac{\partial h^{(1)}}{\partial\phi} = \frac{1}{f_*^2} \frac{\partial f_*}{\partial\phi} \int_f^{f_*} b\Gamma_{12} H_2 df' \tag{5.4.39}$$

where it is important to note that the derivative of the variable upper limit of the integral in (5.4.37) makes no contribution since the integrand is zero at $f = f_*$. Since:

$$\frac{\partial f_*}{\partial\phi} = H_2 \frac{\partial q_2^{(0)}}{\partial\phi} = H_2 \frac{f}{h_1^{(0)}\left[H_2 - h_1^{(0)}\right]} \frac{1}{2} \frac{\partial h_1^{(0)2}}{\partial\phi} = \frac{w_E}{\left[1 - \frac{f}{f_*}\right]} \frac{f_*^2 f}{\beta\gamma_1 H_2^2} R\cos\theta \tag{5.4.40}$$

we obtain:

$$v_2^{(1)} = \frac{\gamma_2}{f} \frac{1}{R\cos\theta} \frac{\partial h^{(1)}}{\partial \phi} = \frac{1}{\left(1 - \frac{f}{f_*}\right)} \frac{w_E}{\beta H_2} \int_f^{f_*} b\,df' \tag{5.4.41}$$

In the subtropical gyre $w_E < 0$, and therefore the sign of (5.4.41) depends on the sign of the cross-isopycnal velocity. If there is heating so that $w_* > 0$, then $b < 0$, and the integral is positive. Thus in the whole region of the adiabatic shadow zone, i.e., east of the line $h^{(0)} = H_2$, the fluid must move northward. This is the eastern branch of the buoyancy-driven recirculation, east of the critical characteristic and calculated numerically by Luyten and Stommel and discussed in Section 5.3.

The flow in this region impinges on the boundary of the shaded region given by the edge of the old shadow zone. Once it crosses this boundary, it must return southward in the shaded sliver in which our solution does not apply.

The flow cannot cross the western boundary of the shaded zone, for that boundary is a streamline of the ventilated flow when buoyancy is included. The situation is shown schematically in Fig. 5.4.3. The northward flow in the old shadow zone is slow and $O(b)$ compared to the wind-driven motion in the ventilated region. It flows northward, and its right boundary is the eastern wall on which $h = H_2$, so that $h^{(1)}$ is zero there. When it meets the eastern boundary of the shaded region, as shown by the streamlines, it must return southward. The western boundary of the shaded sliver also has $h = H_2$ so that the total depth across any section, i.e., $AA'$, resembles the cross section shown above the plan view. $h$ has a minimum at the point $P$ which is the intersection of the line $AA'$ with the eastern edge of the sliver. It then rises to the value $H_2$ on the western edge of the sliver. The entire geostrophic transport recirculates across the section, going northward to the east of the point $P$ in the old shadow zone and flowing southward to the west of the point $P$ within the sliver.

The northward transport in the eastern branch is $O(b)$ since the velocity is $O(b)$ and the width of the region is $O(1)$. The return southward flow takes place in a region of width $O(b)$ and therefore the southward velocity must be $O(1)$ to balance the transport. Since the velocities are $O(1)$, the weak heating is unable at $O(1)$ to affect the southward flow in the sliver. Thus in the sliver the southward flow conserves, to lowest order, potential vorticity so that $f/h_2 = \hat{Q}_2(h)$ in the southward flow. Thus the dynamics in this southward branch of the recirculation is dynamically similar to the ventilated flow in the sense that $q_2$ is preserved to lowest order for weak heating. However, the functional relation between potential vorticity and total depth, $h$, for this flow is determined along the eastern boundary of the sliver as the buoyancy-driven flow impinges on that boundary with known values of $h$ and $h_2$. In particular,

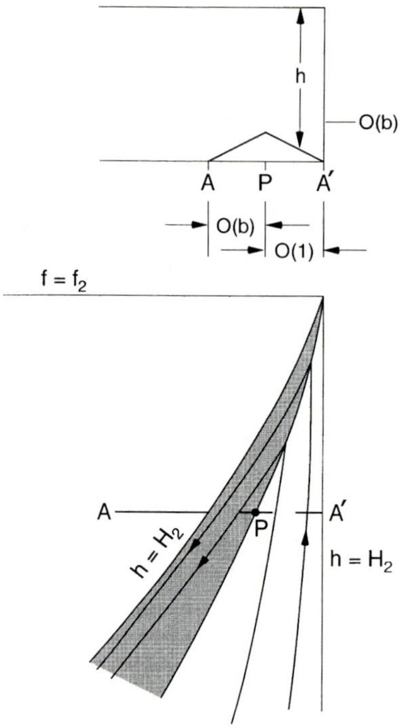


**Fig. 5.4.3.** Schematic presentation of the recirculation in the eastern unventilated zone. *Below*, a plan view of the circulation. Northward moving streamlines in the old, adiabatic shadow zone impinge on the narrow sliver opened up between the adiabatic shadow zone and the eastern edge of the ventilated region. In this region (*shaded*) a swift southward return flow takes place. *Above* cross section of the base of layer 2 across the line $AA$.' In the eastern region of $O(1)$ width a weak $O(b)$ flow is directed northward and is returned as a swift flow with $O(1)$ velocity in the narrow (*shaded*) sliver whose width is $O(b)$. The northward flow is a direct circulation, and the southward flow is indirect

the streamline which flows northward along the eastern boundary of the basin in the old shadow zone reaches the apex of the region at $f = f_2$ before entering the sliver. It enters the sliver, then, with potential vorticity $f_2/H_2$. This means that it flows along western boundary of the sliver in its return to the south since that curve is the line $q_2 = f_2/H_2$.

The buoyancy-driven circulation in the unventilated region is divided geographically and dynamically in two subregimes not immediately evident in the numerical solutions. The eastern portion of the flow consists of a direct circulation, i.e. northward motion, forced by the heating. The western limb of the recirculation consists of southward flow and is an indirect circulation. For small heating this flow conserves potential vorticity and is forced only indirectly by the heating. Its motion is determined by matching to the influx into the sliver of the northward flow driven by the heating. The potential vorticity of the return flow is determined by the influx while the flow itself is forced largely by the wind. That is, once the structure of the flow is determined by potential vorticity conservation, the Sverdrup relation determines the amplitude of the velocity of the return flow.

Another qualitative feature of the unventilated recirculation can be inferred from the solution (5.4.38) which obtains in the eastern portion of the buoyancy-driven recirculation when $b$ is constant. It also yields a qualitatively useful picture of the solution as long as $b$ does not vary rapidly over the region of the northward flow. The important qualitative feature of the solution is the fact that $h^{(1)}$ is a function only of $f/f_*$. That is, from (5.4.27), $h^{(1)}$ is a function only of $h_2^{(0)}$. This means that to order $b$, $h_2$ is a function only of $h$. Thus, the isolines of $h$ and $h_2$ coincide, and, as we have noted above, this is the signature of a region in which the circulation is thermodynamically direct, and in which the horizontal nonlinear advection of density is negligible. The vertical velocity and the cross-isopycnal velocity are in the same direction. This also implies that the nonlinear advective terms in (5.3.8), proportional to $\nabla h \times \nabla h_2$, are identically zero so that:

$$\begin{aligned} \beta v_2 h_2 &= f w_* \\ \beta v_1 h_1 &= f(w_E - w_*) \end{aligned} \qquad (5.4.42a,b)$$

The northward transports in each layer are then decoupled and given in terms of the Ekman pumping and cross-isopycnal velocities. The transport in layer 2 in this region of the old shadow zone is driven entirely by the buoyancy forcing and is decoupled from the wind-driven circulation. Layer 2, in this region acts as a one-layer model driven by $w_*$ rather than Ekman pumping. If $b$ is constant, the argument can be extended to values of $b$ of arbitrary magnitude (Pedlosky 1986).

It should be reiterated that the formal solutions discussed in this chapter do not specify the mechanism that gives rise to the cross-isopycnal flux. The flux is a flow across the quasipermanent large-scale density surfaces, and therefore even baroclinic instabilities, which transfer heat by eddy processes across the large-scale density gradients, could also be considered responsible for the flux. The inability to specify clearly the mechanism is a real weakness of buoyancy-driven theories. Although they provide much insight into the alterations that might be expected from the adiabatic theory of the circulation structure, the inability to specify $w_*$ with the clear connection to external forcing that is the hallmark of the Ekman pumping imbues the solutions with a very provisional, suggestive quality. In the absence of convincing parameterizations of the nonadiabatic processes below the mixed layer the solutions in the present chapter should be viewed as only the first step in investigating of the role of buoyancy forcing of the general circulation.

## References

Chen, L.G. and Dewar W.K. 1993: Intergyre communication in a three-layer model. *J. Phys. Ocean*, **23**, 855–878.

Dewar, W.K. 1987: Planetary shock waves. *J. Phys. Ocean.*, **17**, 470–482.

Korteweg, D.J. and de Vreis G. 1895: On the change in form of long waves advancing in a rectangular channel, and on a new type of long stationary waves. *Phil. Mag.* , **39**, 422–433.

Luyten, J. and Stommel H. 1986a: A beta-control of buoyancy-driven geostrophic flows. *Tellus*, **38A**, 88–91.

Luyten, J. and Stommel H. 1986b: Gyres driven by combined wind and buoyancy flux. *J.Phys. Ocean.*, **16**, 1551–1560.

Luyten, J. R., Stommel H. and Wunsch C. 1985: A diagnostic study of the Northern Atlantic subpolar gyre. *J. Phys. Ocean.*, **15**, 1344–1348.

McDowell, S., Rhines P. and Keffer T. 1982: North Atlantic potential vorticity and its relation to the general circulation. *J. Phys. Ocean.*, **12**, 1417–1436.

Pedlosky, J. 1984: Cross-gyre ventilation of the subtropical gyre: an internal mode in the ventilated thermocline. *J. Phys. Ocean.*, **14**, 1172–1178.

Pedlosky, J. 1986: The buoyancy and wind-driven ventilated thermocline. *J. Phys. Ocean.*, **16**, 1077–1087.

Rhines, P.B. and Schopp, R. 1991: The wind-driven circulation: quasi-geostrophic simulations and theory for nonsymmetric winds. *J. Phys. Ocean.*, **21**, 1438 –1467.

Rhines, P.B. and Young, W.R. 1982: A theory of the wind-driven circulation. I, Mid-ocean gyres. *J. Mar. Res.*, **40** (*Supplement*), 559–596.

Schopp, R. 1988: Spinup toward communication between large oceanic subpolar and subtropical gyres. *J. Phys. Ocean.*, **18**, 1241–1259.

Schopp, R. and Arhan, M. 1986: A ventilated mid depth circulation model for the eastern North Atlantic. *J. Phys. Ocean.*, **16**, 344–357.

# 6 Equatorial Dynamics of the Thermocline: The Equatorial Undercurrent

## 6.1 Introduction

The Coriolis parameter vanishes at the equator, and this distinguishes the dynamics of the equatorial zone from the oceanic dynamics at higher latitudes. Although the Coriolis acceleration due to the tangential component of the earth's rotation vector remains different from 0 at the equator, for the large scales of motion under consideration in this chapter its effect is negligible. Whereas the horizontal pressure gradient outside the tropical zone is balanced by the Coriolis acceleration, its absence at the equator leaves the pressure gradient free to produce large relative accelerations.

The geostrophic approximation is the dynamical starting point and foundation of all the theories of the ocean circulation discussed in the preceding chapters, and its breakdown at the equator signifies that the theoretical results previously obtained cannot be carried directly to the equator. The theories must be modified before the equatorial zone can be included in our overall picture of the oceanic circulation.

The modification occurs on the level of the vorticity balance as well as the momentum balance. The parameter which measures the size of the relative acceleration with respect to the Coriolis acceleration is the Rossby number, defined in (1.2.2) as:

$$R_0 = \frac{U}{fL} \tag{6.1.1}$$

where $L$ is the scale of the motion and $U$ that of the velocity, such that $U/L$ is a measure of the relative vorticity. Near the equator:

$$f = 2\Omega \sin\theta \approx 2\Omega\theta \approx \frac{2\Omega}{R}(R\theta) = \beta y \tag{6.1.2}$$

where $y = R\theta$ is the distance poleward from the equator and where $\beta$ is the equatorial value of the $\beta$ parameter, $2\Omega/R$.

Suppose a motion which straddles the equator has the meridional scale $L$. Then the characteristic value of $f$ for the motion would be $\beta L$. Thus, in the equatorial zone:

$$R_0 = \frac{U}{\beta L^2} \quad . \tag{6.1.3}$$

However, this parameter is, from (1.4.4), also the measure, $\varepsilon$, of the ratio of the relative vorticity *gradient* to the planetary vorticity *gradient*:

$$\varepsilon = \frac{U}{\beta L^2} \tag{6.1.4}$$

Therefore, we can expect that when the distance to the equator, $L$, becomes small enough so that the advection of relative momentum becomes important, i.e., when $R_0$ is $O(1)$, and $L$ is $O(\sqrt{U/\beta})$, this also corresponds to the emergence of the advection of relative vorticity becoming as dynamically significant as the advection of planetary vorticity. We thus expect to lose the geostrophic approximation to the momentum equation simultaneously with the breakdown of the Sverdrup balance. This implies that a full reformulation of the dynamics in the vicinity of the equator is required.

The aspect of the oceanic general circulation of the equatorial region that is perhaps the most striking is the existence of the Equatorial Undercurrent (EUC). Although discovered in the nineteenth century, the EUC was essentially forgotten and was rediscovered in modern times only in 1952. The reason for our initial forgetfulness and the continued ignorance of its existence for so long is related to its subsurface nature. The current is hidden from direct view by a surface current normally flowing westward. The undercurrent itself exists in all three major equatorial oceans as a thin ribbon of water with a width of about 200 km, at depths of the order of 100–200 m, flowing eastward against the prevailing trade winds with speeds of the order of 100 cms/s. These speeds are as large as those of any of the major western boundary currents, and the current which traverses the entire length of each of the Pacific, Atlantic, and Indian Oceans is one of the marvels of the general circulation of the ocean.

Figure 6.1.1 shows cross sections of the current's velocity, temperature, and salinity. The strong eastward velocity has its core at this longitude (155°W, in the Pacific) about 150 m below the surface and has a vertical thickness of about 100 m. The weaker surface velocity is to the *west*, i.e., in the direction of the prevailing wind. The isotherms, as seen in the cross section, become shallow as they approach the equator in the region outside the EUC, as we would expect from the theoretical models of the thermocline discussed in Chapter 4, but in the region of the current they stretch apart, and this bowed shape of the density surfaces is characteristic of the current wherever it is seen. The salinity field acts rather as a tracer of the motion, and the tongues in each hemisphere stretching to the equator suggest that subtropical water flows towards the equator at depth and enters the current.

The zonal velocity in Fig. 6.1.1 has been calculated using the geostrophic approximation. This seems a strange thing to do at the equator, but as we see below, geostrophy for the *zonal* component of the flow is an adequate approximation. This is underscored by a comparison of direct velocity mea-

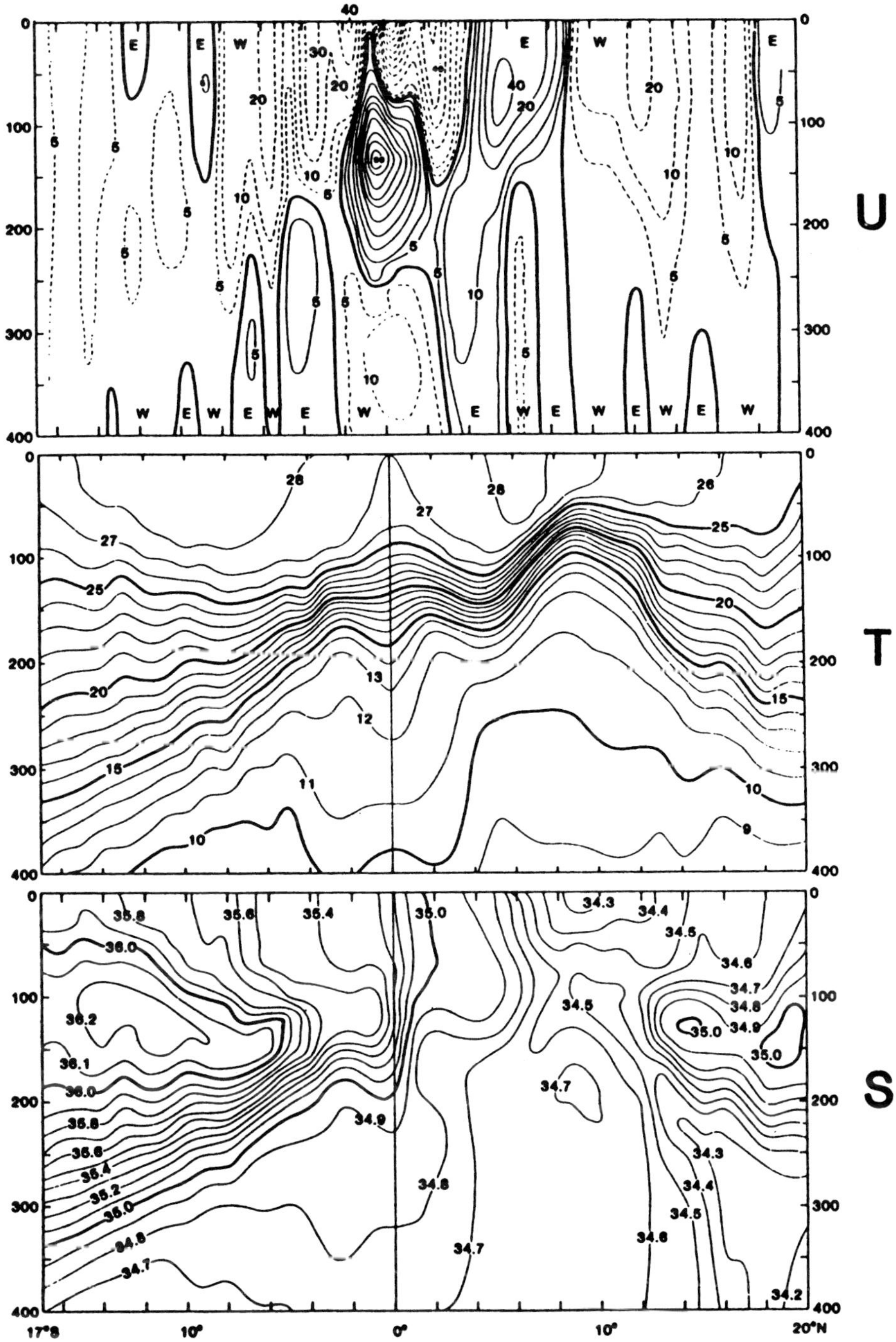


**Fig. 6.1.1.** Meridional cross sections of geostrophic zonal velocity, temperature and salinity at 155°W in the region between Hawaii and Tahiti. Note the bowing of the isotherms in the region of the EUC. The current is contained between the 15 and 25 °C isotherms. Note also the tongue of high-salinity water extending to the EUC, especially from the South Pacific. (From Wyrtki and Koblinsky 1984)

surements, shown in Fig. 6.1.2, with the geostrophic measurements of the preceding figure. The agreement is very good indeed.

The current structure in the equatorial Atlantic is very similar. Figure 6.1.3 shows the temperature and velocity and temperature profiles in the Atlantic at 28°W and the Pacific at 140°W. The similarity is striking.

Figures of latitude-depth cross sections emphasize the two-dimensional structure of the current in the meridional plane. The flow, however, is three-dimensional. The equatorial thermocline, just as the thermocline in mid-latitudes, slopes strongly upward to the east and the EUC, which is embedded

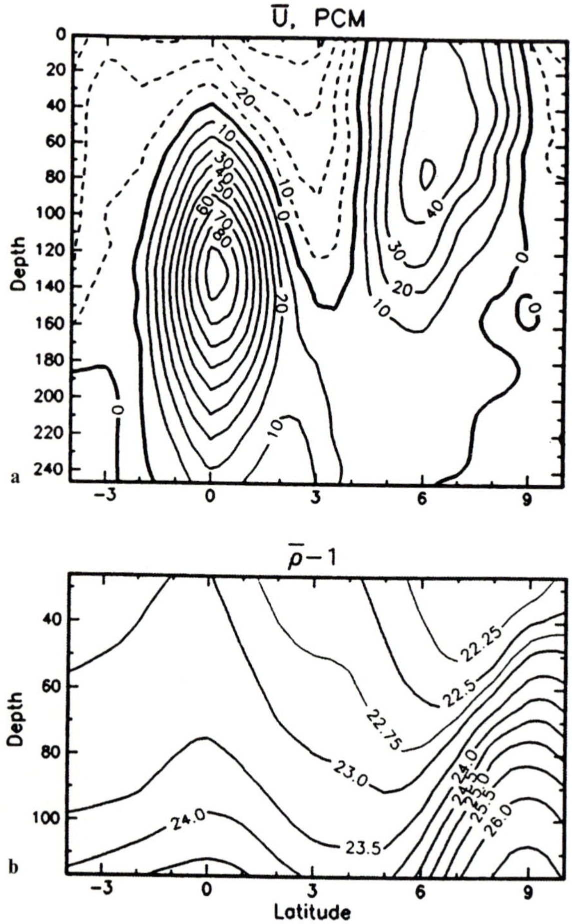


**Fig. 6.1.2.** **a** Contours of zonal velocity in the EUC measured directly with a profiling Doppler current meter in the Pacific at the same longitude as Fig. 6.1.1. Note the excellent agreement between the actual and geostrophic velocities. **b** The density field measured in the same section. Note that the meridional density gradient vanishes at the equator. (From Johnson and Luther 1994)

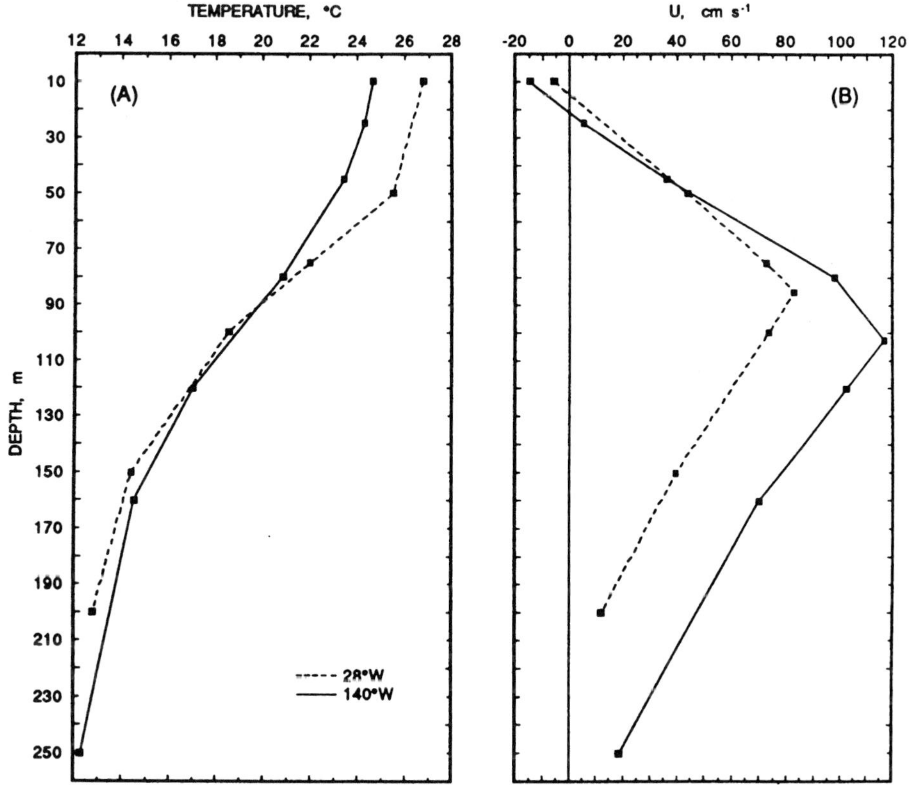


**Fig. 6.1.3.** Temperature and zonal velocity profiles from the Atlantic and Pacific oceans. In each case the measurements represent 2-year means. (From Halpern and Weisberg 1989)

in the equatorial thermocline, flows upwards with the isopycnals. Figure 6.1.4 shows the profile of the EUC at 150°W and further east at 110°W. Points on the profile of the same temperature are joined by lines illustrating both the rise of the isotherms towards the east but also show the rise of the current's core as it flows eastward. As we see below, the three-dimensional character of the current is a vital part of its dynamics. The upward slope of the isotherms is a basin-wide feature of the equatorial thermocline. Figure 6.1.5 presents the zonal sections along the equator in the Pacific and Atlantic, showing the sharp rise (when properly scaled) of the isotherms. A similar climatology of the 14 °C isotherm in the Pacific, shown in Fig. 6.1.6, whose surface forms the base of the EUC, exhibits the same variation. This rises from nearly 200 m in the western Pacific to less than 100 m in the eastern Pacific.

The narrowness of the current and its arrowlike progress across the ocean basin should not lull us into believing that the current exists in isolation from the rest of the ocean at higher latitudes. Indeed temperatures at the core of the EUC (see Fig. 6.1.3, 6.1.4) are low enough (≈18–20 °C) that the origin of the

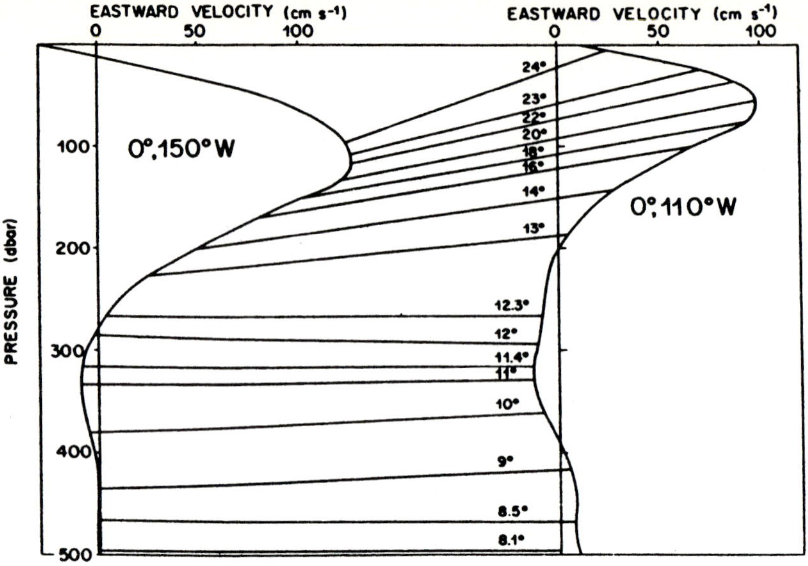


Fig. 6.1.4

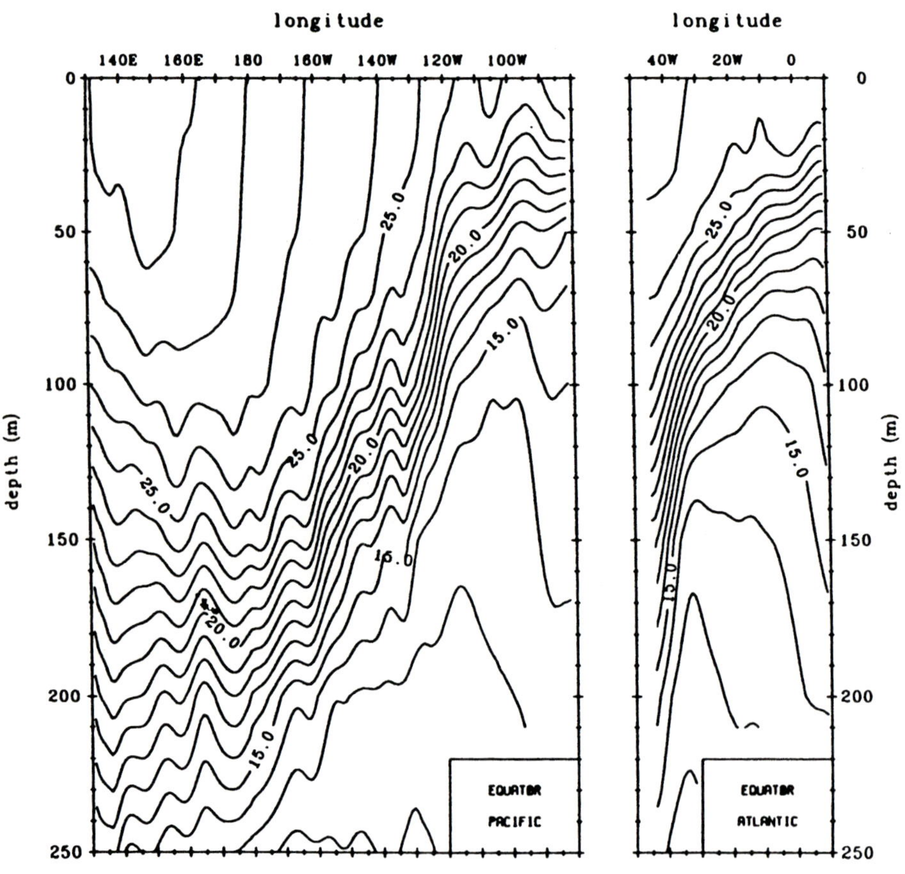


Fig. 6.1.5

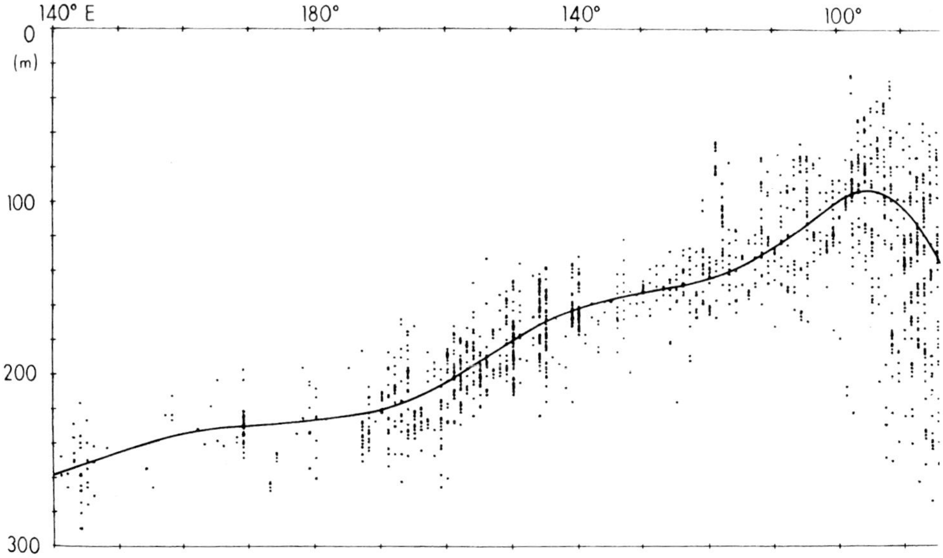


**Fig. 6.1.6.** Depth of the 14 °C isotherm constructed from bathythermograph records whose scatter is shown by *dots*. (From Meyers 1979)

water in the core is certainly water from the subtropical gyres. Careful analyses of the source waters using traditional water properties such as salinity or tracers such as tritium (e.g., Fine et al., 1981, Fine 1987; Tsuchiya et al. 1989,) in the Pacific are quite clear on this point. The implication is that the existence of the EUC and its structure is linked to the dynamics of the thermocline circulation at higher latitudes.

This provides conceptual symmetry to an investigation of the transformation of the dynamics of the subtropical gyre as the equator is approached. The extension of the midlatitude theories to the equator is required for us to complete the theory of the thermocline circulation. At the same time this extension, carrying the dynamics to the equator, provides the dynamic link between the equator and the midlatitudes which is necessary to explain the EUC.

Early theoretical models of the EUC, among the most distinguished being the studies of Charney (1960) and Charney and Spiegel (1971), concentrated on the equatorial band and treated the current as if it were fundamentally isolated from its surroundings. These models ignored the zonal variation of the current and were exquisitely and unrealistically sensitive to the value of the vertical mixing coefficient. Obtaining realistic profiles in these models requires the

◀

**Fig. 6.1.4.** Zonal velocity profiles at 150°W and 110°W at the equator with isotherms of potential temperature illustrating the upward slope of both the current and the isotherms along the equator. (From Bryden and Brady 1985)

**Fig. 6.1.5.** Equatorial sections of annual mean temperature of the Pacific (*left panel*) and Atlantic (*right panel*). Note the sharp rise to the east of the isotherms containing the EUC. (From Wacongne 1990)

presence of substantial dissipation throughout the water column. This, as we see below, is neither necessary theoretically nor appropriate observationally and is related to the isolation of the model current from its surroundings and its treatment as a fundamentally two-dimensional phenomenon.

The diagnostic observational study of Bryden and Brady (1985) is particularly illuminating in emphasizing the link between the equatorial and higher latitude ocean regimes. Bryden and Brady present a diagnostic model of the tropical Pacific based on a series of observations in the eastern Pacific at 110°W and 150°W. Noting that the equatorial zone is well known for being a region of strong upwelling, they stressed the three-dimensional character of the motion. In the past the upwelling at the equator had been pictured largely as a two-dimensional motion in the meridional plane in which poleward motion in each hemisphere in the upper mixed layer, driven by the westward Trade Winds, is balanced by equatorward geostrophic motion beneath the mixed layer. The circuit in this two-dimensional picture is closed at the equator by strong upwelling across the isothermal surfaces. This upwelling across the isothermal surfaces was considered responsible for the bowing of the isotherms observed in the depth-latitude cross sections. Bryden and Brady's analysis of the motion led them to suggest instead that the vertical motion at the equator is a more complex three-dimensional cell in which deep motion approaches the equator and then rises *along isopycnal surfaces* as it moves eastward in the EUC until, in the eastern part of the ocean, the cold water follows the isopycnals to enter the mixed layer, producing the eastern pools of cold water observed in each ocean basin at the equator. The departure of the streamlines from the sloping isopycnals is slight and they estimated the cross-isopycnal velocity near the core of the EUC to be only about one tenth of the vertical velocity itself. To a first approximation the fluid glides up the isopycnal surfaces, conserving density. The motion is fundamentally three-dimensional, and the zonal velocity of the EUC forms a limb of a strongly slanted upwelling cell at the equator.

The conservative nature of the flow has two important and related consequences. First, if density and temperature are nearly conserved, it implies that the explanation for the density structure of the equatorial thermocline must be determined, in analogy with the midlatitude thermocline, as an advective process in which the density at the equator is determined by the density of the fluid carried to the equator from the surrounding region. The current is then not an isolated phenomenon. Second, it implies that the *dynamics* is conservative, and in particular that a conservative theory of the EUC circulation is apt. Conservative dynamics means the fluid has a strong memory of its dynamical properties (such as potential vorticity), and the lasting memory produces a link between regions which are spatially separated.

Of course, the fact that density is conserved is suggestive but not sufficient to mean that the dynamics is completely conservative, for now the vorticity dynamics is not likely to simply be the inviscid Sverdrup vorticity equation. Vertical mixing of momentum and vorticity as well as momentum and vorticity advection might *a priori* be important at the equator. In this chapter we discuss theories which assume fundamentally that a purely conservative model of the dynamics is the appropriate, order one, model for the EUC dynamics, and we ignore vertical mixing of momentum as well as density at lowest order. It is interesting to note that recent measurements (Johnson and Luther 1994) do indicate that vertical mixing is probably *not* significant at the level of the core of the undercurrent.

We therefore take as a starting point the conservative model as a null hypothesis in which the ideal fluid theory of the thermocline, appropriately modified and extended to the equator, is suggested as the fundamental explanation of the EUC. In this view the link between the regimes of the midlatitude thermocline and the equatorial current system is fundamental in explaining equatorial motion.

It is important to emphasize that this linkage is suggested for the time mean, general circulation. The isolation of the equatorial zone on short time scales, over which the equatorial band acts as a wave guide for Kelvin and Rossby waves is well known (see Philander 1990; Moore and Philander 1977, for excellent reviews of equatorial wave dynamics). The trapping scale for low-frequency Rossby waves is shown in these reviews to be of the order of $c_K(k/\omega\beta)^{1/2}$, where $c_K$ is the Kelvin wave speed, and $k$ is the zonal wave number of the wave whose frequency is $\omega$. Note that for low frequencies the "trapping" becomes very feeble. On general circulation scales we are interested in $\omega \approx 0$, and in spite of the equator's effectiveness as a wave guide for higher frequency waves it thus is clear that for steady motions the linkage with higher latitudes is to be naturally expected.

## 6.2 Physics of the EUC: Preliminaries

The solution for the layer thicknesses in midlatitude for the ventilated thermocline, i.e., (4.4.18) in the two moving layer model, gives a prediction for each of the layer thicknesses which is well behaved as the equator is approached. This, of course, does not mean that the solution is correct. In fact, at the equator it is not.

For the case when the wind stress is purely zonal and independent of longitude the Ekman pumping is given by:

$$w_E = -\frac{\partial}{R\cos\theta\,\partial\theta}\left(\frac{\tau\cos\theta}{\rho_0 f}\right) \tag{6.2.1}$$

where $\tau$ is the zonal wind stress. It follows then that the function $D_0^2$ is given by:

$$D_0^2 = \frac{2R(\phi_e - \phi)}{\gamma_2\rho_0}\left[\sin\theta\frac{\partial\tau}{\partial\theta} \quad \frac{\tau}{\cos\theta}\right] \tag{6.2.2}$$

so that, from (4.4.18):

$$h^2 = \frac{2R(\phi_e - \phi)\{\sin\theta(\partial\tau/\partial\theta - \tau/\cos\theta)\}/\rho_0\gamma_2 + H_2^2}{1 + \Gamma_{12}(1 - f/f_2)^2} \tag{6.2.3}$$

where $\Gamma_{12} = \gamma_1/\gamma_2$. Note that $h$ is independent of the magnitude of the Earth's rotation!

In the limit where $\theta$ goes to zero the dominant term in (6.2.3) is proportional to the stress rather than its curl (whereas the opposite is generally true at midlatitudes) and:

$$h^2 \to \frac{2R(\phi_e - \phi)}{\gamma_2(1 + \Gamma_{12})}(-\tau/\rho_0) + \frac{H_2^2}{(1 + \Gamma_{12})}. \tag{6.2.4}$$

Since $\tau$ is negative in the equatorial region of the trade winds, the thermocline as predicted by the limit of the midlatitude theory deepens westward at a rate which depends on the value of the wind stress *near* the equator. The fact that $h$ is well behaved as $\theta$ goes to zero is a misleading criterion to judge the validity of the solution. If we use (6.2.4) to calculate the meridional velocity, $v_2$, we find that:

$$v_2 = \frac{\gamma_2}{fR\cos\theta}\frac{\partial h}{\partial\phi} = \frac{\tau}{h\rho_0 f(1 + \Gamma_{12})}. \tag{6.2.5}$$

The velocity thus becomes singular as the equator is approached, and this is a sign that the physics of the midlatitude model is inadequate to produce a valid solution at the equator.

If the meridional velocity in layer 1 is similarly calculated, and the transports of the two layers are added together, we obtain for the geostrophic meridional transport as $\theta \to 0$:

$$v_1 h_1 + v_2 h_2 = \frac{\tau}{\rho_0 f} \tag{6.2.6}$$

which is equal and opposite to the meridional Ekman layer transport. Thus the geostrophic transport carried to the equator balances to lowest order the Ekman transport at each latitude and longitude. The flux towards the equator in the region beneath the mixed layer is *not* related to the Sverdrup transport, which may indeed be northward in the equatorial regions, but is instead related to the stress and is equatorward as long as $\tau < 0$.

Although (6.2.4) is not valid on the equator, a valid equatorial solution must smoothly match it as the asymptotic limit of the equatorial solution. The flux of submixed layer flow to the equator at the subtropical flanks of the equatorial zone is determined by the midlatitude, ventilated thermocline solution. The connection between the equatorial and midlatitude thermocline circulations is manifested by these matching conditions.

One might explain the emergence of the EUC in the following way. Off the equator the wind blowing to the west produces an east-west pressure gradient as manifested by the westward deepening of the thermocline as in (6.2.4). The westward stress piles up water in the western part of the basin leading to higher pressure in the west. That is, it produces a pressure *force* to the east. Where the Coriolis acceleration is dominant, the response to the pressure gradient is a geostrophically balanced equatorward velocity. On the equator itself, where $f$ vanishes, there can be no such geostrophic balance, and the pressure gradient accelerates fluid down the gradient, producing a strong eastward current along the equator.

This conventional verbal explanation for the EUC is fundamentally incomplete, as the following counterexample demonstrates. Consider a model for the undercurrent which satisfies the content of the verbal argument just set forth. The model is shown schematically in Fig. 6.2.1. A homogeneous layer of fluid is driven by a wind stress at its upper surface. The wind stress is to the west and is balanced overall by a pressure force to the east. It is assumed for simplicity that the properties of the flow are independent of longitude, and that the velocity is only in the zonal direction. Thus, in the absence of longitudinal accelerations the balance of forces in the layer is, assuming a simple model for vertical mixing:

$$0 = -\frac{1}{\rho R \cos\theta} \frac{\partial p}{\partial \phi} + A_v \frac{\partial^2 u}{\partial z^2} \tag{6.2.7}$$

leading to the solution satisfying the no-slip condition on $z = 0$:

$$u = \frac{1}{\rho A_v R \cos\theta} \frac{\partial p}{\partial \phi} z \left( \frac{z}{2} - H \right) + \frac{\tau}{\rho A_v} z \tag{6.2.8}$$

Since the overall pressure force balances the applied wind stress over the width, $L$, of the basin:

$$\tau L = HL \frac{1}{R\cos\theta} \frac{\partial p}{\partial \phi} \tag{6.2.9}$$

so that:

$$u = \frac{\tau}{2\rho A_v} \frac{z^2}{H} \tag{6.2.10}$$

The flow in this simple solution is everywhere in the direction of the applied surface stress. For a westward wind stress the current would be everywhere moving to the west with no undercurrent to the east.

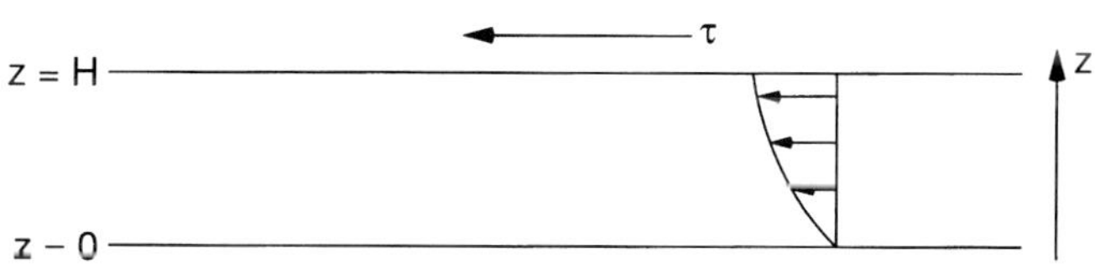


**Fig. 6.2.1.** Simple one-layer model driven by a westward wind stress. The flow is everywhere to the west

This solution is not presented as a realistic model of the equatorial zone, but it does contain all the elements of the verbal argument that has been proposed above as an "explanation" of the undercurrent. It fails to produce an undercurrent because at every level in $z$ the eastward pressure force is balanced by a frictional stress, and there is no eastward acceleration. The value of the solution in the current discussion is its demonstration that it is the vertical *distribution* of pressure that matters. In particular, what is required for the undercurrent is the penetration to depths below the surface of an *unbalanced* pressure gradient available to drive an eastward flow.

An "explanation" for the undercurrent can not logically start with an unbalanced pressure force as if it were an externally imposed force. It is part of the solution to be determined. Appealing to an unbalanced pressure force at the equator to explain the EUC would be no different than an "explanation" of the subtropical gyre which presents it as a response to a high pressure center in the ocean produced by the wind. The theory must, first of all, explain the distribution of the pressure field, and then the velocity field follows. Indeed, in midlatitudes the theory for the circulation is a theory for the height, or pressure fields. Such a theory in midlatitudes comes from uniting the geostrophic relation with potential vorticity dynamics. An analogous development must occur at the equator in order to derive the pressure field that is *consistent* with the undercurrent although at the equator the geostrophic approximation is no longer relevant.

An illuminating discussion of the way in which potential vorticity dynamics could explain the undercurrent was first presented by Fofonoff and Montgomery (1955). They suggested that columns of fluid in subsurface layers approach the equator to replace flow driven away from the equator in the surface mixed layer, and that the fluid approaching the equator conserves its potential vorticity. If a fluid column starts its journey to the equator at a latitude $\theta_0$ where relative vorticity is unimportant, for each fluid column:

$$\frac{f+\zeta}{h}=\frac{f_0}{h_0} \tag{6.2.11}$$

where the 0 subscript refers to variables evaluated at $\theta=\theta_0$. In the region of the undercurrent the zonal velocity is many times greater than the meridional velocity, and the latitude scale is very small compared to the zonal scale, so that momentarily using Cartesian coordinates:

$$\zeta=\frac{\partial v}{\partial x}-\frac{\partial u}{\partial y}\approx-\frac{\partial u}{\partial y}, \quad f=\beta y \tag{6.2.12}$$

and we will also define the parameter $y_0$ by the relation $f_0=\beta y_0$.

Although it is certainly not actually true for fluid columns approaching the equator in the ocean, Fofonoff and Montgomery examined the simple case in which the layer thickness is constant for a fluid column along its path to the equator so that $h=h_0$ Then (6.2.12) reduces to:

$$\frac{\partial u}{\partial y} = \beta(y - y_0) \tag{6.2.13}$$

The shear of the flow becomes negative as the column moves south of its initial position, and relative vorticity is produced as the planetary vorticity is diminished. Integrating (6.2.13) and insisting that u is also zero (i.e., negligible) at the starting latitude where $y = y_0$, yields:

$$u = \frac{\beta}{2}(y - y_0)^2 \tag{6.2.14}$$

which yields an eastward velocity at the equator where $y = 0$:

$$u(y = 0) = \frac{\beta y_0^2}{2}. \tag{6.2.15}$$

With $\beta = 2 \times 10^{-13}$ cm$^{-1}$s$^{-1}$, and choosing $y_0$ to be the width of the observed undercurrent, i.e. $y_0 = 200$ km, (6.2.15) yields as an estimate for $u(y = 0)$ of 40 cm/s, which is rather low. If $y_0$ were chosen to be 300 km, the equatorial speed would be 90 cm/s and closer to the values observed.

Although the solution of Fofonoff and Montgomery is valuable in illustrating how potential vorticity conservation produces an eastward current at the equator, the sensitivity to the starting value, $y_0$, is unacceptable. The starting latitude for the calculation is clearly arbitrary. What (6.2.15) really yields is a *relation* between the lateral scale and the equatorial speed. A complete theory must also yield the value for $y_0$ unconnected with the arbitrary choice of the starting latitude for the integration.

The form of the solution is also inadequate. If we consider the EUC as an inertial boundary current produced at the equator by the emergence of higher order physics due to the breakdown of the geostrophic balance, the profiles of fields such as the relative vorticity should smoothly join to the extra-equatorial fields. Instead, in the Fofonoff-Montgomery solution, shown in Fig. 6.2.2, the relative vorticity abruptly starts its linear increase at the arbitrary latitude $\theta_0$.

It is important to recognize the distinction between potential vorticity conservation and angular momentum conservation in the argument leading to (6.2.15). Angular momentum is not conserved since a zonal pressure gradient is implicitly required for the consistency of (6.2.14). This can be seen by using (6.2.13) in the zonal momentum equation. Thus the momentum equation:

$$u\frac{\partial u}{\partial x} + v\left(\frac{\partial u}{\partial y} - \beta y\right) = -\frac{1}{\rho}\frac{\partial p}{\partial x} \tag{6.2.16}$$

becomes:

$$v\beta y_0 = \frac{\partial}{\partial x}\left[\frac{p}{\rho} + \frac{1}{2}u^2\right]. \tag{6.2.17}$$

The southward flow feeding the undercurrent and producing it by carrying fluid elements which exchange planetary for relative vorticity is possible only if the total Bernoulli function, $p/\rho + u^2/2$ is a function of longitude. A ring of

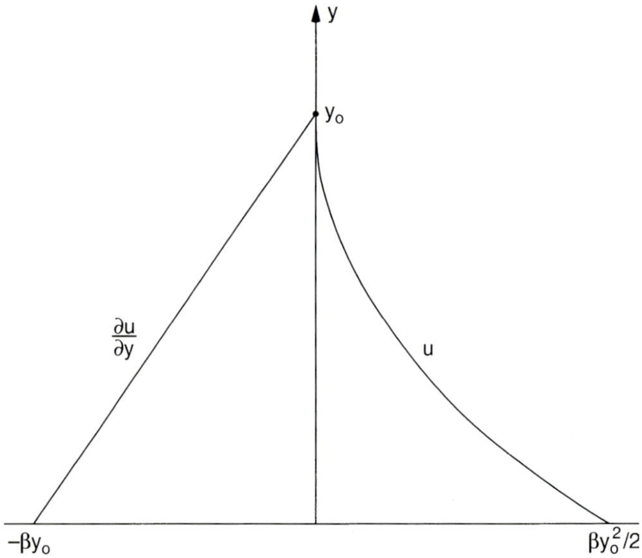


**Fig. 6.2.2.** Fofonoff-Montgomery (1955) solution for the EUC. The velocity and its shear are shown. Note that the shear increases linearly in magnitude from the starting latitude of the calculation where $y = y_o$. The shear reaches an algebraic minimum of $-\beta y_0$ at the equator where the speed is $y_0^2/2$

fluid that conserved angular momentum, i.e. in the absence of a zonal gradient of the Bernoulli function (as would be the case for axially symmetric motions) would arrive at the equator with a *westward* flow equal to $-\beta y_0^2/2$. It is potential vorticity conservation and not angular momentum conservation which is the key to explaining the undercurrent. A complete theory, in the process, must also describe the zonal variations of velocity and pressure consistent with the current.

## 6.3 An Inertial Theory of the Equatorial Undercurrent

### *Equations of Motion and Scaling*

As in the case of the midlatitude thermocline circulation we consider a model consisting of layers of fluid, each with constant density. For simplicity, a two moving layer model is considered, as shown in Fig. 6.3.1.

There is one useful simplification that can be made immediately. Near the equator the metric term, $\cos\theta$, in the equations of motion in spherical co-ordinates is very nearly constant, i.e.:

$$\cos\theta = 1 + O(\theta)^2 \tag{6.3.1}$$

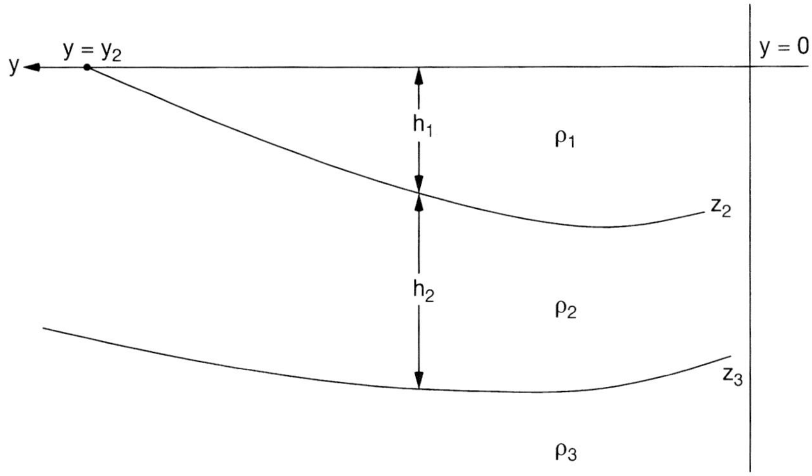


**Fig. 6.3.1.** Two moving layer model for the EUC. The model's outcrop line is at $y = y_2$ and the equator is at $y = 0$

so that writing:

$$x = R\phi \qquad y = R\theta \tag{6.3.2a,b}$$

allows meridional and zonal derivatives to be written:

$$\frac{1}{R\cos\theta}\frac{\partial}{\partial\phi} = \frac{\partial}{\partial x} + O(\theta)^2 \qquad \frac{1}{R}\frac{\partial}{\partial\theta} = \frac{\partial}{\partial y} \tag{6.3.3a,b}$$

where the error in (6.3.3.a) is of order $(\ell/R)^2$ and where $\ell$ is the order of the width of the equatorial current. That is, we can use Cartesian coordinates throughout in the equations of motion while replacing $f$ with $\beta y$ to the same order of accuracy, i.e., of order $(\ell/R)^2$.

Thus in layer $n$ the steady momentum equations can be written following the development leading to (4.2.5) as:

$$u_n\frac{\partial u_n}{\partial x} + v_n\frac{\partial u_n}{\partial y} - \beta y v_n = -\frac{1}{\rho_0}\frac{\quad_n}{\partial x} + F_n(x) + w_*(z_n)\frac{\{u_n - u_{n-1}\}}{h_n}\Theta(-w_*(z_n)) + w_*(z_{n+1})\frac{\{u_{n+1} - u_n\}}{h_n}\Theta(w_*(z_{n+1})) \tag{6.3.4a}$$

$$u_n \frac{\partial v_n}{\partial x} + \frac{\partial v_n}{\partial y} + \beta y u_n = -\frac{1}{\rho_0}\frac{\partial p_n}{\partial y} + F_n^{(y)} + w_*(z_n)\frac{\{v_n - v_{n-1}\}}{h_n}\Theta(-w_*(z_n)) + w_*(z_{n+1})\{v_{n+1} - v_n\}_n \Theta(w_*(z_{n+1})) \tag{6.3.4.b}$$

where $F_n^{(i)}$ is the frictional force in the $n$th layer in the ith direction.

The equation of mass conservation for the $n$th layer is:

$$\frac{\partial(h_n u_n)}{\partial x} + \frac{\partial(h_n v_n)}{\partial y} = -\left[w_*(Z_n) - w_*(Z_{n+1})\right] . \tag{6.3.5}$$

Again, $w_*(z_n)$ represents the mass flux leaving layer $n$ across its upper surface at $z = z_n(x, y)$.

For the two-layer model the dynamical part of the pressure gradient can be related to the layer thicknesses by the hydrostatic relation so that:

$$\frac{1}{\rho_0}\frac{\partial p_2}{\partial x} = \gamma_2 \frac{\partial h}{\partial x}$$
$$\frac{1}{\rho_0}\frac{\partial p_1}{\partial x} = \gamma_2 \frac{\partial h}{\partial x} + \gamma_1 \frac{\partial h_1}{\partial x} \tag{6.3.6a,b,c}$$
$$h = h_1 + h_2$$

The reduced gravity is defined by:

$$\gamma_n = \left(\frac{\rho_{n+1} - \rho_n}{\rho_0}\right) g . \tag{6.3.7}$$

To derive the extension of the dynamics at midlatitude to the equator it is helpful at this point to introduce scales for the variables and write the equations in nondimensional form.

Observations show the undercurrent as a thin ribbon of fluid flowing smoothly across each of the ocean basins. The scales of the current are thus vastly different in the zonal and meridional directions, and we must take that into account. Let $\ell$ be the length scale in the $y$ direction and $L$ be the length scale in the $x$ direction. The ratio $\ell/L \ll 1$. For $\ell = 200$ km and $L = 2000$ km the ratio is 0.1. We thus scale the independent variables as,

$$x = Lx'$$
$$y = \ell y' \tag{6.3.8a,b}$$

where primed variables are nondimensional.

The scale for the zonal velocity is chosen to be U and from the continuity equation we expect that the scale for the meridional velocity should be $U\ell/L$. Thus we introduce nondimensional velocities as:

$$u = Uu'$$
$$v = U\frac{\ell}{L}v'. \tag{6.3.9a,b}$$

The horizontal pressure gradient is of the same order as the Coriolis acceleration in the transition zone between the equator and midlatitudes. The Coriolis acceleration goes to zero right at the equator, but the level of the pressure is set by the forces in the transition region. The pressure gradient at the equator is balanced by relative accelerations, but the order of magnitude of the pressure itself at the equator remains unchanged even though its meridional gradient may increase. At a distance $\ell$ from the equator, for the pressure gradient to be of the same order as the Coriolis acceleration, the scale of the pressure, $P$, should be such that $P/\ell = O(\rho\beta\ell U)$ . Thus the appropriate nondimensionalization of the pressure is:

$$p = \rho_0 U\beta\ell^2 p'. \tag{6.3.10a}$$

Suppose that $H$ is the vertical scale of the motion. The layer thicknesses are then scaled as:

$$h = Hh',\ \ h_1 = Hh_1' \tag{6.3.10b}$$

The characteristic scale for the vertical velocity is $W = UH/L$, and it is this scale against which the cross-isopycnal velocity is measured.

Inserting these scalings into the momentum equations (6.3.4a,b) and *dropping the primes from the dimensionless variables*, we obtain:

$$\begin{aligned}\frac{U}{\beta\ell^2}\left\{u_n\frac{\partial u_n}{\partial x} + v_n\frac{\partial u_n}{\partial y}\right\} - yv_n = &-\frac{\partial p_n}{\partial x} + F_n^{(x)}\frac{L}{U\beta\ell^2}\\ &+\frac{w_*(z_n)}{W}\frac{U}{\beta\ell^2}\frac{\{u_n - u_{n-1}\}}{h_n}\Theta(-w_*(z_n))\\ &+\frac{w_*(z_{n+1})}{W}\frac{U}{\beta\ell^2}\frac{\{u_{n+1} - u_n\}}{h_n}\Theta(w_*(z_{n+1}))\end{aligned} \tag{6.3.11a}$$

$$\begin{aligned}\frac{U}{\beta\ell^2}\frac{\ell^2}{L^2}\left\{u_n\frac{\partial v_n}{\partial x} + v_n\frac{\partial v_n}{\partial y}\right\} + yu_n = &-\frac{\partial p_n}{\partial y} + F_n^{(y)}\frac{1}{U\beta\ell}\\ &+\frac{w_*(z_n)}{W}\frac{U}{\beta\ell^2}\frac{\ell^2}{L^2}\frac{\{v_n - v_{n-1}\}_n}{hn}\Theta(-w_*(z_n))\\ &+\frac{w_*(z_{n+1})}{W}\frac{U}{\beta\ell^2}\frac{\ell^2}{L^2}\frac{\{v_{n+1} - v_n\}}{h_n}\Theta(w_*(z_{n+1}))\end{aligned} \tag{6.3.11b}$$

All variables in (6.3.11a,b) are nondimensional *except for the cross-isopycnal velocities* which appear divided by the characteristic scale for the vertical velocity, $W$, and the *frictional forces* which are scaled by the estimate of the pressure force in each equation.

### *Scales for the Undercurrent*

At the equator the Coriolis acceleration vanishes, and by hypothesis the relative acceleration must enter to balance the pressure gradient. The largest relative acceleration (by a factor $(\ell/L)^2$) is the acceleration of the zonal velocity in (6.3.11a). For it to balance the pressure gradient we must choose the relationship between $U$ and $\ell$ such that:

$$U = \beta\ell^2. \tag{6.3.12}$$

This is precisely the relation obtained from the Fofonoff and Montgomery solution, and it arises here for the same reason, i.e., the need to balance relative acceleration with the Coriolis acceleration or, equivalently, to balance the relative vorticity against the planetary vorticity in the equatorial zone. Of course, right at the equator the Coriolis acceleration vanishes, but in the equatorial zone it is still present, although with reduced effect, and is of the same order as the relative accleration.

It is difficult to give a clear estimate of the size of the dissipative terms in the momentum equations. If, however, we regard $F_n^{(x)}$ as due to horizontal mixing of the zonal momentum we might estimate $F_n^{(x)}$ by:

$$F_n^{(x)} = A_H\left[\frac{\partial^2 u}{\partial y^2} + \frac{\partial^2 u}{\partial x^2}\right] \tag{6.3.13}$$

which would lead to an estimate of the last term in (6.3.11 a) as:

$$F_n^{(x)} = O\left(A_H \frac{U}{\ell^2}\frac{L}{U\beta\ell^2}\right) = \frac{A_H}{\beta\ell^3}\left(\frac{L}{\ell}\right) \tag{6.3.14}$$

We recognize the first ratio in (6.3.14) as ratio of the Munk boundary layer thickness to the EUC meridional scale, all to the third power. Unless the horizontal mixing is very large, the mixing term is negligible. Using the scales for $\ell$ and $L$ previously noted, $A_H$ would have to be larger than $1.6 \times 10^8 \text{cm}^2/\text{s}$ for horizontal mixing to enter the $x$ momentum equation. We can comfortably assume that $A_H$ is not this unrealistically large and therefore assume that horizontal mixing can be neglected at least to the lowest order. A similar analysis can be applied to the frictional term with regard to vertical mixing. If the contribution of vertical mixing to the frictional force in the zonal direction can be represented by:

$$F_n^{(x)} = A_v\frac{\partial^2 u}{\partial z^2} = O\left(A_v\frac{U}{H^2}\right) \tag{6.3.15}$$

where H is the vertical scale of the motion, the ratio occurring in the last term of (6.3.11a) would be of the order:

$$F_n^{(x)} \frac{L}{\beta \ell^3 U} = O\left(A_v \frac{L}{\beta \ell^2 H^2}\right) \tag{6.3.16}$$

For vertical length scales of the order of 100 m this would require a vertical mixing coefficient of the order of 400 $cm^2s^{-1}$, which is much larger than any present estimates at the level of the EUC core. Johnson and Luther (1994) report estimates of $A_v$ between 1 $cm^2s^{-1}$ at the EUC core to values of the order of 50 $cm^2s^{-1}$ near the surface.

Thus ignoring the second term on the right side of (6.3.11a) gives us, as the equatorial approximation to the zonal momentum equation:

$$\begin{aligned}\left\{u_n \frac{\partial u_n}{\partial x} + v_n \frac{\partial u_n}{\partial y}\right\} - y v_n = &- \frac{\partial p_n}{\partial x} \\ &+ \frac{w_*(z_n)}{W} \frac{\{u_n - u_{n-1}\}}{h_n} \Theta(-w_*(z_n)) \\ &+ \frac{w_*(z_{n+1})}{W} \frac{\{u_{n+1} - u_n\}}{h_n} \Theta(w_*(z_{n+1}))\end{aligned} \tag{6.3.17}$$

Note again that this balance occurs since $U = \beta \ell^2$.

The mixing of momentum by cross-isopycnal fluxes is still explicitly present in (6.3.17). Its relative importance is measured by the ratio of $w_*$ to $W = UH/L$. We examine below the case in which this ratio is negligible, as suggested by the diagnostic study of Bryden and Brady, but the effect of the cross-isopycnal momentum mixing is retained for now as part of the general formulation of the equatorial equations.

Even though the variations in the zonal direction are slow compared to the variations in the meridional direction, the zonal advection is as important as the meridional advection in the momentum balance. The weak zonal variation of $u_n$, when multiplied by the large zonal velocity, is of the same order as the strong meridional variation of $u_n$, multiplied by the weak meridional velocity. Thus a two-dimensional theory, in which the zonal variation is ignored, is basically inconsistent with the natural scaling for the equatorial current system.

If we examine the momentum equation for the meridional direction (6.3.11b), we note that the choice $U = \beta \ell^2$, which is imposed by the requirement that the advection of zonal momentum enter to repair the singularity in $v_n$ that would otherwise occur at the equator, implies that the advection terms in (6.3.11b) are of the order $(\ell/L)^2$ and hence are negligible compared to the meridional pressure gradient. This includes the momentum advection by the cross-isopycnal velocity. Similarly, when proper attention is paid to the fact that the frictional term, $F_n^{(y)}$ is proportional to $v_n$ instead of $u_n$, it follows that this dissipative term is of the order $(\ell/L)(\delta_M/\ell)^3$ compared to the pressure gradient. Thus only if the mixing is large enough that the Munk layer thickness

$\delta_M = (A_H/\beta)^{1/3}$ is very much greater than the width of the undercurrent would the mixing of momentum enter the meridional momentum balance. Therefore we obtain the rather surprising result that the *zonal velocity remains in geostrophic balance* in the equatorial region for the EUC.

Although the Coriolis parameter becomes very small as the equator is approached, $v_n$ is so small in comparison to $u_n$ due to the latitudinal narrowness of the current, that the meridional advection is simply too feeble to upset the geostrophic balance for the zonal velocity, even at the equator. It is important to note that it is the smallness of the ratio $(\ell/L)^2$ that preserves geostrophy for the zonal velocity at the equator. The flow is therefore *semi-*geostrophic, i.e., it has one component of the flow in geostrophic balance. This is the swift zonal component, in the direction along which the current is extended, which remains in geostrophic balance. Thus to $O[(\ell/L)^2]$ (6.3.11b) reduces to:

$$yu_n = -\frac{\partial p_n}{\partial y}. \tag{6.3.18}$$

This implies that on the equator where $y$ vanishes the meridional pressure gradient also vanishes. Reference to Fig. 6.1.2b shows that the density field very nearly has a zero gradient in the meridional direction at the equator, which is consistent with the hydrostatic balance and the geostrophic balance implied by (6.3.18) .

The hydrostatic relation relates the scales of the pressure and the depth. Using the relation (6.3.6) and the scaling for $p_n$ and $h$ given by (6.3.10):

$$H = \frac{\beta\ell^2 U}{\gamma_2} = \frac{\beta^2\ell^4}{\gamma_2}. \tag{6.3.19}$$

The scaling relations (6.3.12) and (6.3.19) are two equations for the three scales $U$, $\ell$, and $H$.

However, an additional and very important condition is that the layer thicknesses must match those of the midlatitude solution as the equatorial zone joins to the midlatitude domain. This matching condition is given by (6.2.4) which determines the scaling for the depth as:

$$H^2 = \frac{\tau_0 L}{\rho_0\gamma_2}. \tag{6.3.20}$$

where $\tau_0$ is the scale of the wind stress in the *matching* region, off the equator, where the equatorial physics blends smoothly to the midlatitude physics of the ventilated thermocline. The matching to the midlatitude circulation determines the scales and structure of the equatorial flow.

With $H$ determined in terms of external parameters, both $\ell$ and $U$ follow from (6.3.19), i.e.:

$$\ell = \left( \frac{\gamma_2 \tau_0 L}{\rho_0 \beta^4} \right)^{1/8}$$

$$H = \left( \frac{\tau_0 L}{\gamma_2 \rho_0} \right)^{1/2} \qquad (6.3.21a,b,c)$$

$$U = \left( \frac{\gamma_2 \tau_0 L}{\rho_0} \right)^{1/4}$$

The scales for $H$ and $U$ are independent of the rotation or $\beta$ effect. It is only $\ell$ that depends on rotation. The scaling in (6.3.21) is a purely inertial scaling, and there are no free, internal parameters. For values of $\tau_0 = 1\,\text{dyn/cm}^2$, $L = 3000$ km, $\gamma_2 = 1\,\text{cm/s}^2$, and $\beta = 2 \times 10^{13}\,\text{cm}^{-1}\text{s}^{-1}$, we obtain $H = 170$ m, $U = 130$ cm/s, and $\ell = 250$ km. The scales are encouragingly appropriate for the undercurrent.

The inertial scaling for the zonal velocity is remarkably insensitive to the magnitude of the wind stress. Since it goes as the 1/4 power of the stress, a reduction in the stress by 1/2 reduces the scale for the velocity by only about 15% and the width of the current by less than 10%. This differs greatly from earlier linear theories of the EUC (e.g., McCreary 1981) in which the strength of the EUC is directly proportional to the wind stress on the equator, and the width of the current is essentially independent of the strength of the stress and instead is proportional to the equatorial deformation radius.

The wind stress must be negative at the equator for the matching condition (6.2.4) to be applicable. Should the westward wind stress be shut off in the matching region, we would anticipate the collapse of the EUC.

## 6.4 The Nondissipative Model

The diagnostic study of Bryden and Brady (1985) of the equatorial circulation at 110°W and 150°W in the Pacific suggests that in the core of the EUC the cross-isopycnal velocity is small with respect to the vertical velocity, i.e., that the ratio $w_*/W$ in (6.3.17) is small. Between the 19 °C and the 20 °C isotherm surfaces, which make a sandwich of the core of the EUC at these latitudes, Bryden and Brady estimate the ratio to be in the range 0.05–0.12. Higher in the water column, well above the core, where the top of the current is bled off and enters the mixed layer, the ratio rises to approximately 0.34. For the core of the EUC and the bulk of the current it therefore seems reasonable to ignore dissipation altogether as a first approximation.

If $w_*$ is ignored in the zonal momentum equation (6.3.17), it may be rewritten:

$$(y+\zeta_n)v_n = -\frac{\partial B_n}{\partial x} \tag{6.4.1}$$

where:

$$\zeta_n = -\frac{\partial u_n}{\partial y}, \quad B_n = p_n + \frac{1}{2}{u_n}^2. \tag{6.4.2a,b}$$

The relative vorticity, $\zeta_n$, is given entirely in terms of the shear in the $y$ direction of the zonal velocity component. The other contributor, which is the shear of the meridional velocity in the $x$ direction is smaller by an amount $(\ell/L)^2$. Similarly, only the kinetic energy associated with the *zonal* motion enters the definition of the Bernoulli function $B_n$. The contribution of the meridional velocity is negligible by the same ratio, $(\ell/L)^2$.

In the absence of cross-isopycnal motion (6.3.5) shows that the horizontal mass flux is nondivergent. Thus, expressed in nondimensional units:

$$h_n\vec{u}_n = \hat{k} \times \nabla\psi_n. \tag{6.4.3}$$

Thus (6.4.1) becomes:

$$q_n\frac{\partial\psi_n}{\partial x} = \frac{\partial B_n}{\partial x} \tag{6.4.4}$$

where the potential vorticity, $q_n$, is given by:

$$q_n = \frac{y - \partial u_n/\partial y}{h_n}. \tag{6.4.5}$$

The geostrophic relation (6.3.18) can as well be written in terms of $B_n$ and $q_n$ as:

$$q_n\frac{\partial\psi_n}{\partial y} = \frac{\partial B_n}{\partial y} \tag{6.4.6}$$

which leads to:

$$\vec{u}_n \cdot \nabla q_n = 0 \tag{6.4.7}$$

obtained by eliminating $B_n$ by cross-differentiation between (6.4.4) and (6.4.6).

Multiplying (6.4.4) by $u_n$ and (6.4.6) by $v_n$ and adding we obtain, with the aid of (6.4.3):

$$\vec{u}_n \cdot \nabla B_n = 0 \tag{6.4.8}$$

so that $B_n$, as $q_n$, is conserved along streamlines or:

$$q_n = q_n(\psi_n), \quad B_n = B_n(\psi_n) \tag{6.4.9a,b}$$

Thus, using (6.4.4) and (6.4.6) it follows that:

$$\frac{dB_n}{d\psi_n} = q_n \tag{6.4.10}$$

and eliminating $\psi_n$ between (6.4.9a,b) yields:

$$q_n = Q_n(B_n) \tag{6.4.11}$$

where $Q_n$ is an arbitrary function.
These relations hold only, we recall, in those layers in which the effects of cross-isopycnal mixing can be ignored.

Of the two layers in motion it is the lower layer, layer 2, not exposed to the direct action of the wind stress or Ekman pumping, in which the potential vorticity and the Bernoulli function is conserved. We focus attention on this layer as that in which the core of the EUC develops.

From the hydrostatic relation (6.3.6a,b) we have in scaled units for the nondimensional pressure and layer thicknesses the relation, after using the scaling relations (6.3.10a,b) and (6.3.19):

$$p_2 = h, \quad p_1 = h + \Gamma_{12} h_1 \tag{6.4.12a,b}$$

where again, $\Gamma_{12} = \gamma_1/\gamma_2$

The statement of conservation of potential vorticity (6.4.11) implies that:

$$\frac{[y - \partial u_2/\partial y]}{h_2} = Q_2\left(h + \frac{1}{2}u_2{}^2\right) \tag{6.4.13a}$$

while geostrophy implies that:

$$\frac{\partial h}{\partial y} = -y u_2. \tag{6.4.13b}$$

In these two equations neither $v_2$ nor $x$ appears. The variables $h$, $h_2$, and $u_2$ are linked by two *ordinary* differential equations in the latitude variable, $y$. These two equations form the heart of the determination of the undercurrent structure, but there are three issues that arise. One of them is that the set (6.4.13a,b) consists of two equations in the three unknowns, $u_2$, $h_2$, and $h$. In midlatitudes the equation for potential vorticity conservation is supplemented by the Sverdrup balance equation relating the layer thicknesses to the Ekman pumping, and this closes the system of equations. The Sverdrup condition is obviously not valid at the equator since it is based on the geostrophic balance for the meridional velocity, and this fails at the equator. The second issue is determination of the function $Q_2$, which is a function of $B_2$ and must be found by matching to the midlatitude thermocline flow. Finally, the dependence only on latitude of (6.4.13a,b) although a technical simplification, avoids the issue of the connection and continuity of the solution from one longitude to another.

The first issue, the relation between $h$ and $h_1$, requires a discussion of the dynamics of the upper layer. In midlatitudes it is natural to assume that although the layer does not preserve potential vorticity, it is in geostrophic balance with the direct effect of the wind stress trapped in the surface Ekman layer. At the equator the Ekman depth $(A_v/f)^{1/2}$ formally grows infinitely large, which means simply that the Coriolis effect is unable to limit the depth of penetration of turbulent mixing of the vorticity put in by the wind stress at the surface. We anticipate that a considerable part of the upper layer is directly affected by turbulent mixing. This is consistent with the diagnoses both of Bryden and Brady (1985) of and Johnson and Luther (1994) which show the effects of vertical mixing penetrate to at least 50 m in depth. This renders impossible any simple, invisicid model for the upper layer dynamics.

In an effort to overcome this difficulty while still retaining the essential simplicity of the inertial model for the undercurrent, Pedlosky (1987) suggested simply avoiding the issue altogether and as an expedient chose $h_1$ at each longitude to be just the value at that longitude of the equatorial limit of the ventilated thermocline solution for $h_1$. That is, within the equatorial region Pedlosky took $h_1$ to be a function only of $x$, and the function was chosen to be the value of $h_1$ determined by the midlatitude solution at an asymptotic matching latitude $y_n$, which is large compared to unity (or $\ell$ in dimensional units). Thus $h_1$ is fixed as

$$h_1(x, y) = h_1(x, y_n) \tag{6.4.14}$$

where $h_1(x, y_n)$ is the limit of the ventilated thermocline solution at $y = y_n \gg 1$. If $y_2$ is the nondimensional position of the outcrop line this yields:

$$h_1(x, y_n) = [1 - y_n/y_2]\frac{\left[-2(x - x_e)\tau + H_2^2\right]^{1/2}}{\left[1 + \Gamma_{12}(1 - y_n/y_2)^2\right]^{1/2}} \tag{6.4.15}$$

In (6.4.15) the stress is nondimensional and has been scaled by $\tau_0$. The assumption has also been made, for simplicity, that the stress is strictly zonal and independent of $x$. The depth $H_2$ is the thickness of the second layer at the eastern boundary scaled by the depth $H$ (6.3.21b) and the nondimensional distance $y_2 = f_2/\beta\ell$ where $f_2$ is the Coriolis parameter at the outcrop line. Thus:

$$h_2(x, y) = h - h_1(x, y_n) \tag{6.4.16}$$

in the equatorial zone, and this assumption reduces the system (6.4.13a,b) to two dependent variables. Note that in this formulation the *geostrophic* zonal flow in layer 1 would be the same as in layer 2 since the absence of a tilt of $h_1$ with latitude would eliminate the zonal thermal wind. This does not mean the zonal velocity in layer 1 must equal that in layer 2 since a westward wind driven drift current in layer 1 also exists.

An alternative approach was later suggested by Pedlosky (1991b). The zonal velocity in the upper layer is observed to be considerably smaller than the velocity in the undercurrent core. This implies that the Coriolis acceleration in the upper layer is much less than that of the core. On the other hand, the scaling for the pressure gradient for the upper layer is the same as for the lower layer and in the absence of any more refined consideration this would lead to a pressure gradient in the upper layer much larger than the Coriolis acceleration there. This implies an unbalanced pressure force at lowest order. Thus for a balance at lowest order to occur the scaled pressure gradient in the upper layer satisfies the inequality:

$$\frac{\partial p_1}{\partial y} = O\left(\frac{U_1}{U_2}\right) \ll 1. \tag{6.4.17}$$

If this were the case, then to lowest the order (6.4.12) would imply that:

$$\frac{\partial}{\partial y}\{h + \Gamma_{12}h_1\} \approx 0 \tag{6.4.18a}$$

or when integrated, and matched to the midlatitude solution at $y = y_n$:

$$h + \Gamma_{12}h_1 = h(x, y_n) + \Gamma_{12}h_1(x, y_n). \tag{6.4.18b}$$

Thus:

$$h_1(x, y) = h_1(x, y_n) + [h(x, y_n) - h(x, y)]/\Gamma_{12}. \tag{6.4.19}$$

The two attempts at specifying $h_1$ in terms of $h$ are very crude. The first (6.4.14) allows no deformation of the interface between the two layers and therefore no shear in the geostrophic zonal velocity. The second (6.4.19) establishes the other extreme. In (6.4.19) the interface between the two moving layers moves so much that there is complete compensation of the pressure in the upper layer so that its geostrophic velocity is zero, and the zonal flow is due entirely to the west wind drift at the equator forced by the wind stress. If the stratification between the first two layers is large, i.e., in the limit of very large $\Gamma_{12}$ the two specifications of $h_1$ become equivalent. This equivalence does not imply that either of the specifications is correct, and the incomplete character of the treatment of the upper layer remains a serious weakness of the simple inertial model of the EUC. However, calculations described below show very little qualitative difference between the solution for the undercurrent in layer 2 between the two choices, and the structure of the current seems insensitive to the details of the specification of $h_1$. Attempts to deal with the dynamics of layer 1 in a more deductive are taken up in Section 6.6.

In either case, once $h_1$ is related to $h$, $h_2$ is also determined in terms of $h$, and (6.4.13a,b) becomes a set of two equations in two unknowns *if* the function $Q_2(B_2)$ is known. The determination of the functional relation between potential vorticity and Bernoulli function becomes the central dynamical issue.

The function $B_2$ is constant on streamlines, and therefore the specification of $Q_2(B_2)$ merely identifies the relation of the potential vorticity with its streamline. For flow in the ventilated portion of the subtropical gyre the potential vorticity is set at the time of subduction. If the flow is conservative that value of potential vorticity is maintained as the flow enters the domain of the equatorial flow. The potential vorticity becomes partitioned differently between its parts near the equator as the contribution of the planetary vorticity becomes weaker, and the relative vorticity becomes important as the streamline approaches the equator. Similarly, since the pressure field is scaled with the Coriolis parameter, its contribution to the Bernoulli function becomes weaker in the equatorial region while the kinetic energy, comparatively negligible at higher latitudes, becomes significant at the equator. Each streamline remains a line of constant Bernoulli function in midlatitude as well as in the equatorial region. It is the relative contribution of the pressure and kinetic energy which alters along the streamline as the equator is approached. Although both the potential vorticity and Bernoulli function change from their simple approximate forms in midlatitude to the more complicated representations that are valid at the equator, the *relation* between the potential vorticity and Bernoulli function is retained if the flow is conservative. Once the relation is fixed in midlatitudes, it is maintained as long as the dynamics is conservative, even in the region of semi-geostrophic equatorial dynamics.

Thus for large (nondimensional) $y$, where the solution merges smoothly with the subtropical interior where relative vorticity is negligible, and the Bernoulli functions is given by the pressure field:

$$q_2 = \frac{y+\zeta}{h_2} \approx y_2, \qquad B_2 = h + \frac{1}{2}u_2^2 \approx h, \qquad y \gg 1. \tag{6.4.20}$$

Thus the general statement of potential vorticity conservation, valid in both midlatitudes and in the equatorial zone, $q_2 = Q_2(B_2)$, becomes in midlatitudes:

$$q_2 = Q_2(B_2) = Q_2(h). \tag{6.4.21}$$

However the functional form of $Q_2$ is already known from the discussion of subduction in Sect. 4.4 where the potential vorticity of the subducted fluid is determined by (4.4.14). In *dimensional units* the function $Q_2$ is given by:

$$Q_2(h) = \frac{f_2}{h}. \tag{6.4.22}$$

In our nondimensional units this becomes:

$$Q_2(h) = \frac{y_2}{h} \tag{6.4.23}$$

where $y_2 = f_2/\beta\ell$.

Thus the functional relation of $Q_2$ with respect to its argument is determined and set, once and for all, by the ventilation and subduction process at the outcrop line of the layer 2, and this relation is carried to the equatorial region. It is only the argument of $Q_2$ itself which changes its form. However, all along the streamline the functional relation between $q_2$ and $B_2$ (6.4.23) must apply and therefore:

$$Q_2(B_2) = \frac{y_2}{B_2}. \tag{6.4.24}$$

Thus in the equatorial region (6.4.13a) becomes:

$$q_2 = \frac{y - \partial u_2/\partial y}{h_2} = Q_2(B_2) = \frac{y_2}{B_2} = \frac{y_2}{h + \frac{1}{2}u_2^2} \tag{6.4.25}$$

Solving for $\partial u_2/\partial y$ and using geostrophy (6.4.13b) gives us the system of equations:

$$\frac{\partial u_2}{\partial y} = y - \frac{y_2 h_2}{h + \frac{1}{2}u_2^2}$$
$$\frac{\partial h}{\partial y} = -yu_2 \qquad (6.4.26\text{a,b,c})$$
$$h_2 = h - h_1$$

where $h_1$ is given by either (6.4.16) or (6.4.19).

The system is a second-order system in $y$ and two boundary conditions are required for its solution. The first condition has already been discussed, namely, that for large values of $y$ $h$ should approach the value determined by the midlatitude solution, as given by (6.2.3). In our dimensionless units this is equivalent to the condition:

$$h \to h_n - \frac{\{-2\int_x^{x_e} \tau dx' + H_2^2\}^{1/2}}{\{1 + \Gamma_{12}(1 - y/y_2)^2\}^{1/2}}. \qquad (6.4.27)$$

In (6.4.27) $y$, which is nondimensional and scaled with $\ell$, is large with respect to unity. For the same reason $y_2$ is also large in scaled units. The ratio $y/y_2$, however, may be small in the matching region between the subtropical and equatorial solution if the layer outcrops far from the equator. Note again that $H_2$ in (6.4.27) is the nondimensional value of $H_2$, i.e., equal to the dimensional layer thickness on the eastern wall, divided by the scaling thickness $H$ as given by (6.3.21b).

The second boundary condition that is required comes from the motion of the fluid at the equator. Fluid in the northern hemisphere always starts its journey to the equator with positive potential vorticity since in mid latitudes the sign of the large scale potential vorticity is determined only by the sign of the Coriolis parameter. Fluid approaching the equator and joining the undercurrent from the southern hemisphere has negative potential vorticity. Fluid which might cross the equator and conserve its potential vorticity and finally exit the undercurrent in the other hemisphere would possess potential vorticity of the "wrong" sign. We take as a reasonable condition on the nondissipative solution that fluid not cross the equator and thus remain in the hemisphere in which its potential vorticity has the same sign as the fluid in which it is embedded. Thus:

$$v_2 = 0 \;, \qquad y = 0. \qquad (6.4.28)$$

This physical condition must be re expressed in terms of the variables of the system (6.4.26).

From (6.4.1) and (6.4.6):

$$(v_2 h_2)q_2 = \frac{\partial B_2}{\partial x}, \qquad (u_2 h_2)q_2 = -\frac{\partial B_2}{\partial y} \qquad (6.4.29\text{a,b})$$

From (6.4.24) this implies that:

$$v_2 h_2 = \frac{\partial \psi_2}{\partial x} = \frac{1}{q_2}\frac{\partial B_2}{\partial x} = \frac{B_2}{y_2}\frac{\partial B_2}{\partial x} = \frac{\partial}{\partial x}\left(\frac{B_2^2}{2y_2}\right) \tag{6.4.30a}$$

and similarly that:

$$u_2 h_2 = -\frac{\partial \psi_2}{\partial y} = \frac{\partial}{\partial y}\left(\frac{B_2^2}{2y_2}\right) \tag{6.4.30b}$$

Thus:

$$\psi_2 = \frac{1}{2}\frac{B_2^2}{y_2} + \text{const.} \tag{6.4.31}$$

At the equator where $v_2$ is zero; (6.4.30) implies that both $B_2$ and $\psi_2$ are constant along the equator. Thus the second boundary condition for the system (6.4.26) is:

$$B_2 = B_0, \qquad y = 0 \tag{6.4.32}$$

where $B_0$ is a constant. That is, the *equator is a streamline of the flow*.

Note that the zonal velocity is *not* specified at large values of $y$. The zonal velocity merges smoothly to its value in the subtropical solution as long as the layer thicknesses themselves smoothly merge. For the inviscid system it is only the velocity normal to the boundary layer that is explicitly matched by the asymptotic connection of the solution for $h$ to the zonal structure of the $h$ field far from the equator.

The specification that the Bernoulli function is constant along the equator weaves together each meridional section into a solution which is continuous with longitude and determines the overall longitudinal structure of the equatorial solution. Although the differerential system (6.4.26) has only $y$ as the independent variable, it is the matching to the structure of the midlatitude thermocline and the value of the Bernoulli function on the equator that render the problem three-dimensional and determine the relation of one longitudinal section with another.

The solution is not completely specified, however, until the *value* of the constant $B_0$ is determined. The inviscid boundary layer solution is in reality a family of solutions which depend on the constant $B_0$. There are several plausible ways in which $B_0$ might be chosen. Consider the total mass flux across a meridional section from the equator to the latitude $y_n$ where $y_n$ is some distance outside the equatorial boundary layer, and which we can think of as the latitude where the equatorial solution merges with the midlatitude solution. From (6.4.30b):

$$\int_0^{y_n} hu\,dy = \frac{B_0^2 - B_2^2(x, y_n)}{2y_2} = \frac{B_0^2 - h^2(x, y_n)}{2y_2} \tag{6.4.33}$$

Let the western boundary of the basin be at $x = 0$. At this location, which is the starting point of the undercurrent , the net eastward mass flux in the interval $(0, y_n)$ is given by:

$$\left(\int_0^{y_n} hu\, dy\right)_{x=0} = \frac{B_0^2 - h^2(0, y_n)}{2y_2}. \tag{6.4.34}$$

If $B_0 > h(0, y_n)$ there is a net eastward mass flux at $x = 0$. The situation is shown schematically in Fig. 6.4.1. The flow entering the undercurrent from the subtropical gyre enters on two distinct paths. The first is composed of interior streamlines that directly reach the equator, such as the streamline marked $\psi_i$ in the figure. It flows westward in the interior and turns eastward in the undercurrent producing no net eastward transport across the line $(0, y_n)$. Other streamlines, such as the one labeled $\psi_b$, strike the western boundary before hitting the equator. A critical point, $y_*$, exists in the boundary layer such that south of $y = y_*$ the boundary layer transport is southward. If the streamline $\psi_b$ strikes the western boundary south of $y_*$, it proceeds to the equator and adds to the eastward mass flux of the undercurrent. If the matching position $y = y_n$ is south of the latitude $y = y_*$, there is a net eastward mass flux in the latitude interval $(0, y_n)$ equal to the southward transport of the western boundary layer at $y = y_n$. Only if $B_0$ is equal to $h(0, y_n)$ is the net zonal mass flux balance across a latitude section. This occurs of course only if $y_n$ coincides with the bifurcation latitude $y_*$, for then (6.4.34) would yield the balance.

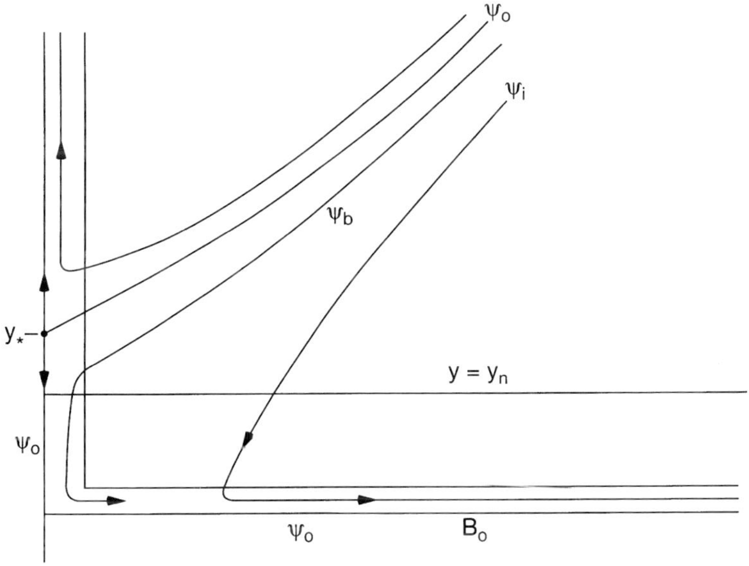


**Fig. 6.4.1.** Connection of the western boundary current and the intitiation of the EUC. The critical streamline $\psi = \psi_0$ intersects the stagnation point of the western boundary current at the latitude $y_*$. The value of the Bernoulli function on the equator is determined by the Bernoulli function on this streamline. If $y_n = y_*$ the net zonal mass flux across the section $(0, y_n)$is zero

This condition on $B_0$ can be understood as follows. Suppose that the western boundary current is fundamentally inertial, at least over the stretch between $y_*$ and the equator. The Bernoulli function would therefore be conserved on each streamline in the boundary current as well as in the interior. Consider the streamline given by $\psi = \psi_0$. This streamline enters the boundary layer at $y = y_*$ and, bifurcating, flows along the western wall to the equator. It is the streamline which forms the inner edge of the boundary layer which then emerges to flow along the equator. Its Bernoulli function is clearly the value $B_0$ that we are seeking to use in constructing the equatorial boundary layer. The general expression for the Bernoulli function is given by (4.2.11b). In dimensionless variables for the layer 2 this is equivalent to:

$$B_2 = h + \frac{1}{2}\left[u_2^2 + v_2^2\right]. \tag{6.4.35}$$

In the interior, outside the western boundary layer and the equatorial current, only the first term is significant. In the EUC the term in $v_2$ is negligible. In the western boundary current, on the other hand, the Bernoulli function is approximately $h + \frac{1}{2}v_2^2$.

On the streamline $\psi_0$ the point on the western boundary at $y = y_*$ is a position where both $u_2$ and $v_2$ are zero since it is a stagnation point. Thus, at this position $B_0$ is equal to $h$ and since the streamline enters the boundary layer where $v_2$ is essentially zero the $h$ field does not change across the boundary layer since in the western boundary layer $v_2$ is in geostrophic balance with $\partial h/\partial x$. Thus for $\psi_0$:

$$B_0 = h(0, y_*) \tag{6.4.36}$$

where $h$ is the depth of layer 2 in the *interior* just outside the western boundary current at the bifurcation latitude. Thus the appropriate value of $B_0$ is the depth of the thermocline at the bifurcation latitude of the western boundary current. Pedlosky (1991a) obtained this result by a detailed consideration of a constant potential vorticity model of the western boundary current, but we see that the argument is far more general. For $B_0$ determined in this way, the net eastward transport in the combined interior and equatorial boundary layer balance if $y_n = y_*$.

Thus if we knew the bifurcation latitude of the western boundary current, the value of $B_0$ would be determined. This requires however, a rather complete theory of the stratified western boundary current which, as we discussed in Chapter 2, is not available. If the circulation is modeled as a single layer, the western boundary layer transport is given by the total interior Sverdrup transport at that latitude (but opposite in sign, of course), i.e., proportional to the curl of the wind stress (or, more precisely, its basin-wide integral if it is a function of longitude). For a multilayer model it is not clear that the bifurcation point in each layer should correspond to the bifurcation point for the total transport. Liu (1994) argues that the vanishing wind stress curl line is an adequate approximation for the bifurcation latitude on the basis of numerical models of the general circulation. McCreary and Lu (1994) present an alternative representation for the bifurcation latitude which in general somewhat departs from the zero wind stress curl line. However, their determination is valid only if the outcrop line occurs far enough south to lie in a region of westward wind stress. Even then determination of the bifurcation point is unfortunately only implicit.

Pedlosky (1987) rather arbitrarily chose a value of $B_0$, and hence implicitly a bifurcation latitude, and examined the nature of the equatorial solution of the set (6.4.26a,b,c). Although the strength of the model's undercurrent depends on $B_0$ the structure of the solution is not very sensitive to its value.

Section 6.5 considers the role of the shadow zone in determining he structure of the solution. In the calculations of Pedlosky (1987) $H_2$ is taken equal to zero. It is assumed that the equator is south of the shadow zone boundary of the lowest thermocline layer and that layer 2 corresponds to an upper layer whose thickness vanishes on the eastern boundary. As we saw in Chapter 4, the presence of a variable mixed layer depth endows even shallow layers in the thermocline with shadow zones, but these are relatively narrow. In the solutions presented by Pedlosky (1987), $H_2$ in the boundary condition (6.4.27) is set equal to zero.

Figure 6.4.2 shows that solution for the case in which $\Gamma_{12} = 1$, $y_2 = 5$ and the matching with the midlatitude solution occurs at $y_n = 2.5$. $B_0$ was chosen to be equal to 1.265, which was equal to $h(0, y_n)$ for the case in which the wind stress is taken to be independent of longitude and equal (in scaled units) to –1 corresponding to a stress of 1 dyn/cm$^2$. The eastern boundary is placed at $x = 1$ and the western boundary at $x = 0$. The relation (6.4.16) is used to relate $h_1$ to $h$.

In the figure panels the solid line represents the zonal velocity in the interval $y \geq 0$. The dashed line represents the shear, $\partial u_2/\partial y$, and the dotted-dashed line is the depth $h$. The dotted curve is the depth of the base of layer 1; note that it is flat, in accordance with (6.4.16).

The zonal velocity at large values of $y$ is *westward* and is determined by the solution. At $x = 0.5$, for example, which is shown in panel b, the zonal velocity far from the equator is –0.05196. With the scaling velocity introduced earlier, $U = 130$ cm/s, this corresponds to a westward flow of 6.7 cm/s. As opposed to the Fofonoff and Montgomery solution the shear is nearly zero in the region outside the undercurrent rather than growing linearly from the matching latitude. Indeed, the westward velocity actually increases in magnitude southward from the starting latitude until at $y \approx 1$ the velocity changes sign and rapidly increases to the east. The solution now has a boundary layer character in which the eastward jet is limited to the domain determined by the internal, physical scale $\ell$ rather than by the matching latitude $y_n$. The maximum eastward velocity is attained on the equator and has a magnitude of order unity in nondimensional units. The current therefore has scales for velocity, width and depth which are in accordance with our original scaling estimates (e.g., the nondimensional variables are order one) and in accordance with observations.

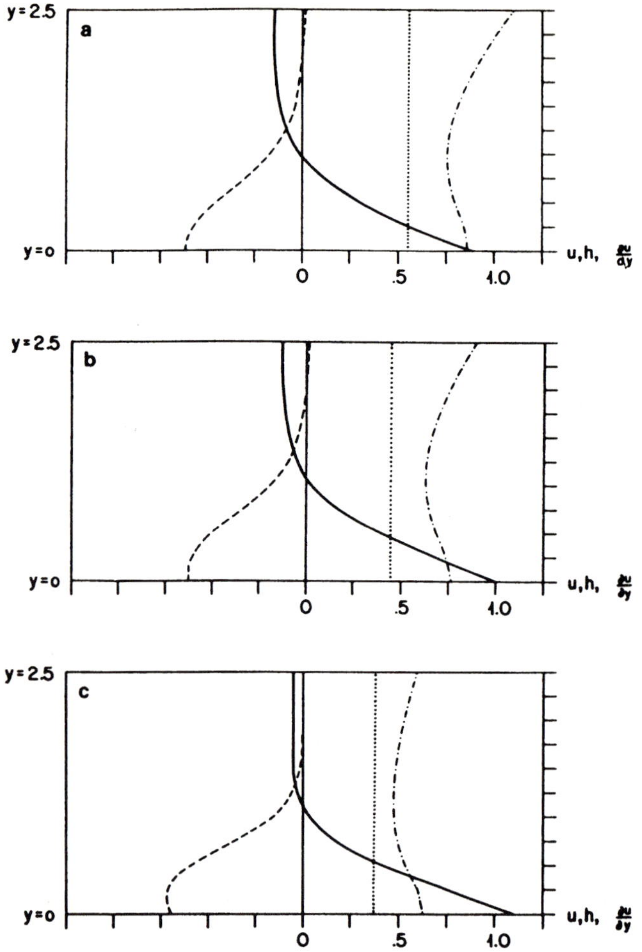


**Fig. 6.4.2.** Solutions of (6.4.26) for $u_2$ (*solid line*), $\partial u_2/\partial y$ (*dashed line*) and $h$ (*dash-dotted*). In this case the wind stress is a constant and $\Gamma_{12} = 1$. The three panels correspond to profiles at $x = 0.25$, 0.50, and 0.75, respectively. $B_0 = 1.265$ and $y_2 = 5$. (From Pedlosky 1987)

The three panels of the figure, at $x = 0.25$, 0.5, and 0.75 show the current accelerating across the basin in response to the continuing flux of fluid entering the current from midlatitudes in the interior of the gyre. On the center line of the current the maximum velocity increases from 0.899 at $x = 0.25$ to 1.28 at $x = 0.75$. Since the Bernoulli function is a constant along the equator, this increase in speed coincides with a decrease in $h$, i.e., with a shallowing of the thermocline along the equator to the east. Figure 6.4.3 shows $h$ along the equator as the solid curve and the thermocline depth at $y = y_n$ as the dotted curve.

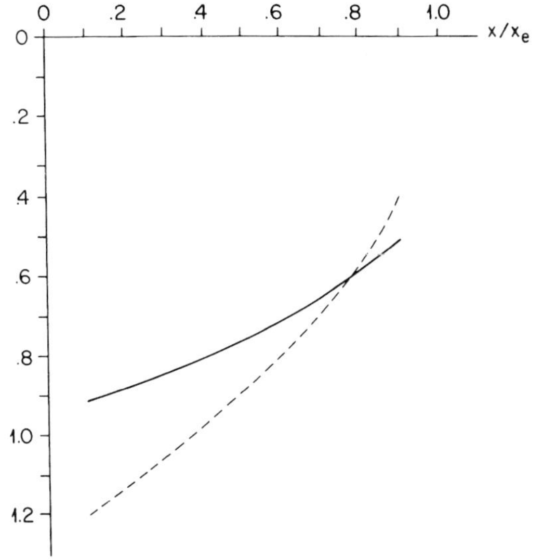


**Fig. 6.4.3.** Base of the moving layer representing the core of the undercurrent shown as a *solid line* at the equator and as a *dashed line* in the matching region at $y = y_n$. (From Pedlosky 1987)

Only the half-plane $y \geq 0$ is shown. The lower half-plane, representing the southern hemisphere, is obtained by reflection. Since $\partial u_2/\partial y$ is not zero on the equator, the velocity profile has a cusp. This is an inevitable consequence of the adiabatic model, which preserves potential vorticity. On the equator where $y = 0$ (6.4.26a) implies that:

$$\frac{\partial u_2}{\partial y}(0) = -y_2 \frac{h_2(0)}{B_0} . \tag{6.4.37}$$

Thus a cusp is present at the equator as long as the thickness of the layer containing the undercurrent differs from zero. Small lateral friction, negligible for determining the overall structure of the current, can smooth away the cusp.

Figure 6.4.4 shows the same profiles of the solution at $x = 0.5$ for the case in which the relation between $h$ and $h_1$ has been chosen to be (6.4.19). The dotted curve is the base of the upper layer, and it is seen to be a reflection of the displacement of the base of the lower layer shown as a solid curve on the right of the figure. The effect of the compensation of the upper layer on the velocity of the undercurrent core is minor. There is a small reduction in the maximum velocity from 1.004 in the example shown in Fig. 6.4.2 to 0.910 in Fig. 6.4.4. Otherwise the qualitative nature of the current is unchanged. Note that the bowing of the isopycnal surfaces, which is also evident in the observations, is due entirely to the tilting of the isopycnal surface required by the thermal wind balance of the zonal velocity. It is not forced by cross-isopycnal motion (there is none in this model) dragging apart the isopycnals.

The isolines of Bernoulli function, $B_2$, are streamlines of the flow, and Fig. 6.4.5 shows some selected streamlines labeled with the value of $B_2$. Along the equator $B_2$ is 1.265, and this is the streamline which emanates from the

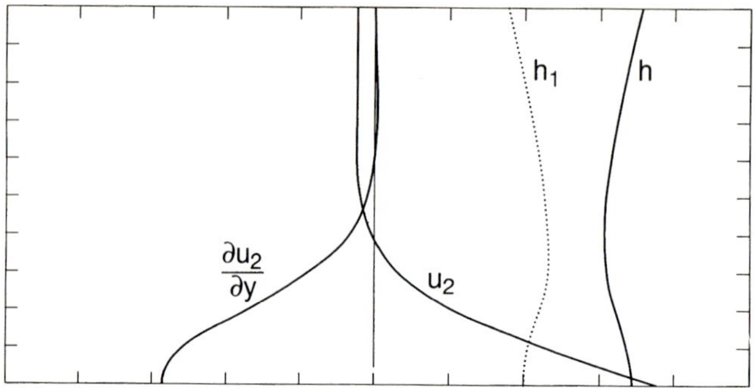


**Fig. 6.4.4.** Profiles of $u_2$, $\partial u_2/\partial y$, $h$, and $h_1$ for the case in which (6.4.19) is used. The parameters are otherwise as in Fig. 6.4.2. The calculation is at $x = 0.5$. The maximum velocity of the eastward velocity is now 0.910

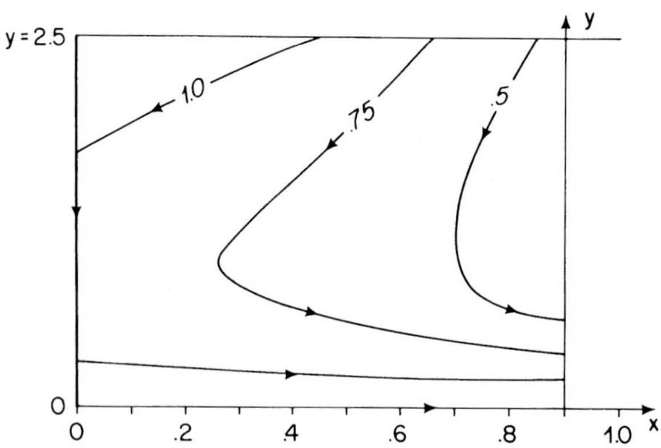


**Fig. 6.4.5.** Several lines of constant $B_2$, i.e., streamlines for the flow calculated in Fig. 6.4.2. (From Pedlosky 1987)

western boundary current. Some streamlines with $B_2$ less than the equatorial value (e.g., $B_2 = 1$) strike the western boundary and flow to the equator in the boundary layer. Streamlines with smaller values of $B_2$, such as $B_2 = 0.75$ flow first southwestward and then turn eastward to feed the undercurrent in the interior adding to its transport and velocity.

Adding more layers and more outcrop lines to improve the vertical resolution of the undercurrent structure does little to alter the over all structure of the velocity field. Figure 6.4.6 shows the result of a calculation of an extended system of four layers in which a system similar to (6.4.26) can easily be derived. The nonlinear boundary value problem leads to a challenging numerical problem but my colleague R. Samelson has succeeded in calculating the velocity profiles which emerge to model the current in layers 2, 3, and 4. The velocity profile decreases monotonically with depth and without reversals of the velo-

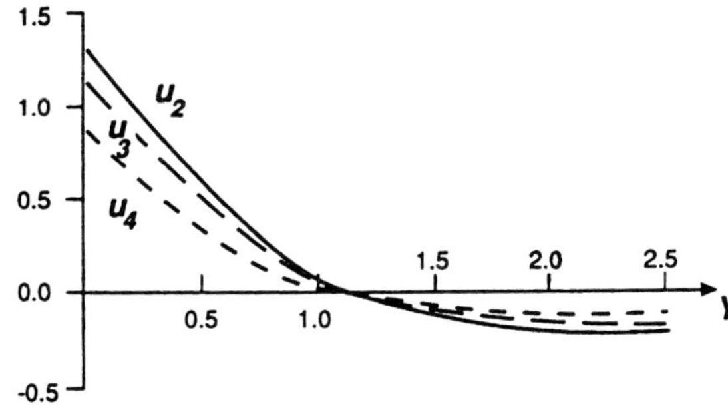


**Fig. 6.4.6.** Results of a four-layer model showing the monotonic decrease of the velocity with depth in the undercurrent solution. (Courtesy of R. Samelson, pers. comm.)

city. Thus the two layer model captures the essence of the core structure in the horizontal plane.

The eastward jet in the present solution continuously increases in strength as it flows eastward; it constantly accelerates as more fluid is added from the interior into its flanks. The solution is unable, as it stands, to describe the observed leveling off of the zonal velocity in the midbasin and its final deceleration as the current terminates on the eastern side of the ocean. Both of these issues are taken up in the following sections. Both the role of the shadow zone and entrainment of the undercurrent into the upper layer must be considered.

## 6.5 Role of the Shadow Zone in the Undercurrent Structure

If layer 2 has a nonzero thickness on the eastern wall, two alterations must be made in the calculation of the undercurrent solution as given in the previous section. The first is quantitative rather than qualitative. The matching condition for large $y$ (6.4.27) has $H_2$ different from zero, and this slightly alters the solution although not its basic character. The more important qualitative effect of having a nonzero thickness on the eastern boundary is the introduction of a shadow zone in layer 2, and its effect on the solution is profound. The issue has been discussed by both McCreary and Lu (1994) and Liu (1994).

The equation for the shadow zone boundary (4.4.22) yields, for the case in which the wind stress is independent of longitude and is directed only in the zonal direction:

$$\frac{2R(\phi_e - \phi)}{\gamma_2 \rho_0}\left[\sin\theta \frac{\partial \tau}{\partial \theta} - \frac{\tau}{\cos\theta}\right] = H_2^2 \frac{\gamma_1}{\gamma_2}\left(1 - \frac{f}{f_2}\right)^2 \tag{6.5.1}$$

In (6.5.1) the variables are dimensional and $\tau$ is the zonal wind stress.

The shadow zone boundary trends southwestward from the outcrop line and may strike the western boundary before it reaches the equator, or it may first strike the equator if the ocean is wide enough. Figure 6.5.1 shows three possible configurations of the flow depending on the position of the shadow zone boundary and its intersection with the equator and/or the western boundary.

Although the dynamics leading to (6.5.1) is not valid at the equator, it holds up to the equatorial zone and hence (6.5.1) gives an accurate picture of the intersection of the ventilated region with the equatorial boundary layer.

Suppose, as in Fig. 6.5.1.a, the shadow zone boundary intersects the equator in the interior before striking the western boundary. The longitude at which this takes place, $\phi = \phi_{eq}$, is calculated form (6.5.1) by letting $\theta \to 0$, i.e.:

$$R(\phi_e - \phi_{eq}) = \frac{\gamma_1 H_2^2}{-2\tau(0)}. \tag{6.5.2}$$

In order for the intersection to occur in the basin it is necessary that $\phi_{eq} > \phi_w$ where $\phi_w$ is the longitude of the western boundary. Therefore the condition

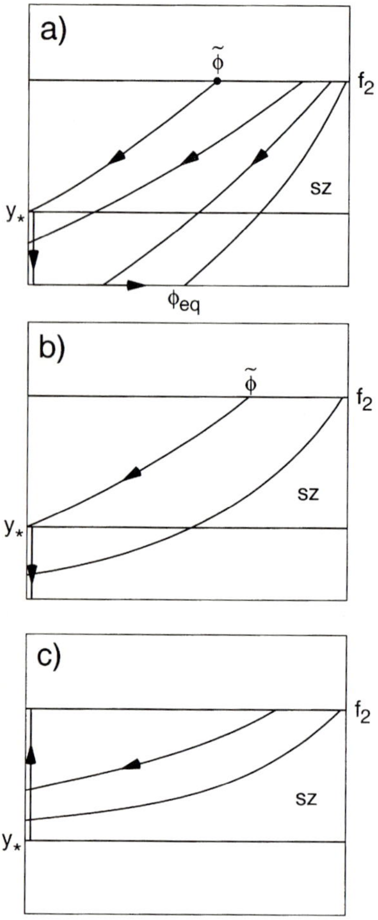


**Fig. 6.5.1a-c.** Position of the shadow zone boundary and its intersection with the western boundary. **a** The shadow zone boundary strikes the equator within the basin, and the undercurrent is fed from both the interior and the western boundary. East of the longitude, $\phi_{eq}$, the transport of the model undercurrent is constant. **b** The shadow zone boundary strikes the western boundary of the basin north of the equator but south of the bifurcation latitude of the western boundary current. The model EUC is fed only from the western boundary current and has uniform transport. **c** The shadow zone boundary strikes the western wall of the basin north of the bifurcation latitude of the western boundary current, and there is no undercurrent

that the shadow zone boundary strikes the equator before reaching the western boundary can be written as:

$$-\tau(0)L_x \geq \frac{\gamma_1 H_2^2}{2} \tag{6.5.3}$$

where $L_x$ is the width of the basin. If the basin is wide enough, or equivalently, if the westward wind stress *at the edge of the equatorial boundary layer* is large enough to satisfy (6.5.3), the shadow zone boundary strikes the equator in the interior in the basin as shown in Fig. 6.5.1a.

When this occurs, the solution in the region *west* of the longitude $\phi = \phi_{eq}$ is qualitatively similar to that discussed in the previous section. The model's EUC is fed from both the western boundary current and the interior and increases in transport and maximum velocity as it flows eastward until it reaches the longitude $\phi_{eq}$. At this longitude, as at every longitude west of this point, (6.4.26) is solved for the current structure matching the subtropical value of $h$ in the region outside the equatorial boundary layer and matching $B_0$ on the equator. This yields the current structure entering the region east of $\phi_{eq}$. In this eastern region the undercurrent is no longer fed by from the interior, and its total transport is constant as a function of longitude as it flows through the equatorial margin of the shadow zone. The structure of the current, however, alters in this region since although the boundary condition for $h$ is independent of longitude in this region, $h_1$ in the shadow zone diminishes eastward. As $h_1$ becomes small compared with $H_2$ even this variation becomes negligible, and the system of equations yields essentially the same boundary value problem at each longitude so that even the structure of the current tends to become independent of longitude. Thus in the region $\phi \geq \phi_{eq}$ the solution of the model's EUC yields a zonal current whose transport is constant. It is important to keep in mind that this leveling off of the transport is a purely adiabatic effect produced by the shielding of the undercurrent from additional mass input after it crosses the shadow zone boundary. In the region east of the intersection of the shadow zone boundary with the equator the depth $h$ of the thermocline base is nearly level, ignoring the small effect of the variation in $h_1$ in determining $u_2$ and $h$ in (6.4.26), and therefore in this region *the eastward pressure force in the lower layer is nearly zero*. The EUC glides eastward in this region as a free, unaccelerating jet.

On the other hand if, as in Fig. 6.5.1b, the shadow zone boundary strikes the western boundary before it reaches the equator and strikes it south of the bifurcation latitude of the western boundary layer, the model's EUC has its transport in layer 2 fixed at the western boundary by the mass flux delivered to the equator in the boundary current. It then flows eastward along the entire equatorial extent of the basin with no change in transport since there is no flow into the undercurrent from the interior.

For the shadow zone boundary to strike the western boundary at or south of the bifurcation latitude where $f = f_*$:

$$L_x\left\{\sin\theta\frac{\partial\tau}{\partial\theta}-\frac{\tau}{\cos\theta}\right\}_{\theta=\theta_*}\geq\frac{\gamma_1 H_2^2}{2}(1-f_*/f_2)^2 \tag{6.5.4}$$

by a simple variation of the argument leading to (6.5.3). If, as Liu (1994) suggests, the bifurcation latitude corresponds to the latitude of zero wind stress curl, then (6.5.4) reduces to the condition (Liu 1994):

$$-\tau(\theta_*)L_x\geq\frac{\gamma_1 H_2^2}{2}(1-f_*/f_2)^2 \tag{6.5.5}$$

which is easier to satisfy than (6.5.3).

If the wind stress in the region of the bifurcation latitude is so small, that the inequality in (6.5.4) or (6.5.5) is not satisfied, the situation shown in Fig. 6.5.1c results. Here, all the flow in layer 2 impinges on the western boundary. None is fed into the EUC in the interior, and none reaches the equatorial zone through the western boundary current. All the flow impinging on the western boundary in layer 2 turns northward and recirculates in the subtropical gyre and is prevented from entering the equatorial zone. There is no equatorial undercurrent in layer 2.

In cases a and b the fluid entering the undercurrent comes originally from fluid subducted along the eastern limb of the outcrop line between the longitude $\tilde{\phi}$ and the eastern boundary. The longitude, $\tilde{\phi}$, can be found by tracing back to the outcrop line the streamline which intersects the western boundary at the bifurcation latitude, i.e.:

$$D_0{}^2(\tilde{\phi},\theta_2)=\left[\frac{D_0{}^2(\phi_w,\theta_*)-H_2^2\Gamma_{12}(1-f_*/f_2)^2}{1+\Gamma_{12}(1-f_*/f_2)^2}\right] \tag{6.5.6}$$

which determines the width of the window of the subducting fluid eventually feeding the undercurrent.

The boundary layer solution for the undercurrent is not sensitive to the value of the wind stress on the equator, but rather indirectly to the wind stress in the matching region through the imposed matching to the subtropical thermocline solution. However, if the wind stress is artificially set to zero in a zone encompassing the matching region, fluid is not delivered to the equator, and the undercurrent does not develop. In this indirect way the undercurrent remains sensitive to the equatorial wind stress even though we can see from the development of the boundary layer problem that once the fluid is delivered to the equator by the extra-equatorial geostrophic flow the acceleration of the fluid to form the undercurrent is essentially a consequence of conservative dynamics rather than local forcing by the wind.

## 6.6 Effect of Entrainment

Observations of the EUC (e.g., Lukas 1986; see also the review by Wacongne 1989) show that in the Pacific, for example, the EUC weakens considerably from the central ocean towards the eastern boundary. This is consistent with the study of Bryden and Brady (1985) who showed the simultaneous rise of the undercurrent core with longitude and the entrainment of fluid from the EUC into the upper surface layer. This entrainment is limited to the upper portion of the current, but as the current shallows this transfer of fluid into the upper layer becomes an increasingly significant fraction of the transport. As each upward sloping density layer reaches the mixed layer, it is peeled off the top of the current and enters the mixed layer, and flows away from the equator driven by the wind as Ekman transport . Layer after layer is thus stripped off the top of the undercurrent, finally producing the observed cold pool of water on the surface in the eastern equatorial oceans as the deeper layers rise to the surface. Pedlosky (1988), following the diagnosis of Bryden and Brady, argued that this entrainment is the principal mechanism for the termination of the undercurrent. This is suggested by a simple consideration of the mass balance in the EUC which quantifies the description just given.

Consider the meridional Ekman transport directed away from the equator at a distance $\ell$ from the equator, i.e., from the flanks of the EUC. This transport, over the basin width $L_x$, is of the order:

$$T_{\text{Ekman}} = \frac{\tau_0}{\rho_0 \beta \ell} L_x \tag{6.6.1}$$

where $\tau_0$ is the scale of the westward directed wind stress.The eastward mass flux in the undercurrent, on the other hand, is of the order:

$$T_{EUC} = UH\ell = \frac{\gamma_2 H}{\beta \ell^2} H\ell = \frac{\tau_0 L_x}{\rho_0 \beta \ell} = T_{\text{Ekman}} \tag{6.6.2}$$

if the scaling relations (6.3.21) are used.

This implies that the transport of the EUC can be completely exhausted by the off-equator mass flux driven by the Ekman transport which is fed, as described by Bryden and Brady (1985), by the entrainment of undercurrent fluid from the upper portion of the EUC. Although the introduction of a shadow zone halts the acceleration of the current by eliminating the east-west pressure gradient, a dissipative mechanism is required to explain its termination.The cross-isopycnal mass flux represents such a nonadiabatic effect, and its inclusion in the dynamics of the undercurrent, as does its effect on the interior discussed in Chapter 5, requires a fundamental alteration of the physics of the model.

Pedlosky (1988) introduced a simple, heuristic model for the effect of the entrainment due to cross-isopycnal mixing which requires only minor alterations in the calculations although introducing fundamental changes in the

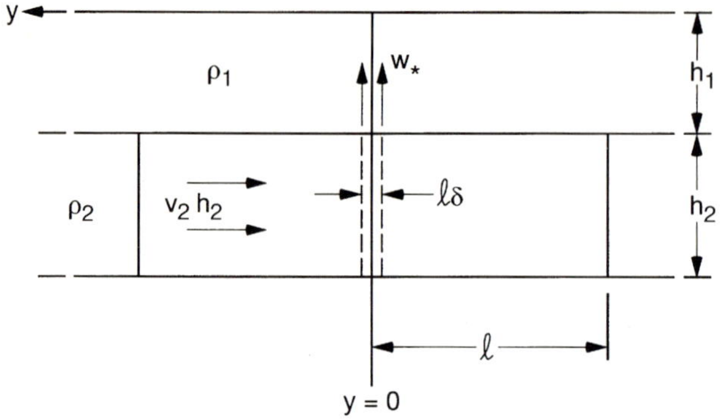


**Fig. 6.6.1.** Schematic presentation of the model of the cross-isopycnal motion in the model. The cross-isopycnal flux is limited to a narrow zone of half-width $\delta\ell$ within the undercurrent of half-width $\ell$

physical problem. Suppose the cross-isopycnal mixing is limited to a very narrow region around the equator, narrow even compared to the EUC width $\ell$, as shown in Fig. 6.6.1. The region has a width $\ell\delta$ where $\delta \ll 1$. If $\delta \ll 1$, the mass balance for this narrow zone must be two-dimensional. That is, integrating the (nondimensional) continuity equation over the latitude band (0,$\delta$) we obtain:

$$\int_0^\delta \frac{\partial u_2 h_2}{\partial x} dy + (v_2 h_2)_{y=\delta} = -\int_0^\delta w_* dy \tag{6.6.3}$$

using the condition that $v_2$ vanishes at $y = 0$. As $\delta$ goes to zero, we obtain in the limit:

$$(v_2 h_2)_{y=0} = -\int_0^\delta w_* dy \equiv -M(x) \tag{6.6.4}$$

where $M(x)$ is the total cross-isopycnal mass flux in the latitude band (0,$\delta$) corresponding to the dimensional width $\ell\delta$.

This model of the cross isopycnal flux is a very artificial one. Although measurements (e.g., Johnson and Luther 1994) suggest increased dissipation near the equator, the restriction of this zone to a small fraction of the undercurrent width is an artifice used to retain the adiabatic dynamics over most of the current's width, since adiabatic dynamics seems adequate to explain the current structure, while allowing the cross-isopycnal flux to affect the transport of the current and its strength at each longitude.

The connection between $M(x)$ and the magnitude of the current follows from the zonal momentum equation in the region just outside $y = \delta$, i.e., at $y = 0^+$. In the absence of friction (6.4.24) and (6.4.29a) yield:

$$h_2 v_2 = \frac{1}{q_2}\frac{\partial B_2}{\partial x} = \frac{1}{2y_2}\partial\frac{B_2^2}{\partial x}. \tag{6.6.5}$$

Thus at the point $y = 0^+$:

$$\frac{1}{2\gamma_2}\frac{\partial B_2^2}{\partial x} = (h_2 v_2)_{y=0^+} = -M(x) \,. \tag{6.6.6}$$

The effect of the concentrated cross-isopycnal flux at the equator is to produce a reduction of the Bernoulli function along the equator proportional to the amount of mass transferred between the layers.

Integrating (6.6.6) from the western boundary where the Bernoulli function may be considered set by western boundary current's value of $B_2$:

$$B_2^2(x, 0^+) = B_0^2 - 2\gamma_2 \int_0^x M(x')dx' \,. \tag{6.6.7}$$

The Bernoulli function on the western boundary, at $x = 0$, is $B_0$. If there were no cross-isopycnal flux, this would be the value of the Bernoulli function all along the equator. The presence of the nonadiabatic mass flux reduces this value with longitude as a function of the strength of the transfer of mass between the layers.

The zonal velocity of the undercurrent satisfies (6.4.29b) so that at each longitude the transport in the undercurrent is equal to:

$$\int_0^\ell u_2 h_2 dy = \frac{1}{2\gamma_2}\left[B_2^2(x, 0^+) - B_2^2(x, \ell)\right]. \tag{6.6.8}$$

The second term in the square bracket on the right side of (6.6.8) is set by the interior, inviscid thermocline solution. As the first term in the bracket is diminished by the cross-isopycnal flux, both the transport in the undercurrent, and its peak velocity weaken with longitude.

The system (6.4.26a,b,c), which holds everywhere outside of the vanishingly thin region of cross-isopycnal flux, can now be solved again with the new boundary condition (6.6.7) applied at $y = 0$ instead of (6.4.32) to which (6.6.7) reduces if $M(x)$ is zero.

In principal, the cross-isopycnal mass flux, which is a measure of the dissipation in the current, should be related to the properties of the current itself. This requires a model for the dissipation. Such a model is discussed below. However, it is useful, as in Chapter 5, to examine the response of the motion to a given distribution of $w_*$ or equivalently to a given longitudinal distribution of $M(x)$.

When the particular density layer is deep, the cross-isopycnal velocity is negligible, as described by the analysis of Bryden and Brady (1985). As the current flows eastward, the uppermost layer is slowly entrained into the mixed layer, exposing the next lowest isopycnal layer to entrainment until, as the jet crosses the basin, each of the layers exhausts its flux into the mixed layer. Only when a particular stratum approaches the upper layer where mixing is important does the cross-isopycnal velocity become significant in that layer. For layers in the core of the undercurrent the effect of the cross-isopycnal flux does not occur until a large portion of the basin has been traversed. In a model in which the undercurrent is represented by a single layer it is possible only crudely to model this behavior by choosing a mass flux function $M(x)$ that is weak in the western part of the basin and grows in strength eastward.

Pedlosky (1988) chose the following distribution for $M(x)$:

$$M(x) = \begin{cases} 0, & 0 \leq x \leq x_b \\ M_0 \left( \dfrac{x - x_b}{x_e - x_b} \right) & x_b \leq x \leq x_e \end{cases} \tag{6.6.9}$$

so that the cross-isopycnal flux is absent until the current flows east of the longitude where $x = x_b$. $M_0$ is *chosen* so that:

$$M_0 = \frac{B_0^2}{y_2(x_e - x_b)} \tag{6.6.10}$$

so that along the equator, from (6.6.7):

$$B_2^2(x, 0^+) = B_0^2 \left[ 1 - \left( \frac{x - x_b}{x_e - x_b} \right)^2 \Theta(x - x_b) \right] . \tag{6.6.11}$$

In (6.6.11) $\Theta(x - x_b)$ is the unit Heaviside function which is zero when its argument, $x - x_b$, is less than zero and equal to unity when $x - x_b \geq 0$. This choice forces $B_2$ to vanish on the equator at the eastern boundary. If, as in the calculations of Pedlosky (1988), the model possesses no shadow zone, the Bernoulli function just outside the boundary current equals $h(x_e, 0)$ given by the ventilated thermocline solution which equals zero if $H_2 = 0$. In this case the transport of the undercurrent is exhausted as the current reaches the eastern boundary.

Figure 6.6.2 shows the profiles of $u_2$, $h$, $h_2$ and $\partial u_2/\partial y$ for the integration of the system (6.4.26) for the case in which (6.6.11) is the boundary condition at $y = 0$ while (6.4.19) has been used to relate $h_2$ to $h$. The longitude $x_b$ is chosen as 0.4 while the eastern boundary is at $x = 1$. Otherwise the parameters are the same as those used to calculate the current in Fig. 6.4.4. Panel a shows the solution at $x = 0.8$ for the case in which there is no cross-isopycnal flux, and therefore $B_2$ on the equator everywhere equals $B_0$, which is equal to 1.265. Note that the maximum velocity, in scaled units, is 1.142. Panel b shows the solution in which (6.6.11) is used. $B_2$ at the equator has now fallen to 0.943 at this longitude, and the maximum velocity of the current at $y = 0$ is similarly reduced from its adiabatic value to 0.881. Otherwise the structure of the current is similar, as might be expected, since the effect of the cross-isopycnal velocity is limited to a narrow region around the equator.

Figure 6.6.3a shows the maximum velocity with and without the cross-isopycnal flux. In the presence of $w_*$ the velocity is reduced significantly. Note, however, that the current continues to accelerate beyond the longitude where the cross-isopycnal velocity begins. The entry of fluid into the current from the sides, tending to accelerate the flow, more than compensates for the deceleration due to $w_*$ until well past the longitude of onset of nonadiabatic effects.

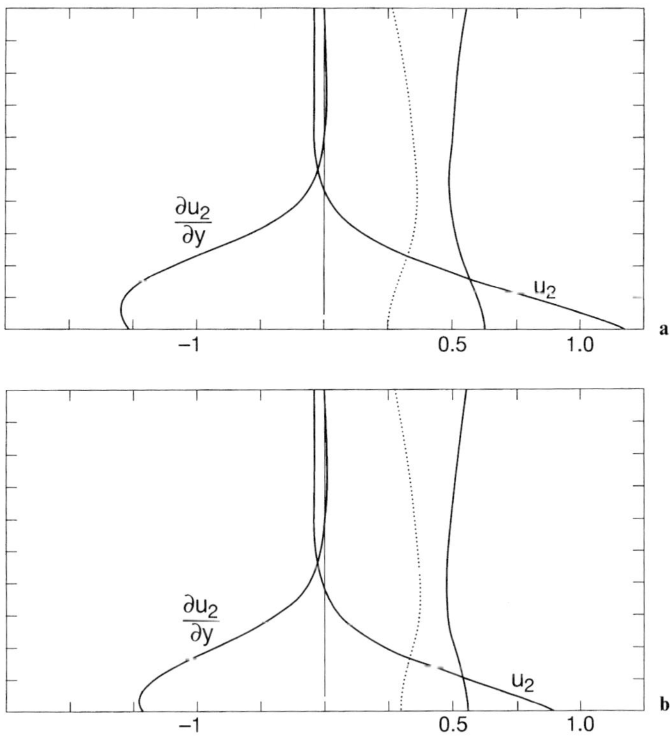


**Fig. 6.6.2.** Profiles of the undercurrent **a** without and **b** with cross-isopycnal flux as specified by (6.6.9) and (6.6.11). Profiles of $u_2$, $\partial u_2/\partial y$, $h$, and $h_2$ are shown. Note that the scales for the nondimensional velocity and depth are shown on the positive $x$ axis while the scale for the shear is shown on the negative $x$ axis

Not until $x \approx 0.7$ does the current begin to decelerate although the entrainment begins at $x = 0.4$. The subsequent fall in the maximum velocity is very rapid and limited mostly to the eastern part of the basin. Figure 6.6.3b shows the layer thicknesses with and without entrainment. The presence of entrainment reduces the thickness of the lower layer and enhances that of the upper layer. In Fig. 6.6.3c the transport of the undercurrent is shown. The difference between the transport with and without entrainment is equal to

$$\Delta T = \int_{x_b}^{x} M\, dx$$

in the interval of entrainment. If the width of the current is defined in terms of the latitude at which the eastward velocity in the EUC is zero, the calculations also show that the width of the current diminishes as a consequence of the entrainment, that is, the current becomes narrower as well as weaker.

The weakest point of this theory is the a priori specification of the cross-isopycnal flux. In an effort to specify $w_*$ in a less arbitrary fashion Pedlosky

a
b
c
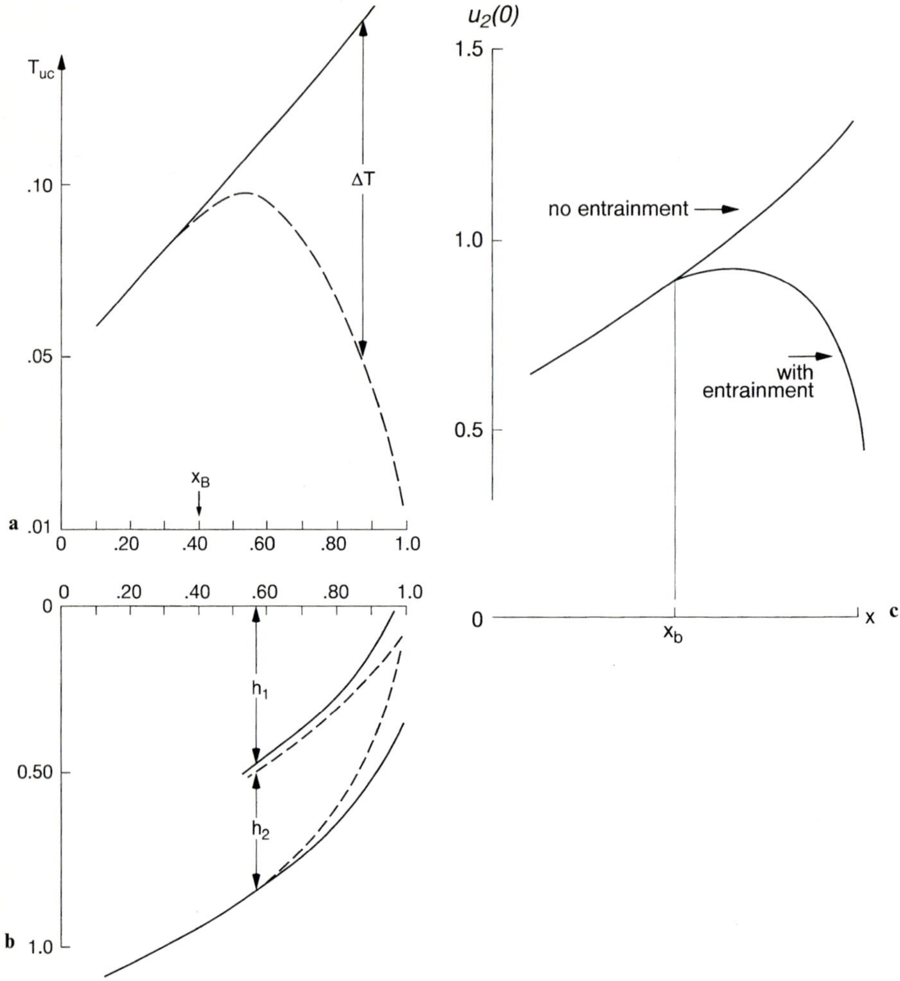


**Fig. 6.6.3.** **a** Velocity in layer 2 on the equator with entrainment (*dotted line*) and without entrainment (*solid line*). **b** Depths of the two layers with (*dotted lines*) and without (*solid lines*). **c** Transport of the current with and without entrainment. The difference in the transport is shown as $\Delta T$ and is equal to the amount of fluid bled off from the undercurrent which enters the upper layer

and Samelson (1989), while retaining the fundamental elements of the model shown schematically in Fig. 6.6.1, introduced explicit dynamics for the upper layer. While retaining (6.4.18) as an approximation for the $y$ momentum equation in the upper layer, they suggested as a simple model for the $x$ momentum equation:

$$-yv_1 = -ru_1 + \frac{\tau}{h_1} - \frac{\partial[h + \Gamma_{12}h_1]}{\partial x}. \tag{6.6.12}$$

The model dynamics of the upper layer, integrated over its depth consists of a balance between the Coriolis acceleration and the surface wind stress, the zonal pressure gradient, and a frictional deceleration proportional to the velocity of the layer itself. This simple representation of the frictional effect is meant to model in the easiest possible manner the observed role of lateral friction (see Johnson and Luther 1994) which should depend only on $u_1$. Far from the equator where this term is dominated by the Coriolis acceleration (6.6.12) consists of balance which combines the Ekman layer transport of the mixed layer and the geostrophic transport of layer 1. Near the equator the separation is no longer valid, and they are lumped together. On the equator the Coriolis acceleration vanishes, and the upper layer zonal current is driven by the imbalance between the zonal pressure gradient and the wind stress. Because of (6.4.18) this pressure gradient is known from the off-equatorial solution of the thermocline problem.

Outside the small region of half-width $\delta$ (in nondimensional units) the upper layer flow is horizontally nondivergent so that a streamfunction can be introduced for the transport, i.e..

$$u_1 h_1 = -\frac{\partial \psi_1}{\partial y}, \qquad v_1 h_1 = \frac{\partial \psi_1}{\partial x} \tag{6.6.13a,b}$$

This allows (6.6.12) to be written as the partial differential equation:

$$y\frac{\partial \psi_1}{\partial x} + r\frac{\partial \psi_1}{\partial y} = -\tau + h_1 P_1(x) \tag{6.6.14}$$

where $P_1(x)$ is the known east west pressure *force* in the upper layer and is equal to $h + \Gamma_{12} h_1$ at the matching latitude outside the equatorial region.

The above equation is linked to the equations in the lower layer through the boundary condition on $\psi_1$ at the equator. The flux of mass across the isopycnals from layer 2 must enter layer 1 in the region of the equator and thus on $y = 0$:

$$h_2 v_2 = \frac{1}{2y_2}\frac{\partial B_2^2}{\partial x} = -h_1 v_1 = -\frac{\partial \psi_1}{\partial x}. \tag{6.6.15}$$

Thus the layers are linked by the condition at the equator:

$$B_2{}^2(x, 0) = -2y_2\psi_1(x, 0) \tag{6.6.16}$$

where it is assumed that both $\psi_1$ and $B_2$ vanish on the eastern boundary.

The characteristics of (6.6.14) are the parabolas:

$$\frac{y^2}{2r} + (1 - x) = \frac{y_e^2}{2r} \tag{6.6.17}$$

each one emanating from a different point, $y_e$, on the eastern boundary. With the starting value $\psi_1 = 0$, and the values of the layer thicknesses known on the eastern boundary, (6.6.16) can be integrated along a small step, $\Delta x$, westward. One of these characteristics strikes the equator at:

$$1 - x = -\Delta = \frac{y_e^2}{2r} \ . \tag{6.6.18}$$

Equation (6.6.18) can be thought of as determining the particular characteristic (i.e., the value of $y_e$) which strikes the equator at this location. The value of $\psi_1$ is determined by integrating along the characteristic and in particular its value on the equator at that longitude is then known. This determines the value of $B_2$ at the equator from (6.6.16). Then the system (6.4.26) can be integrated in latitude at this longitude which is one step in $x$ west of the eastern boundary. This determines $h$ and $h_1$ along the latitude corresponding to the first zonal step from the eastern boundary. With the layer thicknesses known on this meridian the equation (6.6.14) can then be recalculated another step westward along the characteristic and the process iterated until the western boundary is reached. At each step the Bernoulli function is determined on the equator.

This method of determining the Bernoulli function on the equator appears inconsistent with the suggestion of the previous section in which the Bernoulli function on the equator is determined in the west by the value carried to the equator by the western boundary current. The inconsistency really reflects the overidealization of each partial attempt to deal with the dynamics. The previous specification of the Bernoulli function most likely overidealizes the conservative nature of the dynamics which should have a limited range of validity near the western boundary before the effect of cross-isopycnal flux become significant. The linear dynamics of the upper layer discussed in this section is probably more relevant to the eastern regime of the upper density layers of the undercurrent. There does not currently exist an analytical model which joins the two regimes.

Figure 6.6.4 shows the results of the calculation of Pedlosky and Samelson (1989) for the nondimensional, equatorial wind-stress distribution:

$$\tau = -(1 - x) \tag{6.6.19}$$

which is meant to correspond to the wind-stress distribution in the equatorial Atlantic which increases in strength westward. The calculations have been done for $r = 1$ which yield a dissipation time for the upper layer equal to $L/U$, i.e., the characteristic zonal advective time.

Panel a shows the streamfunction of the lower layer. Fluid enters the undercurrent largely in the western basin, both from the western boundary current and from the interior, and rises into the surface layer as it flows eastward. Panel b shows the streamfunction of the upper layer. The flow emerges from the equatorial upwelling region and flows northwestward in a broader pattern than the equatorward directed flow of the lower layer. Panel c presents the isopleths of $u_2$. The undercurrent speed is nearly constant along the equator as the entrainment balances the acceleration due to the convergence of fluid from the interior. The thinning and narrowing of the current, rather than its deceleration, yields the necessary reduction in transport until, very near the eastern boundary, there is a rapid reduction in the current speed as we ob-

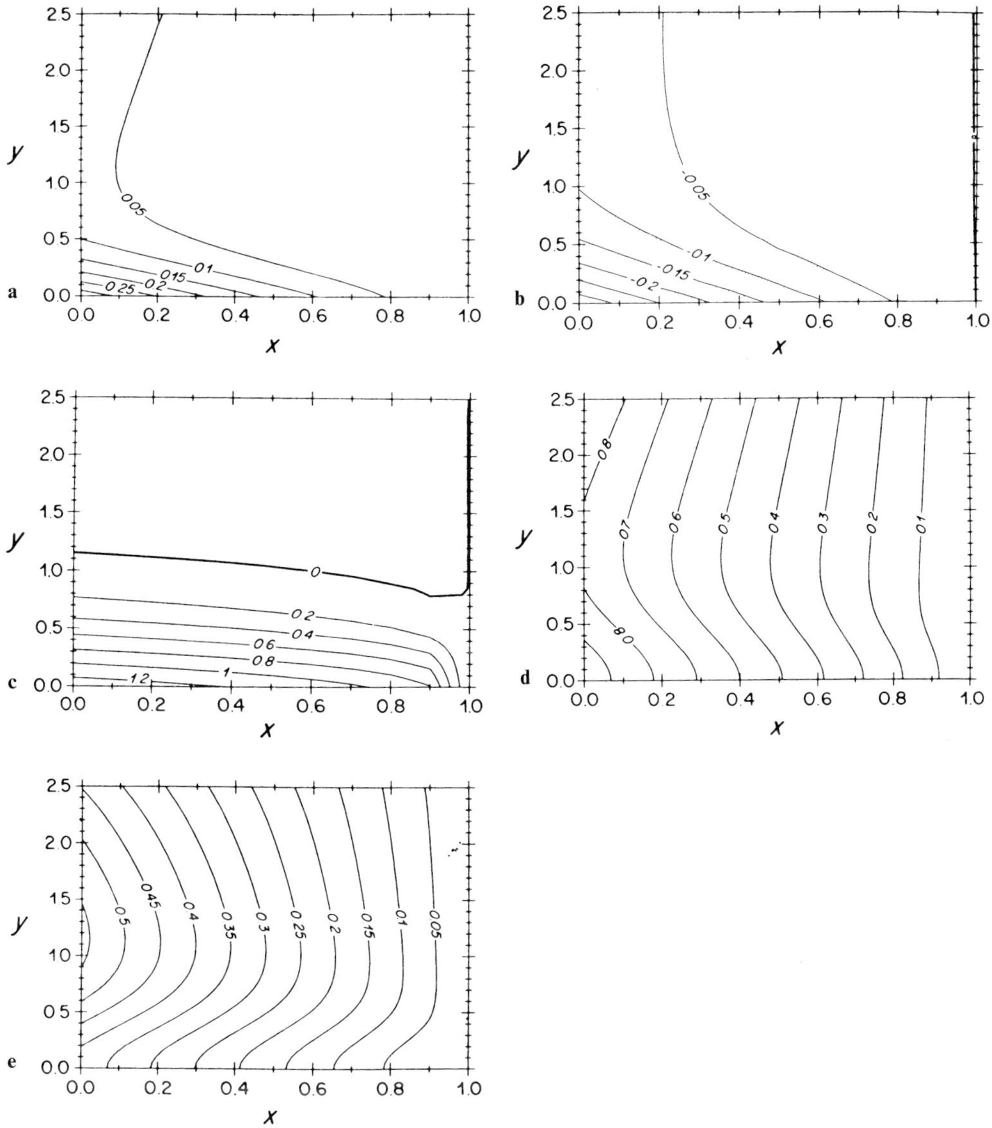


**Fig. 6.6.4a-e.** Two-layer coupled model. **a** Upper layer streamfunction, $\psi_1$. **b** Lower layer streamfunction, $\psi_2$. **c** Contours of $u_2$. **d** $h$. **e** $h_1$. The calculation is for $r = 1$

served in the heuristic model in which $M(x)$ is specified. Panels d and e show $h$ and $h_1$, respectively, demonstrating the bowing of the isopycnals at the equator and the surfacing of the current as it progresses eastward.

The most interesting aspect of the solution, and in view of our earlier remarks perhaps the most reliable part as well, is the rapid decline in the strength of the undercurrent near the eastern boundary. Although the solution requires the undercurrent to be depleted by the time it reaches the eastern boundary, the brusqueness with which this takes place is remarkable and reflects the increasing influence of the entrainment on the shallower structure of the current in the eastern part of the basin.

## 6.7 Numerical Models

The connection between the dynamics of the EUC and the subtropical thermocline is clearly demonstrated by the representation of the EUC as an inertial boundary-layer extension to the equator of the solution for the midlatitude thermocline. The midlatitude thermocline and the EUC are two overlapping elements of a single, complete picture of the general circulation. Nevertheless, the proposed boundary-layer solutions themselves are incomplete and do not yield a completely closed circulation. Only plausibility arguments have been used to link the equatorial flow with the western boundary layer, for example. Similarly, the termination of the undercurrent by entrainment into the upper mixed layer is as much a hypothesis as a definitive result of the theory since the entrainment process is not adequately represented in the simple theories presented. One way to examine the consistency of the hypothesized dynamical elements of the conceptual model is to construct a numerical model whose dynamics is sufficiently complete to allow us to check at least the self-consistency of these individual elements. Of course, care must be taken that the numerical model itself is adequate, and in particular it must be able to describe both the equatorial and midlatitude circulations well enough to persuasively examine their linkage.

The early pioneering model of Cane (1979) is especially significant in that it carefully emphasized the importance of the zonal variation of the flow for the dynamics. Its vertical structure, however, consisted of a layer of a single density. The layer is heuristically subdivided into an upper mixed layer communicating with a single laminar layer below. With the added limitation of the model to a latitude band of 15° on either side of the equator, it is difficult to use the model to examine the connection of the equatorial flow with the midlatitude thermocline. Nevertheless, the model is successful in generating an equatorial boundary layer form for the EUC whose dynamics is largely inertial, and whose velocity satisfies the scaling relation (6.3.12). It is an important indication that the ideas of Fofonoff and Montgomery and the inertial theory outlined in Section 6.4 are relevant to the problem in which the finite zonal extent of the basin is explicitly considered.

A more recent numerical model constructed for the expressed purpose of examining the link between the equator and the midlatitude regions has been presented by McCreary and Lu (1994). The model consists of two moving layers over a resting abyss. The model is an interesting hybrid model containing elements of the standard layer models as well as aspects of conventional continuous models. In particular, the temperature (i.e., density) of each layer is allowed to vary horizontally *within each layer*. It is *assumed* that the density remains independent of $z$ within the layer. The horizontal velocity, if it is geostrophic, therefore varies linearly with $z$. The advection of the density by the *geostrophic* thermal wind is identically zero. If the velocity is geostrophic then, the depth-dependent advective terms in the density equation for each layer are zero, and the assumption of purely horizontal variations of temperature within the layer is consistent with the density equation, assuming the sources of heating are uniform over each layer's thickness. However, as (6.3.17) shows, it is unlikely that the meridional velocity is geostrophic in the equatorial region, and the assumption of a $z$-independent temperature field is somewhat inconsistent there. Nonlinear advection by the ageostrophic velocity produces a $z$-dependent density field. A similar problem arises in the zonal momentum equation due to the nonlinear advection of momentum.

Heuristic entrainment and detrainment laws, governing the transfer of mass between the upper two layers, allow the model to mimic the process of subduction in midlatitudes. Indeed, when the thermodynamics of the model is idealized as adiabatic, McCreary and Lu are able to reproduce the ventilated thermocline solution of Luyten et al. (1983) for the subtropical part of the circulation. The solution in the equatorial region is found numerically *without* the boundary layer assumptions used in previous sections of this chapter. The model domain is large enough to easily contain both the midlatitude circulations of subpolar and subtropical gyres as well as the equatorial flow.

The model, in spite of its inconsistencies in the equatorial region, is thus an ingenious heuristic representation of the circulation. Its ability to close the circulation linking the midlatitudes and the equator, without a priori assumptions of boundary layer matching, makes it a useful vehicle for examining the connection between the two regions. The reader is referred to the paper for further details of the model's formulation.

The wind stress in the "standard" model calculations is chosen to have a negative curl over the region of the subtropical gyre and to become independent of latitude over a broad tropical region straddling the equator. Thus in the region outside the anticipated equatorial boundary layer we expect a subsurface geostrophic flow to balance the poleward meridional Ekman flux. Experiments with more complex stress patterns, in which the stress vanishes in the equatorial domain, are also considered.

Figure 6.7.1 shows the velocity fields as calculated for the upper and lower layers. McCreary and Lu superimposed on the results of the numerical calculation the boundaries of the shadow zone (emanating from the point labeled $y_d$ on the eastern edge of the basin) and the critical streamline connecting the outcrop line with the bifurcation latitude of the western boundary current which occurs at the point labeled $y_b$ in panel b. For the parameters of this calculation the shadow strikes the western boundary at $y_e$ well outside the equatorial boundary layer region but south of the bifurcation latitude. It is

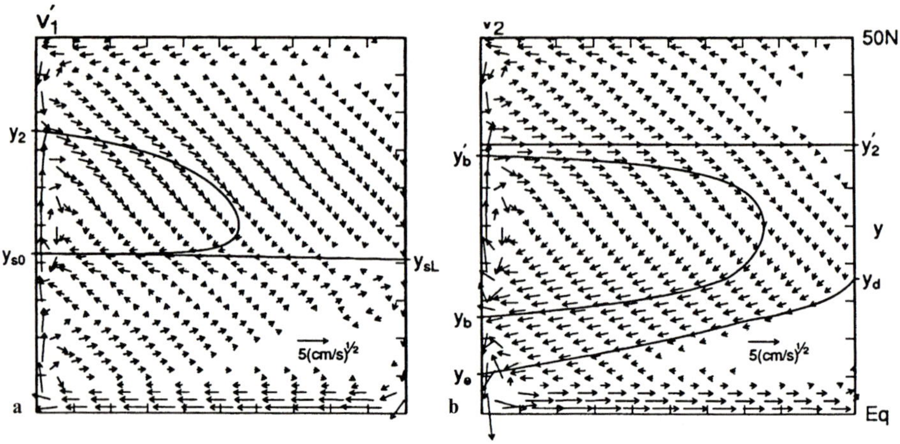

**Fig. 6.7.1a,b.** Circulation in the two-layer model of McCreary and Lu (1994). In the figure the velocity fields in the upper **a** and lower **b** moving layers are shown. *Solid lines* (**b**), analytically derived boundaries of the shadow zone and criticial streamline linking the bifurcation latitude of the western boundary current with the outcrop line at $y = y_d$. In the calculation the EUC is fed from the subtropical gyre through the western boundary current

remarkable how accurately these boundaries, estimated by the analytical theory, continue to guide the pathways of the flow. We can observe in layer 2 how the flow passes from the subduction latitude, $y_d$, and continues into the equatorial zone through the western boundary current after the current bifurcates at $y_b$. In this case, as would be anticipated by the results of Section 6.5, the EUC is fed entirely by the western boundary current.

In Fig. 6.7.2 the same model has been run with a shallower second layer, i.e., $H_2$ is smaller, so that the shadow zone boundary is shifted far enough eastward to strike the equator before hitting the western boundary. The numerical calculations again show the EUC being fed by fluid from the western boundary current but now also by the interior in agreement with the model of the midlatitude/equatorial linkage. The pathway from the midlatitude gyre to the equator is clear, and the EUC exists as the equatorward branch of the ventilated thermocline circulation.

McCreary and Lu (1994) experimented with various wind-stress distributions. They demonstrated that if the wind stress in the equatorial region is expunged, so that fluid cannot reach the equator, the EUC simply disappears, clearly emphasizing the connection between the two regimes. The equatorial wind stress is important as a mechanism for delivering fluid to the equator, where primarily inertial processes are then responsible for producing the EUC.

Perhaps the most complete numerical model of the equatorial region has been constructed by Philander and Pacanowski (1984). Their model is an extension to the equator of the Bryan and Cox (1967) model. It is a primitive equation model (i.e., no significant approximation is made in the horizontal

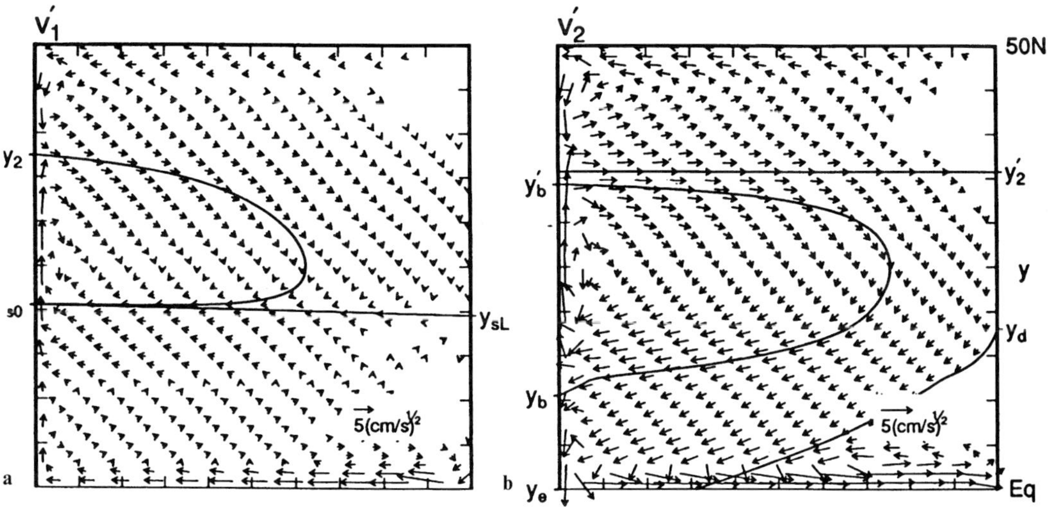


**Fig. 6.7.2a,b.** As in Fig. 6.7.1 except that the parameters of the calculation yield a shadow zone boundary which strikes the equator within the basin. In this case the EUC is fed from the subtropical gyre through the interior as well as the western boundary current. (From McCreary and Lu 1994)

momentum equations while the vertical equation of motion is the hydrostatic balance). It contains 27 levels in the vertical and its horizontal resolution is 1° in longitude by 1/3° in latitude. It has a fairly substantial horizontal mixing coefficient ($A_H = 2 \times 10^7$ cm$^2$/s), and its vertical mixing coefficient depends on the Richardson number, Ri $= \left[N^2/|\partial\vec{u}/\partial z|^2\right]$ in such a way that in regions of low static stability or high shear the mixing rate is increased. The model includes the subtropical gyre in both hemispheres.

Philander and Pacanowski used their model to simulate both the mean and time-dependent circulations of the equatorial zone, and the model results agree favorably with observations, although the model undercurrent is both weaker than observed in the simulation for the Atlantic and also suffers a too rapid loss of fluid by entrainment into the mixed layer. The equatorial thermocline as computed in the model gradually decays due to anomalous warming, and this is related to deficiencies in the model's representation of the link with the cold deep water of the subtropical gyre. Nevertheless it is to date the most distinguished and successful numerical simulation of realistic equatorial conditions.

The relative completeness of the model naturally also produces the spectre of complexity, and only recently have analyses of the model's results focused on the processes of the kind discussed in this chapter. Rather surprisingly, not until the very interesting and revealing analysis of Wacongne (1989) was it possible to describe and diagnose the equatorial dynamical balances of the model.

Figure 6.7.3a shows the model results in a section along the equator in response to the stress field. The zonal velocity, vertical velocity, temperature and salinity are shown in the remaining panels. In the model the rather weak EUC appears to terminate rather far from the eastern boundary, and this is a

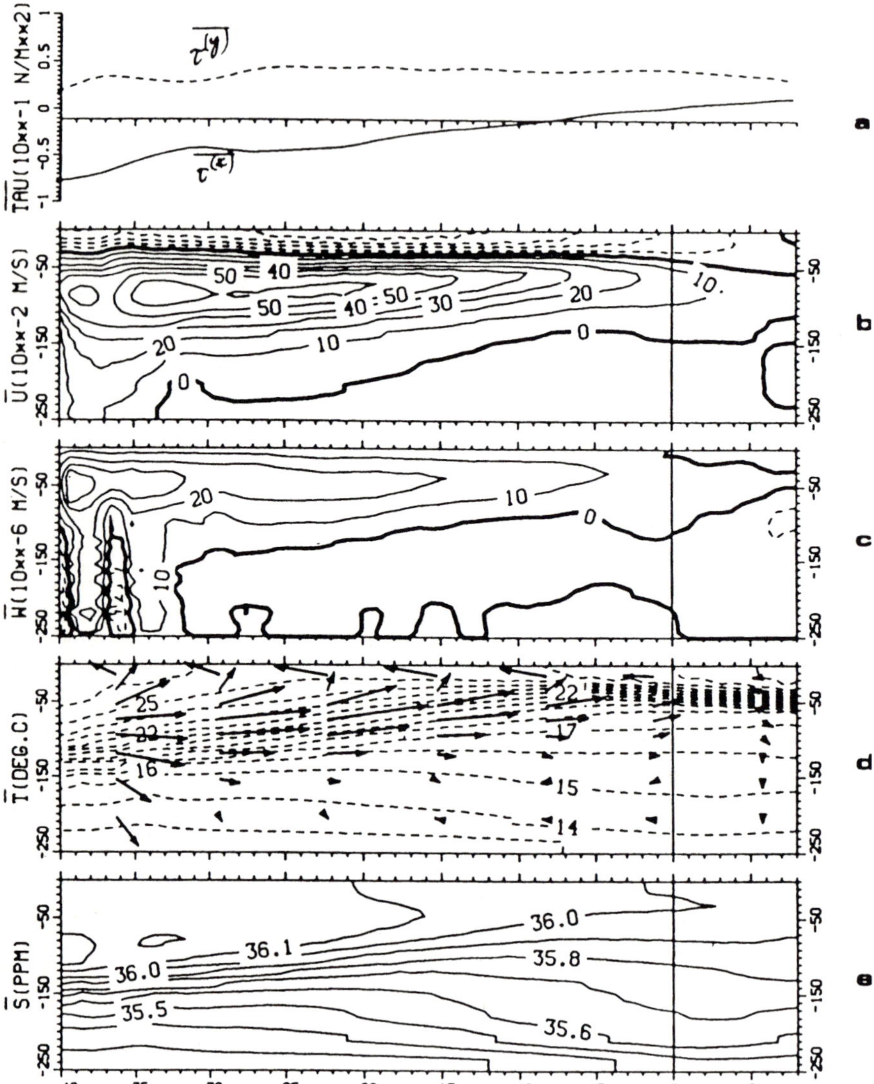

**Fig. 6.7.3a-e.** Results from the Philander and Pacanowski (1984) model applied to the equatorial Atlantic. **a** Zonal and meridional wind stress used in the calculation. **b** Isopleths of zonal velocity in the equatorial zonal plane. **c** Isopleths of vertical velocity. **d** Isotherms in the zonal section with selected velocity vectors *(u, w)* superimposed. **e** Salinity field. (From Wacongne 1989)

sign of the rather large dissipation affecting the current, which also is responsible for its artificially low speed. In panel c the velocity vectors in the $x$-$z$ plane are superimposed on the isotherms, and we observe a rather stronger than expected cross-isopycnal flux except at the very core of the current. In spite of these discrepancies with regard to observations the model is dynamically consistent and reproduces the overall length and velocity scales seen in observations.

Wacongne carefully analysed the balances in the zonal momentum equation, and her results are shown in Fig. 6.7.4. In the upper panel the zones of eastward flow are shown in white, with the region of the equatorial thermocline superimposed as cross-hatching. The core of the model EUC is shown by the dot-dashed heavy line while a heavy dashed line indicates a boundary below which both the zonal pressure gradient and the vertical mixing of momentum are negligible.

A heavy dark line in panel a delineates the region (2) in panel b in which the momentum equation reduces to a balance between the zonal pressure gradient and the eastward acceleration. In this region the simple adiabatic, inertial model of the EUC discussed in Section 6.4 is applicable. As the core of the undercurrent flows eastward it enters a region, labeled (3) in which the horizontal pressure gradient vanishes, and in which the eastward acceleration is small. This is reminiscent of the nonaccelerating region east of the intersection of the shadow zone boundary with the equator discussed in Section 6.5. Above this region vertical mixing penetrates, and the strong vertical mixing is responsible for the deceleration of the current. Indeed, in the center of the basin the core of the model undercurrent falls beneath the thermocline, and here the otherwise free jet is restrained by horizontal mixing.

The numerical model describes a rather more dissipative EUC than either observations or the inertial theory suggest, but it does indicate the regime boundaries anticipated from the simple theories.

This has been further emphasized by the very illuminating analysis of the particle trajectories in the model by Liu et al. (1994) who used the Philander and Pacanowski model in an idealized domain but with similar parameter settings and domain sizes as in the study of Wacongne (1989). Figure 6.7.5 shows the streamline pattern (indicated by the arrows and the dot-dashed lines) and the isolines of potential vorticity (light solid lines) on three density surfaces corresponding to (a) the region somewhat above the core of the EUC, (b) the region of the core, and (c) the region beneath the undercurrent core. We see that water in the core (panel b) reaches the EUC from the interior of the subtropical gyre along paths that nearly conserve potential vorticity in the interior. The fluid reaches the EUC along a path through the western boundary current. The model current receives little water along its flanks in the core, and its acceleration on this surface is limited, and this is consistent with the analysis of Wacongne. Water somewhat above the core flows into the current from both the western boundary layer and the interior although its path in the interior is a contorted zig–zag curve between the subtropical gyre and the EUC. In the

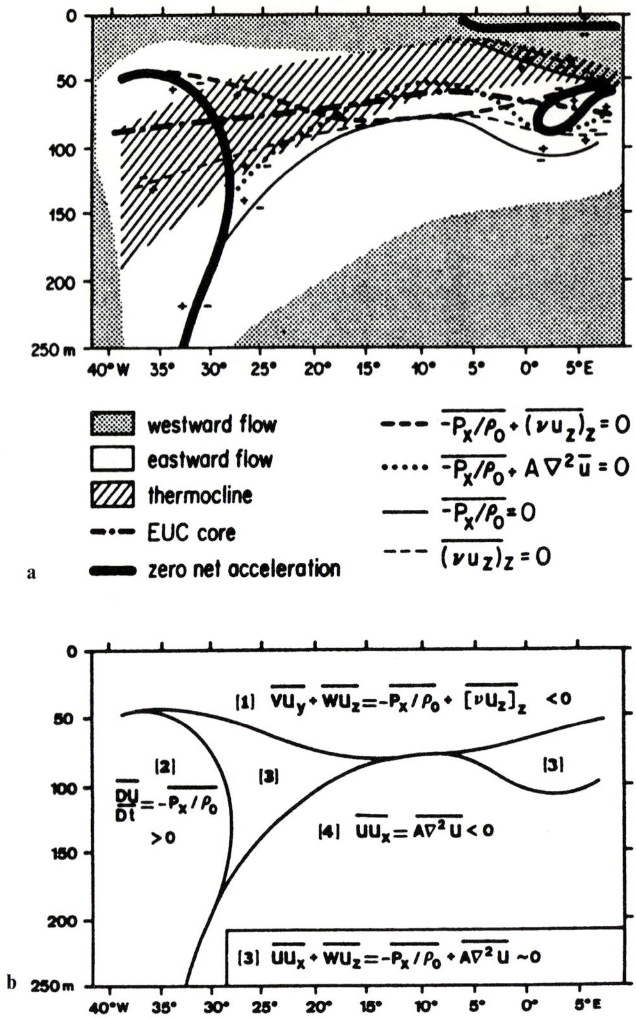


**Fig. 6.7.4a,b.** "Cartoon" showing schematically the balances in the zonal momentum equation along the equator in a vertical plane as manifested in the Philander and Pacanowski model. (From Wacongne 1989)

model most of the water in the EUC travels a considerable distance from deep within the subtropical gyre and its tendency to carry its midlatitude values of potential vorticity to the EUC emphasizes the dynamic link between them. The general circulation of the equatorial regions is therefore an integral part of the overall thermocline circulation, and in particular its potential vorticity is determined by advection to the equatorial region by flow in the subtropical gyre.

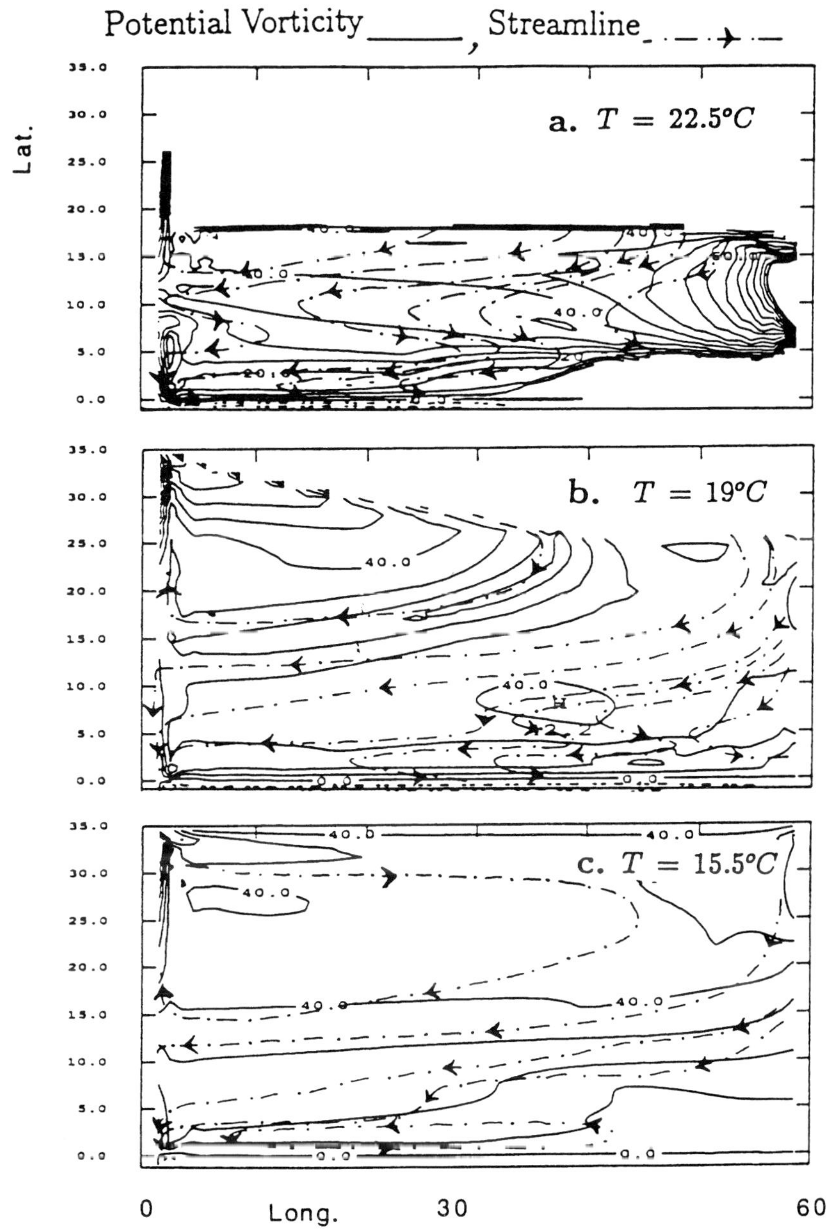


**Fig. 6.7.5a-c.** Potential vorticity (*solid lines*) and the streamlines (*dot-da curves with arrows*) at three levels in the Philander-Pacanowski model showing the connection between the subtropical gyre and the equatorial circulation. Levels at 22.5, 19, and 15.5 °C, corresponding to levels above, at and below the core of the EUC. At the level of the core flow in the subtropical gyre flows to the equatorial region largely conserving potential vorticity and enters the EUC from both the western boundary current and the interior. In the calculation shown the former path accounts for most of the EUC transport. (From Liu et al. 1994)

## References


Bryan, K. and Cox M.D. 1967: A numerical investigation of the oceanic general circulation. *Tellus*, **XIX**, 54–80.

Bryden, H.L. and Brady E.C., 1985: Diagnostic study of the three-dimensional circulation of the upper equatorial Pacific Ocean. *J. Phys. Ocean.*, **15**, 1255–1273.

Cane, M.A., 1979: The response of an equatorial ocean to simple wind stress patterns: II Numerical results. *J. Mar. Res.*, **37**, 253 –299.

Charney, J.G., 1960: Non-linear theory of a wind-driven homogeneous layer near the Equator. *Deep Sea Res.*, **6**, 303–310.

Charney, J.G., and Spiegel S.L., 1971: Structure of wind-driven equatorial currents in homogeneous oceans. *J. Phys. Ocean.*, **1**, 149–160.

Fine, R.A., 1987: The penetration of tritium into the tropical Pacific. *J.Phys. Ocean.*, **17**, 553–564.

Fine, R.A.,Reid J.L., and Ostlund H.G., 1981: Circulation of tritium in the Pacific Ocean. *J. Phys. Ocean.*, **11**, 3–14.

Fofonoff , N.P. and Montgomery R.B., 1955: The Equatorial undercurrent in the light of the vorticity equation. *Tellus*, **7**, 518–521.

Halpern, D. and Weisberg R.H., 1989: Upper ocean thermal and flow fields at 0°, 28° W (Atlantic) and 0°, 140° W (Pacific) during 1983–1985. *Deep Sea Res.*, **36**, 407–418.

Johnson, E.S. and Luther D.S., 1994: Mean momentum balance in the upper and central equatorial Pacific Ocean. *J. Geophys. Res.*, **99**, C4, 7685–7705.

Liu, Z., 1994: A simple model of the mass exchange between the subtropical and tropical ocean. *J. Phys. Ocean*, **24**, 1153–1165.

Liu, Z., Philander S.G.H. and Pacanowski R.C., 1994: A GCM study of subtropical- tropical upper ocean mass exchange. *J. Phys. Ocean.* **24**, 2606–2623.

Lukas, R, 1986: The termination of the equatorial undercurrent in the eastern Pacific. *Prog. Ocean.*, **16**, 63–90.

Luyten, J.R., Pedlosky J. and Stommel H., 1983: The ventilated thermocline. *J. Phys. Ocean.*, **13**, 292–309.

McCreary, J.P. Jr., 1981: A linear stratified ocean model of the equatorial undercurrent. *Phil. Trans. Royal Soc. London*, **298**, 603–645.

McCreary, J.P. Jr., 1985: Modeling equatorial circulation. *Ann. Rev. Fluid Mech.*, **17**, 359–409.

McCreary, J.P. Jr., and P. Lu, 1994: Interaction between the subtropical and equatorial circulations: The subtropical cell. *J. Phys. Ocean.*, **24**, 466–497.

McCreary, J.P. Jr. and Yu Z., 1991: Equatorial dynamics in a 21/2-layer model. *Prog Ocean.* , **29**, 61–132.

Meyers, G., 1979: Annual variation in the slope of the 14° C isotherm along the equator in the Pacific Ocean. *J. Phys. Ocean.*, **9**, 885–891.

Moore, D.W. and Philander S.G.H., 1977: Modelling of the tropical oceanic circulation . In *The sea*, **6**, Eds. Goldberg E.D., McCave I.N., O'Brien J.J. and Steele. J.H. *Wiley Interscience*, New York, pp 319–361.

Pedlosky, J., 1987: An inertial theory of the equatorial undercurrent. *J. Phys. Ocean*, **17**, 1978–1985.

Pedlosky, J., 1988: Entrainment and the termination of the equatorial undercurrent. *J. Phys. Ocean.*, **18**, 880–886.

Pedlosky, J., 1991a: The link between western boundary currents and equatorial undercurrents. *J. Phys. Ocean.*, **21**, 1553–1558.

Pedlosky, J., 1991b: Theoretical developments in ocean circulation theory. *Environmental dynamics series* , **I.**, CNR/Istituto Veneto di Scienza, Lettere ed Arte. Summer School on Environmental Dynamics, Venetzia, Italia, pp 124.

Pedlosky, J., and Samelson R.M., 1989: Wind forcing and the zonal structure of the equatorial undercurrent. *J. Phys. Ocean.*, **19**, 1244–1254.

Philander, S.G., 1990: El Nio, La Nia, and the southern oscillation. *Academic Press*, New York, 293pp.

Philander, S.G., and Pacanowski R.C., 1984: Simulation of the seasonal cycle in the tropical Atlantic Ocean. *Geophys. Res. Letters*, **11**, 802–804.

Tsuchiya, M., Lukas R., Fine R.A., Firing E., and Lindstrom E., 1989: Source water of the Pacific equatorial undercurrent. *Prog. Ocean.*, **23**, 101–147.

Wacongne, S., 1989: Dynamical regimes of a fully non linear stratified model of the Atlantic equatorial undercurrent. *J. Geophys. Res.*, **94**, 4801–4815.

Wacongne, S., 1990: On the difference in strength between Atlantic and Pacific undercurrents. *J. Phys. Ocean.*, **20**, 792–799.

Wyrtki, K. and Koblinsky B., 1984: Mean water and current structure during the Hawaii to Tahiti shuttle experiment. *J. Phys. Ocean.*, **14**, 242–254.

# 7 Abyssal Circulation

## 7.1 Introduction

The base of the thermocline at a depth of roughly 1 km represents the lower boundary of the domain of the vigorous ocean circulation driven by the wind-stress and the large-scale distribution of heating and cooling. Below this region lies a vast volume of fluid extending to depths typically of 4–5 km although the depth diminishes to about 2 km in the vicinity of the great midocean ridge systems. This region is everywhere very cold. In the world ocean the potential temperature is less than 4 °C virtually everywhere below 2 km and in most areas is closer to 2 °C at this depth (e.g. Levitus 1982) and the temperature falls typically to 0–2 °C at 4 km. This *abyssal* region of the ocean is so vast that Worthington (1981) attributes an average temperature of only 3.51 °C to the entire world ocean, so great is the domain of the cold water – in spite of the widespread areas of surface temperatures of the order of 20 °C .

As Warren (1981) remarks in his excellent review article, the implication of the widespread presence of cold water in the region below the thermocline, even in the tropical ocean, was appreciated nearly 200 years ago by Count Rumford who inferred from this fact the existence of a global, deep circulation. Waters of such low temperatures can be formed by cooling of the ocean by the atmosphere only in polar regions. The presence at other latitudes of such cold water implies a large-scale deep circulation, the *abyssal circulation*, which carries the water formed in polar regions to the rest of the ocean. Cold water flows from the polar regions to fill the deep ocean basins, from which the water must eventually rise to the surface and, heated to the observed surface temperatures, must then flow poleward to replace the water which has sunk to the bottom, forming an endless global cell of motion.

The sources of the cold water are found in both polar regions. In the Weddell Sea and the Ross Sea of the Antarctic region cold, saline water sinks to the bottom along the continental shelf to begin its journey northward. At the same time, water from the high-latitude North Atlantic spills through a complex series of passages past Greenland, Iceland and the British Isles and produces the North Atlantic Deep Water that travels southward in the region beneath the thermocline. Estimates of the transport of deep water from each source vary, but each polar region is thought to be responsible for an input into the major basins of between 10–20 $10^6 \times m^3/s$ (i.e., 10–20 sverdrups). Not all of this water has been produced at the surface by cooling. A fraction (see below) is estimated to be water entrained during the sinking as well as entrained laterally as the source water enters the deep basins. In the North Atlantic, for example, McCartney and Talley (1984) have estimated a formation rate of North Atlantic Deep Water of 13–14 sverdrups. In both polar regions the zones of sinking are quite narrow, and the convective sinking itself is difficult to observe directly. The sinking instead is largely inferred from the distribution of ocean properties such as temperature, salinity, and especially oxygen.

The midocean velocities expected in response to this ocean-filling advent of cold water are very low, about an order of magnitude smaller than the wind-driven circulation. At the lowest order these small velocities allow us to calculate the wind-driven circulation without paying much attention to the abyssal circulation which *in the context of the thermocline theory* is negligible.

Nevertheless, the abyssal circulation is of interest in its own right because (a) it occupies such a large proportion of the ocean by volume where, small as is the velocity, it *is* the general circulation, and (b) due to the large volumes involved, its transport is commensurate with the thermocline transport. It is also of considerable importance because of its role in determining the deep temperature, which in turn is responsible in conjunction with the surface heating for determining the background stratification of the ocean. Although considerations of our planet's climate do not enter into the discussion of this book, it should also be intuitively apparent that the massive flux of cold water from the poles to equatorial regions, balanced as it must be by warmer surface water flowing poleward to replace it, must play a key role in the earth's heat balance.

How the abyssal circulation takes place was a subject of considerable early controversy, and the reader is referred to Warren's essay (1981) for a fascinating historical review. Because of the difficulty in making the required long-term direct measurements of currents in the abyss, the deep circulation is still much less well constrained by observations than that of the upper ocean. Therefore, although it occupies the major part of the ocean by volume, the limited theoretical and observational understanding of the deep circulation is reflected here by the nearly inverse proportion of its representation in this book.

In spite of the difficulty in making direct velocity measurements, the measurement of tracer fields and standard hydrographic measurements of the temperature and salinity reveal a circulation which is anything but simple. Figure 7.1.1 from Pickard and Emery (1982) shows meridional cross sections of temperature, salinity, density, and oxygen in the Atlantic Ocean. Especially in the salinity field the layered structure of the circulation is evident. Cold Antarctic Bottom Water extends northward in a giant wedge overlain by North Atlantic Deep Water whose southward motion is inferred from the property fields. Above the North Atlantic water is another tongue of Antarctic water, the Antarctic Intermediate Water, which extends northward into the northern

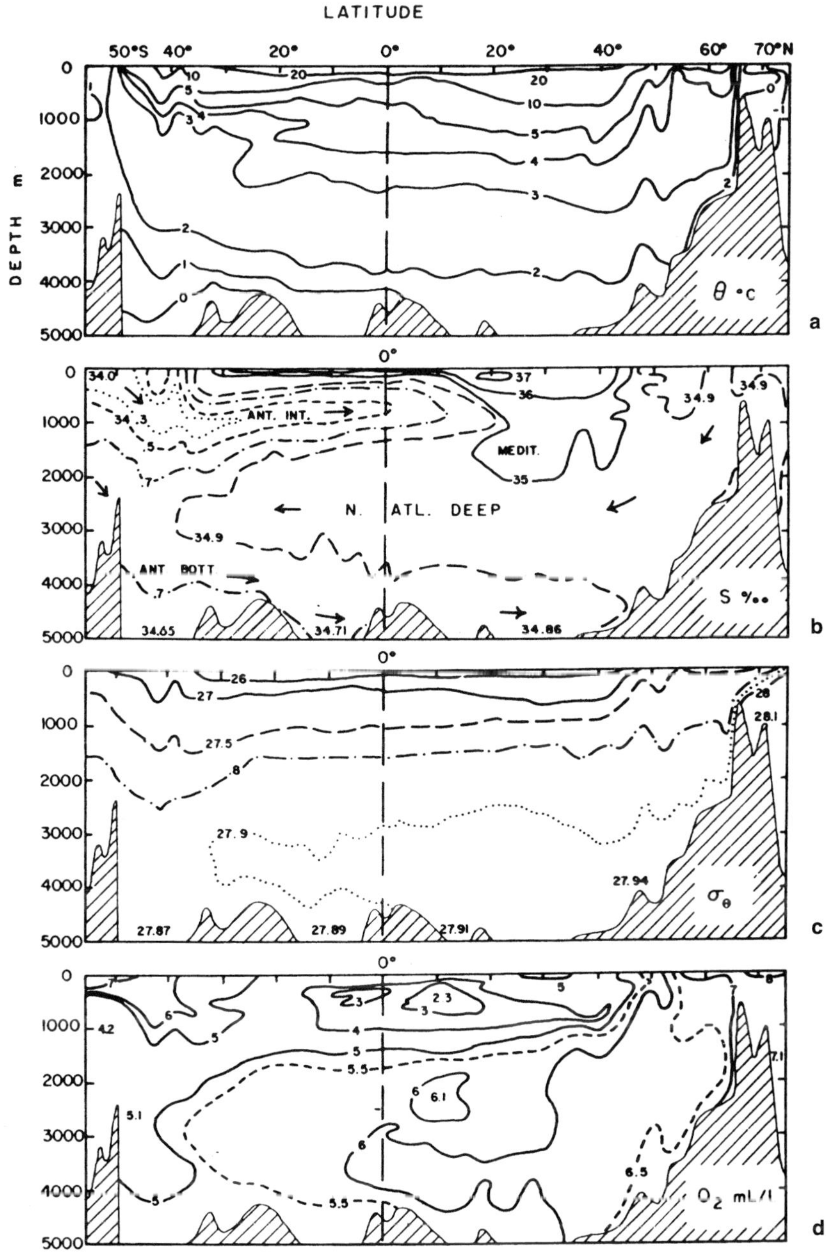


**Fig. 7.1.1.** North-south sections of temperature (**a**), salinity (**b**), density (**c**), and oxygen (**d**) in the Atlantic ocean in the western basin of the ocean. Note the evidence in the salinity field for the layered character of the circulation with tongues of water from sources in the Antarctic appearing to flow northward while being separated by southward flowing North Atlantic deep water. (From Pickard and Emery 1982)

hemisphere. The tongues of the oxygen maxima are particularly important in identifying the poles as the sources of the deep water.

It is virtually impossible on the basis of direct observations to sketch the horizontal structure of this layered circulation with confidence, but the signal of the circulation is easily seen in a complex series of western boundary currents responsible both for carrying part of the flow from the polar sources and for closing the abyssal circulation occurring in the ocean interior. Figure 7.1.2 from Warren (1981) shows the signal in the hydrographic fields of potential temperature, salinity, silica, and oxygen of the western boundary currents associated with the Antarctic bottom water and the southward flowing North Atlantic deep water above. These cross sections also emphasize the strong topographic features at the depths of the abyssal circulation where the Mid-Atlantic Ridge is seen to separate the Atlantic basin into separate subbasins.

The bottom topography evidently has a strong influence on the flow. Observations of deep flows (e.g., Warren and Owens 1988; McCartney et al., 1991; Friedrichs and Hall 1993) each demonstrate the important role of topography in shaping the structure of the deep circulation. In particular the passages from one subbasin to another through breaks in the midocean ridge systems strongly condition the shape of the flow. Figure 7.1.3 from McCartney et al. (1991) shows a bulging, isolated pool of water with potential temperature below 1.8 °C trapped on the eastern side of the gap in the Mid-Atlantic Ridge formed by the Vema fracture zone at 11°N in the Atlantic. Friedrichs and Hall (1993) associate this region with a recirculating gyre which they infer is fed through the Vema fracture zone in the deep water, as shown in Fig. 7.1.4.

Thus the abyssal circulation possesses simultaneously global scales associated with the overall response to the production of cold water in polar regions and its subsequent spread though the abyss, as well as subplanetary scale motions as a consequence of the complex bottom topography of the ocean, of which the Mid-Atlantic Ridge and its system of gaps and faults is a typical example.

The modern theory of the abyssal circulation truly begins with a series of remarkable papers by Stommel (1958) and Stommel and Arons (1960a,b). In fact, the dynamical theory which they developed is actually a straightforward application of Sverdrup theory as described in Chapter 1, but its contextual framework is very original. The application and synthesis of the familiar ideas of Sverdrup theory in the context of the abyssal circulation, driven by prescribed sources of deep water in polar regions, leads to very counterintuitive results, and the theory in this form is familiarly called the Stommel-Arons theory.

A principal component of the Stommel-Arons theory and all subsequent theories is the notion that the water that rises out of the abyss to replace the water which has sunk at the poles does so over broad lateral scales. The water that sinks does so in local, semienclosed regions of the polar oceans, for example, as in the Weddell Sea, and the evidence of the sinking can be read in

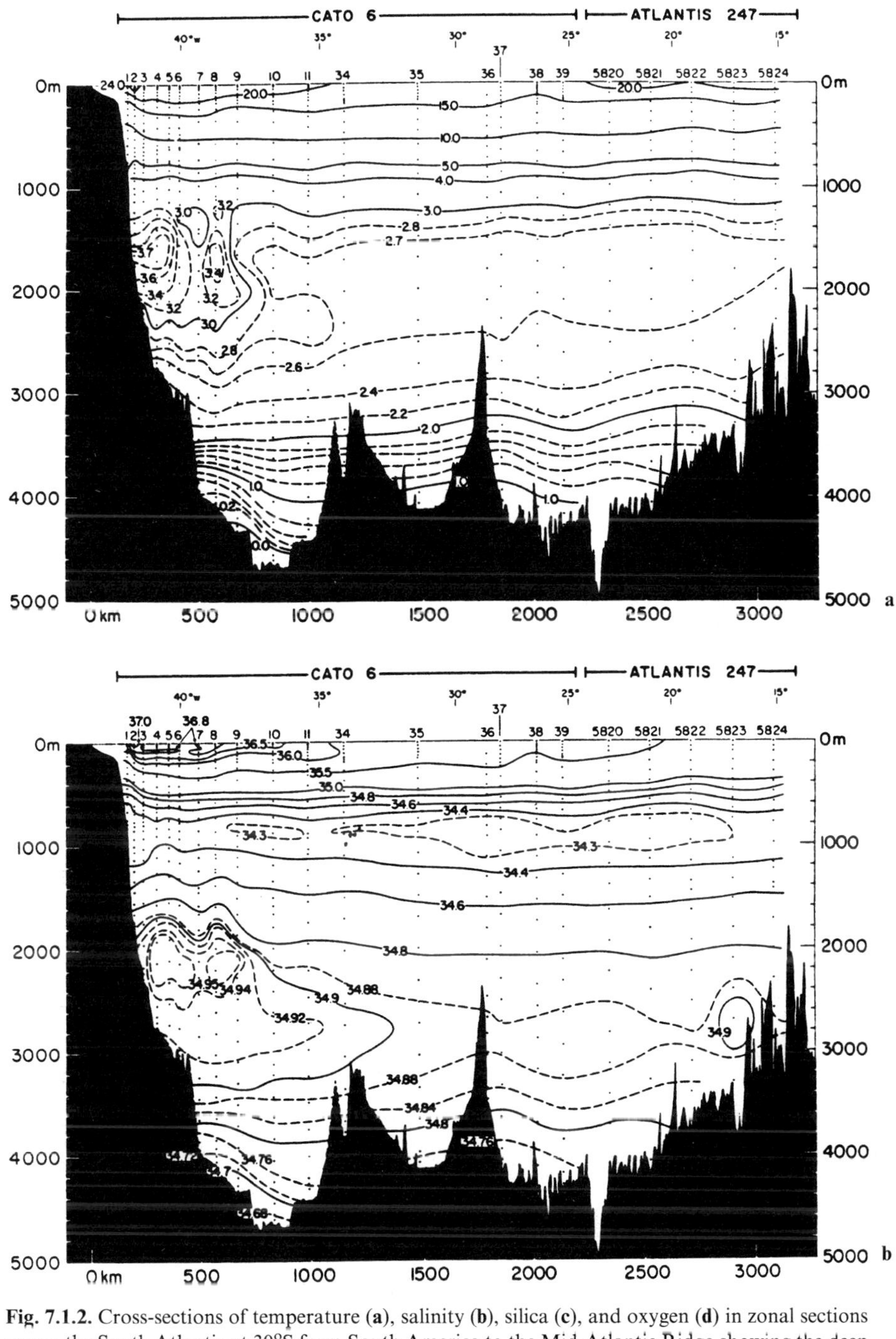


**Fig. 7.1.2.** Cross-sections of temperature (**a**), salinity (**b**), silica (**c**), and oxygen (**d**) in zonal sections across the South Atlantic at 30°S from South America to the Mid-Atlantic Ridge showing the deep western boundary currents associated with Antarctic Bottom Water and, at around 2000 m, the southward flowing North Atlantic Deep Water. (From Warren 1981)

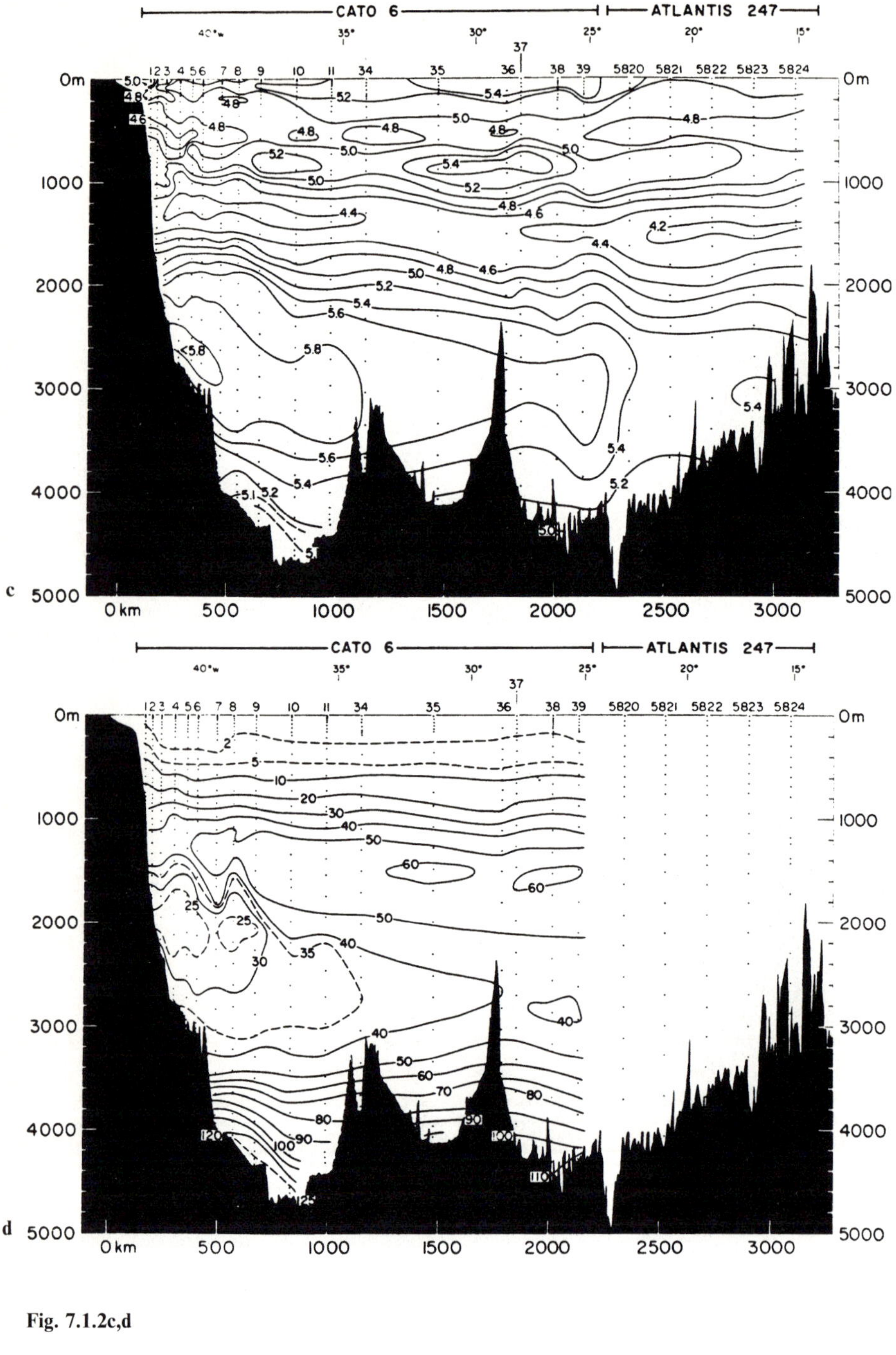
CATO 6
ATLANTIS 247
40°W
35°
30°
25°
20°
15°
0m
1000
2000
3000
4000
5000
0 km
500
1000
1500
2000
2500
3000
c
d

Fig. 7.1.2c,d

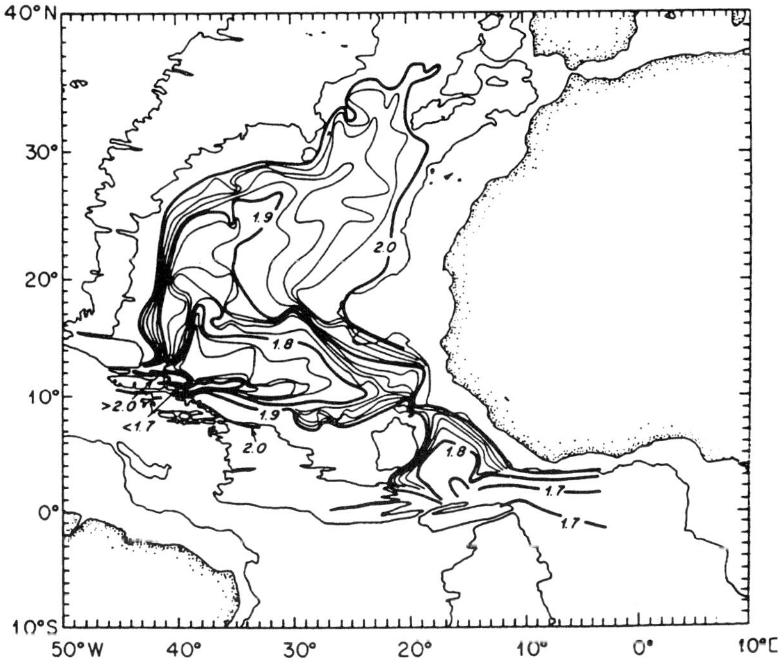


**Fig. 7.1.3.** Contours of bottom potential temperature in the North Atlantic in the region of the Vema channel at 11°N. The contour interval is 0.02 °C. Note the isolated pool of water colder than 1.8 °C to the east of the Vema channel. (From McCartney et al. 1991)

the property fields in these locations. If the water which rises in compensation were to rise in equally localized zones, these regions too could be observed in the distribution of the property fields. Such regions have not been seen, and, as in the Sherlock Holmes story in which the absence of a dog's bark was vital in unraveling a mystery, the absence of observations of narrow rising regions is similarly vital in implying that the rising motion is broad and widespread, rendering it so slow as to be unobservable directly.

At the time that the Stommel-Arons theory was first formulated the oceanic thermocline was pictured fundamentally as a balance between the downward diffusion of heat from the ocean surface and the upward advection of cold water from the abyss. The temperature equation was conceived as a full balance between advection and diffusion of the form:

$$\vec{u} \cdot \nabla T + \mathrm{w}\frac{\partial \mathrm{T}}{\partial \mathrm{z}} = \kappa \frac{\partial^2 \mathrm{T}}{\partial \mathrm{z}^2} \tag{7.1.1}$$

where the first term represents horizontal advection of the temperature. Near the base of the thermocline where the velocities and temperature gradients are

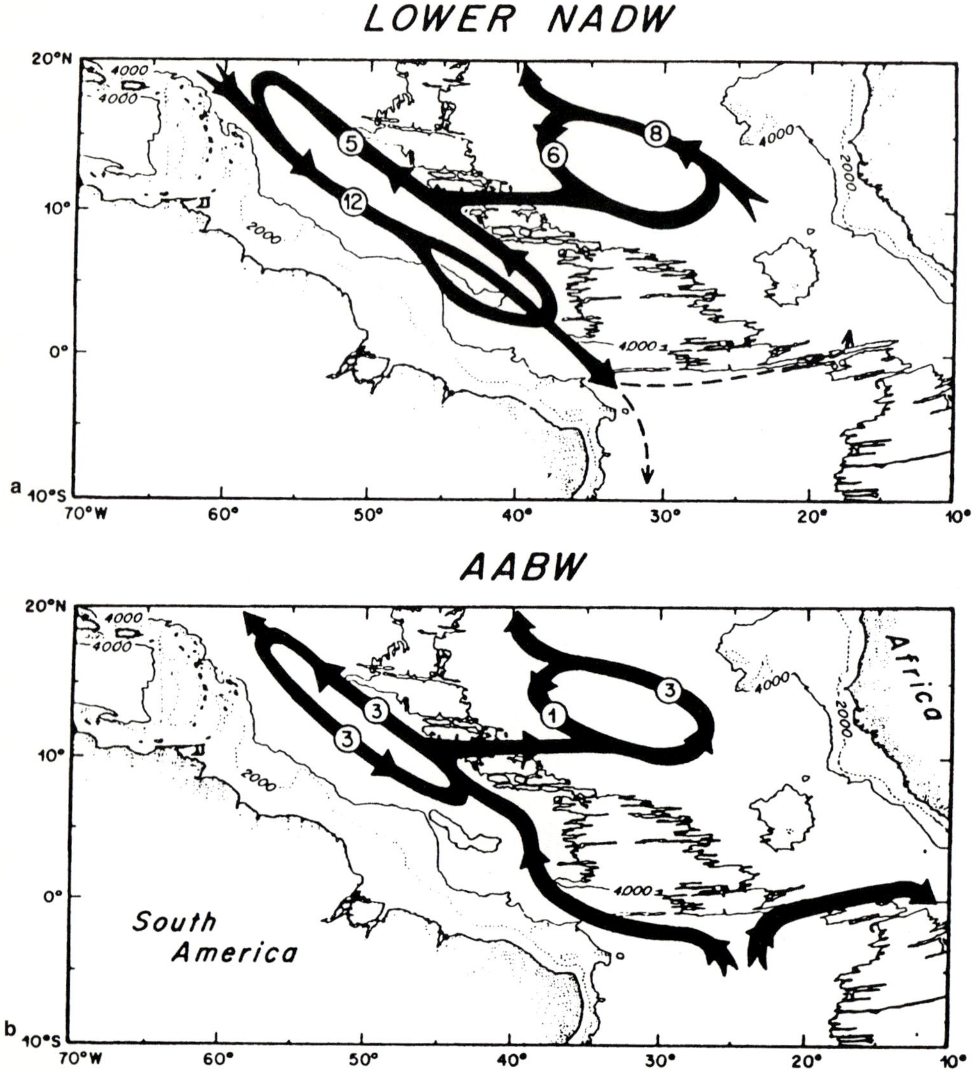


**Fig. 7.1.4.** Schematic circulation patterns in the lower part of the Atlantic circulation, i.e., in the lower North Atlantic Deep Water (**a**) and the Antarctic Bottom Water in the region of the Vema channel (**b**) showing the cyclonic circulation in the region east of the Mid-Atlantic Ridge as inferred by Friedrichs and Hall (1993)

weak, such solutions of the continuous equations for the thermocline which existed suggested the asymptotic balance (see Pedlosky 1986):

$$w_\infty \Re = \kappa \frac{\partial \Re}{\partial z} \tag{7.1.2}$$

where:

$$\Re = \kappa \frac{\partial T}{\partial z} \tag{7.1.3}$$

is the downward heat flux. In this form (7.1.2) is valid even if $\kappa$ is a function of $z$.

Thus, at the base of the thermocline a vertical velocity is produced by the requirement that the rising cold water balance the downward heat diffusion in the steady state. This leads to an estimate of the vertical velocity at the base of the thermocline:

$$w_\infty = \frac{\kappa}{d} \tag{7.1.4}$$

where $d$ is the vertical scale of variation of the heat flux. This vertical velocity, acting on the abyss in a manner analogous to the way in which the Ekman pumping drives the thermocline, was taken by Stommel and Arons as the driving mechanism on the global scale for the abyssal circulation.

Our view of the thermocline, as developed in Chapters 3 and 4, is more advective than diffusive. The results of recent experiments by Ledwell et al. (1993; see also Osborn 1980; Gregg and Sanford 1988) have been referred to above as consistent with this view of the negligible role of diffusion in the dynamics of the thermocline. Nevertheless, it is conceivable that the balance (7.1.2) obtains *at the base of the thermocline* where the velocities are weak even if the presence of diffusion in (7.1.1) is not relevant in the main body of the thermocline.

The overall area of the oceans is roughly $3 \times 10^8$ km$^2$. To obtain a flux of 20 Sverdrups from the abyss to the thermocline to replace the waters sinking into the abyss, an average value for $w_\infty$ would have to be $0.66 \times 10^{-5}$ cm$^2$/s. For a vertical scale of the abyssal motion, $d$, of the order of 1 km, this would suggest an ocean wide average for $\kappa$ of 0.66 cm$^2$/s. This is several times larger than the estimate by Ledwell et al. (1993) of 0.11 cm$^2$/s. As we see below, the abyssal circulation, as predicted theoretically, involves as considerable amount of *recirculation*, involving fluid which does not participate in the vertical overturning cell, but which revolves horizontally in response to the upwelling $w$. The source of near surface water in the North Atlantic, for example, is considerably enhanced by mixing during its journey from the sinking region through the overflows into the North Atlantic basin. Estimates suggest that about half of the deep flow entering the basin is due to entrainment (Warren 1981; Price and Barringer 1994). If the source strength is thus reduced by even half, say, the estimate of the required $\kappa$ becomes considerably closer to the directly measured value of the mixing parameter in the thermocline. There is of course a good deal of uncertainty in all these numbers, and it would be unwise to insist on any particular numerical estimate, but it does seem at least reasonable to consider as apt the original suggestion of Stommel and Arons, i.e., that the thermocline-induced $w_\infty$ is the driving mechanism for the large-scale motion in the abyssal interior. This forms a fundamental presumption of the development which follows.

It is important to note that the theory developed earlier in Chapters 3 and 4 for the dynamics of the thermocline assumes that the effect of the abyssal motion is negligible. The weak velocity of the abyssal motion was neglected in calculating the slope of the deep density surfaces in the thermocline. In addition, the Sverdrup balance was applied to only the depth interval of the thermocline. This is equivalent to the condition that $w_\infty$ is negligible compared to $w_E$ so that the Sverdrup condition can be applied to only the thermocline, and this seems apt on the basis of the above estimates. As noted in Chapter 1, this is also equivalent to assuming that the characteristic *transport* of the abyss is locally small compared to the wind-driven transport, and it is certainly not clear from observations that this is true. If the vertically integrated transports in the oceanic abyssal *interior* were of the same order as the wind-driven transports, the separation of the dynamics of the abyss and the thermocline become problematic, and the two regions must then be considered simultaneously. The theory for such a combined dynamics is currently lacking. Instead, in most of this chapter the thermocline-induced, deep vertical velocity is assumed to be determined only by the thermocline's dynamics, independently of the abyssal motion, and is responsible for driving the interior abyssal motion. Furthermore, it has recently been suggested that the upwelling into the thermocline may indeed be geographically limited. Observations of uniformly low values of the mixing coefficient over most of the oceanic interior are matched by observations of very strong mixing near deep topographic features (Toole et al.1994). It is quite possible then that the overall balance suggested by Stommel and Arons is valid, only that $w_\infty$ is in fact geographically limited. Note that this would imply a generally weak interaction between the dynamics of the abyss and thermocline dynamics over that broad part of the interior where $w_\infty$ is negligible. The effect of such a geographically limited $w_\infty$ is taken up in Section 7.7.

## 7.2 The Stommel, Arons, and Faller Experiment

The theoretical consequences of the application of Sverdrup theory to source driven flow, as would be relevant for the abyss, seemed so nonintuitive to Stommel and Arons themselves that they, in collaboration with Faller, carried out an elegant laboratory experiment to check some of the more surprising results. We invert the process here for didactic purposes and start with an analysis of the experiment (Stommel et. al. 1958).

The experimental situation is shown schematically in Fig. 7.2.1. A pie-shaped basin contains water, and it is spun about a vertical axis at a rate $\Omega$ which was about 1.05 $s^{-1}$ in the experiment. The basin has an apex angle, $\theta_0$, which in the actual experiments was 60°. The water, of density $\rho_0$, has a depth

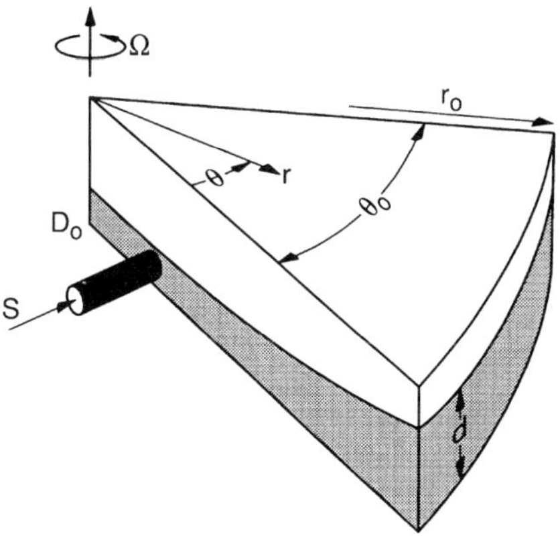


**Fig. 7.2.1.** Schematic representation of the Stommel et al. (1958) experiment. A pie-shaped sector of radius $r_0$ and sector angle $\theta_0$ spins at a rate $W$ about a vertical axis. The sector is fed by a source of strength $S$, and the fluid (*shading*) has a free surface that slopes upward with increasing radius due to the centrifugal effect of the rotation. The source slowly feeds the fluid in the basin, and the free surface of the fluid slowly rises. The position of the source is variable and sinks can also bleed fluid away from the basin

$d$; in the experiments the average depth was about 10 cm. The radius of the sector is $\rho_0$ and this was about 100 cm in the experiments. A localized source of water, shown as entering a lateral boundary in the figure, delivers fluid to the tank at a rate of $S$ cm$^3$/s.

If fluid is in the tank, and there is no source, the depth of the fluid, $d$, in the steady state when there is no motion in the system rotating with the tank satisfies:

$$d = D_0 + \frac{\Omega^2 r^2}{2g}. \tag{7.2.1}$$

This means that the background thickness field, before relative motion occurs, increases radially. Thus the background potential vorticity, $2\Omega/d$, increases inwardly from the rim to the apex of the sector. This endows the sector with an analog to the planetary $\beta$ effect, and the reader can think of the apex of the sector as representing "north."

When the source, $S$, feeds fluid into the sector, the fluid in the sector moves in response. If the source strength is weak enough, that is, if:

$$\frac{S}{\Omega \theta_0 D_0 r_0^2} \ll 1 \tag{7.2.2}$$

the volume change per revolution is slight and the circulation provoked by the source is quasisteady. Indeed, the parameter in (7.2.2) may also be thought of as the relevant Rossby number for the experiment, and under the conditions of the inequality the motion in sector geostrophic except, perhaps, in boundary currents.

The total depth field of the fluid in motion is defined as:

$$d = D_0 + \frac{\Omega^2 r^2}{2g} + \eta \tag{7.2.3}$$

where $\eta$ is the perturbation in the free surface elevation. This corresponds to a hydrostatic pressure field:

$$p = \rho_0\left(\frac{\Omega^2 r^2}{2} - gz\right) + \rho_0 g\eta. \tag{7.2.4}$$

The momentum and mass equations for the single layer of depth $d$ become, after (7.2.4) is used for the pressure and friction is ignored:

$$\begin{aligned}&\frac{\partial\vec{u}}{\partial t} + \vec{u}\cdot\nabla\vec{u} + 2\vec{\Omega}\times\vec{u} = -g\nabla\eta\\ &\frac{\partial\eta}{\partial t} + \nabla\cdot(\vec{u}\eta) + \vec{u}\cdot\hat{r}\left[\frac{\Omega^2 r}{g}\right] + \left\{D_0 + \frac{\Omega^2 r^2}{2g}\right\}\nabla\cdot\vec{u} = 0\end{aligned} \tag{7.2.5a,b}$$

where $\hat{r}$ is a unit vector in the radial direction. In (7.2.5a) a portion of the pressure field balances the centrifugal force, and only the term in $\eta$ is left to drive the motion.

When the source strength is small enough that (7.2.2) is satisfied, solutions for $\eta$ can be conveniently written in the form:

$$\eta = \eta_0\frac{t}{T} + \pi(r,\theta,t) \tag{7.2.6}$$

where $\eta_0/T$ is a constant to be determined. The free surface rises uniformly as the basin fills with water added by the source. Then (7.2.5 a,b) becomes:

$$\begin{aligned}&\frac{\partial\vec{u}}{\partial t} + \vec{u}\cdot\nabla\vec{u} + 2\vec{\Omega}\times\vec{u} = -g\nabla\pi\\ &\frac{\partial\pi}{\partial t} + \left\{D_0 + \frac{\Omega^2 r^2}{2g} + \eta_0\frac{t}{T} + \pi\right\}\nabla\cdot\vec{u} + \vec{u}\cdot\left[\hat{r}\frac{\Omega^2 r}{g} + \nabla\pi\right] = -\frac{\eta_0}{T}.\end{aligned} \tag{7.2.7a,b}$$

When (7.2.2) is satisfied, the relative velocities are weak compared to the rotational velocity of the system, a fact which can be checked *a posteriori*.To the lowest order the velocity outside any boundary layers in which nonlinearity or friction might be important, is in geostrophic balance:

$$\vec{u} = \frac{g}{2\Omega}\hat{k}\times\nabla\pi \tag{7.2.8}$$

where $\hat{k}$ is a vertical unit vector. Note that the term in (7.2.6) which is linear in t does not yield a geostrophic velocity. If (7.2.8) is used in (7.2.7b) we obtain as a steady solution for $\pi$:

$$\frac{\partial\pi}{\partial\theta} = \frac{2\eta_0}{\Omega T}. \tag{7.2.9}$$

If we think of the right hand wall of the sector as the "eastern" boundary of the sector (remember that the potential vorticity of the fluid is increasing towards the apex of the sector), the interior velocity should satisfy the condition that the azimuthal (zonal) velocity vanish there, or equivalently that $\pi$ should be constant there. This holds unless there are sources of mass put in directly at the eastern boundary. In such cases the interior zonal velocity must accept the mass input with obvious alterations in the determination of $\pi$.

Thus except for the case of eastern sources $\pi$ is a linear function of $\theta$ only. Therefore the only geostrophic velocity is $u$, the radial velocity, and this is given by:

$$u = -\frac{g}{r\Omega^2}\frac{\eta_0}{T}. \tag{7.2.10}$$

The velocity in the interior of the sector is always directed towards the apex *regardless of the position of the source*. The radial (outward) mass flux is equal to:

$$T_1 = \int_0^{\theta_0} dru\, d\theta = -\frac{g}{\Omega^2}\theta_0 d\frac{\eta_0}{T} \tag{7.2.11}$$

and its radial dependence depends only on $d(r)$.The mass balance for the sector as a whole requires that the rate at which the volume increases is given by the source strength $S$, i.e.:

$$\frac{\eta_0}{T}\frac{r_0^2}{2}\theta_0 S. \tag{7.2.12}$$

which determines $\eta_0/T$ in terms of $S$. The radial velocity is then related to the source by:

$$u = -\left(\frac{2gS}{\Omega^2 r_0{}^2\theta_0}\right)\frac{1}{r}. \tag{7.2.13}$$

The interior velocity arises as a response to the overall rise in the level of the water in the basin and not to the position of the source of the water. It always flows towards the apex and is analogous to the Sverdrup flow in the ocean interior, which responds only to the interior Ekman pumping. Indeed, if we define the equivalent $\beta$ as:

$$\beta_{\text{eff}} = \frac{2\Omega}{d}\frac{\partial(d)}{\partial r} = \frac{2\Omega^3 r}{gd} \tag{7.2.14}$$

then the ratio $2\Omega/r\beta_{eff}$ is given by:

$$\frac{2\Omega}{r\beta_{eff}} = \frac{gd}{\Omega^2 r^2} \tag{7.2.15}$$

which gives the ratio of the flow driven in the interior by vortex tube stretching with respect to the flux driven directly by the source strength.

On the other hand, it is clear that water must flow away from the apex somewhere, and we can anticipate that this occurs in a "western" boundary layer, with transport $T_w$, as shown in Fig. 7.2.2. Note that $T_w$ is measured as positive for transport towards the apex. To determine the transport $T_w$ it is necessary only to construct the mass balance for that part of the sector contained within a radius r from the apex where $r \leq r_0$. The total source strength $S$ can be divided into several subsources, some of which, as $S_0$, lie within the subregion under consideration and others, such as $S_1$, lie outside the region.

The mass into this subregion consists of the interior flow, the boundary layer flow and the source $S_0$. The sum of these yields the rate of increase of the volume in the subdomain. Thus, to balance mass:

$$T_w + S_0 - T_I = \frac{\eta_0}{T}\frac{r^2\theta_0}{2}. \tag{7.2.16}$$

Recall that $T_I$ is positive for outward flow. The right side of (7.2.16) represents the rate of increase of the volume in the pie-shaped sector with radius $r$. If (7.2.12) and (7.2.13) are used, we obtain:

$$T_w = -S_0 - S\frac{2gD_0}{\Omega^2 r_0^2} \tag{7.2.17}$$

which *is independent of r*. This equation for the transport in the western boundary layer of the system is of course an hypothesis. The solution for the boundary layer has not been found so that (7.2.17) is in fact a consistency condition rather than a bona fide solution for the boundary layer. Similarly, the solution is not complete until the boundary layer required on the *rim* at $r = r_0$ has been found in order that the radial velocity can be brought to rest there. These are important aspects of the total solution that are technically difficult to complete and the experiment is analyzed below under the assumption that such boundary-layer elements of the complete solution can be found. The analysis of Chapter 2 shows that this is a risky assumption in general, but if the source strength is weak enough these boundary layers are linear frictional currents which can always accept the role required for them by

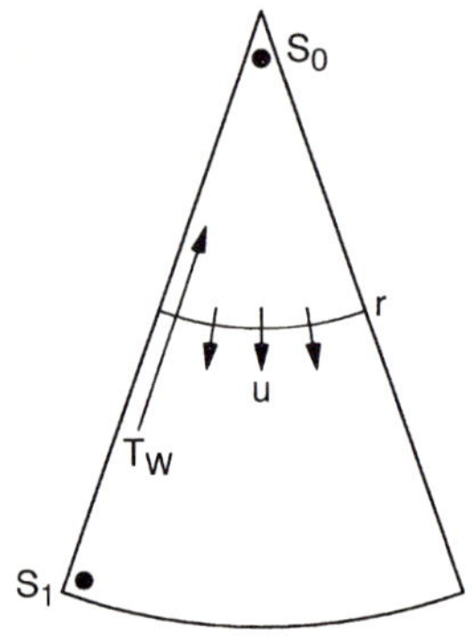


**Fig. 7.2.2.** Schematic plan view of the Stommel et al. (1958) experiment. Sources $S_0$ and $S_1$, such that $S = S_0 + S_1$, are placed at the apex and rim of the sector. A western boundary current has a transport $T_w$, measured positive flowing towards the apex. Mass balance conditions determine the strength of $T_w$

the boundary conditions and mass balance. Hence there is an additional implicit limit on the magnitude of the forcing by the sources.

There are several interesting cases that Stommel, Arons, and Faller considered in their experiment.

Suppose first, for example that the full source of fluid is placed at the apex, i.e., that $S_1 = 0$ and $S_0 = S$ (Fig. 7.2.3a). Then:

$$T_w = -S\left\{1 + \frac{2gD_0}{\Omega^2 r_0^2}\right\}. \tag{7.2.18}$$

The boundary layer flows "southward," away from the source with a mass flux larger than the source strength by an amount $S(2gD_0/\Omega^2 r_0^2)$ which is the measure of the interior mass flux flowing *towards* the source due to the interior stretching of vortex columns. That is, in addition to the flow directly related to the fluid emerging from the source, a substantial *recirculation* of the flow is also taking place, driven in the interior by vortex tube stretching. The boundary current flowing away from the source must carry the combined flux of the source *and* the recirculation of the interior flow which, on entering the

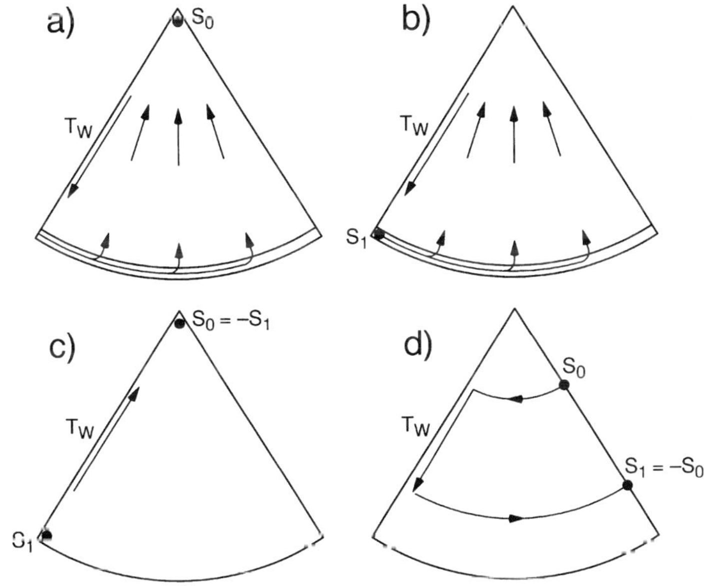


**Fig. 7.2.3a-d.** Schematic presentation of four experiments with different locations for the sources. **a** A single source $S_0 = S$ is placed at the apex. The interior flow is towards the source, and $T_w$ is the boundary layer transport moving away from the source with a transport greater than the source. **b** The source is placed at the outer rim. There is no net source and no interior flow. The flow moves from the source at the rim to the sink at the apex. **d** A source is placed on the "eastern" boundary at one radius, and a sink of equal strength is placed at a larger radius. The flow moves from source to sink on the indirect path though the western boundary layer as shown

apex, is turned southward towards the outer rim to complete the circuit of recirculation. The mass flux of the western boundary current is therefore only an indirect measure of the source strength. While proportional to S, it depends on the parameter $(2gD_0/\Omega^2 r_0^2)$. In the Stommel, Arons and Faller experiment this parameter was order one and so the boundary-current transport is therefore about double the source strength.

If instead, the source of fluid enters at the rim as shown in Fig. 7.2.3b, $S_0 = 0$ and $S_1 = S$. The interior flow is not altered at all, of course. However, now the western boundary current flows southward, *towards* the source, with a strength:

$$T_w = -S\left(\frac{2gD_0}{\Omega^2 r_0^2}\right) \tag{7.2.19}$$

whose value can be greater or less than the source strength, depending on the size of the potential vorticity gradient produced by the slope of the free surface. The interior flow (measured positive outwards) has a transport at the outer rim which is inward, fed by a boundary current along the rim, and this interior transport is greater in magnitude than the source strength and equal to:

$$T_I = -S\left[1 + \frac{2gD_0}{\Omega^2 r_0^2}\right]. \tag{7.2.20}$$

The excess of the interior transport over the source strength $S$ is therefore exactly compensated by the flow towards the source of the western boundary current.

A third interesting possibility is when there is no net source strength at all. The water in the sector is driven by a source at the rim and a sink at the apex (or viceversa) such that $S = 0$ and $S_0 = -S_1$, as shown in Fig. 7.2.3 c. In this case the interior flow is zero and the flow is entirely in the boundary layer flowing from source to sink with a transport, from (7.2.17):

$$T_w = -S_0 = S_1 \tag{7.2.21}$$

Finally, imagine, still with the overall source strength equal to zero, that the source and sink are both placed on the "eastern" boundary, as shown in Fig. 7.2.3d. Since transport boundary layers on the eastern wall are not permitted by geostrophic dynamics, the flow cannot simply stream from source to sink along the eastern boundary. The interior radial velocity, for $S = 0$, is also predicted to be zero. Zonal velocities, on the other hand, are permitted by (7.2.9) even when $S$ (or equivalently $h_0$) is zero. They were previously ruled out since the zonal velocity at the eastern edge of the sector was assumed zero. If however, fluid is injected there, the interior flow on the eastern rim, must match that of the injected fluid since there is no boundary layer along the eastern rim and the interior solution satifies the boundary condition on the zonal velocity there. Hence $\pi$ is a function of r along the eastern boundary in order to produce a geostrophic zonal flow to match the source input. Since $\pi$ is independent of $\theta$, it follows that when $S = 0$ the zonal velocity independent of $\theta$. Fluid can proceed across the sector only as shown in Fig. 7.2.3d. After reaching the western boundary it can travel along the western boundary to the radius of the sink, and retracing its path in azimuth, it leaves the sector by flowing eastward at constant radius to the sink. This is perhaps the most bizarre of all the cases. The constraints placed on the motion by the potential vorticity dynamics in the interior and the absence of an eastern boundary layer forces the flow into an immensely indirect and intricate path to flow from source to sink.

It is even now astonishing to realize that these dynamical ideas achieved full confirmation in the experiments of Stommel, Arons, and Faller (1958), and the reader is referred to their original paper both for a complete discussion of the experimental evidence and for the original sense of wonder at the power of the constraint imposed by the ambient potential vorticity gradient which yields such remarkable flow patterns.

With this experimental confirmation in hand Stommel and Arons were emboldened to apply their ideas to planetary motions in the ocean's abyss.

## 7.3 Stommel-Arons Theory: Abyssal Flow on the Sphere

Stommel and Arons (1960 a,b) extended the dynamical ideas of the previous section to describe a theory for the abyssal motion on a spherical earth. They envisioned the entire ocean as a two-layer model with localized sinking at both poles from the upper to lower layer and a widespread return flow from the lower layer, which represents the abyss, into the upper layer, which is meant to represent the thermocline driven by the wind. The return flow has a vertical velocity, $w_0$, specified over the area of the ocean basin at the interface between the two layers, i.e., at the top of the abyssal layer. This distributed sink acting on the abyss provides the driving mechanism for the interior flow. As in the analysis of the Stommel, Arons, and Faller experiment, western boundary currents are added as needed to satisfy mass balance considerations.

In this form the abyss is represented as a single layer, and therefore at each geographical location the abyss moves in a single direction. As we have noted above, the evidence from tracer and hydrographic data implies that motion in the abyss is strongly baroclinic, with interleaving of water masses moving in quite different directions. The Stommel-Arons model is clearly unable to deal with this vertical structure and should be considered as a model only for the vertical average of the abyssal flow. The authors themselves were concerned about this restriction. (I am indebted to Arnold Arons for a fascinating and enlightening conversation about the historical development of the Stommel-Arons theory). However, they considered that the model presented an important first step to which refinements could be added later. Similarly, the model treats the abyss as a layer of fluid with a flat bottom. Topography is ignored, and, again, Stommel and Arons were clear that they expected the pronounced bottom topography of the abyssal ocean basins to cause "deviations" from the simple theory which they presented. In spite of these reservations the theory still exists today as the fundamental building block of our understanding of the motion of the deep water.

For the layer representing the abyss, the motion is geostrophic. Thus in spherical coordinates:

$$fu = -\frac{1}{R}\frac{\partial P}{\partial \theta}, \qquad fv = \frac{1}{R\cos\theta}\frac{\partial P}{\partial \phi}. \tag{7.3.1 a,b}$$

where the variables have their usual meanings, except that we introduce $P = p/\rho$ where $\rho$ is the constant density of the abyssal layer in the model. $R$ is the earth's radius and $\phi$ and $\theta$ are longitude and latitude, respectively, and $u$ and $v$ are the zonal and northward velocities. The continuity equation for the incompressible fluid of the model abyss is:

$$\frac{1}{R\cos\theta}\frac{\partial(v\cos\theta)}{\partial\theta} + \frac{1}{R\cos\theta}\frac{\partial u}{\partial\phi} + \frac{\partial w}{\partial z} = 0. \tag{7.3.2}$$

Eliminating the pressure from (7.3.1 a,b) and using (7.3.2) yields the Sverdrup vorticity equation:

$$\beta v = f\frac{\partial w}{\partial z} \tag{7.3.3}$$

Note that since the scales are planetary in extent, $f$ and $\beta$ are variable, i.e.:

$$f = 2\Omega\sin\theta, \quad \beta = \frac{2\Omega\cos\theta}{R}. \tag{7.3.4 a,b}$$

The abyss is a single layer of fluid which, in the absence of motion is idealized as having a uniform thickness $H$. At the top of this layer a vertical velocity, $w_0$, is specified which drives the interior flow. As discussed in Section 7.1, this velocity is identified with the flow entrained into the base of the thermocline, $w_\infty$, owing to the balance achieved there between diffusion and upward advection of temperature. It is important to keep in mind that in this theory this vertical velocity is not due to thermal anomalies induced only by the motion in the abyss itself but rather is due primarily to dynamics in the adjacent thermocline.

Since the density in the layer is constant, the horizontal geostrophic motion is depth independent and therefore the vertical integral of (7.3.3) over the depth $H$ yields:

$$\beta v = \frac{f w_0}{H} \tag{7.3.5}$$

or:

$$v = \tan\theta \frac{w_0 R}{H} \tag{7.3.6}$$

The meridional velocity is always *poleward* as long as $w_0$ is positive. Thus in the southern hemisphere the flow is southward while it must be northward in the northern hemisphere. As long as $w_0$ is finite at the equator, it follows that $v$ *must vanish on the equator*. There can therefore be no interior flow across the equator. Any flow crossing the equator must occur in regions of different dynamics where the geostrophic Sverdrup balance is upset, i.e., in the western boundary current.

From (7.3.1b) and the condition that $P$ be constant on the eastern boundary of the ocean basin at $\phi = \phi_e$ :

$$P = -\frac{2\Omega R^2}{H}\int_{\phi}^{\phi_e} w_0(\phi, \theta)\sin^2\theta d\phi'. \tag{7.3.7}$$

The zonal velocity follows from (7.3.1a):

$$u = \frac{1}{\sin\theta}\frac{R}{H}\frac{\partial}{\partial\theta}\left[\int_{\phi}^{\phi_e} w_0(\phi, \theta)\sin^2\theta d\phi'\right] \tag{7.3.8}$$

in which form it is valid even for eastern boundaries which are not meridians, i.e., for which $\phi_e = \phi_e(\theta)$.

If, for simplicity, the velocity $w_0$ is taken to be independent of horizontal position, the algebra is considerably simplified. For a barotropic model such as this there is no substantial qualitative effect in considering $w_0$ constant, although when baroclinic effects are considered, as in Section 7.7, the geographical distribution of $w_0$ becomes important. For now, however, we take $w_0$ as a constant. Thus, for example:

$$u = 2w_0\cos\theta\frac{R}{H}(\phi_e - \phi) \tag{7.3.9}$$

so that the zonal flow is always eastward.

Consider as a first example, the ocean basin composed of the spherical sector shown in Fig. 7.3.1. It stretches from the equator to the pole, and its meridional boundaries are at $\phi = \phi_e$ on the eastern and $\phi = \phi_w$ on the western boundary. A source of fluid meant to represent the sinking of cold water at the pole in the North Atlantic is shown at the apex of the sector and has a magnitude $S_0$. The streamlines of the interior flow are already known since, as in the experiment described in the previous section, the interior flow is completely independent of the position of the sources.

The mass balance for the subsector north of an arbitrary latitude $\theta$, shown by the dotted line in the figure, is found by balancing the sum of the interior mass flux, the western boundary current's transport, and the source's contribution, $S_0$, all into the sector, against the flux out of the abyss due to

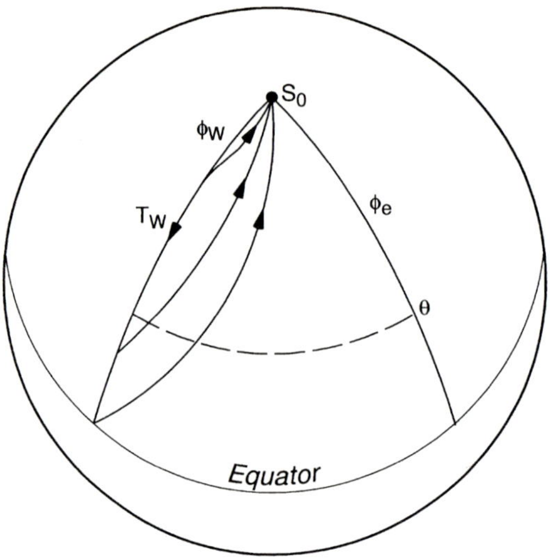


**Fig. 7.3.1.** Circulation in a spherical sector driven by the source $S_0$ in the apex of the basin and uniform upwelling at the upper surface of the abyss. To determine the transport in the western boundary current a mass balance is made for the area north of the arbitrarily chosen latitude $\theta$

the vertical motion through the interface into the thermocline. Thus the northward flux of the interior velocity $T_I$ is, using (7.3.6):

$$T_I = \int_{\phi_w}^{\phi_e} vH \cos\theta \, R \, d\phi' = w_0 R^2 \sin\theta(\phi_e - \phi_w) \tag{7.3.10}$$

while the fluid entering the thermocline and leaving the abyss in that subsector is:

$$T_{\text{out}} = \int_{\theta}^{\pi/2} \int_{\phi_w}^{\phi_e} w_0 R^2 \cos\theta \, d\phi \, d\theta' = w_0 R^2 (\phi_e - \phi_w)[1 - \sin\theta]. \tag{7.3.11}$$

If $T_w$ is the *northward* transport of the western boundary current, and $S_0$ is the source strength, the mass balance implies that:

$$T_w + T_I + S_0 = T_{\text{out}} \tag{7.3.12}$$

or, using (7.3.10) and (7.3.11):

$$T_w = -2w_0 R^2 (\phi_e - \phi_w) \sin\theta + \left[ w_0 R^2 (\phi_e - \phi_w) - S_0 \right]. \tag{7.3.13}$$

If the mass is balanced over this northern hemispheric sector:

$$S_0 = \int_0^{\pi/2} \int_{\phi_w}^{\phi_e} w_0 R^2 \cos\theta \, d\phi d\theta = w_0 R^2 (\phi_e - \phi_w). \tag{7.3.14}$$

In this case, where the sector is isolated from the rest of the ocean, the two terms in the square bracket in (7.3.13) exactly balance and:

$$T_w = -2S_0 \sin\theta. \tag{7.3.15}$$

The western boundary layer transport is southward for all $\theta \geq 0$. At the pole the transport is exactly twice the source strength since the boundary current must carry southward the fluid issuing from the source as well all of the interior flow impinging on the apex. This interior flux, which comprises a basinwide recirculation, has a northward mass flux at the apex equal to the source strength . Thus at the apex half the mass flux of the boundary layer comes from the source and half from the recirculating interior flow. The interior is fed from the western boundary current as the current flows southward, as indicated in Fig. 7.3.1. The transport of the western boundary current falls to zero as the current approaches the equator. During its flow southward the western boundary current loses a total mass flux equal to $2S_0$. One half of this flux is recirculated in the interior and ends up at the apex to return through the western boundary current. The other half is lost through the interface to the thermocline and is directly replaced in the western boundary layer at the apex by the source.

Of course there is no reason why the sector need be isolated from the other hemisphere. That is, there is no a priori reason why the balance in (7.3.14) need apply if there are sources or sinks of fluid in the other hemisphere. The interior flow is still zero at the equator but the transport of the western boundary current at the equator, from (7.3.13) would be:

$$T_w(\theta = 0) = w_0 R^2 [\phi_e - \phi_w] - S_0. \tag{7.3.16}$$

If the upwelling to the thermocline is greater than the source strength at the pole, there must be an import of fluid across the equator from the south, and $T_w$ is positive. If the source strength is larger than the upwelling, the reverse is true, and the boundary current flows across the equator from the northern to southern hemispheres. This is likely to be the situation in the North Atlantic.

On the other hand, if we consider a similar sector in which there is no source at the apex, all the water must be imported from the southern hemisphere. In such a case $S_0$ would be zero and $T_w$ is given by:

$$T_w = w_0 R^2 (\phi_e - \phi_w)[1 - 2\sin\theta] \tag{7.3.17}$$

instead of (7.3.15). This might be the case in a simple model of the North Pacific which is thought to have no local sources of deep water (Warren 1981). The interior flow is the same as in the previous example. However, now at the equator the boundary current is positive as the current delivers the necessary flow to the northern hemisphere sector. As the current flows northward, its mass is exhausted as the flow through the upper surface of the abyss into the thermocline draws fluid from the boundary layer. In fact at 30°N the transport vanishes. North of this latitude the boundary current actually moves *southward*, driven by the cyclonic recirculation of the interior flow.

This important example emphasizes the fact that the direction of the interior flow and the boundary layer flow can be directly opposed to that which might be naively inferred from the knowledge of the source position and the characteristics of the water at any location with respect to the source. Thus, for example, the water in the western boundary current north of 30°N has its source in the southern hemisphere. However, because of the strongly recirculating character of the flow it is moving southward towards the source. A flux equal to $w_0 R^2(\phi_e - \phi_w)$ enters the northern hemisphere across the equator in the western boundary current and enters the interior south of 30°N. Half of this flux is absorbed into the thermocline in the area south of 30°N. At 30°N, where $T_w = 0$, the other half of the original mass flux, i.e. 0.5 $w_0 R^2(\phi_e - \phi_w)$ enters the northern portion of the sector through the interior and upwells into the thermocline. However, a flux equal to the full flow across the equator reaches the apex of the sector, i.e., a flux of $w_0 R^2(\phi_e - \phi_w)$. This flux is fed by the recirculation in the western boundary current in the region north of 30° which carries at the apex a flux, $w_0 R^2(\phi_e - \phi_w)$, southward. Hence the maximum flux in the boundary current at the pole is *double* the transport that entered the region north of 30°. The recirculation is twice as strong as the flow involved in the mass circuit from the source in the western boundary layer at the equator to the distributed sink in the northern portion of the gyre. The ability of the source/sink flow to create strong recirculations tends to decouple the local flow direction from the origin of the source water. This is true both for the interior and the boundary layer.

Consider now an ocean basin bounded between the latitudes $\theta_s$ and $\theta_n$ and the longitudes, as before, $\phi_e$ and $\phi_w$. Again, let $w_0$ be geographically uniform so that (7.3.10) applies. Suppose that a source of strength $S_n$ is placed in the northern sector at the northern boundary, and a source $S_s$ is placed at the southern boundary of the basin. The longitude of each of the sources is arbitrary. The situation is shown in Fig. 7.3.2 and presents a unifying and generalizing arrangement of the previous two examples.

The total mass flux upwelling in the basin is:

$$\begin{aligned}\int_{\phi_w}^{\phi_e}\int_{\theta_s}^{\theta_n} w_0 R^2 \cos\theta\, d\theta\, d\phi &= w_0 R^2(\phi_e - \phi_w)\left[\sin\theta_n - \sin\theta_s\right]\\ &= S_s + S_n \qquad (7.3.18)\end{aligned}$$

which relates the upwelling amplitude to the *total* source strength.

Consider the upwelling into the thermocline in the subregion north of the latitude $\theta$. The flux leaving the abyss is given by the first line in (7.3.18) with $\theta_s$ replaced with $\theta$. The mass balance for this region consists of balancing this sink against the fluid that enters in the western boundary layer, the fluid delivered by the source $S_n$ and the interior flow (7.3.6) integrated across the latitude $\theta$. When the balance is made the result requires a western boundary layer transport:

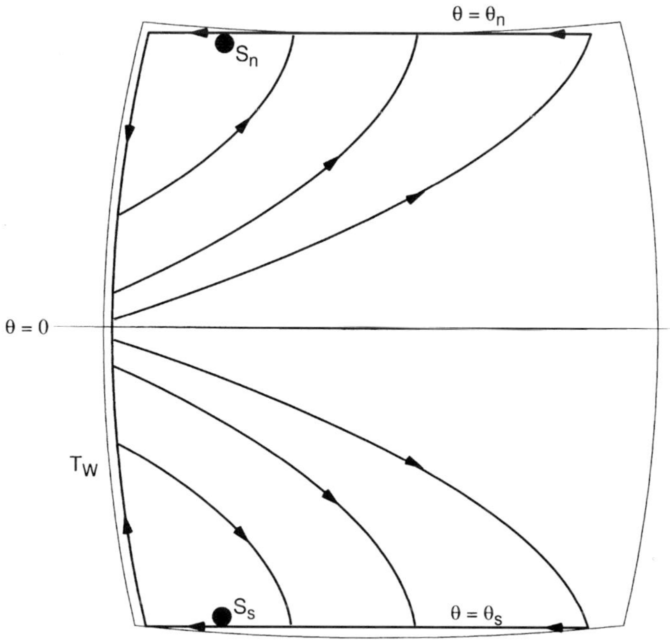


**Fig. 7.3.2.** Enclosed ocean basin whose northern and southern boundaries are $\theta_n$ and $\theta_s$, respectively. The basin typically straddles the equator. A source of strength $S_n$ lies at the northern boundary, and a source $S_s$ is placed at the southern boundary. The inferred circulation is as shown for the case where the two sources are equal, and where $\theta_n = -\theta_s$. Note the northern and southern boundary layers required to close the circulation provoked by the interior flow

$$T_w = w_0 R^2(\phi_e - \phi_w)\left[\sin\theta_n - 2\sin\theta\right] - S_n. \tag{7.3.19}$$

The upwelling velocity is related to the sum of the source terms, i.e.:

$$S_s + S_n = w_0 R^2(\phi_e - \phi_w)\left[\sin\theta_n - \sin\theta_s\right]. \tag{7.3.20}$$

When, for example, $S_n$ is zero the entire deep water source is in the south. Again, this might be considered a model of the situation in the Pacific, which lacks a deep water source in the northern hemisphere. In this case:

$$T_w = w_0 R^2(\phi_e - \phi_w)\left[\sin\theta_n - 2\sin\theta\right] \tag{7.3.21}$$

which vanishes where:

$$\theta = \sin^{-1}\left\{\frac{\sin\theta_n}{2}\right\}. \tag{7.3.22}$$

Thus the position of the point where the western boundary layer flow changes direction is a function only of the latitude of the position of the northern boundary of the basin. If $\theta_n$ is 70° instead of 90° the position of the stagnation point in the western boundary layer moves only slightly, i.e., to 28°N from 30°N. If $\theta_n$ becomes small on the other hand (e.g., the Indian Ocean), the latitude where the boundary current changes direction becomes $1/2\theta_n$. If, instead, the two sources are equal, i.e., if $S_n = S_s = S$:

$$T_w = w_0 R^2(\phi_e - \phi_w)\left\{\frac{\sin\theta_n + \sin\theta_s}{2} - 2\sin\theta\right\}. \tag{7.3.23}$$

If the southern boundary of the region lies in the southern hemisphere, $\sin\theta_s < 0$. If the domain is symmetric, i.e., if $\theta_s = -\theta_n$ the first term in the final bracket of (7.3.23) will vanishes, and the boundary-layer transport vanishes only on the equator. In this case, where both the source strength and the geometry of the two hemispheres are perfect mirror images of each other, there is no interchange across the equator in either the interior or boundary layer. The situation is shown in Fig. 7.3.2. Note the need in general for northern and southern boundary layers required to close the flow produced by the circulation in the interior.

When the sources and each of the hemispheric basins are mirror images, the two hemispheres are isolated from each other. When the symmetry is broken, either by unequal source strengths or by dissimilar basin geometries in the two hemispheres, there results a flux across the equator in the western boundary current, and the two hemispheres are in communication. This lack of symmetry is certainly the more realistic situation and therefore we expect to see cross equatorward transport in the western boundary layer in all basins.

A further illuminating example presented by Stommel and Arons in their original paper is shown in Fig. 7.3.3. A hemispheric sector is partly blocked by a barrier at the longitude $\phi^*$ which stretches from the apex to a latitude $\theta^*$. The barrier is meant to represent a continent, and it is of interest to see its effect on the deep circulation driven by a source, $S_0$, placed to the south of the continent. The source can be thought of either as the actual location of production of deep water or as the point of entry of fluid from the western boundary layer into the basin.

The interior *meridional* flow is still given everywhere by (7.3.6). Consider the region north of $\theta = \theta^*$. There are two subsectors, one on each side of the barrier. The total upwelling into the thermocline for the entire region north of $\theta$ is given by (7.3.11). For the subsector to the right of the continent and in the region north of the latitude $\theta = \theta^*$, the total mass balance for the region $\phi \geq \phi^*$ is:

$$T_w^* + w_0 R^2(\phi_e - \phi^*)\sin\theta = w_0 R^2(\phi_e - \phi^*)[1 - \sin\theta] \tag{7.3.24}$$

so that the boundary layer transport, $T_w^*$, on the eastern side of the continent is given by:

$$T_w^* = w_0 R^2(\phi_e - \phi^*)[1 - 2\sin\theta] \tag{7.3.25}$$

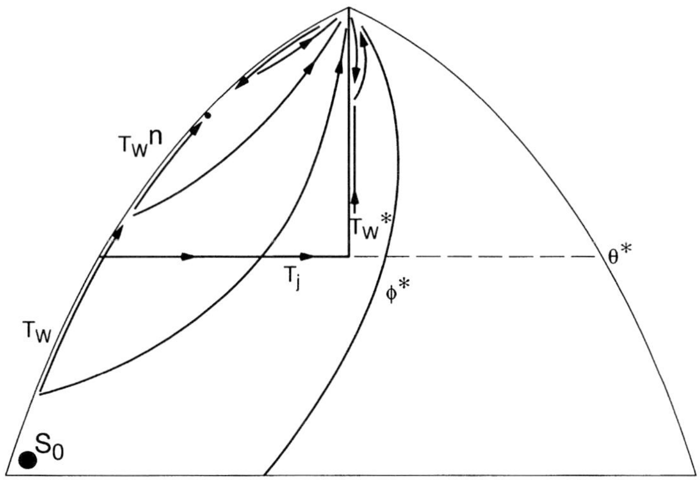


**Fig. 7.3.3.** Ocean sector is split by a barrier at longitude $\phi^*$ which extends from the apex at 90°N southward to a latitude $\theta^*$. The presence of the barrier produces a splitting of the western boundary layer transport in the southern region into two tributary currents, one on each western boundary in the region north of $\theta^*$ joined by a zonal jet with transport $T^j$

which is the same as (7.3.17) with $\phi_w$ replaced by $\phi^*$. Thus, the circulation in the eastern subbasin is exactly the same as if the western subbasin did not exist except that the transport of the western boundary layer is reduced because the size of the interior domain and hence the recirculating interior transport is reduced.

For the western subbasin the meridional flow is still given by (7.3.6). However, the zonal velocity is now given by (7.3.9) with $\phi_e$ replaced by $\phi^*$, which forms the eastern boundary of this subbasin. If we now repeat the analysis that leads to (7.3.17) for the western subbasin, we obtain for the transport of the western boundary current of the *western* basin in the region *north* of $\theta^*$:

$$T_w^n = w_0 R^2(\phi^* - \phi_e)[1 - 2\sin\theta]. \tag{7.3.26}$$

If (7.3.25) and (7.3.26) are compared to (7.3.17) we see that the *sum* of the two boundary-layer transports on the two western boundaries of the subbasins north of $\theta^*$ exactly equals the boundary-layer transport on the western side of the full basin in the absence of the continent. The introduction of the continent has split the western boundary layer transport into two tributary currents whose combined transport matches smoothly to the transport of the single western boundary current which exists in the region south of $\theta^*$, and which is still given by (7.3.17). This implies that the western boundary layer transport at the western boundary of the full sector suffers an abrupt decline as the current passes the latitude of the southern tip of the continent. The western boundary passes the latitude of the southern tip of the continent. The western boundary layer transport falls from the value given by (7.3.17) to that given by (7.3.26). Since the interior meridional transport is continuous across the latitude $\theta^*$, this implies that the drop in transport is effected by a narrow jet, as shown in Fig. 7.3.3 which carries the difference between the transports given by (7.3.17) and (7.3.26), and which becomes the western boundary layer transport at the southern tip of the continent. The eastward transport, $T_j$, in the zonal jet connecting the two western boundary currents is:

$$\begin{aligned} T_j &= T_w(\theta^*) - T_w^n(\theta^*) = T_w^*(\theta^*) \\ &= w_0 R^2(\phi^* - \phi_w)\left[1 - 2\sin\theta^*\right] \end{aligned} \tag{7.3.27}$$

Note that the latitudes where the boundary layer transport vanishes are the same in each subdomain if $\theta^*$ lies south of 30°N. The presence of a barrier splitting the basin, as would the presence of a continent such as South America, produces a zonal jet linking the boundary currents of the two western boundaries at the latitude where one of them terminates at the tip of the continent.

Since the problem is linear, combinations of the examples described in this section can be used to describe a complex picture of the total circulation of the abyss of the global ocean. The first attempt to do so in a systematic way was due to Stommel (1958) in a classic "letter to the editors" in which he boldly outlined a scheme for the abyssal circulation based on the ideas described here. The schematic circulation that he proposed is shown in Fig. 7.3.4. The numbers

**Fig. 7.3.4.** Abyssal circulation proposed by Stommel (1958)

associated with circulation were refined in a later paper (Stommel and Arons 1960b), but the pattern of flow remains similar to that shown in the figure (see also Kuo and Veronis 1970).

The original figure of Stommel's is shown because it highlights an important point. The patterns of flow described in the theory assume that the ocean floor is flat. As noted earlier, the presence of the great oceanic ridge systems is expected to disturb the low patterns considerably. The ridge system is indicated in Stommel's figure by the shaded areas, but the streamlines are drawn as if the barriers to the deep flow presented by the ridges were not present. The flow in the figure moves with serene indifference to the presence of the topographic barriers. In Section 7.5 we take up the question of the effect of the ridges and their gaps on the Stommel-Arons circulation. However, first we consider in more detail the circulation associated with isolated sources and sinks within the domain of the basin.

## 7.4 Dipole Circulation Associated with Isolated Sources

In Section 7.2 we described an experimental situation in a rotating pie-shaped sector in which a source and sink are placed at different radii on the eastern boundary of the sector, and argued on the basis of mass balance considerations that the flow of mass from source to sink takes place in purely zonal jets which are connected to the western boundary current. This picture, in which the flow moves zonally *from* the source and then *to* the sink, is only a partial picture of the circulation. In this section we reexamine the situation more carefully within the context of the Stommel-Arons theory on the sphere.

Consider the situation shown in Fig. 7.4.1. A source of finite extent has a total strength $S_+$ and feeds the abyss by pumping fluid downward in the zone indicated by the upper circle. South of the source lies a sink with an equal and opposite total strength, i.e., $S_- = -S_+$. Outside of these zones the vertical pumping velocity $w_0$ is zero. What is the circulation produced by such a distribution?

The meridional velocity is given by (7.3.6) and is zero except in the regions of the source and sink. In the region of the source, where $w_0 < 0$, the meridional flow (assuming the source is in the northern hemisphere) is southward while under the sink where $w_0 > 0$, the meridional flow is northward as indicated in the figure. Elsewhere the interior flow can only be zonal and given by (7.3.8) while the pressure anomaly is given by (7.3.7). Note that the pressure disturbance and zonal velocity are both zero east of the source and sink. The pressure anomaly is also zero west of the source and sink *outside* the latitude bands containing the source and sink. This has important implications. If the zonal flow from the source to the western boundary were unidirectional, for example, strictly westward, the pressure west of the source would have to increase monotonically across the latitude interval of the source

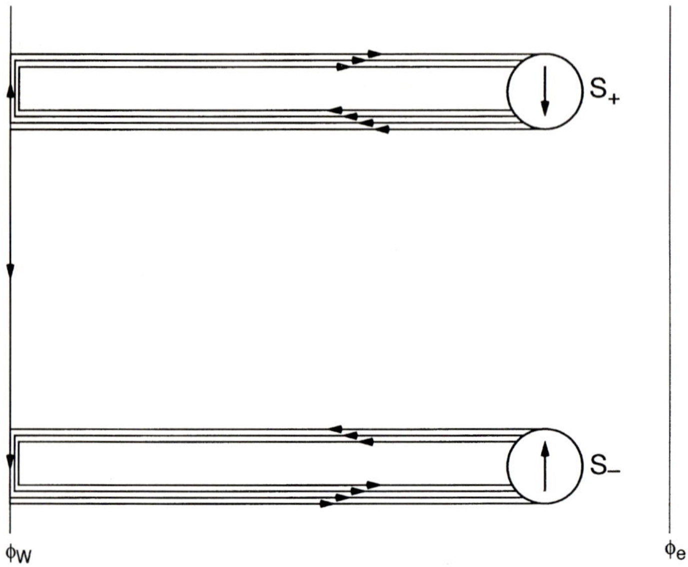


**Fig. 7.4.1.** Schematic circulation pattern produced by a localized source of strength $S_+$ set off some distance from the western boundary. South of the source there is a sink of strength $S_- = -S_+$. The source and sink are finite in extent. The interior flow is limited to the latitude bands of the source and sink and consists of dipolar jets positioned to the west of both the source and sink. Meridional flow exists only at the source and sink or in the western boundary current. The two limbs of each dipolar flow are slightly different in strength, and the difference is just equal to the mass issuing from the source which travels in the western boundary current to the sink

and could not continuously match the region of zero pressure anomaly outside the latitude strip containing the source. Thus the flow cannot be simply unidirectional. The meridional pressure gradient is, from (7.3.7):

$$\frac{1}{R}\frac{\partial P}{\partial \theta} = -\int_{\phi}^{\phi_e} \frac{\cos\theta}{\beta H}\frac{\partial}{\partial \theta}\{f^2 w_0\} d\phi' \tag{7.4.1}$$

where the fact that $\cos\theta/\beta$ is a constant has been used in deriving (7.4.1). $w_0$ is localized in latitude. In the zone of the source $w_0$ is negative, falling from zero at the southern edge of the source region, reaching a minimum somewhere in the latitude interval of the source, and returning to zero as the northern edge of the source region is reached. In this region the pressure anomaly is positive, and it too is limited to the region where $w_0$ is nonzero. The meridional pressure *gradient* is therefore positive in the first latitude interval where $f^2 w_0$ is algebraically decreasing, and where the pressure is increasing northward from its zero starting value. The meridional pressure gradient is negative in the remaining, northern part of the source interval where the pressure is approaching zero from its positive maximum. The geostrophic zonal flow is therefore *bidirectional*, i.e., of a dipolar character. As shown in the figure, it is positive in the northern subinterval where the pressure gradient is negative and negative (westward) in the southern region where the pressure gradient is positive. This is consistent with the closure of the meridional flow under the source as shown in the figure.

The zonal velocity itself is determined from (7.4.1) by geostrophy and is given by (7.3.8). Although the meridional integral of the pressure gradient is exactly zero, the same is not true for the zonal velocity because of the factor involving the Coriolis parameter. Thus we may write (7.3.8) in the revealing form:

$$u = \frac{1}{f}\int_{\phi}^{\phi_e} \frac{R\cos\theta}{\beta H}\frac{\partial}{R\partial\theta}(f^2 w_0)\,d\phi'. \tag{7.4.2}$$

Although the total pressure drop in the region of negative pressure gradient is equal to the pressure rise in the region of positive pressure gradient, the Coriolis parameter in the region of the positive pressure gradient is smaller (since it is farther south) than is the zonal velocity in the region of negative pressure gradient in the north where the flow is eastward and $f$ is larger. Thus the zonal flow in the westward branch of the dipole flow is slightly larger than the flow in the eastward branch. The difference in the two flows is exactly equal to the mass added to the flow in its southward excursion in the region of the source before it turns westward. This can be seen by rewriting (7.4.2) as:

$$uH = \int_{\phi}^{\phi_e} R\cos\theta\,d\phi'\left[w_0 + \frac{1}{\beta R}\frac{\partial}{\partial\theta}(f w_0)\right]. \tag{7.4.3}$$

The ratio of the second term in the square bracket in (7.4.3) to the first is of the order of $R/L_y$ where $L_y$ is the north-south scale of the source region. If the source has a scale much less than the planetary scale, it is the second term which dominates locally. This term integrates to zero when (7.4.3) is integrated over the source region (recall that $\cos\theta/\beta$ is constant) and gives rise to a large dipolar, bidirectional component of the zonal flow with no net zonal flux. The smaller part of the flow is given by the first term, and this is the only term which survives when an integral over the entire source is made. Thus, integrating from the latitude $\theta_1$ to the latitude $\theta_2$,

$$\int_{\theta_1}^{\theta_2} uH\,R d\theta = \int_{\theta_1}^{\theta_2}\int_{\phi}^{\phi_e} w_0 R^2\cos\theta\,d\theta\,d\phi' + \left.\int_{\phi}^{\phi_e} w_0\frac{f}{\beta}R\,d\phi'\right\}_{\theta=\theta_1}^{\theta=\theta_2} \tag{7.4.4}$$

where the last term in (7.4.4) is the difference between the zonal integral in the last term evaluated at the two latitudes $\theta_1$ and $\theta_2$.

If the two latitudes lie outside of the latitude interval occupied by the source, i.e., if the integral is over the entire source region, the last term vanishes. The first term increases as $\phi$ moves westward until $\phi$ is just west of the source. The integral then ceases to grow and remains constant. Thus for all

$\phi \leq \phi_{ws}$, the western longitudinal extent of the source, the *total* zonal flux is constant, westward, and equal to:

$$\int_{\theta_1}^{\theta_2} uHR\,d\theta = \iint\limits_{\substack{\text{source}\\ \text{area}}} w_0 R^2 \cos\theta\, d\phi\, d\theta = S_+. \tag{7.4.5}$$

Just enough more fluid flows westward in the southern branch of the bidirectional jet than flows eastward in the northern branch to deliver to the western boundary layer a net amount of flow equal to the flux of mass issuing from the source. This is shown schematically in the figure where four streamlines move westward while three streamlines return to the source. The difference of one unit is provided by the source. In the western boundary layer the recirculating portion (three units in the figure) turn northward to close the recirculating, dipolar cell. The smaller portion of one unit, representing the flux $S_+$, turns southward in the western boundary layer where it flows southward with no change in transport until it reaches the latitude band of the sink. In this region the arguments given above for the source can be applied directly to the flow in the latitude interval of the sink. Again, in the southern limb of the dipolar flow to the sink the smaller Coriolis parameter provides more flow to the sink than issues from it, and this excess is provided by the southward flux in the western boundary current which terminates at the southern edge of the latitude band containing the sink.

It is true that there is a net mass flux from source to sink through zonal jets and the western boundary current. However, focusing on the mass balance alone can be somewhat misleading as the results of the analysis shows. By far the largest velocities are not connected to the requirements of simple mass flux but are provoked by the reaction of the flow to the presence of vortex tube stretching in the presence of a powerful potential vorticity constraint. The dominant flow, in terms of its magnitude, is the recirculating flow, and, once again, its direction at any location is not simply related to its position relative to its source. In the western boundary current, for example, in the latitude band of the source, most of the water moves *northward* even though the global mass budget requires a southward boundary layer transport of the fluid from the source to feed the sink in the south. In the regions of both the source and the sink the strong recirculation implies that slightly less than half of the flow moves towards the source and away from the sink.

The dipolar jets, with their relatively strong lateral shears, especially when the north-south scale becomes small, are likely to be unstable, and the idealized rectilinear flow is surely only an idealization of the more complex flow that is more likely to result. Something of the kind can be inferred from the original photographic evidence of the Stommel, Arons, and Faller (1958) experiment in the case where the source and sink flow, analyzed, here was attempted. The reader sees in the original illustration of the experiment a large billowing ink cloud issuing westward from the source. It is not possible to reconstruct the flow structure occurring in the experiment, but it is plausible that the dipolar flow predicted here had become unstable.

The recirculations produced by the sources driven by the pumping out of the abyss can also be found when the source is produced by the flow drawn through a gap in the eastern boundary of a subregion of the oceanic basin. This is the subject of the next section.

## 7.5 Effect of Ridges and Gaps

As described in Section 7.1, the ocean basins at depth are actually divided into subbasins by the enormous subsurface mountain ranges produced by the geological processes of seafloor spreading. The midocean ridge systems are found in all oceans, and the peaks of the ridges rise to within 2 km of the ocean surface. If the ridge systems were unbroken, they would represent impenetrable barriers to the abyssal circulation and divide the basin into sets of subplanetary scale vessels in each of which the Stommel-Arons theory could be applied separately. However, the ridges are fractured at many places along the length of the ridge and these deep gaps offer passage for the water of one subbasin to communicate with the neighboring subbasin. The restriction of the interbasin deep flow to these gaps can be expected to have a significant influence on the circulation patterns constructed using the Stommel-Arons theory. To examine this in more detail, consider the model shown in Fig. 7.5.1. To simplify matters to their essentials a cartesian coordinate system is used.The discussion follows that of Pedlosky (1994b).

An abyssal basin with eastern and western boundaries at $x_e$ and $x_w$, respectively, is divided in two by a north-south ridge system at $x = x_t$. The basin lies between latitudes of $y_n$ and $y_s$ to the north and south where the basin is also closed by solid boundaries although this is not necessary to complete the solution. The ridge is broken, here for illustration, into three segments. The latitude interval for the jth segment is $y_{\text{js}} \leq y \leq y_{\text{jn}}$, for $j = 1, 2, 3$. There are two gaps between the three segments through which the fluid can leak from one basin to the next as well as gaps between the northern extent of the ridge and the northern basin boundary and the southern limit of the ridge and the southern boundary. These last two gaps can also be thought of as representing gaps in a more prolonged ridge system.

The flow is on the $\beta$ plane and governed by the equation for the geostrophic streamfunction:

$$\psi = \frac{p}{\rho_0 f_0}. \tag{7.5.1}$$

Because of the relatively small scales that are produced in the circulation by the presence of the ridges and gaps, a quasi-geostrophic theory is applied in which

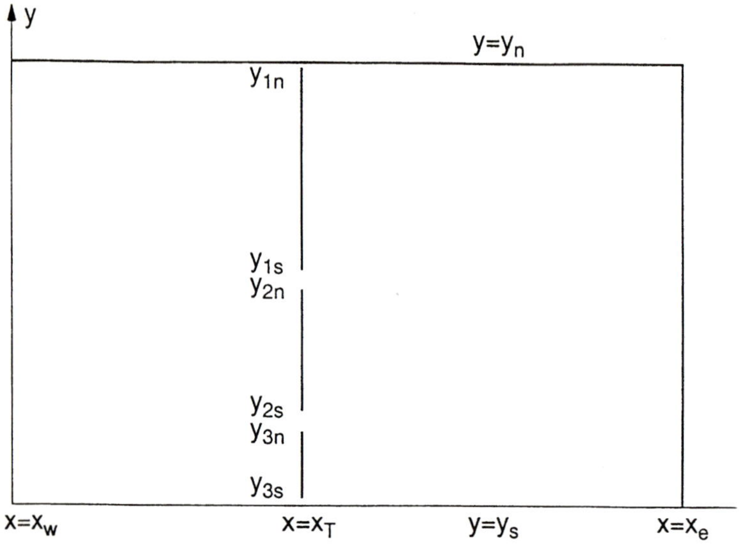


**Fig. 7.5.1.** Abyssal basin with its idealized ridge system. The ridge system, at $x = x_t$, has gaps at the locations shown. The jth ridge element extends from $y = y_{js}$ to $y = y_{jn}$

the vorticity equation is written in terms of the geostrophic streamfunction (7.5.1).

The pumping velocity $w_0$ is an arbitrary function of latitude and longitude. However, the longitudinal dependence is relatively inconsequential for the present argument. The dependence on latitude is important, and it is imagined that the pumping velocity would, in the *absence* of the ridge, produce a large scale closed gyre. That is, it is anticipated that the interior streamfunction has a minimum within the basin representing the center of a large cyclonic gyre of the type anticipated in the Stommel-Arons theory.

The presence of the ridge produces boundary-layer structures on the eastern side of the ridges (the western side of the eastern subbasin). To describe these boundary layers with maximum simplicity a bottom friction model is used so that the entire dynamics is governed by the vorticity balance:

$$\beta v = \frac{f w_0}{H} - r\left(\frac{\partial v}{\partial x} - \frac{\partial u}{\partial y}\right) \tag{7.5.2}$$

where $r$ is a parameter which measures the effect of bottom friction. The bottom friction is imagined to be negligible outside western boundary layers which requires that:

$$\frac{r}{\beta(x_e - x_t)} \ll 1. \tag{7.5.3}$$

When the velocities are written in terms of the geostrophic streamfunction (7.5.2) becomes:

$$\beta \frac{\partial \psi}{\partial x} = \frac{f w_0}{H} - r\nabla^2\psi. \tag{7.5.4}$$

The last term in (7.5.4) is negligible in the interior by virtue of (7.5.3). In the region *east* of the ridge the interior solution which satisfies the boundary condition that $\psi$ is zero at $x = x_e$ is the Stommel-Arons solution:

$$\psi = \psi_I = -\int_x^{x_e} \frac{f w_0(x', y)}{\beta H} dx'. \tag{7.5.5}$$

Since the interior solution is a balance between the $\beta$ effect and the pumping by the flow into the thermocline, information in the solution propagates strictly westward so that the interior solution east of the ridge is unaware of the presence of the ridge. The solution (7.5.5) is exactly the same as if the ridge were absent. The streamfunction must be a constant on the eastern boundary of course, and this constant is chosen to be zero without any loss of generality. The streamfunction must also be constant on each of the ridge segments so that no fluid flows through the unbroken portion of the ridge. The value of the constant on each ridge generally differs, however, from zero. The difference between its value on each ridge segment and its value on the eastern wall is a measure of the meridional mass flux, at the location of the ridge segment, in the interval between the ridge and the eastern boundary. As flow leaks into or out of the eastern basin through the gaps, this northward mass flux changes from segment to segment. The difference in the value of the streamfunction from one segment to the next measures the net flux of fluid that flows through the gap between them.

To determine these constants the dynamics of the boundary layer on each ridge segment must be examined. On each segment there is a boundary layer involving a balance between the $\beta$ term and the bottom-friction term in (7.5.4). The boundary-layer solution which satisfies (7.5.4) for small $r$ and which yields a constant value of $\psi$ on the ridge is, for the $j$th segment:

$$\psi_j = \psi_I(x, y) - \left[\psi_I(x_t, y) - \Psi_j\right] e^{-(\xi/\delta)}. \tag{7.5.6}$$

The solution (7.5.6) is valid on the eastern side of each ridge segment. In (7.5.6):

$$\xi = (x - x_t), \qquad \delta = \frac{r}{\beta}. \tag{7.5.7 a,b}$$

The constant $\Psi_j$ in (7.5.6) depends on which ridge segment is considered and to this point it is arbitrary. The determination of $\Psi_j$ is the key to the solution which determines the degree to which the subbasins communicate.

In the latitude intervals between the ridge segments, where the information flow from the eastern boundary is uninterrupted by the presence of a ridge segment, the original interior solution (7.5.5) remains valid. West of the ridge system but within the latitude band of each segment the interior Sverdrup dynamics must hold up to each ridge segment, so that the interior solution west of the ridge is given in the jth latitude band by:

$$\psi_j = -\int_x^{x_t} \frac{f w_0}{H} dx' + \Psi_j \tag{7.5.8}$$

The integral in (7.5.8) is taken from any point west of the $j$th ridge segment and integrated to $x_t$, the position of the ridge, where the value of $\psi$ is $\Psi_j$.

To determine the constant $\Psi_j$ we return to the momentum equation. The form of the momentum equation consistent with (7.5.2) is:

$$f\hat{k} \times \vec{u} = -\frac{\nabla p}{\rho} - r\vec{u}. \tag{7.5.9}$$

The bottom friction is represented in the momentum equation by a body force which acts as a simple drag on the flow. An alternative formulation, in which the friction is limited to the Ekman layer, and the velocity in the lower boundary layer is considered separately, gives the same final result. It is also possible to use lateral mixing of momentum and vorticity instead of bottom friction and also obtain the same result as that which follows below (Pedlosky 1994b).

Suppose $\hat{t}$ is a tangent vector to the "island" comprising the $j$th ridge segment. If (7.5.9) is dotted with $\hat{t}$ and then integrated around the island's boundary:

$$\begin{aligned}\oint_C \hat{t} \cdot (\hat{k} \times \vec{u}) f\, d\ell &= \oint_C (\hat{t} \times \hat{k}) \cdot \vec{u} f\, d\ell = \oint_C \hat{n} \cdot \vec{u} f\, d\ell \\ &= -\oint_C \frac{\nabla p}{\rho} \cdot \hat{t}\, d\ell - r \oint_C \vec{u} \cdot \hat{t}\, d\ell \end{aligned} \tag{7.5.10}$$

where $\hat{n}$ is the vector normal to the ridge segment and $d\ell$ is the differential line element along the contour $C$ which coincides with the boundary of the segment. Since the normal velocity on the ridge segment must vanish, the term on the left hand side of (7.5.10) must vanish. The first term on the right side of (7.5.10) must also vanish since it is the closed line integral of a perfect differential. This leaves as a condition on the solution:

$$\oint_C \vec{u} \cdot \hat{t}\, d\ell = 0 \tag{7.5.11}$$

for all $r \neq 0$.

The circuit $C$ runs counterclockwise around the ridge segment. The condition (7.5.11) states that there can be no net tangential force exerted along the ridge segment by the flow. In the case of the ridge the dominant term in (7.5.11) comes from that part of the contour where the tangential velocity is the greatest. On the eastern side of the ridge there is a boundary layer with a tangential velocity of the order of $\psi/\delta$, and this is much greater than the velocity on the western side of the ridge where there is no boundary layer, and the velocity is of the order $\psi$. Hence the condition (7.5.11) is equivalent, within the boundary-layer approximation $\delta \ll x_e - x_t$, to the condition,

$$\int_{y_{js}}^{y_{jn}} v\,dy = 0 \tag{7.5.12}$$

applied to the boundary-layer solution on the *eastern* side of the jth ridge segment.

If (7.5.6) is used to evaluate $v$ on the ridge segment at $x = x_t$, where $\xi = 0$:

$$\frac{1}{\delta}\int_{y_{js}}^{y_{jn}} \left[\psi_I(x_t, y) - \Psi_j\right] dy = 0 \tag{7.5.13}$$

or since $\psi_j$ is a constant:

$$\Psi_j = \frac{1}{(y_{jn} - y_{js})}\int_{y_{js}}^{y_{jn}} \psi_I(x_t, y)\,dy. \tag{7.5.14}$$

The constant value of the streamfunction on the $j$th ridge segment is just the *average* value of the interior streamfunction on the eastern side of the ridge. The result is identical to that of Godfrey (1989) although the derivation differs in detail. The result is not special to the form of the dissipation used. As long as the dynamics is linear, and the friction term in the boundary layer can be parameterized by some differential operator only in $x$, the boundary layer solution is of the form:

$$\psi = \psi_I + \left[\Psi_j - \psi_I(x_t, y)\right] F(\xi/\delta). \tag{7.5.15}$$

The function $F(\xi/\delta)$ describes the boundary layer structure in longitude in the region east of the ridge, and it has been normalized in (7.5.15) so that $F(0) = 1$. Clearly (7.5.6) is a special case of (7.5.15). If the momentum equation is once more integrated around the ridge, we obtain:

$$\oint_C \vec{\Im} \cdot \hat{t}\, d\ell = 0 \tag{7.5.16}$$

where $\vec{\Im}$ is the frictional force in the momentum equation. Note that (7.5.16) would be valid even if the nonlinear advection of momentum were considered. What would probably no longer be true in the presence of strong nonlinearity is the simple form of the boundary-layer structure suggested in (7.5.15) in which the amplitude of the boundary-layer correction at the wall depends linearly on the difference between $\Psi_j$ and the interior streamfunction times a function only of longitude. For the linear problem, however, such a form is to be expected, and if we again use the fact that the dominant friction force occurs on the eastern flank of the ridge segment where the boundary layer is located, (7.5.16) becomes:

$$\int_{y_{js}}^{y_{jn}} \Im^{(y)} dy = 0. \tag{7.5.17}$$

$\Im^{(Y)}$ is the meridional component of the frictional force. This frictional force is proportional to a derivative of order n of the meridional velocity and $n$ is 0, 2, or an even higher, integer depending on whether bottom friction, lateral friction, or some higher order type of friction must be used in the parameterization of friction. Thus on the boundary of the $j_{\text{th}}$ ridge segment:

$$\Im^{(y)} \propto \left[\Psi_j - \psi_I(x_t, y)\right] F^{(n+1)}(0) \tag{7.5.18}$$

where $F^{(n)}$ is the $n_{th}$ derivative of $F$.

Then (7.5.17) can be satisfied only if (7.5.14) holds. The condition that $\Psi_j$ be equal to the average value of the interior streamfunction is therefore a rather general result. It does depend on the general form of the boundary-layer solution (7.5.15) being valid, and nonlinearity may upset this if strong recirculations develop at the tip of the segments, for example, so that the boundary-layer approximation, allowing $y$ derivatives to be neglected with respect to $x$ derivatives, fails. We saw in Chapter 2 that this can happen if the boundary-layer motion becomes strongly inertial, and this places an implicit limit on the magnitude of the velocity for which (7.5.14) holds. For the relatively weak flows of the abyss this does not seem to be a significant concern.

Note that since the constant $\Psi_j$ is the average of the local interior streamfunction on the ridge segment, the constant changes from one segment to another.The boundary-layer velocity along the ridge is, from (7.5.6):

$$v = \frac{1}{\delta}\{\psi_I(x_t, y) - \Psi_j\}e^{-\xi/\delta}. \tag{7.5.19}$$

If $\psi_I(x_t, y) > \Psi_j$, i.e., if the interior streamfunction at a given latitude along the ridge segment is greater than its *average value along the segment,* the boundary-layer flow is northward; if the reverse is true, the boundary-layer flow is southward. Therefore the direction of the boundary-layer flow on each ridge segment is determined only by the character of the interior streamfunction in the neighborhood of the ridge segment. Since $\Psi_j$ is the average value of $\psi_I$ along the ridge, there must be at least one latitude where $\psi_I$ equals its average and thus at least one latitude where the bracket in (7.5.19) vanishes. Since the zonal velocity is zero everywhere on the ridge, this means there must be at least one stagnation point along the ridge where the total velocity vanishes. If the ridge segment is long enough, there may be two such stagnation points.

Consider the situation shown in Fig. 7.5.2. The ridge has no gaps except at the northern and southern tips of the ridge near the northern and southern boundaries. That is, the three ridge segments are fused into a single segment with gaps between only $y_n$ and $y_{1n}$ and $y_{1s}$ and $y_s$ The solid line in the figure is a schematic of the latitude distribution of $\psi_I$ along the ridge. It is everywhere

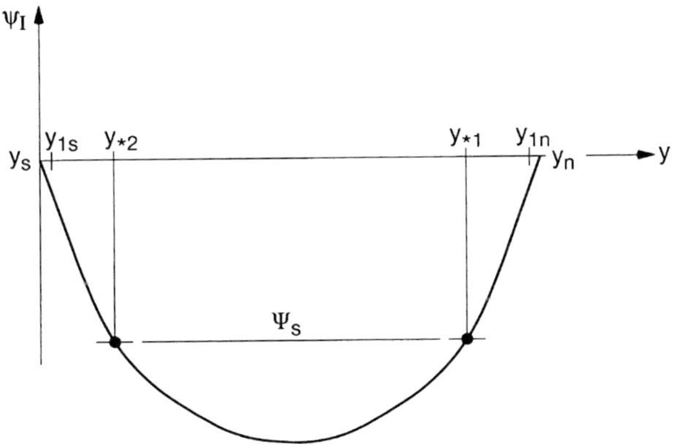


**Fig. 7.5.2.** Curve of $y(x_t, y)$ as a function of $y$. The horizontal line labeled $\Psi_1$ is the constant value of the streamfunction on the ridge. The stagnation points of the boundary layer flow occur at $y_1^*$ and $y_2^*$

negative and has a single minimum, representing a cyclonic cell that would fill the basin with an unbroken Stommel-Arons gyre if there were no ridge. The constant $\Psi_1$ is determined by the integral condition (7.5.14), and its value is indicated by the horizontal line in the figure which intersects the curve of $\Psi_I(x_t, y)$ at two locations. These correspond to the latitudes $y_1^*$ and $y_2^*$ which are the stagnation points of the flow along the ridge.

In Fig. 7.5.3 the streamline pattern is shown for this ridge system for the case where:

$$\frac{f}{\beta H} w_0 = \frac{f_0}{\beta_0 H} W_0 \sin\left(\pi \frac{y}{y_n}\right) \tag{7.5.20}$$

and where the origin in $y$ has been placed at $y_s$. The geostrophic streamfunction in the figure has been scaled with $f_0 W_0 x_e / \beta_0 H$. The ridge is placed at $x_t = 0.51 x_e$. For this configuration the value of the streamfunction on the ridge, given by (7.5.14) is $\Psi_1 \approx -0.34$. This yields two stagnation points at $y \approx 0.24\ y_n$ and y $y \approx 0.76\ y_n$. These points are labeled $y_2^*$ and $y_1^*$ respectively, and are shown in the figure. The streamline in the interior which corresponds to the value $\Psi_1$ joins the two stagnation points and encloses a region in which the flow is completely encircled by this streamline and recirculates in a pool anchored to the eastern side of the ridge as shown in Fig. 7.5.3. Fluid in this pool does not communicate with the fluid in the western basin but remains in the recirculating pool driven by the local upwelling into the thermocline. A small flux, of order $(\beta x_e / f)\psi_I$ enters the pool to replace the fluid entering the thermocline, but this is negligible within the $\beta$-plane quasi-geostrophic approximation.

In the solution east of the ridge (7.5.5) in this case yields:

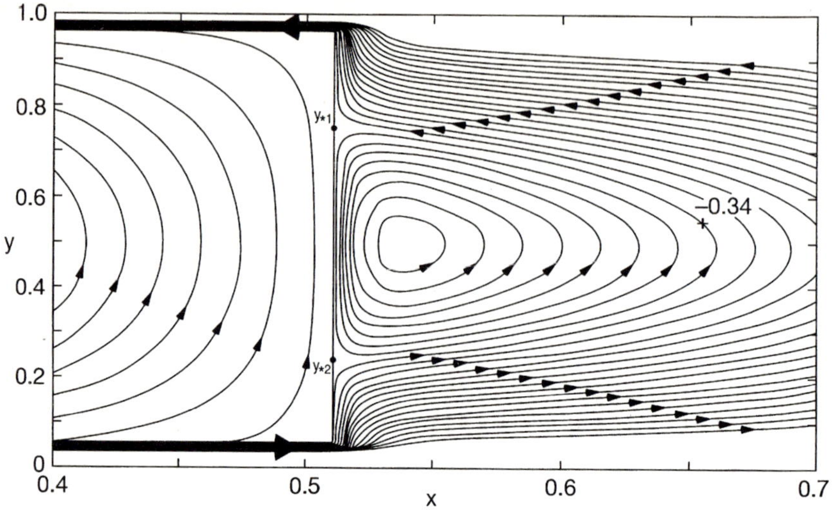


**Fig. 7.5.3.** Streamline pattern in the case where the ridge has gaps only near the termini at $y = y_{1n}$ and $y = y_{1s}$. The flow develops a recirculation zone whose extent is determined by the two stagnation points on the ridge, $y_1^*$ and $y_2^*$ and the streamline, $\psi I = \Psi_1$ which joins them

$$\psi_I = -\frac{f_0}{\beta_0 H} W_0 x_e \left[1 - \frac{x}{x_e}\right] \sin\left(\pi \frac{y}{y_n}\right). \tag{7.5.21}$$

This implies that an amount of fluid equal to:

$$-\psi_I(x_t, y_n/2) = \frac{f_0 W_0 x_e}{\beta_0 H}\left[1 - \frac{x_t}{x_e}\right] \tag{7.5.22}$$

would *in the absence of the ridge* enter the eastern basin in the south and exit back to the western basin in the north. With the previous scaling for the streamfunction, and the position of the ridge at $x_t = 0.51x_e$, this yields a nondimensional mass flux across the ridge position equal to 0.49 units of transport *when there is no ridge*. The streamfunction at the outer edge of the recirculation zone corresponds to a value of $\psi \approx -0.34$ and since the streamfunction is zero on the eastern boundary, this implies that a flux equal to 0.34 enters through the southern gap and leaves through the northern gap. Even when the gap is as narrow as in this case where $y_{1n} = 0.98\ y_n$ and $y_{1s} = 0.02$, the flow from the eastern to the western basin is 68% of the unblocked value. The rest of the transport recirculates within the eastern basin in the trapped cell east of the ridge.

If instead of a long unbroken ridge, whose length is great enough to span the distance from one stagnation point to the other, the ridge is broken into smaller fragments, the situation is as shown in Fig. 7.5.4. With the same upwelling velocity as given by (7.5.20) and thus the same interior streamfunc-

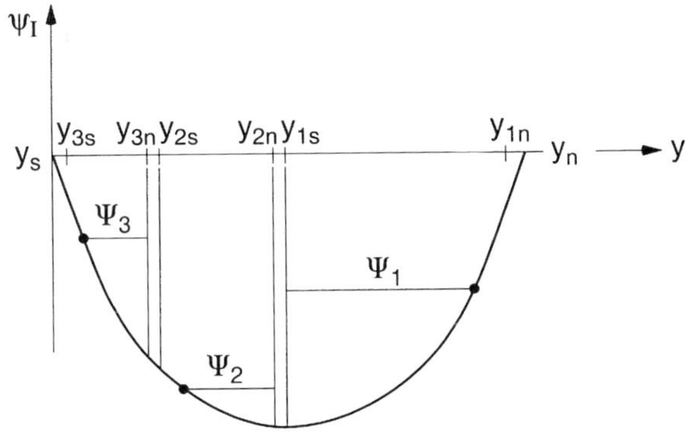


**Fig. 7.5.4.** Curve showing the interior streamfunction on the ridge for the case where the ridge is broken into three segments. The constant values of the streamfunction on each segment, $\Psi_j, j = 1, 2, 3$ are shown by the horizontal line segments. They intersect the curve of $\psi_I(x_t, y)$ in only a single point on each segment

tion in the eastern basin as given by (7.5.21) the values of the constant streamfunctions on the three segments are found to be $\Psi_j = -[0.34, 0.445, 0.25]$ for $j=[1, 2, 3]$. With these values of $\Psi_j$ the direction of the transport on each segment is determined by (7.5.19) and the critical streamlines joining the stagnation points to the interior flow are also determined. Figure 7.5.5. presents the situation schematically. Since each of the $\Psi_j$ are different from each other, no two of the stagnation points are joined by the same streamline. Instead, the streamline emanating from one stagnation point intersects another ridge segment and joins its western boundary layer and leaves the eastern basin to flow into the western basin. None of the recirculation streamlines emanating from the stagnation points encloses a completely isolated pool. Each of the recirculating regions is instead fed by *zonal jets* which trail westward from each ridge gap. The transport through each ridge gap is just equal to the difference of the constant stream function value on neighboring ridge segments. For example, in Fig. 7.5.5, the tightest recirculation is bounded by the streamline $\psi = \Psi_2$ and consists of fluid which enters the gap between the first and second ridge segments and executes a compact cyclonic gyre before leaving through the same gap through which it entered. On the other hand, fluid entering the gap between the second and third ridge segments leaves either through the gap between the northern-most segment and the northern boundary, or through the gap between the first and second segments. The fluid between the streamlines $\Psi_2$ and $\Psi_3$, for example, leaves through the northernmost gap. Note that continuity requires that the flow through the gap between segments 1 and 2 be bi-directional while the flow through the other gaps is unidirectional. The full streamline pattern is shown in Fig. 7.5.6.

Since the streamfunction for the region east of the ridge is determined independently of the presence of the ridge, the flow pattern in the eastern basin

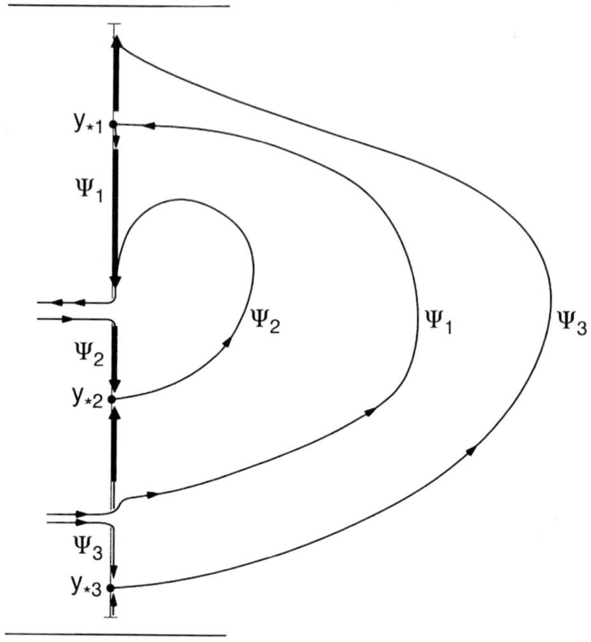


**Fig. 7.5.5.** Schematic presentation of the stagnation points and the critical streamlines which join them to the flow and their reattachment to the jets feeding the eastern basin through the western boundary layers

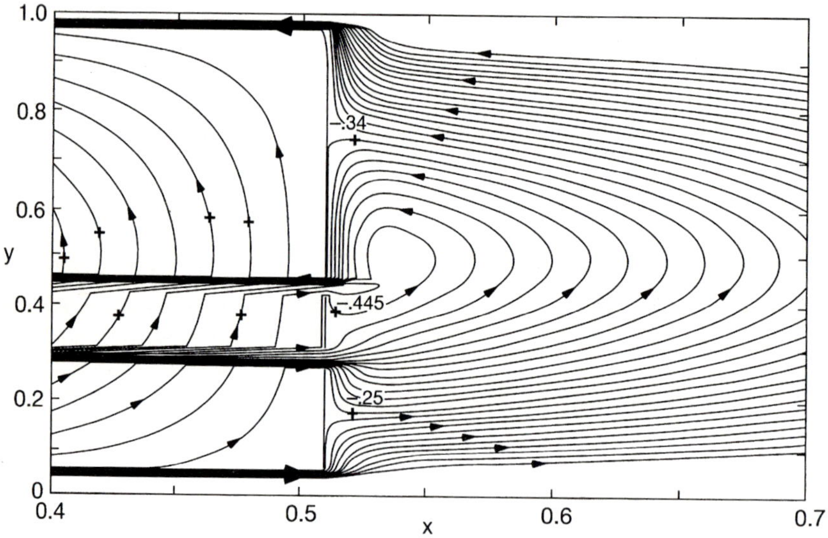


**Fig. 7.5.6.** Full streamline pattern for the stagnation point structure shown in Fig. 7.5.5

is independent of the ridge. The fate of the flow when it impinges on the ridge is determined by the stagnation point structure, and the flow through the gaps in the ridge is completely determined in turn, by the eastern, interior streamfunction and the position of the gaps (or, equivalently, on the value of the constant streamfunction on each segment). Thus the flow is drawn through the gaps to satisfy the conditions already determined in the eastern basin rather than being forced through the gaps by the flow in the western basin. The zonal jets are forced in the western basin as a reaction to the mass drawn though the ridge gaps in the manner already familiar from the discussion of the Stommel-Arons theory in Section 7.4.

Pedlosky (1994b) suggested that these recirculations, anchored to the ridges and their gaps, are responsible for the subplanetary recirculations seen in the data, as described in Section 7.1 (e.g., Fig. 7.1.4). Extension of the theory to a stratified model, in the context of a two-layer model (Pedlosky 1994a) shows that the presence of the recirculation is strongest within the layers actually blocked by the ridge, but that the recirculation extends upward, at least weakly, into the unblocked layer above the ridge.

## 7.6 Establishment of the Abyssal Circulation

The Stommel-Arons theory relates the structure of the abyssal circulation to a widespread upwelling into the thermocline. The horizontal structure of the upwelling is fixed a priori in the model. This, with the specification of the position of the sources of deep water, determines the boundary-current transports required to close the circulation as well. The distribution of the upwelling into the thermocline is really, of course, not completely independent of the motion of the waters of the abyss, and this dependence can be expected to alter the horizontal structure of the upwelling and the resulting horizontal circulation. At the same time, baroclinic effects can be expected also to be important in determining the form of the circulation, for, as we noted in Section 7.1, the circulation as observed has a strong baroclinic character. One important feature of the baroclinic circulation is that the way in which information is signaled involves slow baroclinic Rossby waves and baroclinic Kelvin waves as well as the rapid barotropic Rossby wave, and this affects both the structure of the steady circulation and the process of adjustment of the circulation to changes in the sources or sink of the flow. A similar dependence was noted in Chapter 3.

A very illuminating model of the abyssal circulation which attempts to extend the ideas of Stommel and Arons along the direction of these concerns has been presented by Kawase (1987). Although the assumption of linearity of the dynamics is maintained, the upwelling from the abyss into the thermocline is coupled to the abyssal motion in a direct way. Indeed, while the interior upwelling in the Stommel-Arons model is independent of the abyssal motion,

Kawase presents a discussion of a dynamics in which the upwelling depends *only* on the motion of the abyss. This represents an opposite extreme from the classical theory. The Stommel-Arons model imagines the upwelling as a response to the wind- and buoyancy-driven thermocline motion. The thermocline dynamics provides an upwelling velocity, $w_0$, at the base of the thermocline, which is considered given. In Kawase's model there is no wind driven circulation and the forcing for the interior circulation must be generated entirely internally, i.e., there is no forcing from the thermocline. As we see below, this places a stronger burden on the abyssal dynamics than is perhaps realistic dynamically, but the theory is very revealing in its delineation of the way in which the circulation is established and controlled by baroclinic wave processes.

The fundamental model is shown in Fig. 7.6.1. The ocean is represented as a two-layer model and, as in the Stommel-Arons theory, the lower layer represents the abyss. The layers in the absence of motion have thicknesses $H_1$ and $H_2$, and the subscript 1 for all variables refers to the upper layer and 2 refers to the lower layer. The *departure* of the interface from its rest position is $h$. A source of deep water is introduced at the northwest boundary of the basin with a strength $Q$, and this represents a flux of water from the upper to the lower layer occurring outside the region to which the dynamics directly applies. The total upwelling through the interface between the two layers must equal $Q$ in the steady state.

Although it is possible to write the equations in spherical coordinates, it is useful to simplify the discussion to its essentials by employing cartesian coordinates. The basin, however, straddles the equator. In fact the basin itself (although not the placement of the sources) is symmetrical about the equator and the equatorial $\beta$ plane approximation is used in which the Coriolis parameter is approximated as $f = \beta y$.

If $p_n^*$ is the *total* pressure in the $n$th layer, then the linear, time-dependent horizontal momentum equations for the two-layer model can be written as:

$$\frac{\partial \vec{u}_n}{\partial t} + f\hat{k} \times \vec{u}_n = -\nabla \frac{p_n^*}{\rho_0} - \kappa \vec{u}_n. \tag{7.6.1}$$

In (7.6.1) the dynamics is linearized as in the Stommel-Arons theory, and the Boussinesq approximation is made so that the density, in the horizontal equations of motion, is replaced by the mean density $\rho_0$. A simple drag law, proportional to the velocity, with coefficient $\kappa$ is introduced to allow the representation of the western and equatorial boundary layers where needed. Otherwise this term is considered negligible.

The hydrostatic approximation to the vertical equation of motion implies that:

$$p_n^* = -\rho_n gz + p_n(x, y, t) \tag{7.6.2}$$

while continuity of total pressure at the interface implies that:

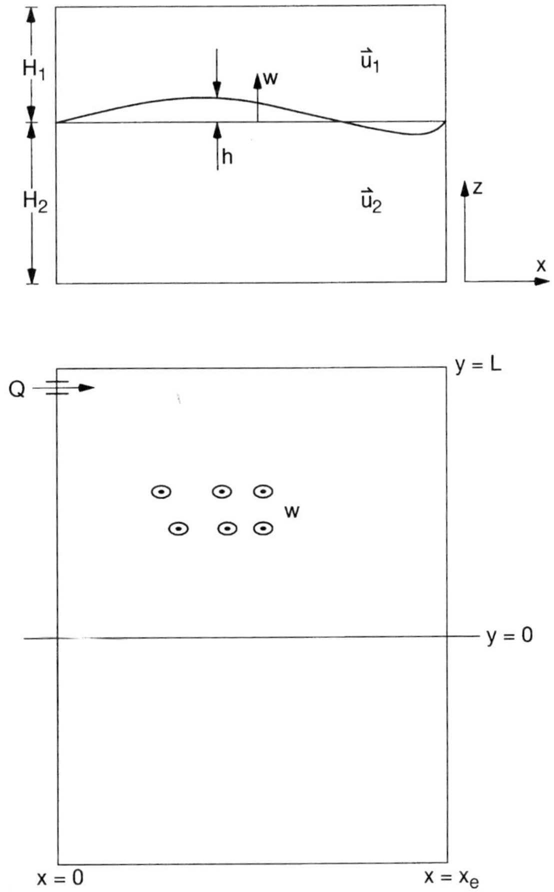


**Fig. 7.6.1.** Two-layer model of Kawase (1987) for the abyssal circulation. A cross-interface velocity w balances the source of deep water $Q$ which enters the abyss in the northwest corner of the basin. The basin straddles the equator and the b-plane approximation is used so that $f = \beta y$

$$p_2 = p_1 + \rho_0 g' h \tag{7.6.3}$$

where

$$g' = \left(\frac{\rho_2 - \rho_1}{\rho_0}\right) g.$$

The linearized equation for mass conservation for the lower layer is:

$$\frac{\partial h}{\partial t} + H_2 \left[\frac{\partial u_2}{\partial x} + \frac{\partial v_2}{\partial y}\right] = -w \tag{7.6.4}$$

where $w$ is the cross-interface velocity at the interface, $z = h$. In the absence of wind driving and Ekman pumping there is no exchange through the upper surface of the model, and therefore the mass conservation equation for the total transport, yields:

$$\nabla \cdot \vec{U}_B = 0 \tag{7.6.5}$$

where:

$$\vec{U}_B = H_1\vec{u}_1 + H_2\vec{u}_2 \tag{7.6.6}$$

so that the total transport is horizontally nondivergent and derivable from a streamfunction, i.e.:

$$\vec{U}_B = \hat{k} \times \nabla\Psi_B. \tag{7.6.7}$$

If the momentum equations for each layer are each multiplied by their respective $H_n$ and added together to form an equation for the barotropic transport, we obtain:

$$\left(\frac{\partial}{\partial t} + \kappa\right)\nabla^2\Psi_B + \beta\frac{\partial \Psi_B}{\partial x} = 0. \tag{7.6.8}$$

Aside from a solution exponentially decaying in time, the only solution of (7.6.8) which satisfies the condition that $\Psi_B$ vanish on the basin boundary is the solution in which $\Psi_B$ and *all the barotropic velocities are zero*. In the Kawase model there is no barotropic mode, and the only circulation pattern is the internal mode driven by the source-sink forcing. There is no forcing of the abyss by the thermocline's wind-driven motion, and in this sense the physics of the model is quite distinct from the Stommel-Arons model.

The vanishing of the barotropic transport implies that:

$$\vec{u}_1 = -\frac{H_2}{H_1}\vec{u}_2. \tag{7.6.9}$$

Subtraction of the two momentum equations yields the governing equation for the baroclinic mode:

$$\left[\frac{\partial}{\partial t} + \kappa + f\hat{k}\times\right](\vec{u}_1 - \vec{u}_2) = g'\nabla h \tag{7.6.10}$$

where (7.6.3) is used to express the difference of the pressure gradients between the two layers. If (7.6.9) is used, (7.6.10) can be expressed entirely in terms of the velocity of the flow in the lower layer. Thus:

$$\frac{\partial \vec{u}_2}{\partial t} + f\hat{k} \times \vec{u}_2 = -g''\nabla h - \kappa\vec{u}_2 \tag{7.6.11}$$

where $g'' = g' H_1/(H_1 + H_2)$

The key assumption of Kawase's model is the connection suggested between the cross-interface velocity w and the displacement of the interface. Kawase takes:

$$w = \lambda h \tag{7.6.12}$$

Kawase (see also Tziperman 1987) argues that (7.6.12) is the appropriate two-layer finite-difference analog of the equation of a continuously stratified model representing a balance between vertical advection and heat conduction. However, it is probably clearer and preferable to accept (7.6.12) on its own terms as a model for the cross-isopycnal flux in the two-layer model. This implies that if the thickness of the lower layer increases, a compensating mass flux into the upper layer from the lower layer tends to act to damp the thickness anomaly and drive the system towards its original state, eliminating the thermal anomaly represented by the interface distortion. In fact, the continuity equation, (7.6.4) now becomes:

$$\frac{\partial h}{\partial t} + H_2\left[\frac{\partial u_2}{\partial x} + \frac{\partial v_2}{\partial y}\right] = -\lambda h \tag{7.6.13}$$

in which the damping of the thermal anomaly associated with the interface displacement is evident. The time scale for this thermal dissipation to restore the interface to its undisturbed value, i.e., the time required to damp out the thermal anomalies is $\lambda^{-1}$.

We are particularly interested in steady solutions of the governing equations (7.6.11) and (7.6.13). In the steady state the global departure of the interface from its rest position is obtained by integrating (7.6.13) over the basin and using the fact that a net flux, equal to $Q$, enters through the lateral boundary:

$$\lambda \iint_A h\,dx\,dy = Q \tag{7.6.14}$$

where $A$ is the total area of the basin. Since the problem is linear, the structure of the solution is independent of the strength of the forcing, and (7.6.14) serves only to determine the amplitude of the motion. The structure is determined by the internal dynamics, and we focus our attention on this.

For diagnostic purposes it is useful to consider the potential vorticity equation. If the vorticity equation is formed from the curl of the momentum equations, and the continuity equation (7.6.13) is used:

$$\frac{\partial}{\partial t}\left[\zeta_2 - \frac{f}{H_2}h\right] + \beta v_2 = -\kappa\zeta_2 + \frac{f}{H_2}\lambda h, \tag{7.6.15}$$

where $\zeta_2$ is the relative vorticity:

$$\zeta_2 = \frac{\partial v_2}{\partial x} - \frac{\partial u_2}{\partial y}. \tag{7.6.16}$$

It is important to note that if both thermal and momentum dissipations are zero, i.e., if both $\kappa$ and $\lambda$ are zero, the poleward velocity in the steady state is *zero* as long as $h$ is bounded. This is particularly important for the role of the thermal dissipation since $\kappa$ is generally considered small nearly everywhere. The vertical velocity at any location depends then on the thermal anomaly produced by the abyssal motion itself. In particular, in the absence of advection in the model, the thermal anomaly introduced into the basin in the region of the source propagates through the basin, and the consequences of the balance of thermal propagation and dissipation determine the spatial structure of the motion field in the model.

From the momentum equation (7.6.11) both $u$ and $v$, the zonal and meridional velocities, can be directly related to the interface height $h$, namely, in a steady state:

$$v = g'' \frac{f \partial h/\partial x - \kappa \partial h/\partial y}{\kappa^2 + f^2}, \qquad u = -g'' \frac{f \partial h/\partial y + \kappa \partial h/\partial x}{\kappa^2 + f^2} \tag{7.6.17}$$

If $\kappa$ is negligibly small, and $f$ is $O(1)$, then (7.6.17) reduces to the geostrophic balance. This holds over most of the basin with the exception of the western boundary current region and the model's equatorial boundary layer. Using (7.6.17) in the mass conservation equation (7.6.13) yields a single equation for h which can be considered the governing equation for the motion. Thus, after some algebra, and remembering that $f = \beta y$, we have for steady state motion:

$$h\frac{\lambda}{H_2} - \beta g'' \frac{\partial h}{\partial x}\left\{\frac{f^2 - \kappa^2}{(f^2+\kappa^2)^2}\right\} - \nabla^2 h\left(\frac{\kappa g''}{(f^2+\kappa^2)}\right) + \frac{\partial h}{\partial y}\frac{2\beta f g'' \kappa}{(f^2+\kappa^2)^2} = 0. \tag{7.6.18}$$

The structure of this rather daunting equation simplifies considerably in the limit of small $\kappa$.

There are three natural length scales in the problem. (a) the basin scale of $O(x_e)$. (b) the Stommel boundary layer scale $\delta_s = \kappa/\beta$ which is the scale of the western boundary current in the model; and (c) a third intrinsic scale, $\delta_R$, which is key to understanding the interior dynamics of the circulation. This scale arises from a balance between the westward propagation of baroclinic Rossby waves and the dissipation of the anomalous height field, $h$, in the wave.The characteristic baroclinic Rossby wave speed for scales larger than the deformation radius, is directed westward and has the magnitude of the order:

$$c_R = \frac{\beta g'' H_2}{f^2}. \tag{7.6.19}$$

In the time scale $\lambda^{-1}$ associated with the decay of thermal anomalies such baroclinic wave disturbances propagated a distance $\delta_R$ , given by:

$$\delta_R = c_R \lambda^{-1} = \frac{\beta g'' H_2}{f^2 \lambda}. \tag{7.6.20}$$

In most cases of interest $\delta_s \ll \delta_R \leq x_e$. Indeed, outside of the western boundary layer, and away from the equator, the terms in $\kappa$ in (7.6.18) are negligible, and the dynamics reduces to a balance of the first two terms of the equation so that almost everywhere the equation for $h$ becomes:

$$\frac{\partial h}{\partial x} - \frac{\lambda f^2}{\beta g'' H_2} h = 0. \tag{7.6.21}$$

This is the steady-state equation for the westward propagation of a baroclinic Rossby wave that is simultaneously being damped at a rate $\lambda$.

This solution decays away from the eastern boundary. In this limit the velocity field is in geostrophic balance since the terms in $\kappa$ in the momentum equations are neglected. Note that if $\lambda$ were zero, the zonal derivative of $h$ would vanish as well, as long as $h$ is bounded. Since the velocity field is in geostrophic balance, this would imply that as long as $h$ remains bounded the meridional velocity would also become small relative to $h$. That is, the solution of (7.6.21) that satisfies the condition that the zonal velocity vanishes on $x = x_e$ is:

$$\begin{aligned} h &= h_0 e^{\left[\frac{(x-x_e) f^2 \lambda}{\beta g'' H_2}\right]} \\ v &= h_0 \frac{f\lambda}{\beta H_2} e^{\left[\frac{(x-x_e) f^2 \lambda}{\beta g'' H_2}\right]} \\ u &= h_0 \frac{2\lambda(x_e - x)}{H_2} e^{\left[\frac{(x-x_e) f^2 \lambda}{\beta g'' H_2}\right]} \end{aligned} \tag{7.6.22 a,b,c}$$

In (7.6.22) $h_0$ is a constant since $h$ must be constant along $x = x_e$ to bring to zero the geostrophic zonal velocity there.

Of course the integral constraint (7.6.14) drives $h$ to behave as $\lambda^{-1}$ so that a meridional velocity exists in the basin interior for small $\lambda$, but it would require very large interface displacements in that limit, which would be inconsistent with the linearization assumption.

According to (7.6.22) the maximum displacement and the maximum horizontal and vertical velocity occur near the eastern boundary and decay away in the scale $\delta_R$. For large $\lambda$, i.e., for $\delta_R \ll x_e$, the disturbance of the interface and the velocities are limited to the region within $\delta_R$ of the eastern wall. Since $\delta_R$ depends inversely on $f^2$, this distance becomes large as the equator is approached, even for large damping, i.e., for large $\lambda$. In this limit (small $\delta_R$) we can expect the disturbance to be confined to a narrow region near the eastern boundary except in the equatorial zone where the disturbance reaches westward and eventually reaches the western boundary. Thus we anticipate that in the limit of large $\lambda$ there is a concentration of the flow in the equatorial zone of the model as a consequence of the privileged ability of the disturbances to propagate *westward* from the eastern boundary near the equator since $c_R$ becomes large as $f$ goes to zero.

In the vicinity of the equator the governing equation (7.6.18) can be simplified by ignoring $x$ derivative with respect to $y$ derivatives in the Laplacian of $h$ to obtain:

$$\frac{h}{\delta_R} - \frac{\partial^2 h}{\partial y^2}\frac{\delta_s}{\left[1+\frac{\kappa^2}{f^2}\right]} - \frac{\partial h}{\partial x}\frac{\left(1-\frac{\kappa^2}{f^2}\right)}{\left[1+\frac{\kappa^2}{f^2}\right]^2} + \frac{2\beta\delta_s/f}{\left[1+\frac{\kappa^2}{f^2}\right]^2}\frac{\partial h}{\partial y} = 0. \tag{7.6.23}$$

A balance can be obtained between the first two terms if the $y$ scale of the equatorial boundary layer is of the order:

$$\delta_{\text{eq}} = \sqrt{\delta_s \delta_R}. \tag{7.6.24}$$

The relative size of the third term with respect to the first two terms is then $O(\delta_R/x_e)$ and can be ignored when this parameter is small. It is exactly in this limit that the equatorial intensification is obtained. This is the limit in which the response in midlatitudes is limited to the eastern boundary and communicates only to the western boundary, where the source is located, through the equatorial region. The fourth term is always small compared to the first two terms as long as $\delta_s\delta_R \ll R^2$, where $R$ is the earth's radius.

The flow from the source at the northwest corner of the basin is directed southward in a western boundary layer of width $\delta_s$ which arises as a balance of the second and third terms in (7.6.18) with the possible presence of the first term if $\delta_R$ should be as small as $\delta_s$. Otherwise the balance is the same as the standard Stommel boundary layer, i.e., in the western boundary layer region:

$$\frac{\partial^2 h}{\partial x^2} + \frac{1}{\delta_s}\frac{\partial h}{\partial x} - \frac{h}{\delta_s \delta_R} = 0 \tag{7.6.25}$$

so that the third term is negligible if $\delta_s \ll \delta_R$.

When these elements of the solution are put together to represent the total solution and the integral condition relating the amplitude of the motion to the source strength is used the solution is complete. Kawase (1987) presents the solution in three interesting parameter limits. The first is when $\delta_s \ll x_e$, and $\lambda$ is large enough that $\delta_R$ is also small compared to the basin scale. In this case we expect the region of motion to be limited to narrow regions near the eastern boundary and along the equator as well as in the boundary layer along the western boundary.

Figure 7.6.2 shows the solution as calculated by Kawase (1987). In panel a the height field $h$ is given while in panel b the velocity vectors are shown. It is important to note that Kawase excluded from this representation of the velocity field vectors greater than 0.25 of the inflow velocity at the source, i.e., these very large velocities have been omitted from the figure. The largest part of the response, as we see from the region of the concentrated height contours is in fact along the western boundary. The fluid moves rapidly southward from the source in the western boundary layer and flows along the equator. At the

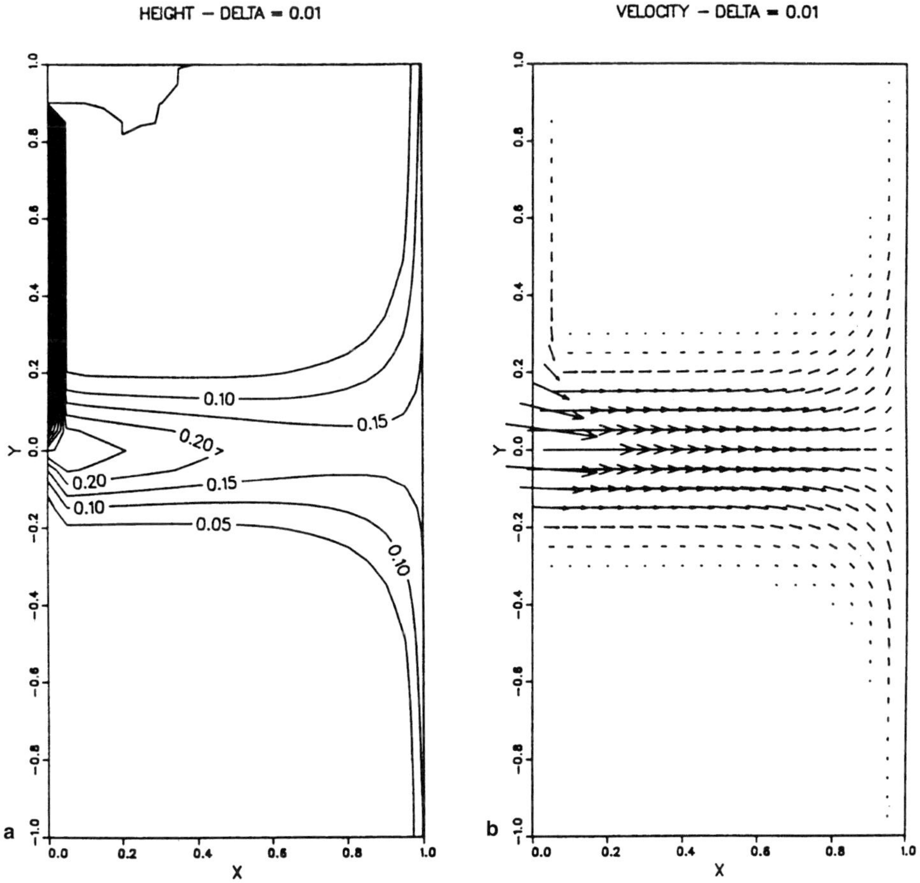


**Fig. 7.6.2.** Analytical solution for h **a** and the velocity vector field **b** for the case where $\delta_B = 0.01\ x_e$. and where $\delta_s = 0.01\ x_e$. Contour intervals for $h$ refer to an arbitrary scaling. Note that in **b** velocity vectors in the rapid western boundary layer are omitted. (From Kawase 1987)

eastern boundary it turns both north and south, and both the motion, and the interface displacement is limited to a narrow zone near the eastern boundary in both hemispheres. The upwelling into the thermocline occurs largely in the region of the order $\delta_R$ near the eastern boundary. Since $\lambda$ is large, the thermal signal originating from the eastern boundary is trapped close to the eastern wall.

On the other hand as $\lambda$ diminishes the thermal signal reaches further across the basin. Figure 7.6.3 shows the solution when $\delta_R$ is of the order of the basin width. The velocity field fills the basin in both hemispheres and the interface displacement is nearly flat at the equator. The thermal anomaly propagating westward from the eastern boundary is everywhere able to reach across the basin before decaying, and there is no special concentration of the motion field

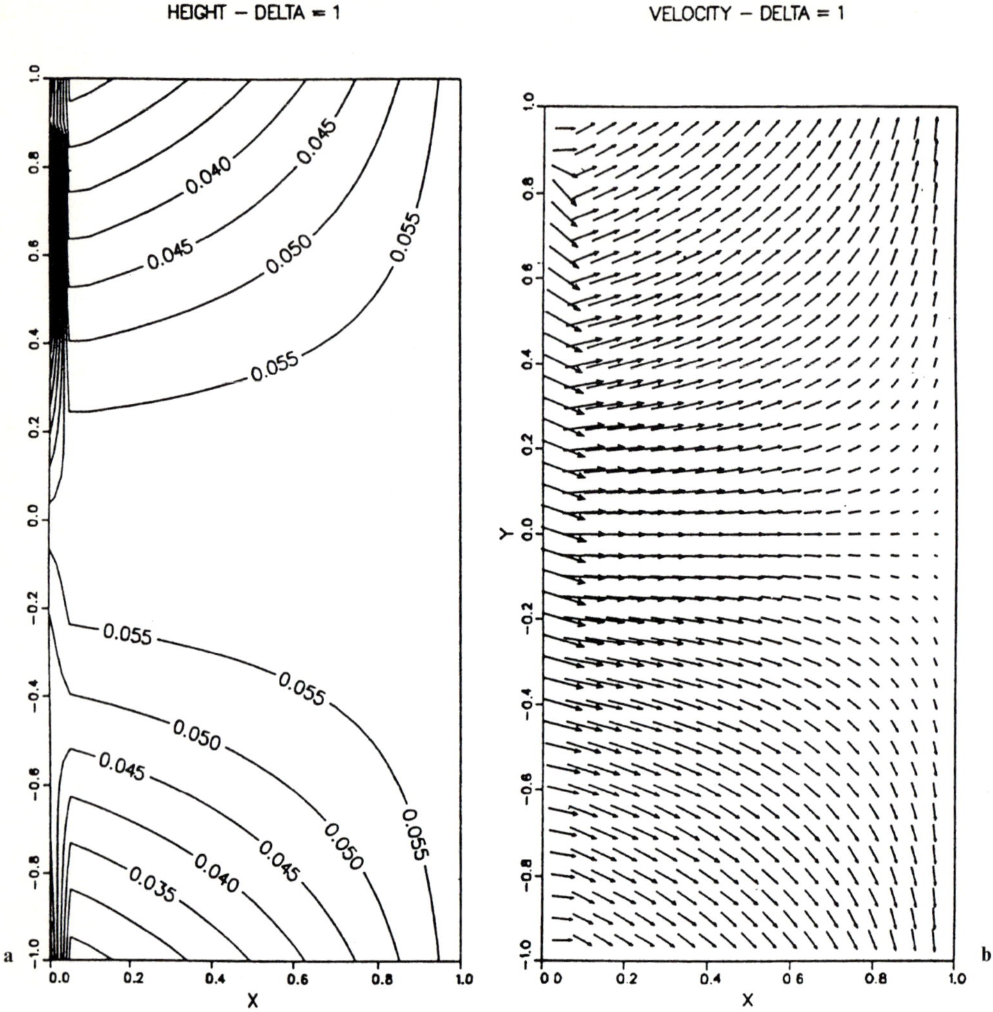


**Fig. 7.6.3** As in Fig. 7.6.2 but for the case where $\delta_B = x_e$. (From Kawase 1987)

or height anomaly at the equator in (7.6.22). The height field, as can be seen from (7.6.22), tends to be relatively uniform at the equator where $f \approx 0$, and this is apparent in panel a of the figure. The absence of the equatorial boundary layer in this case is due to the ability of the anomalies to penetrate across the basin at midlatitudes *as well as* on the equator so that the equatorial zone is no longer a privileged path for westward propagation.

The role of the eastern boundary as a source of thermal anomaly for the interior is made even clearer by the fascinating time-dependent calculations carried out by Kawase (1987). Using a numerical model in which the linear shallow water equations could be integrated in time for the same basin and source configuration as discussed above, Kawase followed the evolution of the flow field from a state of rest driven by the turning on the source in the northwest corner of the basin. Figure 7.6.4 shows the evolution of the height field for the case where $\delta_R/x_e = 0.3$ and $\delta_s/x_e = 0.0134$. In the four panels of the figure the height field is shown at 5, 10, 20, and 40 days after the source is switched on. The development of the flow is striking. A coastal Kelvin wave with characteristic speed $(g''H_2)^{1/2}$ moves rapidly down the western boundary away from the source with a southward fluid flow developing behind it. In the second panel the Kelvin wave disturbance leaves the western boundary, as it must, since it cannot travel southward in the southern hemisphere on the western boundary. It starts to travel across the basin, trapped in the equatorial wave guide. Upon reaching the eastern boundary, the signal splits evenly due to the symmetry of the basin and the thermal disturbance, i.e. the displacement of the interface, and moves poleward in each hemisphere. In the final panel the disturbance along the eastern boundary enters the interior and propagates westward as a baroclinic Rossby wave. Its penetration into the interior is well estimated by the scale $\delta_R$, which we recall is a function of latitude. The propagation continues until the steady solution is established as shown in Fig. 7.6.5 which agrees well with the analytical solution shown in Fig. 7.6.3.

It is particularly important to note how the anomaly introduced at the western boundary enters the interior to establish the steady solution. It does so in a very indirect way in which the propagation of Kelvin and Rossby waves is instrumental in the establishment of the flow. Since the Kelvin waves move only along the boundary and the equator , the interior flow is established by the propagation of baroclinic Rossby waves inward from the *eastern* boundary where the anomaly in the height field is set up by the Kelvin wave signal that has carried the anomaly signal first along the equator and then poleward in each hemisphere. The ability of the Rossby waves to penetrate westward is what determines the structure of the interior flow and also the final equatorial structure, even as here, when the source of the anomaly is in the west. The dynamics is very nonlocal, and the response to the forcing is conditioned by the wave properties of the flow so that even the nature of the steady solution becomes clear only by considering the underlying wave dynamics.

As Fig. 7.6.5 shows, the interior response in each hemisphere is symmetric, with the upwelling into the thermocline being equally divided between the two hemispheres. This implies that half of the source flow crosses the hemisphere in the western boundary layer to upwell in the southern hemisphere. The western boundary current flows from a source in one hemisphere to the other hemisphere in which there is no source but only the distributed sink of fluid entering the thermocline. In agreement with the Stommel-Arons theory [see equation (7.3.17)] the western boundary current has a stagnation point at half the distance from equator to pole (at 30° in spherical coordinates) where the boundary current transport changes direction.

The role of the dissipation and wave propagation is particularly clear in the Kawase model, and this renders it particularly illuminating. Indeed, the

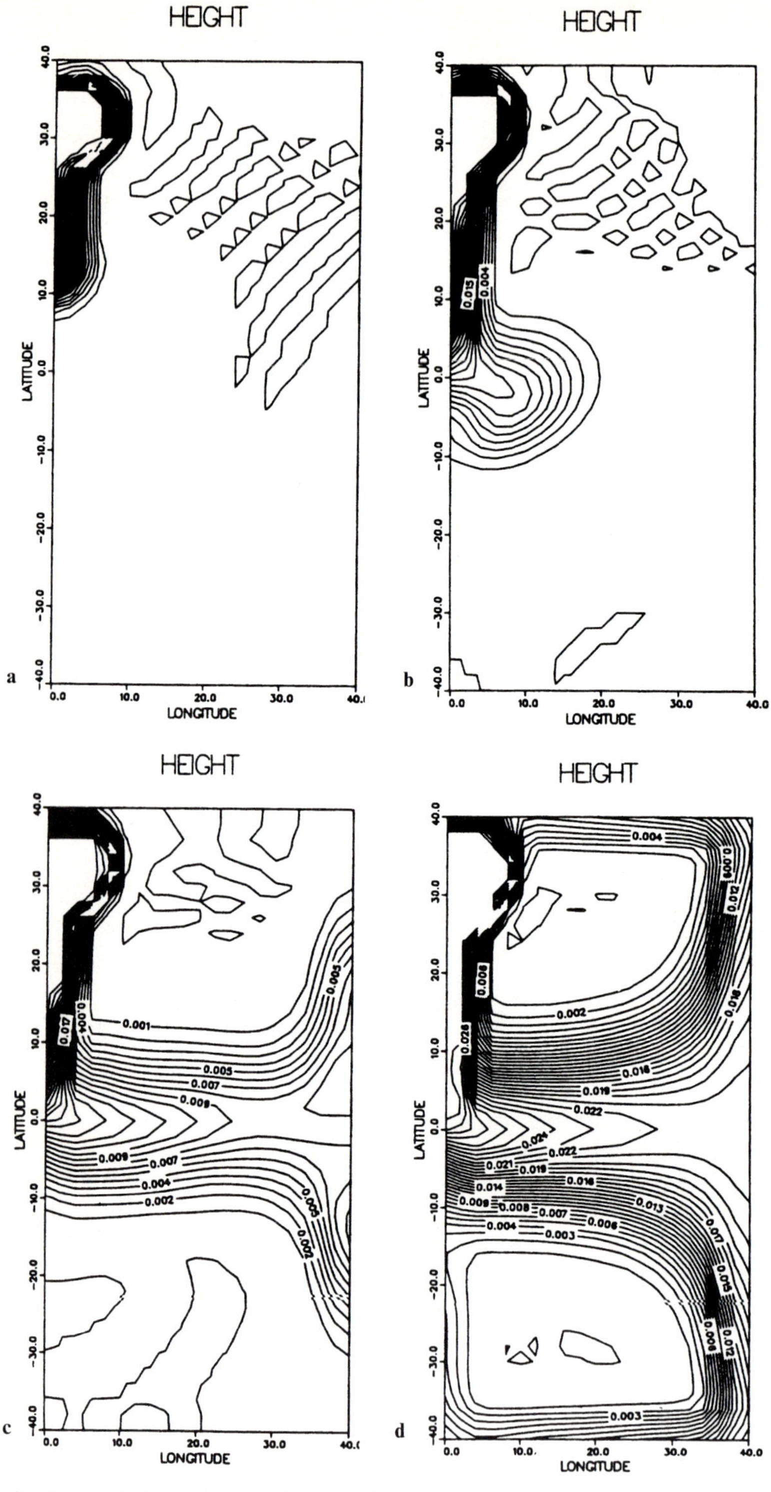


**Fig. 7.6.4a–d.** Establishment of the height field from a state of rest in response to a source in the northwest corner of the basin. Days 5 (**a**),10 (**b**), 20 (**c**), and 40 (**d**) from the intitation of the flow. $\delta_B = 0.3\ x_e$. (From Kawase 1987)

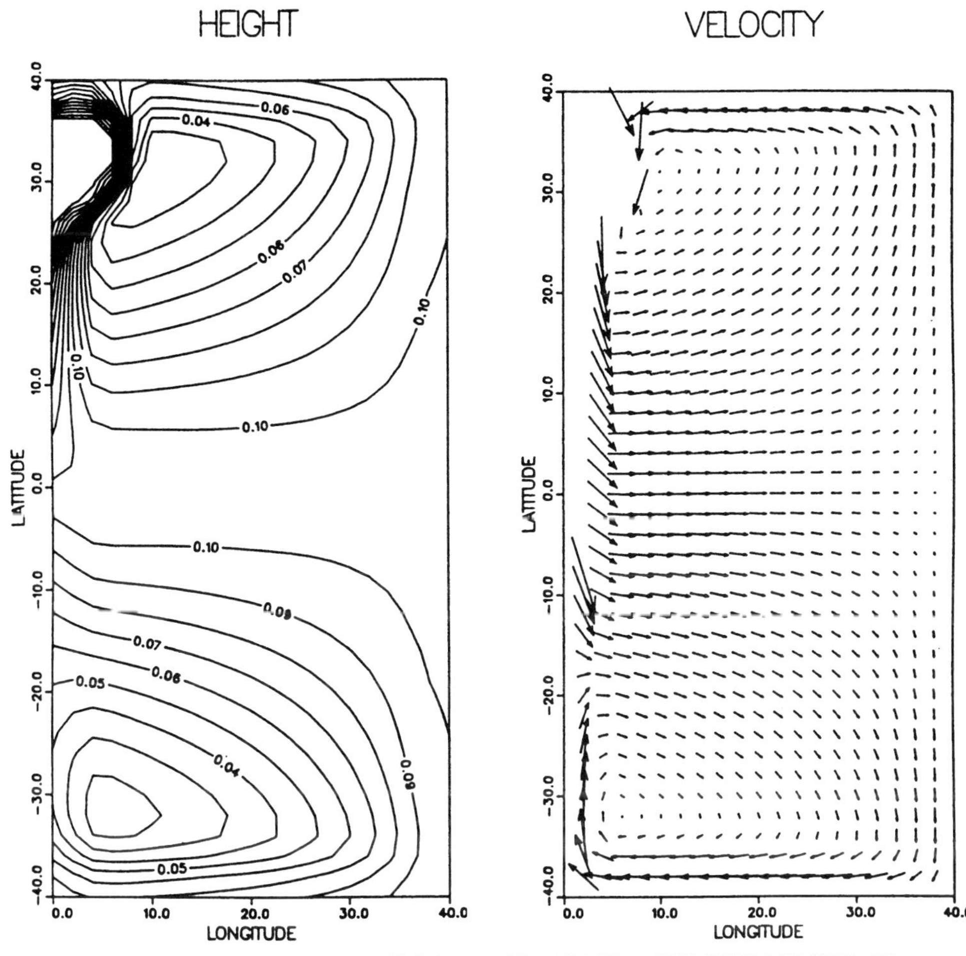


**Fig. 7.6.5.** Final steady state at day 200 of the calculation shown in the previous Fig. 7.6.4. (From Kawase 1987)

restriction that fluid in the abyss can upwell only where the abyss itself induces an anomaly actually allows the height field to act as a tracer of the wave perturbations, either in the spin-up phase or in the final steady state where the Rossby wave field is frozen in a balance between westward propagation and local dissipation. This role of the Rossby wave remains important even when the restriction of the upwelling to satisfy (7.6.12) is relaxed. The next section examines a fully baroclinic model to describe the layered structure of the abyss. Although the upwelling is determined a priori as in the Stommel-Arons theory, the baroclinic wave dynamics discussed by Kawase becomes the key in understanding the determination of the vertical structure of the abyssal flow.

## 7.7 Vertical Structure of the Abyssal Circulation

### *Formulation*

The Stommel-Arons theory represents the abyssal region as a single, homogeneous layer in which the horizontal velocity is independent of depth. It follows that at each geographical location the flow must have only a single direction. Yet, as we noted in Section 7.1 (see Fig. 7.1.1), the motion in the abyss is layered with interleaving tongues of water from different polar sources apparently flowing in opposite directions. In order to explain this behavior a baroclinic model of the abyss is needed in which the vertical structure of the abyssal flow is determined by the dynamics of the abyss rather than being imposed by the specification of small number of layers to represent the flow. The results of the previous section in which the model of Kawase is discussed suggest that the baroclinic dynamics of the large-scale Rossby waves is essential in understanding the behavior of the interior abyssal flow outside the western boundary current. It was observed there that the propagation of signals from the source region follows an intricate, indirect path so that the thermal anomalies actually enter the interior from the *eastern* boundary and only then move westward to put the interior into motion. In fact, the results of the classical Stommel-Arons theory prepare us for the indirect relationship between the direction of the motion in the interior and the position of the sources. This is even more evident when the flow is baroclinic and thus has more degrees of freedom in the vertical.

At the same time, instead of restricting the driving of the abyssal region to areas in which the abyss develops its own thermal anomalies, in this section we return to the starting point of the Stommel-Arons theory in which the thermocline dynamics specifies a vertical velocity at its base which drives the abyss. It is likely that some combination of the two models of the driving would be most realistic, but for simplicity we choose the Stommel-Arons approach in the expectation that limiting the domain of motion to those regions which are self-driven by the abyssal anomalies is overly restrictive.

In the Stommel-Arons theory the upwelling velocity into the thermocline, $w_o$, is often taken *for illustrative purposes* to be a constant. For the barotropic model this presents little loss of generality. As noted in our earlier discussion, Stommel and Arons were keenly aware that $w_0$ is probably not spatially constant and suggested in fact that since theoretical considerations suggest it to be inversely proportional to the depth scale of the thermocline, it should be largest in eastern and equatorial regions and certainly not be spatially uniform. The combination of non-uniform upwelling and baroclinic wave dynamics is seen to yield an intriguing suggestion about the relative importance of the baroclinic to barotropic response of the abyss to the upwelling into the thermocline.

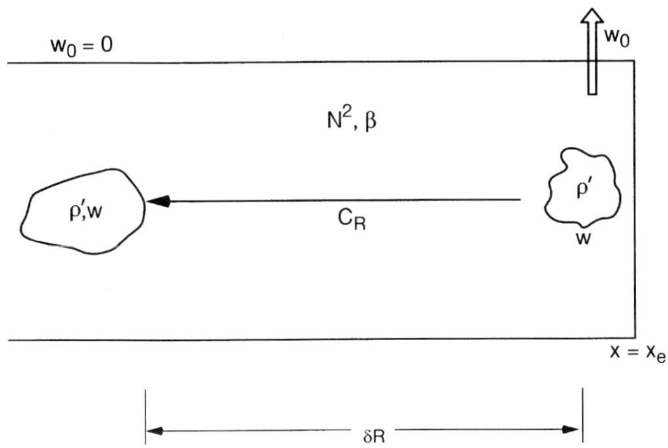


**Fig. 7.7.1.** Schematic diagram showing a zonal cross section of a stratified abyssal region driven by an upwelling velocity, $w_0$, which is limited to the eastern portion of the basin.The abyss has a background buoyancy frequency, $N$. The thermal anomaly produced by the upwelling propagates westward with the Rossby wave speed $c_R$ and extends a distance $\delta_R$ westward as determined by a balance between propagation and dissipation of the anomaly. In the western region the anomaly arrives at a location where $w_0$ is zero, but where the local vertical velocity, $w$, in the body of the fluid produced by the local density anomaly is different from zero

The physical essentials of the idea can be grasped by considering a simple, extreme conceptual example. Suppose, as in Fig. 7.7.1, the upwelling into the abyss is concentrated *entirely* on the eastern boundary as shown in the figure. Suppose there is a background stratification measured by the buoyancy frequency $N$. The upwelling in the eastern portion of the basin produces a density anomaly locally which propagates westward as an ensemble of baroclinic Rossby waves each with a different vertical structure. Each Rossby wave extends westward a distance $\delta_R$ which is appropriate to its vertical mode structure. This distance arises as a balance between the propagation westward as a Rossby wave and the decay of the anomaly through thermal dissipation. The westward extent of the anomaly reaches into a region in which $w_0$ is zero (this can always be arranged conceptually by limiting the region in which $w_o$ is different from zero to a narrow zone in the east). In this region the density anomaly carried by the wave is nonzero and forces, locally, an accompanying vertical velocity in consequence of the distortion of the density surfaces. The vertical velocity in turn implies a meridional velocity as a consequence of the Sverdrup vorticity equation:

$$\beta v = f \frac{\partial w}{\partial z} \tag{7.7.1}$$

whose validity is independent of the stratification. When (7.7.1) is integrated over the entirely depth of the abyss:

$$\beta \int_{\text{abyss}} v\, dz = f w_0 \tag{7.7.2}$$

assuming that the bottom of the abyss is flat or that the vertical velocity is negligibly small there. If $w_0 = 0$ in the western region which the anomaly has reached by Rossby wave propagation, then locally the integral of the meridional velocity must vanish although $v$ itself, from (7.7.1), differs from zero. That is, *the barotropic portion of the abyssal flow can be locally negligible compared to the baroclinic part if the upwelling velocity varies strongly in the horizontal.* The flow must therefore be layered, at least in the western region of the basin. If a thermal anomaly is produced on the eastern boundary of the basin, say, by the Kelvin wave propagated from the source region or from a region of upwelling further south rather than by local upwelling, the flow is layered even in the eastern basin if the upwelling is weak there as well. Thus the layering of the abyssal flow can be associated with the spatial nonuniformity of the abyssal upwelling.

Pedlosky (1992) and Edwards and Pedlosky (1995) studied a continuously stratified, linear model of the abyssal circulation to examine this phenomenon. In particular, the model is used to examine the relation between the structure of the abyssal flow in both the interior and the boundary layer and the vertical structure of the sources on the perimeter of the basin representing the sources of deep water in the abyss. As might be expected from the indirect nature of the forcing of the interior, the relationship between the vertical structure of the input sources and the vertical structure of the interior flow is a complex one.

The model dynamics consists of geostrophic balance for the velocities outside the region of the western boundary current as well as the hydrostatic approximation. The total density field, as in the quasi-geostrophic models of Chapter 3, is represented in terms of a background field which is independent of horizontal position plus a weak anomaly which varies laterally, namely:

$$\rho_{\text{total}} = \rho_0 + \rho_v(z) + \rho_H(x, y, z). \tag{7.7.3}$$

The basic restriction of the model is that

$$\rho_H \ll \rho_v \ll \rho_0. \tag{7.7.4}$$

Thus the density is nearly constant and has a weak vertical variation which determines a background buoyancy frequency:

$$N = \left( \frac{-g}{\rho_0} \frac{\partial \rho_v}{\partial z} \right)^{1/2} \tag{7.7.5}$$

and an even weaker lateral variation of density that drives a baroclinic motion through the thermal wind relation. Thus, again using Cartesian coordinates for simplicity:

$$\rho_0 f v = \partial p / \partial x$$

$$\rho_0 f u = -\partial p/\partial y$$

$$\rho_H g = -\partial p/\partial z. \qquad (7.7.6\,\text{a,b,c})$$

The density appearing in (7.7.6c) is the density anomaly $\rho_H$ since $p$ refers to the horizontally varying pressure field. The thermal wind relation implies that:

$$f\frac{\partial v}{\partial z} = -\frac{g}{\rho_0}\frac{\partial \rho_H}{\partial x}, \quad f\frac{\partial u}{\partial z} = \frac{g}{\rho_0}\frac{\partial \rho_H}{\partial y}. \qquad (7.7.7\,\text{a,b})$$

In (7.7.6) and (7.7.7) $f = f_0 + \beta y$.
The condition of incompressibility of the abyssal flow is:

$$\frac{\partial u}{\partial x} + \frac{\partial v}{\partial y} + \frac{\partial w}{\partial z} = 0 \qquad (7.7.8)$$

from which (7.7.1) follows by eliminating the pressure in the momentum equations.

The density equation is a balance between advection and the diffusion of density (or equivalently temperature and salinity) by small-scale turbulence. If the assumption that the laterally varying density anomaly is small with respect to $\rho_v$ is used, the density equation reduces to the linear balance:

$$w\frac{\partial \rho_v}{\partial z} = \kappa\frac{\partial^2 \rho_H}{\partial z^2}. \qquad (7.7.9)$$

Here the assumption is made that it is vertical mixing of the density anomaly which is important in the interior of the basin rather than horizontal mixing. The density equation can be rewritten in terms of the pressure as:

$$wN^2 = \frac{\kappa}{\rho_0}\frac{\partial^3 p}{\partial z^3}. \qquad (7.7.10)$$

On the lower boundary, i.e., at $z = -H$, the vertical velocity is zero, while at the top of the abyss, at $z = 0$, the vertical velocity is given by $w_0(x, y)$, thus, with (7.7.10):

$$\frac{\partial^3 p}{\partial z^3} = 0, \qquad z = -H$$

$$\frac{\partial^3 p}{\partial z^3} = \frac{\rho_0 N^2}{\kappa} w_0, \qquad z = 0 \qquad (7.7.11\text{a,b})$$

The density anomaly on both the bottom at $z = -H$ and at the top of the abyss are taken to be zero, although the latter condition can easily be relaxed. However, the simplest model is one in which, as in the Kawase model, the temperature anomalies are driven internally by the response to the upwelling and the sources rather than externally imposed density forcing. Thus:

$$\frac{\partial p}{\partial z} = 0, \quad \text{on } z = 0, -H \tag{7.7.12a,b}$$

From (7.7.1) and (7.7.6a):

$$f\frac{\partial w}{\partial z} = \frac{\beta}{\rho_0 f}\frac{\partial p}{\partial x}. \tag{7.7.13}$$

When the equation for the density (7.7.10) is used to eliminate $w$, we obtain the final equation for the pressure field in the abyss:

$$\frac{\partial^4 p}{\partial z^4} - \frac{N^2 \beta}{f^2 \kappa}\frac{\partial p}{\partial x} = 0. \tag{7.7.14}$$

In deriving (7.7.14) the simple case in which $N^2$ is independent of $z$ is considered here. Variable buoyancy frequency can be considered with no qualitative alteration in the physics. In fact (7.7.14) is the analog in the continuously stratified case of (7.6.21). That is, (7.7.14) describes the balance between the westward propagation of the pressure field in the Rossby wave and the dissipation due to thermal mixing which diminishes the anomaly as it propagates westward.

On the eastern boundary of the basin the geostrophic zonal velocity is taken to be zero. That is, the only deep water sources that are considered are on either the northern, southern, or western boundaries. On the eastern wall where there can be no boundary current involving higher order dynamics than those included in (7.7.14) the boundary condition must be:

$$\frac{\partial p}{\partial y} = 0. \tag{7.7.15}$$

Note that this only implies that the pressure on the eastern boundary must be independent of latitude. It may still be a nontrivial function of $z$. There is in general a density anomaly produced on the eastern boundary by the process of Kelvin wave propagation around the rim of the basin, as described in the the preceding section, which establishes a density anomaly, independently of $y$, on $x = x_e$ . This eastern boundary density perturbation must be determined to obtain the complete interior solution.

On the other boundaries the velocity must match the velocity of the sources. However, on the other boundaries the velocity normal to the boundary are generally not in geostrophic balance, and additional terms representing horizontal mixing of momentum and/or density must be added to the formulation of the physics. We suppose that on these other boundaries:

$$\vec{u} \cdot \hat{n} = -s(\vec{r}, z) \tag{7.7.16}$$

where $s$ represents the flux of deep water into the basin that depends on both depth, $z$, and position, $\vec{r}$, along the boundary $C$.

This completes the mathematical formulation of the problem. However, some general insight into the solution can be obtained by simple scaling considerations. In particular we can estimate the ratio of the baroclinic velocity, induced by the density gradients, to the barotropic velocity that must still satisfy the Stommel-Arons integral relation (7.7.2). Thus we can estimate the vertical average, or barotropic, component of the flow from (7.7.2) as:

$$V_{\text{barotropic}} = \frac{f w_0}{\beta H}. \tag{7.7.17}$$

The baroclinic velocity can be estimated from the thermal wind relation as:

$$V_{\text{baroclinic}} = \frac{g}{f}\frac{H}{\rho_0}\frac{\partial \rho}{\partial x} = \frac{g}{f}\frac{H}{\rho_0}\frac{\Delta_\rho}{x_e}. \tag{7.7.18}$$

However, the density anomaly, $\Delta\rho$, can be related to the vertical velocity through the density equation, (7.7.9), i.e., assuming that $w$ is $O(w_0)$:

$$\frac{\Delta_\rho}{\rho_0} = \frac{w_0 N^2 H^2}{g\kappa} \tag{7.7.19}$$

so that:

$$V_{\text{baroclinic}} = \frac{w_0 N^2 H^3}{f\kappa x_e}. \tag{7.7.20}$$

Therefore, the ratio of the barotropic to baroclinic velocity is of the order of:

$$\frac{V_{\text{baroclinic}}}{V_{\text{barotropic}}} = \frac{\beta N^2 H^4}{\kappa f^2 x_e} \tag{7.7.21}$$

If $\beta = 2 \times 10^{-13}\text{cm}^{-1}, N = 10^{-13}\text{s}^{-1}, H$ (for the scale of the motion)1 km, $\kappa = 0.1\text{cm}^2\text{s}^{-1}, f = 10^{-4}s^{-1}$, and $x_e = 3000$ km this parameter is about 70. A larger value of the diffusion coefficient or a weaker background stratification reduces the ratio, but the implication is clear. The baroclinic velocity is large compared with the vertical average, and on the basis of the physical argument given above we can expect *the Stommel-Arons barotropic velocity to be a relatively small residual obtained by vertically averaging the large baroclinic velocity*.

An important constraint on the interior solution which allows us to determine the density anomaly on the eastern boundary arises from the inability of the boundary layers to transfer a *net* vertical mass flux locally. This inability follows from the balance that occurs in the density equation in the boundary layer. Each variable in the western boundary layer can be represented as the sum of the interior solution (subscript I) and a boundary layer *correction* denoted by a tilde, e.g.:

$$\rho_H = \rho_I + \tilde{\rho}. \tag{7.7.22}$$

The correction variables vanish outside the boundary layers and are used to satisfy the boundary conditions, for example, of no slip and no heat flux, at the side walls.

The density equation for the correction to the density field in the western boundary layer is easily shown to be:

$$-\tilde{w}\frac{\rho_0 N^2}{g} = \kappa_H \frac{\partial^2 \tilde{\rho}}{\partial x^2} \tag{7.7.23}$$

where $\kappa_H$ is the coefficient of lateral diffusion of the density anomaly. If (7.7.23) is integrated across the boundary layer:

$$\int \tilde{w}\, dx = \frac{\kappa_H g}{N^2 \rho_0}\frac{\partial \tilde{\rho}}{\partial x}(0) = -\frac{\kappa_H f}{N^2}\frac{\partial \tilde{v}}{\partial z}(0). \tag{7.7.24}$$

The integral in (7.7.24) is taken over the entire boundary layer width. The integral of the right side leaves us with the correction to the density gradient at the wall since by definition the correction function vanishes outside the boundary layer. Since the *long-shore* velocity can be shown to remain in geostrophic balance, the thermal wind relation yields the second step in (7.7.24). The long-shore velocity in the correction field is larger than the interior flow in the western boundary current, and therefore the right side of (7.7.24) represents, to the lowest order, the density gradient and vertical shear of the total meridional velocity. If either $v$ is zero on the side wall (i.e., if no-slip condition on $v$ is used), or if the side wall is thermally insulated, the right hand side of (7.7.24) vanishes, and it follows that the *net* vertical mass flux in the side wall boundary layers is negligibly small. Thus at each $z$ level the total mass flux which enters the basin through the sources, $s(\vec{r}, z)$, must enter the interior *at that depth* although not necessarily at the same geographical position along the boundary as the source. Thus, integrating the continuity equation for the *interior* motion:

$$\iint_A \left(\frac{\partial u}{\partial x} + \frac{\partial v}{\partial y}\right) dA = -\oint_C \vec{u} \cdot \hat{n}\, d\ell = \oint_C s(\vec{r}, z)\, d\ell \tag{7.7.25}$$

where $A$ and $C$ represent the area and perimeter of the basin. The last equality in (7.7.25) follows from the inability of the boundary layers to take the mass flux entering the basin and transport it to a different depth. Since the variables in (7.7.25) are the interior velocities which are in geostrophic balance:

$$\begin{aligned}\iint_A \left(\frac{\partial u}{\partial x} + \frac{\partial v}{\partial y}\right) dx\, dy &= -\iint_A \frac{\beta v}{f}\, dx\, dy = -\iint_A \frac{\beta}{\rho_0 f^2}\frac{\partial p}{\partial x}\, dx\, dy \\ &= -\int_{y_s}^{y_n} dy \frac{\beta}{\rho_0 f^2}[p(x_e) - p(x_w)].\end{aligned} \tag{7.7.26}$$

Thus:

$$\int_{y_s}^{y_n} \frac{\beta}{\rho_0 f^2}[p(x_e) - p(x_w)]dy = \oint_C s(\vec{r}, z)d\ell \tag{7.7.27}$$

where $x_e$ and $x_w$ are the eastern and western boundaries of the basin, and $y_s$ and $y_n$ are the southern and northern extremes of the basin. Note that the constraint (7.7.27) involves only the perimeter average of the input sources, i.e., the interior pressure drop across the basin involves only:

$$S(z) = \oint_C s(\vec{r}, z)d\ell. \tag{7.7.28}$$

### *Technique of Solution*

The boundary conditions (7.7.11 a,b) and (7.7.12 a,b) as well as the even number of z derivatives in (7.7.14) suggest that the solution of the interior problem might be usefully represented in terms of the cosine expansion:

$$p(x, y, z) = \sum_{n=0}^{\infty} P_n(x, y) \cos\left(\frac{n\pi(z + H)}{H}\right) \tag{7.7.29}$$

where:

$$P_n = \frac{1}{C_n}\int_{-H}^{0} p \cos\left[\frac{n\pi(z + H)}{H}\right]dz \qquad C_n = \begin{cases} H & n=0 \\ H/2 & n>0 \end{cases} \tag{7.7.30}$$

The horizontal velocities have a similar cosine expansion, while the density and vertical velocity have an expansion in sine functions instead. The expansion for these variables can all be obtained from the pressure field. It is important to note that because of the inhomogeneous condition (7.7.11 b), the equation for $P_n$ *cannot* be obtained by inserting (7.7.29) directly into (7.7.14) and differentiating term by term. Instead (7.7.14) must be multiplied by $\cos[n\pi(z + H)/H]$ and integrated over the depth range $(-H, 0)$. When this is done, we obtain as the governing equation for $P_n$:

$$\frac{\partial P_n}{\partial x} - \alpha_n P_n = \frac{(-1)^n}{C_n}\frac{\rho_0 f^2}{\beta} w_0(x, y) \tag{7.7.31}$$

where:

$$\alpha_n = \left(\frac{n\pi}{H}\right)^4 \frac{\kappa f^2}{\beta N^2}. \tag{7.7.32}$$

Equation (7.7.31) describes the variation with longitude of the amplitude of each vertical mode of the solution. The $n = 0$ mode is the barotropic mode and carries the vertically averaged abyssal transport in the interior. The modes with $n \geq 1$ are the baroclinic modes, and they possess no vertically averaged transport. Since $\alpha_n = 0$ for $n = 0$, the barotropic internal mode satisfies a local relation between the longitudinal pressure gradient, and therefore the meridional velocity, and the forcing, i.e.:

$$\frac{\beta}{\rho_0 f}\frac{\partial P_0}{\partial x} = \beta v_0 = f\frac{w_0}{H}. \tag{7.7.33}$$

The zero-subscripted variables on the left side of (7.7.33) represent the barotropic mode, and when (7.7.33) is compared with (7.3.5), we see that the barotropic mode satisfies the Stommel-Arons dynamics for the interior. Indeed, with the entire problem broken into vertical modes, the lowest mode, corresponding to $n = 0$, is exactly the same as the barotropic model of Stommel and Arons. The barotropic mode can thus be calculated independently of the baroclinic modes. Thus all the results of Section 7.3 can be carried over directly to the barotropic mode of the stratified model. This assumes of course that effects such as nonlinearity or bottom slope do not enter to mix the vertical modes and cause them to interact. The reader is referred to Pedlosky and Chapman (1993) for an example of the role of topography in mixing the baroclinic and barotropic modes and the way in which the baroclinic modes can become net transporters of mass in the presence of a bottom slope.

For the baroclinic modes, $n \geq 1$, the solution of (7.7.14) for the pressure and velocity fields is nonlocal, i.e.:

$$P_n = P_n(x_e)e^{-\alpha_n(x_e - x)} - \frac{(-1)^n \rho_0 f^2}{C_n \beta}\int_x^{x_e} w_0 e^{-\alpha_n(x' - x)}dx' \tag{7.7.34}$$

The pressure and its spatial derivatives depend on the integral of the forcing along the latitude y between the point $x$ and the eastern boundary at$x = x_e$. The solution also depends on the value of the pressure anomaly $P_n(x_e)$ on the eastern boundary. For $n = 0$ this adds an irrelevant constant to the solution but for the baroclinic modes this produces a pressure anomaly which varies in both the $x$ and $y$ directions (recall that $\alpha_n$ is a function of $y$ through its dependence on $f$). On $x = x_e, P_n(x_e)$ represents a pressure anomaly that is independent of $y$ and a function of $z$ only through the dependence of $P_n$ on its index $n$. Thus it represents a density anomaly on the eastern boundary that induces interior pressure perturbations as a consequence of the westward Rossby wave radiation by the long baroclinic Rossby waves. The presence of the density anomaly on the eastern boundary is due ultimately to the disturbance of the density field by the input sources which propagates around the basin and establishes a steady state density anomaly on the eastern boundary. As fluid enters the basin at a depth $z$ and rises to leave through the upper boundary, its upward motion can take place only if there is sufficient diffusion of density to allow the fluid to adjust to the background density gradient according to (7.7.9). Since there is a net upwelling there is a net displacement of the density surfaces and in particular an adjustment of the density on the eastern boundary.

This adjustment, or equivalently $P_n(x_e)$, can be determined from the mass balance constraint (7.7.27). If each term in (7.7.27) is represented in a cosine series:

$$\int_{y_s}^{y_n} dy \frac{\beta}{\rho_0 f^2} [P_n(x_e) - P_n(x_w)] = S_n \tag{7.7.35}$$

where $S_n$ is the Fourier component of the *perimeter integral of the input sources*, i.e.:

$$S_n = \frac{1}{C_n} \int_{-H}^{0} S(z) \cos\left[\frac{n\pi(z+H)}{H}\right] dz. \tag{7.7.36}$$

$S(z)$ is given by (7.7.28) and is the total mass flux per unit of depth interval that enters the basin over the entire perimeter of the basin, and $S_n$ is the $n_{th}$ Fourier cosine amplitude of the depth distribution of all the sources around the perimeter. It is important to note that $S_n$ is independent of the placement of the sources along the boundary; it recognizes the distribution of the total source strength only with $z$. This steady-state constraint is the result of the rapid passage of the Kelvin waves. On the scale of the geostrophic motions the boundary adjustment is instantaneous, and the integral condition (7.7.35) holds as soon as the barotropic Rossby wave crosses the basin and the Sverdrup balance is established.

If (7.7.34) is used in the constraint (7.7.35), $P_n(x_e)$ is determined, that is:

$$P_n(x_e) \int_{y_s}^{y_n} dy \frac{\beta}{f^2 \rho_0} \left[1 - e^{-\alpha_n(x_e - x_w)}\right] + \int_{y_s}^{y_n} \int_{x_w}^{x_e} dy\, dx' \frac{(-1)^n}{C_n} w_0 e^{-\alpha_n(x' - x_w)} = S_n. \tag{7.7.37}$$

Thus:

$$\frac{P_n(x_e)}{\rho_0} = \frac{S_n}{I_n} - \frac{(-1)^n}{C_n I_n} \int_{y_s}^{y_n} \int_{x_w}^{x_e} w_0 e^{-(x' - x_w)} dx' dy \tag{7.7.38}$$

where $I_n$ is a number that depends only on the geometry, background stratification of the basin, and the vertical diffusion through the parameter $\kappa$, i.e.:

$$I_n = \int_{y_s}^{y_n} \frac{\beta}{f^2} \left(1 - e^{-\alpha_n(x_e - x_w)}\right) dy. \tag{7.7.39}$$

When $n = 0$ the parameter $\alpha_n = 0$, and $I_n$ is also 0. Thus for $n = 0$ the integral condition (7.7.37) places no constraint on the irrelevant constant $P_0(x_e)$ and instead is a restatement of the overall mass balance of the basin:

$$S_0 H = \int_{y_s}^{y_n} \int_{x_w}^{x_e} w_0 dx\, dy. \tag{7.7.40}$$

On the other hand, the baroclinic structure of the eastern pressure anomaly is not a simple reflection of the overall source-sink balance but instead is a consequence of the Kelvin wave propagation of the perturbation induced by the baroclinicity of the source. As (7.7.38) demonstrates, this depends on the overall source structure and the distribution of the upwelling in the interior. The anomaly itself on the eastern boundary is the heritage, in the steady state, of the prior passage of the Kelvin wave carrying information about the source strength.

Now that the pressure anomaly on the eastern boundary is determined, the total solution may be written, using (7.7.38) in (7.7.34) as:

$$\begin{aligned}\frac{P_n(x, y)}{\rho_0} &= \frac{S_n e^{-\alpha_n (x_e - x)}}{I_n} \\ &\quad - \frac{(-1)^n}{C_n} \frac{e^{\alpha_n (x_e - x)}}{I_n} \int_{y_s}^{y_x} \int_{x_w}^{x_e} w_0(x', y) e^{-\alpha_n (x' - x)} dx' dy \\ &\quad - \frac{(-1)^n f^2}{C_n \beta} \int_x^{x_e} w_0(x', y) e^{-\alpha_n (x' - x)} dx'.\end{aligned} \tag{7.7.41}$$

The first two terms of the solution for $P_n$ spring from the density anomaly on the eastern boundary while the last term on the right side is due to the direct forcing of the interior by the upwelling. Both effects induce Rossby wave propagation westward. Again, note that the interior flow is indifferent to where on the boundary the sources of deep flow are added. The flow could enter anywhere along the northern, western, or southern boundaries without altering the solution in (7.7.41) as along as the perimeter-averaged input is identical.

### *Structure of the Circulation*

A question of particular interest can now be investigated directly. Namely, what is the relationship between the vertical structure of the solution and the vertical structure of the sources? That is, does the interior flow reflect directly the vertical form of the input flow?

Consider first the case in which $w_0$ is independent of $x$ but may still depend on $y$. We argued above that the $x$ dependence of $w_0$ may lead to layering. Is this also true of $y$ dependence of the upwelling? The $x$ integrals in the expression for $P_n$, (7.7.41), can be performed directly if $w_0$ is independent of $x$. Using the fact that $\alpha_n \beta / f^2$ is independent of $y$:

$$\frac{P_n}{\rho_0} = \frac{S_n}{I_n} e^{-\alpha_n(x_e - x)} - \frac{(-1)^n}{C_n} e^{-\alpha_n(x_e - x)} \frac{f^2}{\beta \alpha_n} \left[ \frac{\int_{y_s}^{y_n} dy\, w_0 \frac{\beta}{f^2} \left(1 - e^{-\alpha_n(x_e - x_w)}\right)}{I_n} - w_0 \right] - \frac{(-1)^n f^2}{C_n \alpha_n \beta} w_0. \tag{7.7.42}$$

If, and only if, $w_0$ is independent of $y$ as well as $x$, do the two terms in the large bracket of (7.7.42) cancel, and the last term becomes an irrelevant constant so that:

$$\frac{P_n}{\rho_0} = \frac{S_n}{I_n} e^{-\alpha_n(x_e - x)} \tag{7.7.43}$$

In this case, *and only in this case,* the vertical distribution of the pressure anomaly and the associated velocity are determined entirely in terms of the vertical structure of the input sources. If, for example, all the baroclinic source input is in vertical mode number $N$, i.e., if $S_n = 0$ for $n \neq N$, then only $P_N$ differs from zero and thus has the same vertical structure as the source. This occurs *only* when the upwelling is completely uniform in space. Thus even when $w_0$ is independent of $x$ but is a function of latitude, the vertical structure of the circulation depends on the vertical modes excited in the Rossby wave field that propagate westward. If, for example, the upwelling is limited to a very narrow region near the southern boundary the Kelvin waves still establish an isolated density anomaly on the eastern boundary which then propagates westward in the form of baroclinic Rossby waves. Their vertical structure depends on $P_n(x_e)$, which, as we see from (7.7.38), depends on the entire upwelling distribution in the basin as well as the sources on the perimeter.

Thus if the interior upwelling depends on either latitude or longitude, the vertical structure of the interior circulation is not determined only by the vertical form of the input of the sources but by the full dynamics of the baroclinic Rossby waves in the interior.

Edwards and Pedlosky (1995) examined the response of this abyssal model to an upwelling distribution of the form:

$$w_0 = K e^{-s(x_e - x)} e^{\,t(y - y_s)} \tag{7.7.44}$$

where $K$ is chosen to satisfy the over all mass balance (7.7.40). The parameters $t$ and $s$ are chosen to control the variation of the upwelling spatially. Thus the larger the parameter $sx_e$, the more concentrated on the basin's eastern boundary is the upwelling, while the larger the parameter $tx_e$, the more the upwelling remains localized near the southern boundary of the basin. Edwards and Pedlosky introduced the parameter $\alpha$ where:

$$\alpha = \frac{\pi^4 f_0 L^2 \kappa}{N^2 H^4}. \tag{7.7.45}$$

and $f_0$ is a characteristic value of the Coriolis parameter in the basin ($10^{-4} s^{-1}$) while $L = x_e - x_w$, is the width of the basin. In terms of $\alpha$:

$$\alpha_n L = \alpha n^4 \frac{f}{f_0} \frac{f}{\beta L}. \tag{7.7.46}$$

In their calculations Edwards and Pedlosky chose $\alpha = 0.01$.

Note that the Rossby modes corresponding to higher n are trapped near the eastern boundary, and therefore the baroclinic modes with low $n$ will dominate in the western basin. With $\alpha = 0.01$ the gravest modes reach across the basin with little damping.

Figure 7.7.2 shows the meridional velocity in the interior solution driven by a source distribution, $S(z)$, shown in panel a. $S(z)$ has the form:

$$S = S_0 \left[ 1 - \cos\left( \frac{2\pi(z + H)}{H} \right) \right] \tag{7.7.47}$$

so that the fluid entering the basin enters at middepth in the abyss in *the perimeter average* and contains Fourier harmonic content in only the zero and second Fourier modes. Panel b shows the meridional velocity profiles at nine zonal locations on the line $y = 0.5L$. At each location a solid vertical line represents the zero line for $v$. The origin of the $x$ coordinate, corresponding to $x_w$ is placed at $x = 0$ while $x$ is scaled with $L$ in the figure. In the example shown in panel *b* the upwelling velocity is spatially uniform, and we can see that the result of (7.7.43) obtains. The vertical structure of the meridional velocity is precisely the same as that of the source with a maximum of $v$ at middepth. The scale of the velocity in the figure is $fKL/\beta H$ where $K$, in (7.7.44), is chosen to satisfy the overall mass balance (7.7.40). The only structural reflection of the Rossby wave dynamics is the slight decay in the amplitude of the response in the west with respect to the eastern basin. In panel c there is instead a substantial variation of the upwelling with $x$ such that $sL = 2$. Now the meridional velocity profiles are no longer similar to the source structure. Instead there is a progression of the structure from east to west. In the east, where there is strong upwelling, the meridional velocity is strongly positive near the surface, reflecting the vortex tube stretching there. There is a minimum at $z \approx H/4$ corresponding to the decay of that stretching. At middepth the effect of the source again becomes important, and the meridional velocity shows a secondary maximum. In the west the meridional velocity changes sign in the upper part of the water column, and the flow is directed equatorward. Since the signal of the upward vertical velocity produced in the east propagates westward, in the west there is a positive $w$ in the body of the fluid. However, since $w$ is very small at the upper boundary of the abyss due to the exponential decay of $w_0$, with increasing distance westward, the vertical velocity must

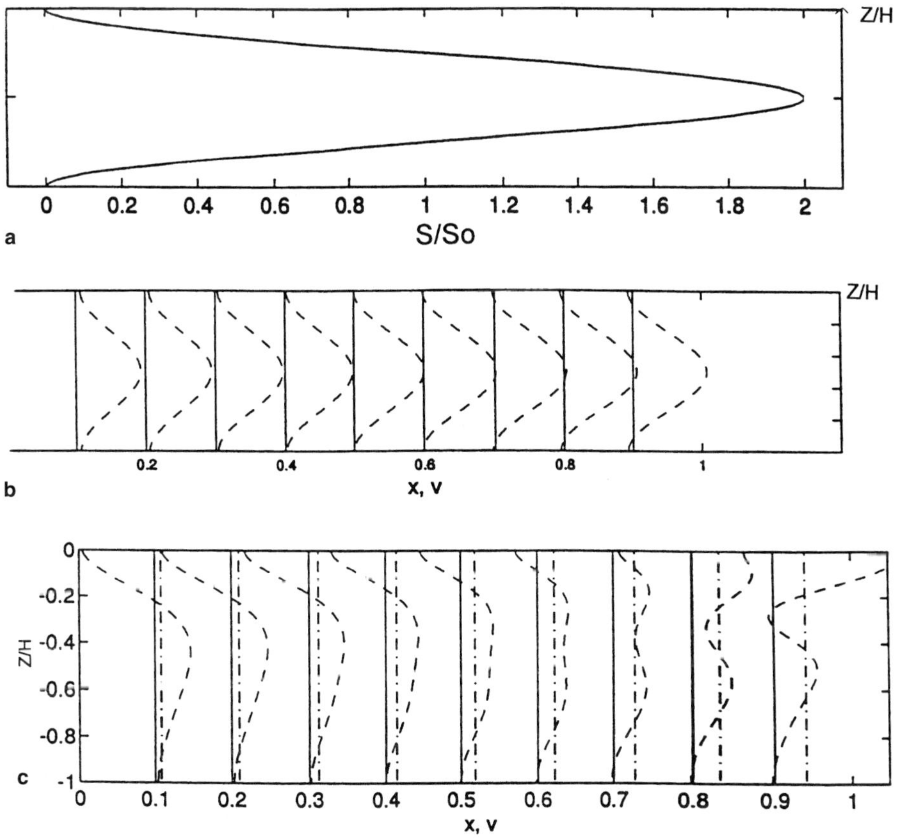


**Fig. 7.7.2a,b.** Response of the abyssal flow as shown in the meridional velocity profile to an integrated source function S(z). $S$ has the form. $S_0\left[1-\cos\left(\frac{2\pi(z+H)}{H}\right)\right]$. The constant $K$ in (7.7.44) is adjusted to satisfy the overall mass balance. **b,c** Meridional velocity profiles at different x locations across the basin. *Solid vertical line,* zero for *dotted line,* meridional velocity profile. The variable x is scaled with L while v is scaled with $fKL/\beta H$. **b** meridonal velocity profiles for the case where the upwelling is spatially uniform ($t=s=0$). **c** The meridional velocity profiles for the case where there is a strong variation of the upwelling velocity in the zonal direction, $s(x_e-x_w)=2$

become nearly zero at the upper boundary. Thus there is vortex tube *compression* near the upper surface in the western region which drives an equatorward flow according to (7.7.1). At middepth the effect of the source is again important, and the flow is poleward. As discussed above, since $w_0$ is essentially zero in the west when sL is sizable, the meridional velocity must very nearly have a zero vertical average, i.e., it must be layered. In Fig. 7.7.2c the dashed-dot line shows the magnitude of the barotropic component of the meridional velocity. It decays westward, of course, in the same way as does $w_0$. At almost all locations the baroclinic velocity far exceeds the barotropic velocity so that the barotropic meridional velocity calculated from Stommel-Arons theory is seen to be a small residual which remains after vertical averaging.

Another particularly revealing example is shown in Fig. 7.7.3. In this example the fluid enters predominantly at the very bottom of the abyss. Here:

$$S = \frac{S_0}{2}\left[1 - \cos\left(\frac{\pi z}{H}\right)\right] \tag{7.7.48}$$

and is shown in panel a of the figure. Panel b shows the response. The meridional velocity profiles are shown again at nine locations in $x$ each corresponding to the position of the origin of the graph of the profile of $v$ . Here again, $t = 0$, while $sL = 2$. As in the previous figure the velocity in the upper portion of the water column reflects the surface forcing and the Rossby wave dynamics. There is poleward flow in the east and equatorward flow in the west. As before the western region is one of almost complete compensation of the velocity field with depth with strong poleward flow at the bottom balancing the equatorward flow in the upper part of the water column. The dashed-dotted line showing the barotropic or vertically averaged velocity is again very small, especially in the west, compared with the baroclinic velocity. There is an important difference between the response to $S(z)$ as given by (7.7.48) and the previous example. In Fig. 7.7.3b the bottom velocity is poleward and quite strong all across the basin. The fluid enters the basin near the bottom and must

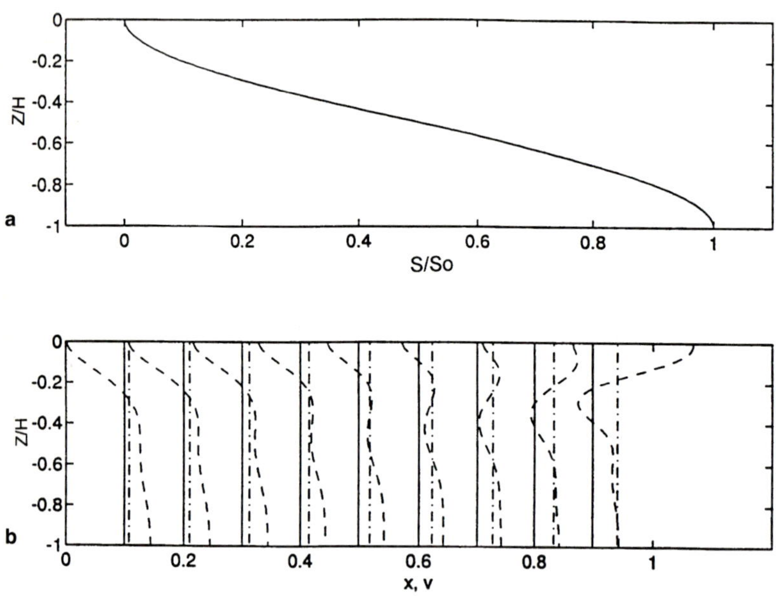


**Fig. 7.7.3a,b.** As in Fig. 7.7.2. **a** Source function $S(z)$ **b** the Meridional velocity profiles across the basin

rise to leave through the interface with the thermocline in order to close the circulation. This means that there must be vortex tube stretching near the bottom of the abyss and thus a poleward velocity. It is important to note that this bottom poleward velocity is in response to the *perimeter average* of the source input. This bottom poleward flow in the interior would occur independently of whether the source were placed on the northern, western, or southern boundary of the basin.

If instead of a zonal variation of the upwelling velocity we allow $w_0$ to depend on y a similar response occurs. In Fig. 7.7.4 the same source function as in Fig. 7.7.2 is used, i.e., $S(z)$ is given by (7.7.47). Only now there is no zonal variation of $w_0$ $(s = 0)$ and instead the upwelling decays exponentially from the southern boundary $(tx_e = 2)$. In this case the upwelling is very small in the north, and for large $y$ the pressure anomaly is due entirely to the westward propagation of the density anomaly which has been established on the eastern boundary. This anomaly is established by the passage of the Kelvin waves in response to the source flow through the basin perimeter and by the disturbance produced in the south by the upwelling taking place there. Since $w_0$ is small in

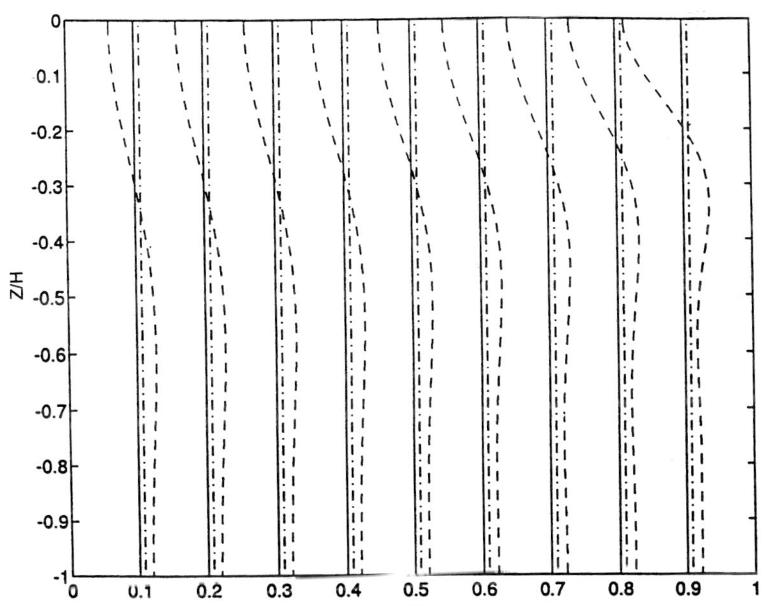


**Fig. 7.7.4.** As in Fig. 7.7.2, but now $sL = 0$ while $tL = 2$, i.e., the upwelling is strongly localized in the southern part of the basin. The velocity profiles for are shown at $y/L = 0.9$. At this latitude the direct forcing of the interior flow is negligible, and the motion is driven almost entirely by the density anomaly on the eastern boundry which has been produced by the Kelvin wave signal carrying the information of the upwelling which occurs to the south. The Rossby wave propagating westward produce a baroclinic flow which strongly exceeds the barotropic Stommel-Arons velocity at all longitudes

the north, the barotropic velocity is small there for all $x$, and we can expect that the meridional velocity has a very small vertical average compared with the baroclinic velocity. Figure 7.7.4 shows the meridional velocity profiles again at zonal locations across the basin at $y = 0.9L$. Again, the dashed-dotted line shows the Stommel-Arons, barotropic component to the velocity and it is seen to be small at each $x$ location with respect to the baroclinic velocity. The flow in the upper part of the water column is now equatorward at each zonal position as vortex compression in the upper part of the water column is induced by the midlevel source, and the small upwelling at $z = 0$ at this latitude.

Thus for the interior velocity field the combination of Rossby and Kelvin wave dynamics leads to a velocity structure at every location which is quite different than that indicated by the sources. The source is important for the structure of the velocity field, especially at the bottom of the water column, but only for spatially featureless $w_0$ is the velocity field's structure dominated by the source structure.

Edwards and Pedlosky (1995) also investigated the boundary layer structure required to close the circulation.They added horizontal mixing of both momentum (in the equation for the flow component parallel to the boundary) and horizontal mixing for density. The reader is referred to their paper for the details of the calculation. The boundary layer structure is complex. The barotropic component of the flow in this linear model always has the lateral structure of the Munk layer while the baroclinic portion of the flow has a double boundary layer structure that depends on the level of stratification. If the Munk layer is thicker than the Rossby deformation radius, the baroclinic structure has an inner scale of the order of the Rossby deformation radius, which exists on all boundaries, and an outer scale, which dominates the baroclinic flow structure, which occurs only on the western boundary and has a characteristic width for the $n$th baroclinic mode equal to $\delta_w = (n^2 f^2 \kappa_h / \beta H^2 N^2)$ (Warren 1981). Edwards and Pedlosky show that the vertical structure of the velocity field in the boundary layer depends not only on the local source structure of the flow entering the basin through the boundary but also on the integrated properties of the source and the upwelling over the entire basin. This occurs because the western boundary current must carry not only the flow entering from the sources but also accept the interior recirculating flow which impinges on the boundary. The boundary layer turns the impinging flow to bring its normal velocity to zero so that the current flows along the boundary. Since the interior flow and its vertical structure depends, as we have seen, on nonlocal properties of the source distribution and the upwelling, the boundary layer flow also has its structure determined by remote forcing.

Thus in both the interior and the boundary layer the flow pattern at any level can be expected to differ significantly from that predicted by the classical Stommel-Arons model although, at least in models with no topography, the barotropic mode which carries the total transport is still given by the Stommel-Arons theory. The layering of the velocity field is not explainable simply in terms of the properties of the source waters of the abyss but depends instead on the vorticity dynamics of the interior where the motion is driven by the upwelling in the interior and the density anomaly on the eastern boundary.

## References


Edwards, C.A. and Pedlosky J, 1995: The influence of distributed sources and upwelling on the baroclinic structure of the abyssal circulation. *J. Phys. Ocean.*, **25**, 2259–2284.

Friedrichs, M.A.M., and Hall M.M., 1993: Deep circulation in the tropical North Atlantic. *J. Mar. Res.*, **51**, 697–736.

Godfrey, J.S., 1989: A Sverdrup model of the depth-integrated flow for the world ocean allowing for island circulations. *Geophys. Astrophys. Fluid Dyn.*, **45**, 89–112.

Gregg, M.C. and Sanford T.B., 1988: The dependence of turbulent dissipation on stratification in a diffusively stable thermocline. *J. Geophys. Res.*, **93**, 12,381–12,392.

Kawase, M. 1987: Establishment of deep ocean circulation driven by deep-water production. *J. Phys. Ocean.*, **17**, 2294–2317.

Kawase, M., 1993. Topographic effects on the bottoom water circulation of the western tropical North Atlantic ocean. *Deep - Sea Res.*, **40**, 1259–1283.

Kawase, M., and Straub D, 1991: Spinup of source-driven circulation in an abyssal basin in the presence of bottom topography. *J. Phys. Ocean.*, **21**, 1501–1514.

Kuo, H.-H. and Veronis G., 1970: Distribution of tracers in the deep oceans of the world. *Deep Sea Res.*, **17**, 29–46.

Ledwell, J.R., Watson A.J. and Law C.S., 1993: Evidence for slow mixing across the pycnocline from an open-ocean tracer release experiment. *Nature*, (Letters), **364**, 701–703.

Levitus, S. 1982: Climatological atlas of the world ocean. *U.S. Government Printing Office*. pp 173.

McCartney, M.S. and Talley L.D., 1984: Warm - to - cold conversion in the northern North Atlantic Ocean. *J. Phys. Ocean.*, **14**, 922–935.

McCartney, M.S., Bennett S.L., and Woodgate-Jones M.E., 1991: Eastward flow through the Mid-Atlantic Ridge at 11° N and its influence on the abyss of the eastern basin. *J. Phys. Ocean.*, **21**, 1089–1121.

Osborn, T.R., 1980: Estimates of the local rate of vertical diffusion from dissipation measurements. *J. Phys. Ocean.*, **10**, 83–89.

Pedlosky, J., 1987: Thermocline theories. In: *General circulation of the ocean*, Eds. Abarbanel H.D.I. and Young W.R., *Springer*, Berlin Heidelberg New York, pp 55–101.

Pedlosky, J., 1992: The baroclinic structure of the abyssal circulation. *J. Phys. Ocean.*, **22**, 652–659.

Pedlosky, J., 1994a: Stratified abyssal flow in the presence of fractured ridges. *J. Phys. Ocean .*, **24**, 403–417.

Pedlosky, J., 1994b: Ridges and recirculations – gaps and jets. *J. Phys. Ocean.*, (in press).

Pedlosky, J., and Chapman D.C., 1993: Baroclinic structure of the abyssal circulation and the role of meridional topography. *J. Phys. Ocean.*, **23**, 979–991.

Pickard, G.L. and Emery W.J., 1982: Descriptive physical oceanography. *Pergamon Press*, Oxford, pp 249.

Price, J.F. and Barringer M.O., 1994, Outflows and deep water production by marginal seas. *Prog. in Ocean.*, **33**, 161–200.

Stommel, H. 1958: The abyssal circulation. *Deep -Sea Res.* (Letters), **5**, 80–82.

Stommel, H., and Arons A.B., 1960a: On the abyssal circulation of the world ocean. I. Stationary planetary flow patterns on a sphere. *Deep - Sea Res.* , **6**, 140–154

Stommel, H., and Arons A.B., 1960b. On the abyssal circulation of the world ocean. II. An idealized model of the circulation pattern and amplitude in oceanic basins. *Deep- Sea Res.*, **6**, 217–233.

Stommel, H., Arons A.B. and Faller A.J., 1958: Some examples of stationary planetary flow patterns in bounded basins. *Tellus*, **10**, 179–187.

Toole, J.M., Polzin K.L., and Schmidt R.W., 1994: Estimates of diapycnal mixing in the abyssal ocean. *Science*, **264**, 1120–1123.

Tziperman, E., 1987: The Mediterranean outflow as an example of deep buoyancy driven flow. *J. Geophys. Res.*, **92**, 14510–14520.

Warren, B.A., 1981: Deep circulation of the world ocean. In: *Evolution of physical oceanography*. Eds. Warren, B.A. and Wunsch C. MIT Press, Cambridge, pp 6–41.

Warren, B.A., and Owens W.B., 1988: Deep currents in the central Subarctic Pacific Ocean. *J. Phys. Ocean.*, **18**, 529–551.

Worthington, L.V., 1981: The water masses of the world ocean: some results of a fine-scale census. In: *Evolution of physical oceanography*. Eds. Warren, B.A. and Wunsch C. MIT Press, Cambridge, pp 42–69.

# Subject Index

Printing: Saladruck, Berlin
Binding: Buchbinderei Lüderitz & Bauer, Berlin